QUICK REFERENCE MANUAL FOR SILICON INTEGRATED CIRCUIT TECHNOLOGY

QUICK REFERENCE MANUAL FOR SILICON INTEGRATED CIRCUIT TECHNOLOGY

W. E. BEADLE
J. C. C. TSAI
R. D. PLUMMER

Editors

A Wiley-Interscience Publication
JOHN WILEY & SONS
New York • Chichester • Brisbane • Toronto • Singapore

Copyright © 1985 by Bell Telephone Laboratories, Incorporated

Published by John Wiley & Sons, Inc.

This publication is designed to provide accurate and
authoritative information in regard to the subject
matter covered. It is sold with the understanding that
the publisher is not engaged in rendering legal, accounting,
or other professional service. If legal advice or other
expert assistance is required, the services of a competent
professional person should be sought. *From a Declaration
of Principles jointly adopted by a Committee of the
American Bar Association and a Committee of Publishers.*

Library of Congress Cataloging in Publication Data:

Main entry under title:

Quick reference manual for silicon integrated circuit
  technology.

  "A Wiley-Interscience publication."
  Includes index.
  1. Integrated circuits—Handbooks, manuals, etc.
2. Silicon—Handbooks, manuals, etc.   I. Beadle, W. E.
II. Tsai, J. C. C.   III. Plummer, R. D.
TK7874.Q38   1984   621.381'73   84-25667
ISBN 0-471-81588-8

Printed in the United States of America

10 9 8 7 6 5 4 3 2 1

CONTRIBUTORS

PREFACE

It has been our observation that workers in the semiconductor industry make extensive use of handy curves and graphs in their work. These design aids often represent data accumulated over several years of experience by an individual or a small group of associates. Often it takes the form of a set of curves or a clever nomograph. Unfortunately, despite the usefulness of such information, it usually has very limited circulation throughout the design community. This manual is a collection of such reference data, gleaned by the authors in their various roles in the design, development, processing, manufacture, and characterization of silicon devices and integrated circuits.

Topics include: properties of silicon, mathematical expressions, measurements, chemical recipes, diffusion, ion implantation, process data, conductivity of diffused layers, properties of p-n and metal-semiconductor junctions, surfaces, MOS, and reliability. A comprehensive table of the physical constants is also provided. Again, the manual emphasizes graphs, charts, nomographs, simple expressions, etc. designed to provide quick answers to questions. The intent is not to educate, explanations are at best terse, rather, the manual is intended to augment the designers basic knowledge with "quick" reference data.

The editors are indebted to the authors for providing their design tools and supplying their energies to this task. Special thanks are given to F. Cohen and D. D. Williams whose thorough work and infinite patience helped shape this large data base into a useful manual and R. T. Cronan for his skillful rendering of the information contained in the figures used throughout this manual. Finally we thank our management, R. L. Batdorf and J. Godfrey for their continuing support of this project.

CONTENTS

QUICK REFERENCE MANUAL FOR SILICON INTEGRATED CIRCUIT TECHNOLOGY

1. PHYSICAL CONSTANTS

TABLE 1-1
RECOMMENDED CONSISTENT VALUES OF THE FUNDAMENTAL CONSTANTS[1]

QUANTITY	SYMBOL	VALUE*	UNCERTAINTY (ppm)
Atomic mass unit	$u = (10^{-3} \text{ kg mol}^{-1})/N_A$	$1.6605655(86) \times 10^{-27}$ kg	5.1
Avogadro's constant	N_A	$6.022045(31) \times 10^{23} \text{ mol}^{-1}$	5.1
Bohr magneton	$\mu_B = e\hbar/2m_e$	$9.274078(36) \times 10^{-24}$ J T^{-1}	3.9
Bohr radius	$a_0 = \alpha/4\pi R_\infty$	$0.52917706(44) \times 10^{-10}$ m	0.82
Boltzmann constant	$k = R/N_A$	$1.380662(44) \times 10^{-23}$ J K^{-1}	32
Classical electron radius	$r_e = \mu_0 e^2/4\pi m_e$ $= \alpha\lambdabar_C$	$2.8179380(70) \times 10^{-15}$ m	2.5
Diamagnetic shielding factor, spherical H_2O sample	$1 + \sigma(H_2O)$	$1.000025637(67)$	0.067
Electron Compton wavelength	$\lambda_c = \alpha^2/2R_\infty$ $\lambdabar_c = \lambda_c/2\pi = \alpha a_0$	$2.4263089(40) \times 10^{-12}$ m $3.8615905(64) \times 10^{-13}$ m	1.6 1.6
Electron g-factor	$g_e/2 = \mu_e/\mu_B$	$1.0011596567(35)$	0.0035
Electron magnetic moment	μ_e	$9.284832(36) \times 10^{-24}$ J T^{-1}	3.9
Electron rest mass	m_e	$0.9109534(47) \times 10^{-30}$ kg $5.4858026(21) \times 10^{-4}$ u	5.1 0.38
Elementary charge	e	$1.6021892(46) \times 10^{-19}$ C	2.9
Faraday constant	$F = N_A e$	$9.648456(27) \times 10^4$ C mol^{-1}	2.8
Fine structure constant, $\mu_0 c e^2/2h$	α α^{-1}	$0.0072973506(60)$ $137.03604(11)$	0.82 0.82
Gas constant	R	$82.0568(26)$ cm^3 atm mol^{-1} K^{-1} $1.98719(6)$ cal mol^{-1} K^{-1}	31 31
Gravitational constant	G	$6.6720(41) \times 10^{-11}$ N m^2 kg^{-2}	615
Josephson frequency-voltage ratio	$2e/h$	$483.5939(13)$ THz V^{-1}	2.6
Magnetic flux quantum	$\phi_0 = h/2e$ h/e	$2.0678506(54) \times 10^{-15}$ Wb $4.135701(11) \times 10^{-15}$ J Hz^{-1} C^{-1}	2.6 2.6
Molar gas constant	R	$8.31441(26)$ J mol^{-1} K^{-1}	31

* The digits in parentheses following a numerical value represent the standard deviation of that value in terms of the final listed digits. For definitions of the symbols shown in this column, see page 1-5.

TABLE 1-1 (Contd)

RECOMMENDED CONSISTENT VALUES OF THE FUNDAMENTAL CONSTANTS

QUANTITY	SYMBOL	VALUE*	UNCERTAINTY (ppm)
Molar standard volume, ideal gas ($T_0 = 273.15$ K, $p_0 =$ 1 atm)	$V_m = RT_0/p_0$	$0.02241383(70)$ m^3 mol^{-1}	31
Muon g-factor	$g_\mu/2$	$1.00116616(31)$	0.31
Muon magnetic moment	μ_μ	$4.490474(18) \times 10^{-26}$ J T^{-1}	3.9
Muon rest mass	m_μ	$1.883566(11) \times 10^{-28}$ kg	5.6
		$0.11342920(26)$ u	2.3
Neutron Compton wavelength	$\lambda_{C,n} = h/m_n c$	$1.3195909(22) \times 10^{-15}$ m	1.7
	$\lambdabar_{C,n} = \lambda_{C,n}/2\pi$	$2.1001941(35) \times 10^{-16}$ m	1.7
Neutron rest mass	m_n	$1.6749543(86) \times 10^{-27}$ kg	5.1
		$1.008665012(37)$ u	0.037
Nuclear magneton	$\mu_N = e\hbar/2m_p$	$5.050824(20) \times 10^{-27}$ J T^{-1}	3.9
Permeability of vacuum	μ_0	$4\pi \times 10^{-7}$ H m^{-1} $= 12.5663706144 \times 10^{-7}$ H m^{-1}	
Permittivity of vacuum	$\epsilon_0 = (\mu_0 C^2)^{-1}$	$8.85418782(7) \times 10^{-12}$ F m^{-1}	0.008
Planck's constant	h	$6.626176(36) \times 10^{-34}$ J Hz^{-1}	5.4
	$\hbar = h/2\pi$	$1.0545887(57) \times 10^{-34}$ J s	5.4
Proton Compton wavelength	$\lambda_{C,p} = h/m_p c$	$1.3214099(22) \times 10^{-15}$ m	1.7
	$\lambdabar_{C,p} = \lambda_{C,p}/2\pi$	$2.1030892(36) \times 10^{-16}$ m	1.7
Proton gyromagnetic ratio	γ_p	$2.6751987(75) \times 10^8$ s^{-1} T^{-1}	2.8
Proton gyromagnetic ratio (uncorrected)	γ_p'	$2.6751301(75) \times 10^8$ s^{-1} T^{-1}	2.8
	$\gamma_p'/2\pi$	$42.57602(12)$ MHz T^{-1}	2.8
Proton magnetic moment	μ_p	$1.4106171(55) \times 10^{-26}$ J T^{-1}	3.9
Proton magnetic moment in Bohr magnetons	μ_p/μ_B	$1.521032209(16) \times 10^{-3}$	0.011
Proton moment in nuclear magnetons	μ_p/μ_N	$2.7928456(11)$	0.38

* The digits in parentheses following a numerical value represent the standard deviation of that value in terms of the final listed digits. For definitions of the symbols shown in this column, see page 1-5.

TABLE 1-1 (Contd)
RECOMMENDED CONSISTENT VALUES OF THE FUNDAMENTAL CONSTANTS

QUANTITY	SYMBOL	VALUE*	UNCERTAINTY (ppm)
Proton moment in nuclear magnetons (uncorrected)	μ_p'/μ_N	2.7927740(11)	0.38
Proton rest mass	m_p	$1.6726485(86) \times 10^{-27}$ kg 1.007276471(11) u	5.1 0.011
Quantum of circulation	$h/2m_e$ h/m_e	$3.6369455(60) \times 10^{-4}$ J Hz^{-1} kg^{-1} $7.273891(12) \times 10^{-4}$ J Hz^{-1} kg^{-1}	1.6 1.6
Radiation constant, first	$C_1 = 2\pi hc^2$	$3.741832(20) \times 10^{-16}$ W m^2	5.4
Radiation constant, second	$C_2 = hc/k$	0.01438786(45) m K	31
Ratio, electron to proton magnetic moments	μ_e/μ_p	658.2106880(66)	0.010
Ratio, muon mass to electron mass	m_μ/m_e	206.76865(47)	2.3
Ratio, muon moment to proton moment	μ_μ/μ_p	3.1833402(72)	2.3
Ratio, proton mass to electron mass	m_p/m_e	1836.15152(70)	0.38
Rydberg constant	R_∞	$1.097373177(83) \times 10^7$ m^{-1}	0.075
Second radiation constant	$C_2 = hc/k$	0.01438786(45) m K	31
Specific electron charge	e/m_e	$1.7588047(49) \times 10^{11}$ C kg^{-1}	2.8
Speed of light in vacuum	c	299792458(1.2) m s^{-1}	0.004
Stefan-Boltzmann constant	$\sigma = (\pi^2/60) \, k^4/\hbar^3 c^2$	$5.67032(71) \times 10^{-8}$ W m^{-2} K^{-4}	125

*The digits in parentheses following a numerical value represent the standard deviation of that value in terms of the final listed digits. For definitions of the symbols shown in this column, see below.

SYMBOL	UNIT	SYMBOL	UNIT	SYMBOL	UNIT
C	coulomb	K	kelvin	s	second
F	farad	kg	kilogram	T	tesla=webers/meter^2
H	henry	m	meter	V	volt
Hz	hertz	mol	mole (gram molecular wt)	W	watt
J	joule	N	newton	Wb	weber

TABLE 1-2
MISCELLANEOUS CONSTANTS

QUANTITY	VALUE
kT/q (300K)	0.0259 volt
(298K)	0.0257 volt
(293K)	0.0252 volt
(273K)	0.0235 volt
1 eV	$1.6021892(46) \times 10^{-19}$ J
1 eV/molecule	23.06 Kcal/mole
Voltage-wavelength product	12398.520(32) eV•Å
Thermochemical calorie	4.184 J

TABLE 1-3
LENGTH CONVERSION FACTORS

UNIT	QUANTITY IN EQUIVALENT UNITS					
	inch	mil	cm	mm	μm	Å
inch	1	10^3	2.54	25.4	2.54×10^4	2.54×10^8
mil	10^{-3}	1	2.54×10^{-3}	2.54×10^{-2}	25.4	2.54×10^5
cm	0.3937	3.937×10^2	1	10	10^4	10^8
mm	3.937×10^{-2}	39.37	0.1	1	10^3	10^7
μm	3.937×10^{-5}	3.937×10^{-2}	10^{-4}	10^{-3}	1	10^4
Å	3.937×10^{-9}	3.937×10^{-6}	10^{-8}	10^{-7}	10^{-4}	1

TABLE 1-4

PHYSICAL CONSTANTS OF Ge AND Si AT 300K[2] [3]

QUANTITY	Ge	Si
Atomic Weight	72.60	28.09
Atoms, Total (cm^{-3})	4.418×10^{22}	4.995×10^{22}
Boiling Point (°C)	2700	2600
Burger's Vector, b (Å)	—	3.74
Coefficient of Thermal Expansion	—	See Figure 2-35
Compliance ($cm^2 \cdot dyne^{-1}$) <111>		5.32×10^{-13} [4]
Crystal Structure	Diamond 8 atoms/unit cell	Diamond 8 atoms/unit cell
Density (ρ) (g/cm^3)	5.32	2.33
Density of Surface Atoms (cm^{-2}) (100) (110) (111)	6.24×10^{14} 8.83×10^{14} 7.21×10^{14}	6.78×10^{14} 9.59×10^{14} 7.83×10^{14}
Dielectric Constant	16.0	11.8
Effective Density of States (see Section 2.01) Conduction Band, N_C (cm^{-3}) Valence Band, N_V (cm^{-3})	1.04×10^{19} 6.0×10^{18}	3.22×10^{19} 1.83×10^{19}
Elastic Constant ($dyne \cdot cm^{-2}$) [5] C_{11} C_{12} C_{44}	1.29×10^{12} 0.483×10^{12} 0.671×10^{12}	1.66×10^{12} 0.639×10^{12} 0.795×10^{12}
Electron Affinity (eV) (111)	—	4.85 (111) [6]
Electron-Effective Mass (at 4K) [7] Longitudinal: (m_1/m_o) Transverse: (m_t/m_o) *Density of States: ($m*/m_o$)	1.58 0.082 0.55	0.98 0.19 1.08
Energy Gap (eV)	0.67	1.12

*See Figure 2-2.   Reprinted by permission of John Wiley & Sons, Inc.

TABLE 1-4 (Contd)

PHYSICAL CONSTANTS OF Ge AND Si AT 300K

QUANTITY	Ge	Si
Ground-State Degeneracy Factor g_D g_A	— —	2 4

Hole-Effective Mass (at 4K) [8]
    Heavy: (m_1/m_0)
    Light: (m_2/m_0)
    *Density of States: m^*/m_0
	
	0.28
	0.044
	0.29
		
		0.537
		0.153
		0.591

Intrinsic Carrier Concentration †
    n_i (cm^{-3})
    $n_i^2 (cm^{-6})$
	
	2.4×10^{13}
	5.76×10^{26}
		
		1.38×10^{10}
		1.90×10^{20}

Intrinsic Debye Length $$L = \left[\frac{kT\epsilon}{2q^2 n_i}\right]^{1/2} \quad (\mu m)$$ $n_i L (cm)^{-2}$	 0.69 1.65×10^9	 28.7 4.0×10^7
Intrinsic Resistivity (ohm•cm)	47	2.3×10^5
Lattice Constant (Å)	5.66	5.43

Lattice Contraction Coefficient, $\beta (cm^3)$ [9]
 	
  For Phosphorus in Lattice
  For Boron in Lattice
	
	—
	—
		
		7.2×10^{-25}
		2.3×10^{-24}

Melting Point (°C)	937	1415
Poisson's Ratio, γ	—	0.27
Refractive Index	4.0	3.4

Scattering Limited Velocity $(cm•s^{-1})$
    Electron
    Hole
	
	6.2×10^6
	5.7×10^6
		See Figure 2-34
		$\sim 1.0 \times 10^7$
		$\sim 8.4 \times 10^6$

Shear Modulus, μ $(dyne/cm^2)$	—	7.55×10^{11}
Specific Heat, C_p [J/ (g • °C)]	0.31	0.7
Thermal Conductivity, K_{th} [w/(cm•°C)]	0.6	1.5[10] See Figure 2-43

Thermal Diffusivity
$$D_{th} = \frac{K_{th}}{\rho C_p} \left(\frac{cm^2}{sec}\right)$$
[7]D_{th} (@ 400K)
    (@ 600K)
    (@ 800K)
    (@ 1000K)
    (@ 1200K)
	
	0.36
	
		0.9
		
		0.52
		0.29
		0.19
		0.14
		0.12

*See Figure 2-3. †See Figure 2-4   Reprinted by permission of John Wiley & Sons, Inc.

TABLE 1-5

PHYSICAL CONSTANTS OF SiO_2, Si_3N_4, Si-O_x-N_y, AND SIPOS AT 300K

STRUCTURE	SiO_2[2]	Si_3N_4	Si-O_x-N_y	SIPOS[11]
	AMORPHOUS	AMORPHOUS	AMORPHOUS	AMORPHOUS
DC Resistivity ($\Omega \bullet cm$):				
@ 25°C	$10^{14} - 10^{16}$	$\sim 10^{14}$	—	$10^6 - 10^{14}$
@ 500°C	—	$\sim 2 \times 10^{13}$	—	
Density ($g \bullet cm^{-3}$)	2.27	3.1	—	$2.2 - 2.33$
Dielectric Constant	$3.8 - 3.9$	7.5	$4.77 - 6.12$	$5.0 - 9.0$
Dielectric Strength ($V \bullet cm^{-1}$)	$\sim 5 \times 10^6$	$\sim 1 \times 10^7$	$\sim 5 \times 10^6$	—
Energy Gap (eV)	~ 8	~ 5.0	—	—
Etch Rate in Buffered HF* (Å/min)	1000	$5 - 10$	$33 - 400$	~ 0
Infrared Absorption Band (μm)	9.3	$11.5 - 12.0$	9.3 and 12.0	$9.0 - 12.0$
Linear Expansion Coefficient (°C^{-1})	5.0×10^{-7}	—	—	
Melting Point (°C)	~ 1700	—	—	
Molecular Weight	60.08			
Molecules/cm^3	2.3×10^{22}			
Refractive Index	1.46	2.05	$1.60 - 1.88$	$2.0 - 3.6$†
Specific Heat [$J/(g \bullet °C)$]	1.0			
Stress in the Film on Silicon (dyne/cm^2)	$2 - 4 \times 10^9$[12][13] compression	$9 - 10 \times 10^9$[14]		$1 - 6 \times 10^9$ both compression & tension
Thermal Conductivity [$W/(cm \bullet °C)$]	0.014	—	—	
Thermal Diffusivity (cm^2/s)	0.006			

* Buffered HF: 34.6% (wt.) NH_4F, 6.8% (wt.) HF, 58.6% H_2O.

† At 589 nm

TABLE 1-6

SIN FILM PROPERTIES FOR 50 kHz AND 13.56 MHz FROM A RADIAL
FLOW PLASMA REACTOR[15]

	50 kHz	13.56 MHz
Deposition rate (Å/min)	130	100
Film uniformity (%)	± 5	± 10
Refractive index	1.93	1.96 2.02
Refractive index uniformity (%)	0.1	1.5
Pinholes density (n/cm^2)	0.5	1.0
Etch Rate (BHF 10-1) (nm/min)	15	30
Stress (10^9 dynes/cm^2)	compressive 3	tensile 2
H content (at %)	18	24
Si/N ratio	.82	.93 — 1.0
Breakdown voltage (x 10^6 V/cm)	6	9

This figure was originally presented at the Spring 1983 Meeting of
the Electrochemical Society, Inc., held in San Francisco,
California.

TABLE 1-7

FREE ENERGY OF FORMATION OF METAL OXIDES AT 500K[16]

OXIDE	$-F°$ (Kcal/mole)*	$-F°$ [(Kcal/g • atom O)]
Al_2O_3	362.1	120.7
Cr_2O_3	240.2	80.1
MoO_2	114.5	57.2
Na_2O	83.0	83.0
NiO	46.1	46.1
SiO_2	187.9	94.0
Ta_2O_5	434.9	87.0
TiO	112.2	112.2
ZrO_2	238.4	119.2

*F°: free energy at the standard state

TABLE 1-8

PHYSICAL CONSTANTS OF SELECTED METALS[17]

METAL	ATOMIC NO.	ATOMIC WEIGHT	DENSITY P (20°C) (g/cm³)	MELTING POINT (°C)	SPECIFIC HEAT Cp at 20°C [cal/(g•°C)]	COEF OF LINEAR THERMAL EXPANSION NEAR 20°C (10⁻⁶/°C)	THERMAL CONDUCTIVITY (10°C) [cal/(cm•°C•s)]	RESISTIVITY (20°C) (μΩ•cm)
Ag	47	107.88	10.49	960.8	0.0559 (0°C)	19.68 (0–100°C)	1.0 (0°C)	1.59
Al	13	26.98	2.699	660	0.215	23.6 (20–100°C)	0.53	2.6548
Au	79	197.0	19.32	1063	0.0312 (18°C)	14.2	0.71 (0°C)	2.35
Be	4	9.013	1.848	1277	0.45	11.6 (25–100°C)	0.35	4
Cr	24	52.01	7.19	1875	0.11	6.2	0.16	12.9 (0°C)
Cu	29	63.54	8.96	1083	0.092	16.5	0.941	1.673
Fe	26	55.85	7.87	1536.5	0.11	11.76 (25°C)	0.18 (0°C)	9.71
Ga	31	69.72	5.907	29.78	0.079	18 (0–30°C)	0.07–0.09 Melting	17.4(a); 8.1(b); 54.3(c)
In	49	114.82	7.31	156.2	0.057	33	0.057	8.37
Mo	42	95.95	10.22	2610	0.066	4.9 (20–100°C)	0.34	5.2 (0°C)
Ni	28	58.71	8.902 (25°C)	1453	0.105	13.3 (0–100°C)	0.22 (25°C)	6.84
Pb	82	207.21	11.36 (rolled)	327.4	0.0309 (0°C)	29.3 (17–100°C)	0.083 (0°C)	20.648
Pd	46	106.7	12.02	1552	0.0584 (0°C)	11.76	0.168 (18°C)	10.8
Pt	78	195.09	21.45	1769	0.0314 (0°C)	8.9	0.165	10.6
Rh	45	102.91	12.44	1966	0.059 (0°C)	8.3	0.21 (17°C)	4.51
Sn	50	118.70	7.2984 (β)	231.9	0.054	23 (Poly, 0–100°C)	0.150 (0°C)	11 White Sn (0°C)
Ta	73	180.95	16.6	2996 ± 50	0.034 (25°C)	6.5	0.13	12.45 (25°C)
Ti	22	47.90	4.507	1668	0.124	8.41	0.037 (50°C)	42
W	74	183.86	19.3	3410	0.033	4.6	0.397 (0°C)	5.65 (27°C)
Zn	30	65.38	7.133 (25°C)	419.5	0.0915	39.7 (20–250°C)	0.27 (25°C)	5.916; 49 (25°C)
Kovar[18]	—	—	8.36	1450	0.105 (0°C); 0.155 (300°C)	4.6–5.2 (30–400°C)	0.0395 (30°C); 0.0485 (300°C); 0.0585 (500°C)	62.7 (100°C)

TABLE 1-9

EUTECTIC COMPOSITION AND EUTECTIC TEMPERATURE OF BINARY ALLOYS OF
Ge AND Si

ELEMENT	Ge				Si			
	ATOM % OF Ge	WT % OF Ge	TEMP °C	REF-PG	ATOM % OF Si	WT % OF Si	TEMP °C	REF-PG
Ag	25.9	19.0	651	19 — 23	10.6	3.0	840	20 — 10
Al	30.3	53.0	424	21 — 38	12.3	12.7	577	21 — 55
As	19.0	18.0	723	19 — 166	59.5	35.5	1073	19 — 180
	—	—	—	—	10.0	4.0	786	19 — 180
Au	27.0	12.6	356	19 — 206	18.6	3.2	370	21 — 103
Bi	<0.1	<0.1	271	21 — 183	1×10^{-8}*	—	271	21 — 198
Ca	—	—	—	—	69.0	61.0	980	19 — 408
Cd	<0.1	<0.1	319	21 — 282	—	—	—	—
Ce	—	—	—	—	81.5	47.0	1240	19 — 461
Co	25.0	29.0	1110	19 — 476	23.0	12.5	1195	19 — 503
	73.0	77	810	19 — 476	77.5	62.0	1259	19 — 503
Cu	36.0	41.0	640	19 — 585	30.2	16.1	802	19 — 631
Fe	34.0	40.8	1130	20 — 327	34.0	20.5	1200	19 — 713
	~64.5	~70.3	845	20 — 327	67.0	50.5	1212	19 — 713
	75.5	80.0	859	20 — 327	73.5	58.0	1208	19 — 713
Ga	5×10^{-3}	—	29.8	21 — 446	1.2	0.5	19	21 — 457
In	0.05	—	157	21 — 478	2×10^{-8}	—	156	21 — 553
Mg	61.0	82.3	680	19 — 765	54.5	58.0	950	19 — 917
	1.15	3.4	635	19 — 765	1.16	1.34	638	19 — 917
Mn	52.5	59.4	697	19 — 767	~68.0	~50.0	1142	20 — 507
	—	—	—	—	21.0	12.0	1040	20 — 507
Ni	62.0	67.0	775	19 — 769	46.0	29.0	964	19 — 1040
Pb	0.02	—	207	21 — 485	5×10^{-8}	—	207	21 — 722
Pt	—	—	—	—	39.0	8.4	983	20 — 624
	—	—	—	—	23.0	4.2	830	20 — 624
Sb	17.0	11.0	590	19 — 773	0.3	7×10^{-4}	629.4	21 — 801
Sn	0.3	0.002	231	21 — 490	1×10^{-5}	—	231.9	21 — 818
Te	49.9	34.4	723	21 — 491	—	—	—	—
	15.0	9.0	375	19 — 776	—	—	—	—
Ti	13.4	19.0	1360	21 — 492	13.7	8.5	1330	19 — 1198
	—	—	—	—	86.0	78.0	1330	19 — 1198
Zn	4.5	5.0	398	21 — 496	—	—	—	—

*Calculated

TABLE 1-10

METAL-SILICON BARRIER POTENTIALS AND METAL WORK FUNCTIONS FOR Ag, Aℓ, Au, Cu, Mg, Ni, Pd, AND Pt CONTACT ON ~0.5 Ω•cm n-TYPE SILICON

PARAMETER*	DEFINITION	Ag	Aℓ	Au	Cu	Mg	Ni	Pd	Pt
V_{int} (V) [22] †	Built-In Voltage	0.507	0.450	0.612	0.518	0.200	0.528	0.618	0.696
Φ_m (eV) [22] †	Metal Work Function	4.31	4.20	4.70	4.52	3.7	4.74	5.0	5.3
$\Phi_{b(I-V)}$ (eV) [6] †	Barrier from I-V Measurements	0.68	0.61		0.73				
Φ_{bc} (eV) [6] †	Barrier from Capacitance Measurements	0.79	0.70	0.82	0.75				
Φ_{bph} (eV) [6] †	Barrier from Photoemission Measurements	0.68	0.61	0.73	0.62				
Φ (eV) [22] ‡	Calculated Barrier Height	0.721	0.699	0.822	0.732	0.417	0.703	0.883	1.022

* See Figure 1-1 for illustration of terms.

† Experimental values from listed reference.

‡ Calculated values from listed reference.

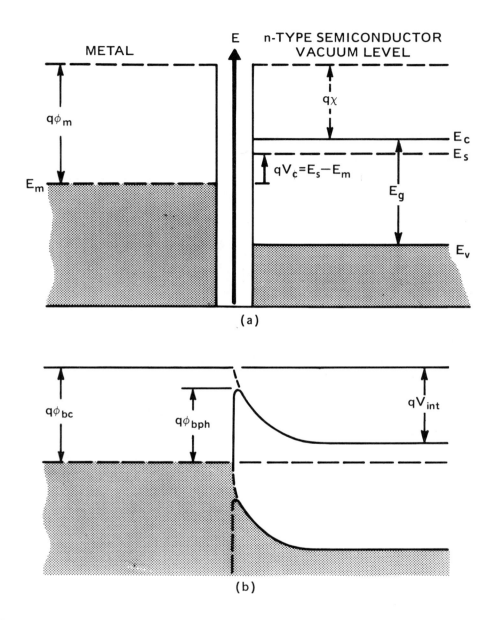

Figure 1-1— Energy Band Diagrams of a Metal and n-Type Silicon
Before (a) and After (b) Being Brought into Contact.

TABLE 1-11

BARRIER HEIGHTS ON n- AND p- TYPE Si

METAL	METAL SILICIDE	$\phi_B(V)$ n-TYPE Si	p-TYPE Si	STRUCTURE	FORMING TEMP (°C)	MELTING TEMP (°C)	REF
Ag		*	0.55				23
Al		*	0.50				23
Au		0.80*	0.30				24
Co	CoSi	0.68	—	Cubic	400	1460	25
	CoSi$_2$	0.65	—	Cubic	450	1326	25
Cr	CrSi$_2$	0.57	—	Hexag	450	1475	25
Cu		*	0.69				23
Hf	—	—	0.9				26
Hf	HfSi	0.4 — 0.5	—	Orthorh	550	2200	25
Ir	IrSi	0.93	—	—	300	—	25
Mg		*					
Mn	MnSi	0.76	—	Cubic	400	1275	25
	Mn$_{11}$Si$_{19}$	0.72		Tetrag	800†	1145	25
Mo	MoSi$_2$	0.55		Tetrag	1000†	1980	25.
Ni	—	0.60*		0.5			24
	Ni$_2$Si	0.7 — 0.75		Orthorh	200	1318	25
	NiSi	0.66 — 0.75		Orthorh	400	992	25, 27
	NiSi$_2$	0.7		Cubic	800†	993	25
Pb	—	—	0.41				23
Pd		*					
Pd	Pd$_2$Si	0.745 ± .015	0.355				28
	Pd$_2$Si	0.72 — 0.75	—	Hexag	200	1330	25
Pt		*					
Pt	PtSi	0.84	0.26	Orthorh	300	1229	25, 29
Rh	RhSi	0.69	0.34	Cubic	300	—	25, 29
Ti	TiSi	0.6	0.6	Orthorh	650	1540	25
V		0.7 ± .04		—			26
W	WSi$_2$	0.65		Tetrag	650	2150	25
Zr	ZrSi$_2$	0.55	0.55	Orthorh	600	1520	25, 29

* See Table 1-10.
† $\leqslant 700°C$ under very clean conditions.

TABLE 1-12

DISTRIBUTION COEFFICIENTS AT THE MELTING
POINTS OF Ge AND Si[30]

ELEMENT	Ge	Si
Ag	4×10^{-7}	—
Al	0.073	0.0020
As	0.02	0.3
Au	1.3×10^{-5}	2.5×10^{-5}
B	17.	0.80
Bi	4.5×10^{-5}	7×10^{-4}
Cd	$>1 \times 10^{-5}$	—
Co	$\sim 10^{-6}$	8×10^{-6}
Cu	1.5×10^{-5}	4×10^{-4}
Fe	$\sim 3 \times 10^{-5}$	8×10^{-6}
Ga	0.087	0.0080
Ge	1	0.33
In	0.001	4×10^{-4}
Li	0.002	0.01
Mn	$\sim 10^{-6}$	$\sim 10^{-5}$
N	—	$<10^{-7}$ *
Ni	3×10^{-6}	—
O	—	0.5
P	0.080	0.35
Pb	1.7×10^{-4}	—
Pt	$\sim 5 \times 10^{-6}$	—
S	—	10^{-5}
Sb	0.0030	0.023
Si	5.5	1
Sn	0.020	0.016
Ta	—	10^{-7}
Te	$\sim 10^{-6}$	—
Tl	4×10^{-5}	—
V	$<3 \times 10^{-7}$	—
Zn	4×10^{-4}	$\sim 1 \times 10^{-5}$

* Uncertain

TABLE 1-13

IMPURITY LEVELS IN Si AND Ge[31]

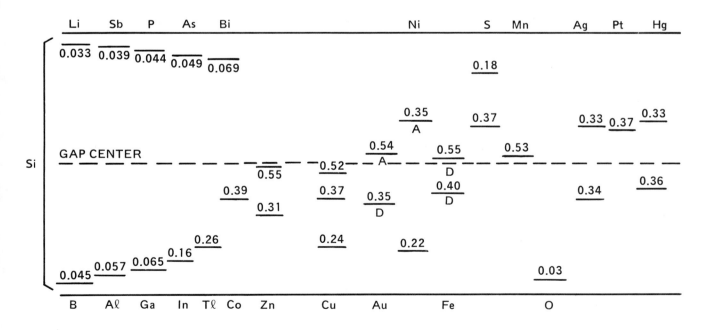

TABLE 1-14

DEEP IMPURITY LEVELS IN Si: MEASUREMENTS BY CONVENTIONAL
TECHNIQUES AND BY SURFACE CAPACITANCE STUDIES*[32]

IMPURITY	LEVEL(S) FROM CONVENTIONAL TECHNIQUES		LEVELS FROM SURFACE CAPACITANCE STUDIES, FAHRNER AND GOETZBERGER (1972) (eV)
	(eV)	FROM SZE (1969) (eV)	
Ag	0.29 (c, a)	0.33 (c, a)	0.36 (c)
	0.32 (v, d)	0.34 (v, d)	0.33 (v)
Au	0.54 (c, a)	0.54 (c, a)	0.30 (c)†
	0.35 (v, d)	0.35 (v, d)	0.83 (c)
	0.033 (v, a)?	—	—
Ba	—	—	0.32 (c)
	—	—	0.52 (v)
Be	0.17 (v, a)	—	0.42 (v)
Cd	0.45 (c, a)	—	0.20 (c)
	0.55 (v, a)	—	0.30 (v)
Co	0.53 (c, a)	—	0.63 (c)
	0.35 (v, a)	0.39 (v, a)	—
Cr	0.41 (c, d)	—	—
Cs	—	—	0.30 (c)
Cu	0.52 (v, a)	0.52 (v, a)	0.26 (c)
	0.37 (v, a)	0.37 (v, a)	0.53 (v)
	0.24 (v, a)	0.24 (v, a)	0.40 (v)
Fe	—	—	0.14 (c)
	—	0.55 (c, d)	0.51 (c)
	0.40 (v, d)	0.40 (v, d)	0.40 (v)
Ge	—	—	0.27 (c)
	—	—	0.50 (v)
Hg	0.31 (c, a)	—	—
	0.36 (c, a)	0.33 (c)	0.35 (c)
	0.33 (v, d)	0.36 (v)	0.39 (v)
	0.25 (v, d)	—	—
In	0.16 (v, a)	0.16 (v, a)	—
Li	0.033 (c, d)	0.033 (c, d)	—
Mg	0.11 (c, d)	—	—
	0.25 (c, d)	—	—
Mn	0.53 (v, d)	0.53 (v, d)	0.43 (c)
	—	—	0.45 (v)

* The letters c, v, indicate whether the energy has been given with respect to the conduction or valence band edges. The letters a, d, show whether the level is generally considered to be an acceptor or a donor.

† Some instability of these levels was observed.

TABLE 1-14 (Contd)

DEEP IMPURITY LEVELS IN Si: MEASUREMENTS BY CONVENTIONAL
TECHNIQUES AND BY SURFACE CAPACITANCE STUDIES*[32]

IMPURITY	LEVEL(S) FROM CONVENTIONAL TECHNIQUES		LEVELS FROM SURFACE CAPACITANCE STUDIES, FAHRNER AND GOETZBERGER (1972) (eV)
	(eV)	FROM SZE (1969) (eV)	
Mo	0.33 (c, d)	—	—
	0.34 (v, d)	—	—
	0.30 (v, d)	—	—
N	0.14 (c, d)	—	—
Ni	0.35 (c, a)	0.35 (c, a)	—
	0.23 (v, a)	0.22 (v, a)	—
O	0.16 (c, d)	—	0.51 (c)*
	0.38 (c, a)	—	0.41 (v)
	—	0.03 (v, a)	—
Pb	—	—	0.17 (c)
	—	—	0.37 (v)
Pd	0.34 (v, a)	—	—
Pt	0.25 (c, a)	—	—
	0.36 (v, a) ?	0.37 (c, a)	—
	0.30 (v, d)	—	—
S	0.18 (c, d)	0.18 (c, d)	0.26 (c)
	0.37 (c, d)	0.37 (c, d)	0.48 (v)
	0.52 (c, d)	—	—
Se	? (c, d)	—	0.25 (c)
	—	—	0.40 (c)
Sn	—	—	0.17 (c)
	—	—	0.37 (v)
Ta	—	—	<0.14 (c)
	—	—	0.43 (c)
Te	0.14 (c, d)	—	—
	? (c, d)	—	—
Ti	—	—	0.21 (c)
Tl	0.26 (v, a)	0.26 (v, a)	0.30 (v)
V	—	—	0.49 (c)
	—	—	0.40 (v)
W	0.22 (c, a)	—	—
	0.30 (c, a)	—	—
	0.37 (c, a)	—	—
	0.34 (v, d)	—	—
	0.31 (v, d)	—	—
Zn	0.55 (c, a)	0.55 (c, a)	0.55 (c)
	0.31 (v, a)	0.31 (v, a)	0.26 (v)

* These oxygen levels broaden at room temperature and disappear after weeks.

Reprinted by permission of John Wiley & Sons, Inc.

TABLE 1-15

THERMAL EXPANSION COEFFICIENT (α) & STIFFNESS PARAMETER $[E/(1-\nu)]$ FOR
SELECTED FILMS, SILICIDES, CONSTITUENT METALS, SILICON,
AND SUBSTRATES OVER THE TEMPERATURE RANGE 20-400°C

FILM	α ppm/°C	$[E/(1-\nu)]$ 10^{12} dynes/cm^2	REFERENCE
Undoped Polysilicon	2.81 ± 0.20	1.5 ± 0.16	33
Doped Polysilicon ($\rho_s \sim 16\ \Omega/\square$	2.95 ± 0.24	1.24 ± 0.16	33
CVD Si_3N_4	1.6	3.7	34
RF Plasma SiN	1.5	1.1	34
CVD BN	1.0	1.3	35
Ti	8.5	4.6	34
$TiSi_2$	14.5	1.0	34
Ta	6.5	3.2	34
$TaSi_2$	16.3	1.1	34
Mo	5.0	4.1	34
$MoSi_2$	14.7	1.0	34
W	4.5	4.8	34
WSi_2	13.7	1.2	34
<100> Silicon*	3.24		36
<100> Silicon		1.81	37
<1102> Sapphire (Al_2O_3)	7.7	5.0	34
Fused Quartz	0.55	0.9	34

* See Figure 2-26 of Section 2 for a plot of α versus temperature for silicon.

REFERENCES

1. E. Richard Cohen, "Fundamental Constants Today," *Research/Development* (March 1974), pp. 32-38.

2. A.S. Grove, *Physics and Technology of Semiconductor Devices.* New York: John Wiley & Sons, 1967, pp. 102-103.

3. S.M. Sze, *Physics of Semiconductor Devices.* New York: John Wiley & Sons, 1969, pp. 57-58.

4. J.W. Corbett, R.S. McDonald, and G.D. Watkins, "The Configuration and Diffusion of Isolated Oxygen in Silicon and Germanium," *J. Phys. Chem. Solids,* Vol. 25 (1964), p. 878.

5. H.J. McSkimin, "Measurement of Elastic Constants at Low Temperatures by Means of Ultrasonic Waves — Data for Silicon and Germanium Single Crystals, and for Fused Silica," *J. Appl. Phys.,* Vol. 24, No. 8 (August 1953), pp. 988-997.

6. A. Thanailakis, "Contacts Between Simple Metals and Atomically Clean Silicon," *J. Phys. C: Solid State Phys.,* Vol. 8 (1975), pp. 655-668.

7. *Semiconductors.* ed. N.B. Hannay, New York: Reinhold, 1959, pp. 324-330.

8. H.D. Barber, "Effective Mass and Intrinsic Concentration in Silicon," *Solid-State Electronics,* Vol. 10 (1967), pp. 1039-1051.

9. N.D. Thai, "Anomalous Diffusion in Semiconductors — a Quantitative Analysis," *Solid-State Electron.,* Vol. 13 (1970), pp. 165-172.

10. "Physical/Electrical Properties of Silicon," *Integrated Silicon Device Technology,* Vol. 5 AD-605-558, under Contract AF 33 (657)-10340, Research Triangle Institute, July 1964, pp. 98-106.

11. H.R. Maxwell, Jr., W.R. Knolle, and R.D. Plummer, private communication.

12. R.J. Jaccodine and W.A. Schlegel, "Measurement of Strains at Si-SiO_2 Interface," *J. Appl. Phys.,* Vol. 37, No. 6 (May 1966), p. 2429.

13. C.M. Drum, J. D. Ashner, and P.F. Schmidt, private communication.

14. T.F. Retajczk, Jr. and W.R. Knolle, private communication.

15. F. Martinet, G. Guegan, and I.C. Jesionka, "Effect of Frequency on Plasma Si N Deposited in a parallel Radial Flow Reactor." This paper was originally presented at the *Spring Meeting, The Electrochem. Soc.,* San Francisco, CA, May 8-13, 1983, (Extended Abstracts, Vol. 83-1), p. 181.

16. M.P. Lepselter, "Beam-Lead Technology," B.S.T.J., *45,* No. 2 (February 1966), p. 247.

17. *Metals Handbook,* 8th ed., Vol. 1, Am. Soc. for Metals (1961).

18. W.H. Kohl, *Handbook of Materials and Techniques for Vacuum Devices.* New York: Reinhold, 1967, p. 426.

19. M. Hansen, *Constitution of Binary Alloys.* New York : McGraw-Hill, 1958.

20. F.A. Shunk, *Constitution of Binary Alloys, Second Supplement.* New York: McGraw-Hill, 1969.

21. R.P. Elliot, *Constitution of Binary Alloys, First Supplement.* New York: McGraw-Hill, 1965.

22. B. Pellegrini, "Properties of Silicon-Metal Contacts Versus Metal Work-Function, Silicon Impurity Concentration and Bias Voltage," *J. Phys. D: Appl. Phys.,* Vol. 9 (1976), pp. 55-68.

23. B.L. Smith and E.H. Rhoderick, "Schottky Barriers on p-Type Silicon," *Solid-State Electron.,* Vol. 14 (1971), pp. 71-75.

24. R.A. Zettler and A.M. Cowley, "p-n Junction-Schottky Barrier Hybrid Diode," *IEEE Trans. on Electron Devices,* Vol. ED-16, No. 1 (January 1969), p. 60.

25. S.M. Sze, unpublished work.

26. A.N. Saxena, "Hafnium-Silicon Schottky Barriers: Large Barrier Height on p-Type Silicon and Ohmic Behavior on n-Type Silicon," *Appl. Phys. Lett.,* Vol. 19, No. 3 (August 1971) pp. 71-73.

27. J.M. Andrews and F.B. Koch, "Formation of NiSi and Current Transport Across the NiSi-Si Interface," *Solid-State Electron.,* Vol. 14 (1971), pp. 901-908.

28. C.J. Kircher, "Metallurgical Properties and Electrical Characteristics of Palladium Silicide-Silicon Contacts," *Solid-State Electron.,* Vol. 14 (1971), pp. 507-513.

29. J.M. Andrews and M.P. Lepselter, "Reverse Current-Voltage Characteristics of Metal-Silicide Schottky Diodes," *Solid-State Electron.,* Vol. 13 (1970), pp. 1011-1023.

30. F.A. Trumbore, "Solid Solubilities of Impurity Elements in Germanium and Silicon," B.S.T.J., *39,* No. 1 (January 1960), pp. 205-233.

31. S.M. Sze and J.C. Irvin, "Resistivity, Mobility and Impurity Levels in GaAs, Ge, and Si at 300°K," *Solid-State Electron.,* Vol. 11 (1968), pp. 599-602.

32. A.G. Milnes, *Deep Impurities in Semiconductors.* New York: John Wiley & Sons, 1973, pp. 14-15.

33. T.F. Retajczyk, Jr., private communication.

34. T.F. Retajczyk, Jr., and A.K. Sinha, private communication.

35. T.F. Retajczyk, Jr. and A.K. Sinha, "Elastic Stiffness and Thermal Expansion Coefficient of BN Films," *Applied Physics Letters,* Vol. 36 (2) (January 1980), pp. 161-163.

36. P.J. Burkhardt and R.F. Marvel, "Thermal Expansion of Sputtered Silicon Nitride Films," *J. Electrochem. Soc.: Solid-State Science* Vol. 116, No. 6 (June 1969), pp. 864-866.

37. W.A. Brantley, "Calculated Elastic Constants for Stress Problems Associated with Semiconductor Devices," *J. Appl. Phys.,* Vol. 44, No. 1 (January 1973), pp. 534-535.

2. PHYSICAL PROPERTIES

LIST OF SYMBOLS

Symbol	Definition
a	Absorption coefficient
C_T	Total phosphorus concentration
E_c	Energy at conduction band edge
E_f	Fermi energy
E_g	Energy gap
E_i	Intrinsic Fermi energy
E_v	Energy at valence band edge
h	Plank's constant
k	Boltzmann's constant
K_{th}	Thermal conductivity
L	Effective (extrinsic) Debye length
m	Electron rest mass
m_l	Light hole mass
m_h	Heavy hole mass
$m^{(N)}$	''Effective density-of-states'' mass
$m^{(N)}{}_{l_c}$	Effective density-of-states mass for conduction-band electrons
$m^{(N)}{}_{l_v}$	Effective density-of-states mass for valence-band holes
m_1	Transverse electron-effective-mass in the conduction band
m_{11}	Longitudinal electron-effective-mass in the conduction band
$<m>$	Average density-of-states mass
n	Equilibrium concentration of electrons
n_a	Un-ionized acceptor density
n_d	Un-ionized doner density
n_i	Intrinsic Fermi level
n_n	Electron concentration in n-type material
n_p	Electron concentration in p-type material
N	Impurity concentration
N_a	Acceptor doping density
N_d	Donor doping density
N_c	Effective density-of-states at the conduction band edge
N_v	Effective density-of-states at the valence band edge
p	Equilibrium concentration of holes
p_n	Hole concentration in n-type material
p_p	Hole concentration in p-type material
q	Electronic charge
T	Absolute temperature
V_D	Drift velocity
ε_a	Acceptor energy level measured relative to the valence band edge
ε_d	Donor energy level measured relative to the conduction band edge
ε_λ	Spectrical emissivity
ρ	Resistivity
μ	Carrier mobility
μ_n	Electron mobility
μ_p	Hole mobility
ν	Number of valleys in momentum space
λ	Wavelength

2.1 Equilibrium Semiconductor Electron and Hole Carrier Concentrations[1]

The equilibrium concentrations of electrons, n, and holes, p, are given by

$$n = N_c \exp\left[-(E_c - E_f)/kT\right] = n_i \exp\left[(E_f - E_i)/kT\right] \qquad 2.1.1$$

and

$$p = N_v \exp\left[-(E_f - E_v)/kT\right] = n_i \exp\left[(E_i - E_f)/kT\right], \qquad 2.1.2$$

where N_c and N_v are the fictitious, but convenient, effective density-of-states at the conduction band and valence band edges, respectively; E_g is the energy gap; E_f is the Fermi energy; n_i is the intrinsic carrier concentration; E_i is the intrinsic Fermi level; k is Boltzmann's constant; and T is the absolute temperature. These quantities are defined in Figure 2-1.

The intrinsic Fermi level relative to the center of the band gap is

$$E_i = kT \ln (N_v/N_c). \qquad 2.1.3$$

The effective density-of-states at the conduction band (valence band) edge is given by

$$N_c, N_v = 2(2\pi mkT/h^2)^{3/2} \left(\frac{m^{(N)}}{m}\right)^{3/2} \qquad 2.1.4$$

or

$$N_c, N_v = 4.84\times10^{15}\, T^{3/2} \left(\frac{m^{(N)}}{m}\right)^{3/2} \text{ in units of cm}^{-3}, \qquad 2.1.5$$

where m is the electron rest mass and $m^{(N)}$ is the appropriate "effective density-of-states" mass for each band.

For the conduction band,

$$m^{(N)}\big|_c = \nu^{2/3} (m_{11} m_1^2)^{1/3}, \qquad 2.1.6$$

where m_{11} and m_1 are the longitudinal and transverse electron effective masses and ν is the number of valleys in momentum space (four for Ge, six for Si).

Hensel et al[2] have reported the following measured values for silicon at 4.2K:

$$m_{11}/m = 0.9163 \pm 0.0004$$

$$m_1/m = 0.1905 \pm 0.0001.$$

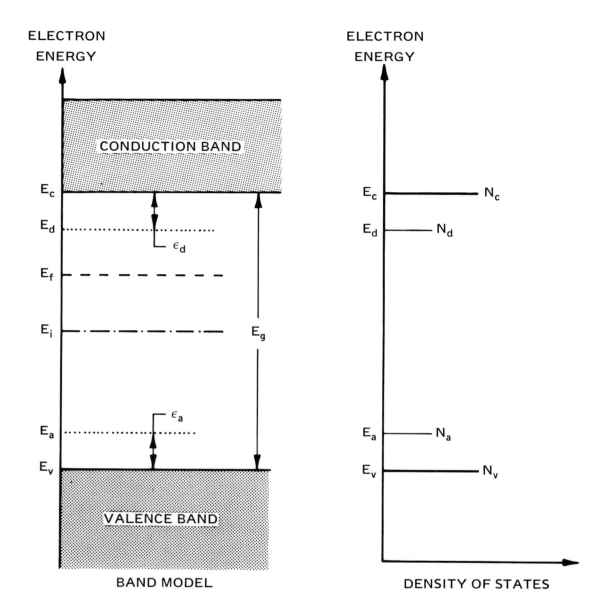

$$n=N_c \exp \left[-(E_c-E_f)/kT\right]=n_i \exp \left[(E_f-E_i)/kT\right]$$

$$p=N_v \exp \left[-(E_f-E_v)/kT\right]=n_i \exp \left[(E_i-E_f)/kT\right]$$

$$E_i=\tfrac{1}{2}(E_c+E_v)+\tfrac{1}{2}kT \ln(N_v/N_c)$$

Figure 2-1 — Band Model and Density-Of-States Diagrams for Hole and Electron Concentrations

From this, the conduction band effective mass (Equation 2.1.6) is

$$\frac{m^{(N)}|_c}{m} = 1.062.$$

The temperature dependence of $m^{(N)}|_c$ is given in Figure 2-2.[3] At 300K, the value is 1.18.

For the valence band, the "effective density-of-states mass" is given in terms of light and heavy hole masses, m_ℓ and m_h, by

$$m^{(N)}|_v = [m_\ell^{3/2} + m_h^{3/2}]^{2/3}. \qquad\qquad 2.1.7$$

Barber[3] reports $m_h/m = 0.537$ and $m_\ell/m = 0.153$ for heavy and light hole effective masses at 4.2K. From Equation 2.1.7, the value for $m^{(N)}|_v/m$ is 0.591. The temperature dependence of $m^{(N)}|_v$ is given in Figure 2-3. [3] At 300K, the value is 0.81.

The hole-electron product (pn) in equilibrium is given by

$$pn = n_i^2 = N_c N_v \exp(-E_g/kT), \qquad\qquad 2.1.8$$

where E_g is the energy band gap.

The geometric mean of the conduction and valence band effective density-of-states masses, $\langle m \rangle$, is referred to as the "average density-of-states effective mass," i.e.,

$$\langle m \rangle = (m^{(N)}|_c \bullet m^{(N)}|_v)^{1/2}. \qquad\qquad 2.1.9$$

From Equations 2.1.4, 2.1.5, and 2.1.8, the intrinsic carrier concentration, n_i, is

$$n_i = 2(2\pi mk/h^2)^{3/2} \left(\frac{\langle m \rangle}{m}\right)^{3/2} T^{3/2} \exp(-E_g/2kT) \qquad\qquad 2.1.10$$

or

$$n_i = 4.84 \times 10^{15} \left(\frac{\langle m \rangle}{m}\right)^{3/2} T^{3/2} \exp(-E_g/2kT) \ (cm^{-3}). \qquad\qquad 2.1.11$$

Note that $\langle m \rangle$ and E_g are implicit functions of temperature in these expressions.

The experimental intrinsic carrier concentration as a function of reciprocal temperature is given in Figure 2-4 for Ge and Si. These curves are empirically fit to the experimental data (450K<T<1100K) of Morin and Maita[4] [5]

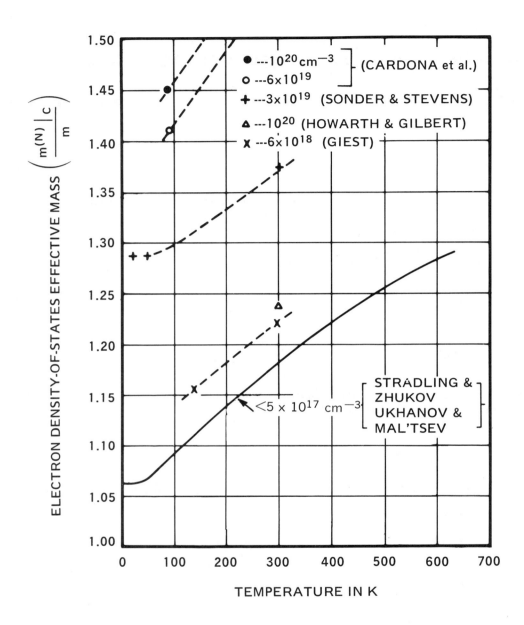

Figure 2-2 — Temperature Dependence of Silicon Electron (Conduction Band) Density-Of-States Effective Mass[3] (Reprinted with permission from Pergamon Press, Ltd.)

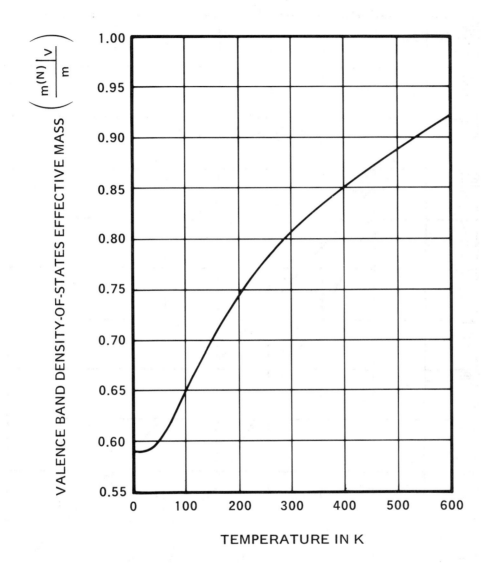

Figure 2-3 — Calculated Temperature Dependence of Silicon Hole (Valence Band) Density-of-States Effective Mass[3] (Reprinted with permission from Pergamon Press, Ltd.)

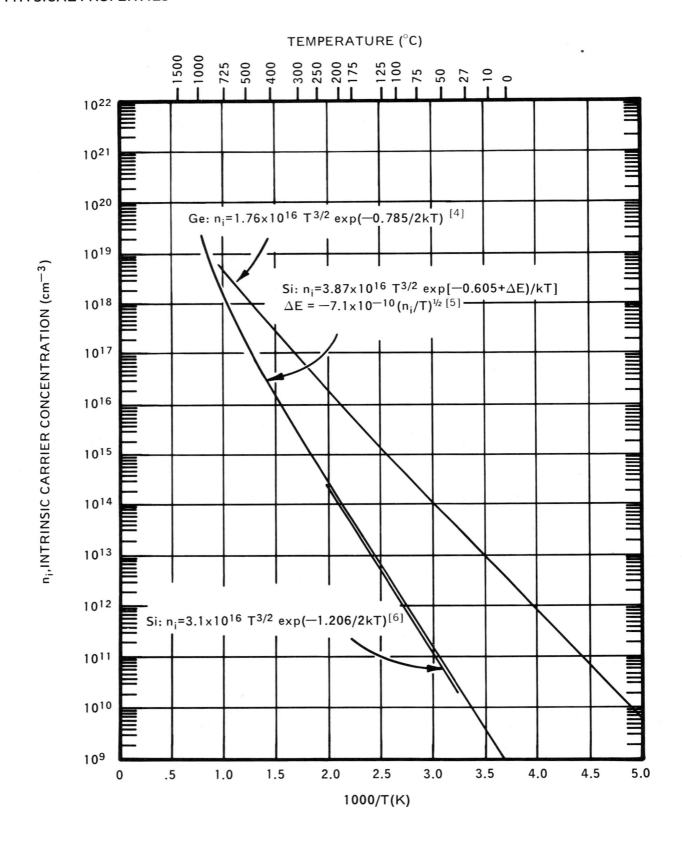

Figure 2-4 —Temperature Dependence of Intrinsic Carrier Concentration for Ge and Si

For Ge

$$n_i = 1.76 \times 10^{16} \, T^{3/2} \exp(-0.785/2kT) \, (cm^{-3}), \qquad\qquad 2.1.12$$

and for Si

$$n_i = 3.87 \times 10^{16} \, T^{3/2} \exp[-(0.605 + \Delta E)/kT]. \qquad\qquad 2.1.13$$

$$\Delta E = -7.1 \times 10^{-10} \, (n_i/T)^{1/2}.$$

Putley and Mitchell [6] made measurements over the temperature range 300-500K and fit their data for Si to

$$n_i = 3.10 \times 10^{16} \, T^{3/2} \exp(-1.206/2kT). \qquad\qquad 2.1.14$$

This curve is also shown in Figure 2-4. Herlet[7] made p-n junction measurements to extend the n_i data down to 250K. His data is in reasonable agreement with equation 2.1.14. Figure 2-5 shows n_i versus T(K) for Ge, Si, GaAs, and GaP.[8] These curves give the overall temperature dependence of n_i for these materials.

2.2   Energy Gap, E_g, and Density-Of-States Effective Mass, $<m>$

Thurmond[8] provides the following data on the temperature dependence of the energy gap, $E_g(T)$, of Si and Ge.

$$E_g(T) = E_g(0) - \alpha T^2/(T+\beta), \qquad\qquad 2.2.1$$

where

	$E_g(0)$ [eV]	$\alpha/10^{-4}$ [eV K^{-1}]	β[K]
Ge	0.7437 ± 0.001	4.774 ± 0.30	235 ± 40
Si	1.17 ± 0.001	4.730 ± 0.25	636 ± 50

In Figure 2-6, $[E_g(T) - E_g(0)]$ is plotted as a function of temperature for Si and Ge.

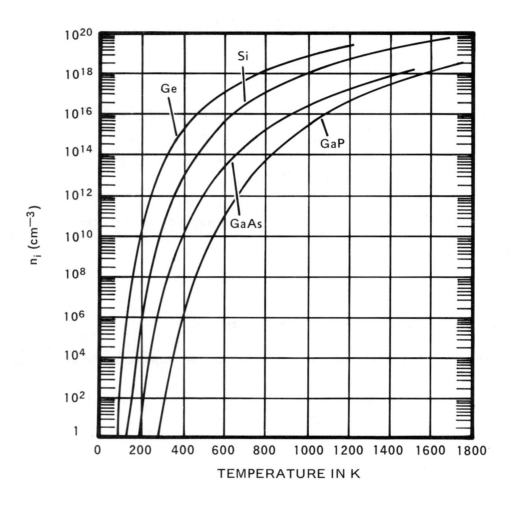

Figure 2-5 — The Intrinsic Carrier Concentration, n_i in cm^{-3}, as a Function of T for Ge, Si, GaAs, and GaP[8] (Reprinted by permission of the publisher, The Electrochemical Society, Inc.)

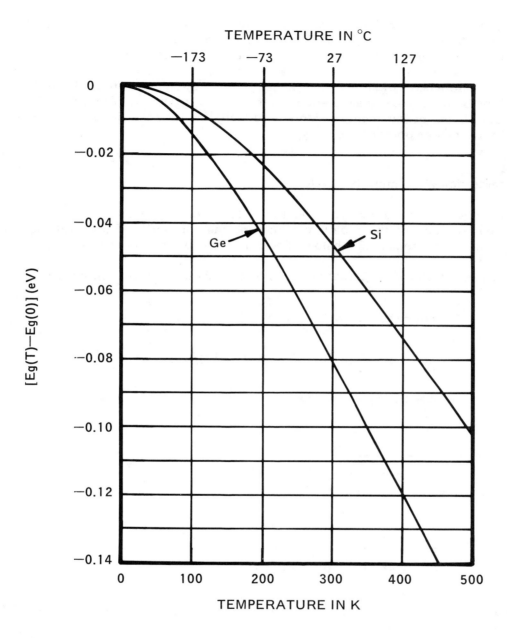

$$E_g(T) - Eg(0) = -\alpha\, T^2/(T+\beta)$$

	$\alpha/10^{-4}$ [eV K^{-1}]	β[K]	Eg(0)[eV]
Ge:	4.774 ± 0.3	235 ± 40	0.744
Si:	4.730 ± 0.25	636 ± 50	1.17

Figure 2-6 —The Difference Between the Energy Band Gap, Eg, at T (K) and at 0 K, as a Function of Temperature for Ge and Si[8]   (Reprinted by permission of the publisher, the Electrochemical Society, Inc.)

Thurmond[8] uses the silicon experimental data for n_i by Morin and Maita[4] [5] and the temperature dependence of E_g given by Equation 2.2.1 and Equation 2.1.9 to calculate $<m>$. His data for silicon is presented in Figure 2-7 along with some published values obtained by other methods. At 300K, $<m>/m$ is 0.98. The Si-M&M curve is based upon the n_i measurements of Morin and Maita.[5] The Si-P&M curve is based upon the n_i measurements of Putley and Mitchell.[6] The Si-B curve is from Barber.[3] The dashed-line extension of the Barber curve is the present estimate of $<m>$ to the melting point of Si. The Ge-M&M curve is based upon the n_i measurements of Morin and Maita[4] ; the solid line is the present estimate of $<m>$ from 0 K to the melting point of Ge.

2.3 Carrier Concentration With Impurity Levels

Equilibrium-free carrier concentrations and the Fermi level for doped semiconductors are determined by the simultaneous solution of the equations describing the appropriate Fermi statistics and electrical neutrality (Equations 2.3.1 through 2.3.5 below). See Figure 2-1 for definitions of symbols.

Electron and hole concentrations are (Equations 2.1.1 and 2.1.2 repeated for convenience):

$$n = N_c \exp \left[-(E_c - E_f)/kT \right] = n_i \exp \left[(E_f - E_i)/kT \right] \qquad \text{2.3.1}$$

$$p = N_v \exp \left[-(E_f - E_v)/kT \right] = n_i \exp \left[(E_i - E_f)/kT \right]. \qquad \text{2.3.2}$$

The number of un-ionized donors, n_d, and acceptors, n_a, are:

$$n_d = \frac{N_d}{1 + \frac{1}{2} \exp \left[(E_c - \epsilon_d) - E_f \right] kT} \qquad \text{2.3.3}$$

$$n_a = \frac{N_a}{1 + \frac{1}{2} \exp \left[E_f - (E_v + \epsilon_a) \right]/kT}. \qquad \text{2.3.4}$$

Here, N_d and N_a are the donor and acceptor doping densities and ϵ_d and ϵ_a are the associated energy levels measured relative to the band edge.

These charged quantities must satisfy electrical neutrality in the crystal, i.e.,

$$n + (N_a - n_a) = p + (N_d - n_d). \qquad \text{2.3.5}$$

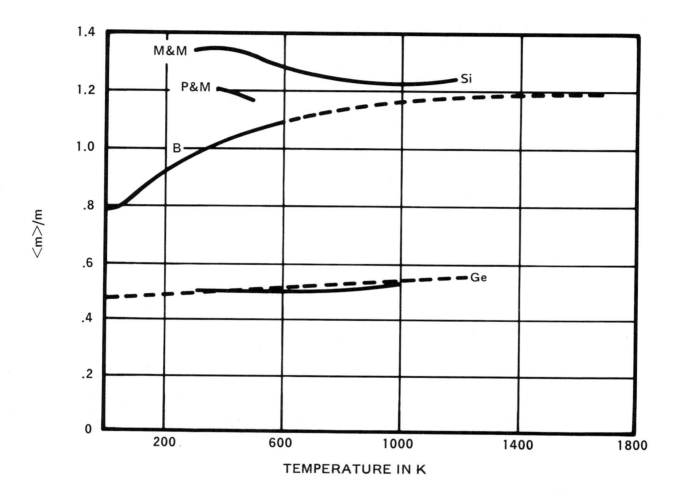

Figure 2-7 — Temperature Dependence of the Density-Of-States Average Effective Mass $\langle m \rangle$ in Units of the Electron Rest Mass, m [8] (Reprinted by permission of the publisher, the Electrochemical Society, Inc.)

The equilibrium np product is n_i^2 therefore, equation 2.3.5 becomes a quadratic in either n or p,

$$n^2 - n\,[(N_d-n_d)-(N_a-n_a)] - n_i^2 = 0 \qquad\qquad 2.3.6$$

$$p^2 - p\,[(N_a-n_a)-(N_d-n_d)] - n_i^2 = 0, \qquad\qquad 2.3.7$$

where n_i is given by Equation 2.1.10.

Figure 2-8 presents the Fermi potential calculated from these equations for n-type (ϵ_d = 0.044eV) and p-type (ϵ_d=0.044eV) silicon as a function of doping and temperature. Figure 2-9 gives the Fermi potential as a function of doping at 300K. The dashed curves assume that the impurity atoms are always ionized independently of Fermi level. In these calculations, the effective mass temperature dependence shown in Figure 2-1 and 2-2 is assumed.

The Fermi potential versus temperature as a function of doping concentration for germanium is given in Figure 2-10.

If the donors and acceptors are fully ionized, Equations 2.3.5 and 2.3.6 become, for n- and p-type semiconductors, respectively:

$$n_n = \frac{1}{2}\left[(N_d-N_a) + \sqrt{(N_d-N_a)^2 + 4\,n_i^2}\right] \qquad\qquad 2.3.8$$

$$p_p = \frac{1}{2}\left[(N_a-N_d) + \sqrt{(N_a-N_d)^2 + 4\,n_i^2}\right] \qquad\qquad 2.3.9$$

and if the net doping $(N_a - N_d)$ is much greater than n_i

$$n_n \sim (N_d - N_a) \qquad\qquad 2.3.10$$

$$p_p \sim (N_a - N_d). \qquad\qquad 2.3.11$$

and the minority carrier densities are:

$$p_n = \frac{n_i^2}{N_d - N_a}$$
2.3.12

$$n_p = \frac{n_i^2}{N_a - N_d} \quad .$$
2.3.13

Equations 2.3.10 through 2.3.13 are usually adequate for practical semiconductor device work. Band-model parameters for silicon are tabulated in Table 2-1 for convenience.

TABLE 2-1

SUMMARY OF BAND—MODEL PARAMETERS FOR SILICON

PARAMETERS	SYMBOL	EQUATION	VALUE (300K)
Conduction Band Effective Density-of-States Mass	$m^{(N)}\vert_c$	2.1.6	1.18
Valence Band Effective Density-of-States Mass	$m^{(N)}\vert_v$	2.1.7	0.81
Effective Density-of-States, Conduction Band	N_c	2.1.5	$3.22 \times 10^{19} (\text{cm}^{-3})$
Effective Density-of-States, Valence Band	N_v	2.1.7	$1.83 \times 10^{19} (\text{cm}^{-3})$
Intrinsic Carrier Concentration	n_i	2.1.13	$1.38 \times 10^{10} (\text{cm}^{-3})$
Energy Gap	E_g	2.2.1	1.12 eV

2.4 Debye Length

Mathematical descriptions of the electric potential and fields at semiconductor surfaces are obtained from solutions of Poisson's equation and contain a factor which has the dimensions of length. This characteristic length is called the effective Debye length L. For small disturbances of the electric potential at the surface, the potential within the semiconductor decays exponentially, from the surface to the bulk value, as a function of depth z in terms of z/L.

For larger disturbances, the functional dependence of the potential is more complex but is still described in terms of the effective Debye length. See Figure 2-11.

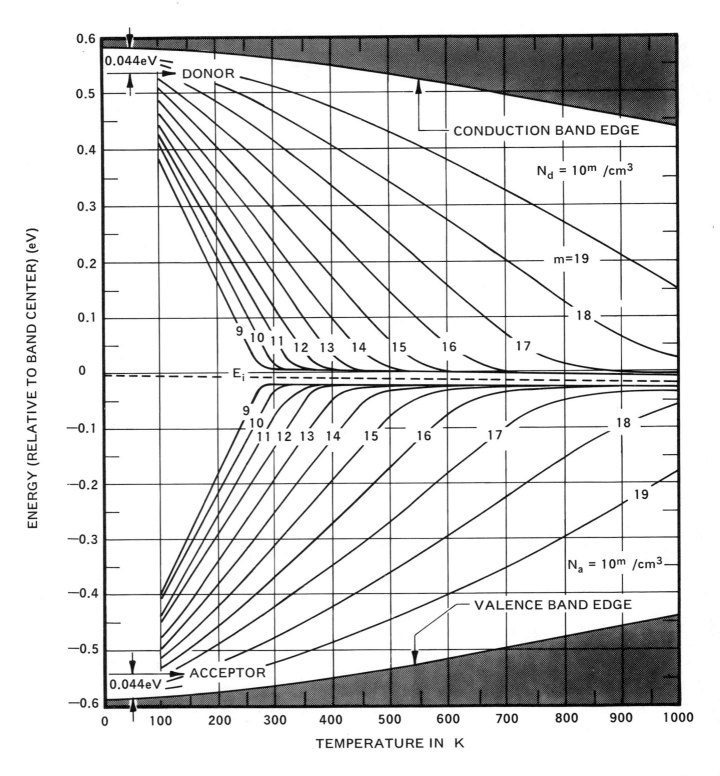

Note: Curves above the center of bandgap assume a donor level at 0.044eV (below the conduction band); curves below bandgap center assume an acceptor level at 0.044eV (above the valence band).

Figure 2-8—Calculated Fermi Potential for Silicon

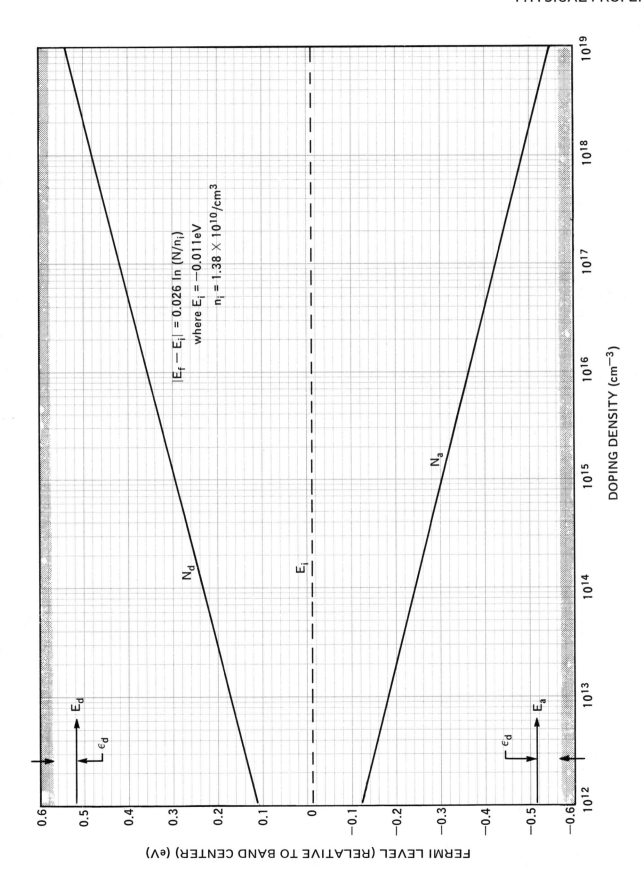

Figure 2-9 — Fermi Level Versus Doping for Si, T = 300K

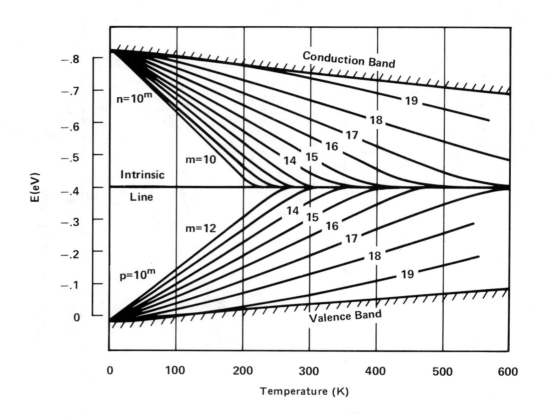

Figure 2-10 — Fermi Level Versus Temperature for Ge[9]
(Reprinted by permission of John Wiley & Sons, Inc.)

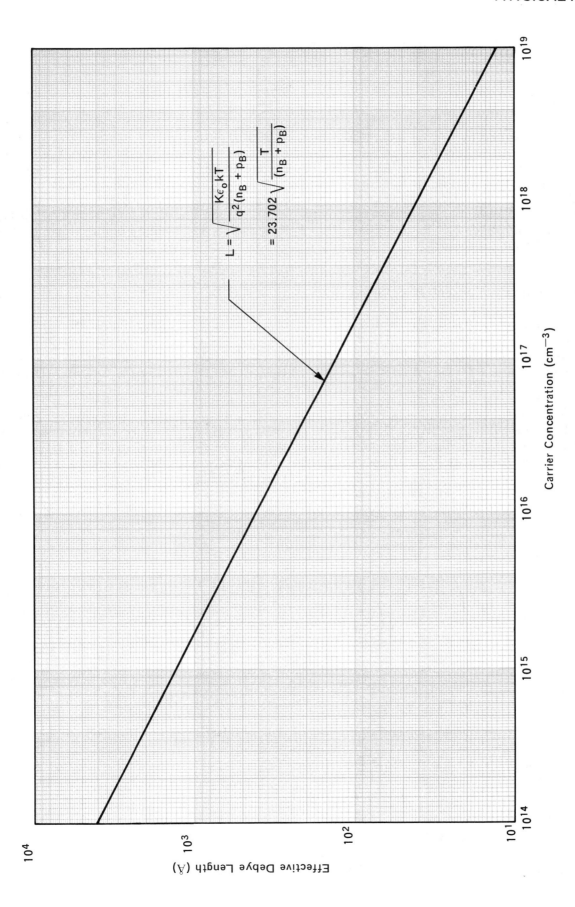

$$L = \sqrt{\frac{K\epsilon_o kT}{q^2(n_B + p_B)}}$$

$$= 23.702 \sqrt{\frac{T}{(n_B + p_B)}}$$

Carrier Concentration (cm^{-3})

Effective Debye Length (Å)

Figure 2-11 — Effective Debye Length for Silicon at 300K [10]

2.5   Resistivity and Mobility Versus Impurity Concentration for Silicon

Introduction

Irvin's curves of resistivity versus impurity concentration have been among the most commonly used conversions by semiconductor engineers. [11] [12] These curves were derived empirically from collections of experimental data measured by various laboratories prior to 1960. Measurements by Wagner [13] showed discrepancies in the impurity concentration range from 10^{16} to 10^{19} cm^{-3} for p-type silicon. For the past few years, the National Bureau of Standards (NBS) has conducted careful concentration and resistivity measurements on bulk, floating-zone grown boron-doped and phosphorus-doped silicon crystals.[14] [15] [16] Data from Esaki and Miyahara[17] and Fair and Tsai[18] for high concentration phosphorus were also used for the resistivity versus phosphorus concentration curve. Data for phosphorus from F. Mousty, P. Ostoja, and L. Passari were also added to the NBS data.[19]   NBS added some imaginary points to improve the accuracy of curve-fitting procedures.

Resistivity Versus Impurity Concentration Curves

The resistivity versus concentration curves at 23°C are shown in Figures 2-12 through 2-18. The p-type curve was determined from boron-doped samples while the n-type curve was determined from phosphorus-doped samples. For easy use, the curves were plotted on 2-cycle log-log paper except Figure 2-12 where extrapolated data are presented on 3 X 3 log-log paper. The measured lower concentration limits were 10^{13} cm^{-3} for both p-type and n-type silicon. The curves shown below 10^{13} cm^{-3} were linear extrapolations and they were not verified from measurement. The high concentration limits were close to 5×10^{20} cm^{-3} for both types of dopants. No experimental data were obtained from grown crystals for impurity concentrations greater than 2×10^{20} cm^{-3}. The few data points at concentrations greater than 2×10^{20} cm^{-3} were obtained from diffused thin-layer samples. In fact, in Irvin's curve for the n-type silicon, the data were from resistivity and neutron activation analysis of phosphorus-diffused samples also.

The NBS curve fitting for boron and phosphorus dopants are given below:

Boron-Doped Silicon

Resistivity-to-concentration conversion at 23°C

$$N = \frac{1}{q\rho}\left(k_{min} + \frac{k_{max}-k_{min}}{1+\left(\dfrac{\rho}{\rho_{ref}}\right)^{\alpha}}\right)$$

where

q = electronic charge 1.6021×10^{-19} C

ρ = resistivity (ohm-cm)

N = carrier concentration = impurity concentration (cm^{-3}) for $N < 10^{20}$ cm^{-3}

k_{min} = 2.13×10^{-3} $\left(\dfrac{V \bullet s}{cm^2}\right)$

k_{max} = 1.947×10^{-2} $\left(\dfrac{V \bullet s}{cm^2}\right)$

ρ_{ref} = 1.833×10^{-2} (ohm-cm)

a = 1.105.

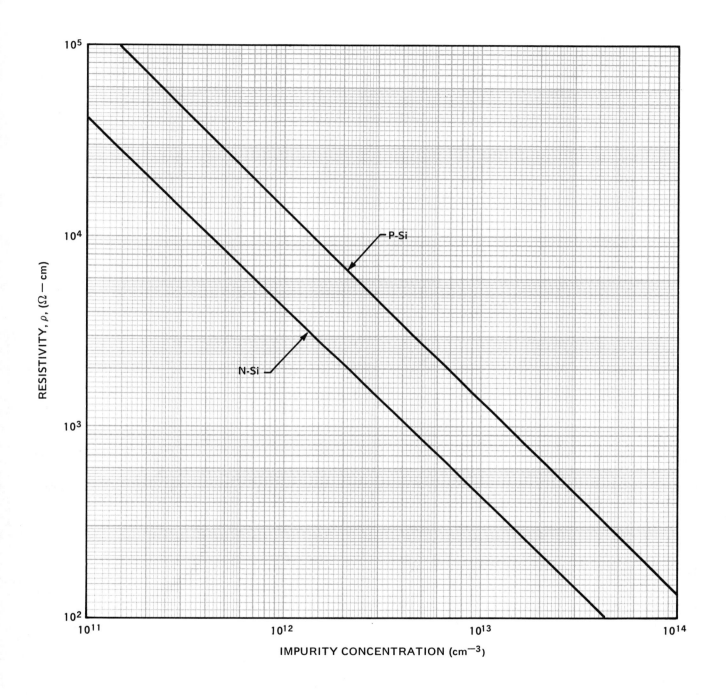

Figure 2-12 — Resistivity Versus Impurity Concentration at 300K

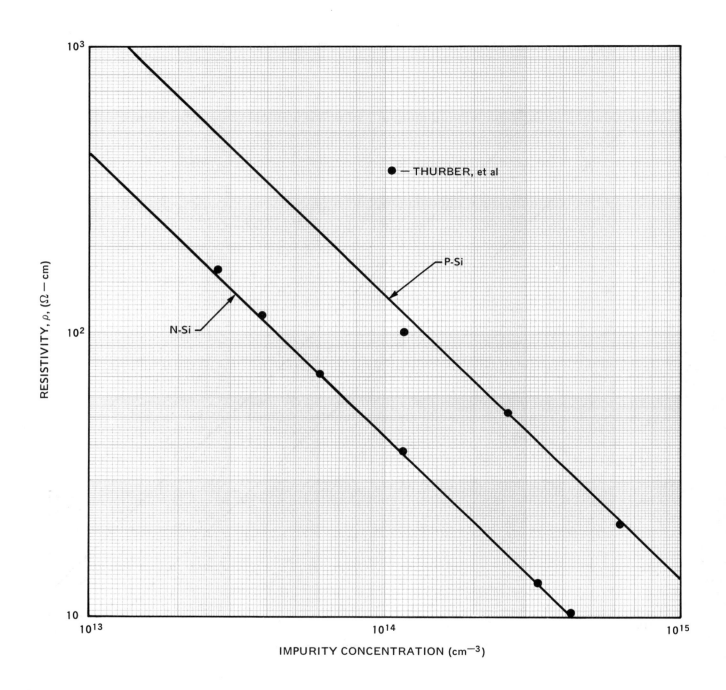

Figure 2-13 — Resitivity Versus Impurity Concentration at 300K

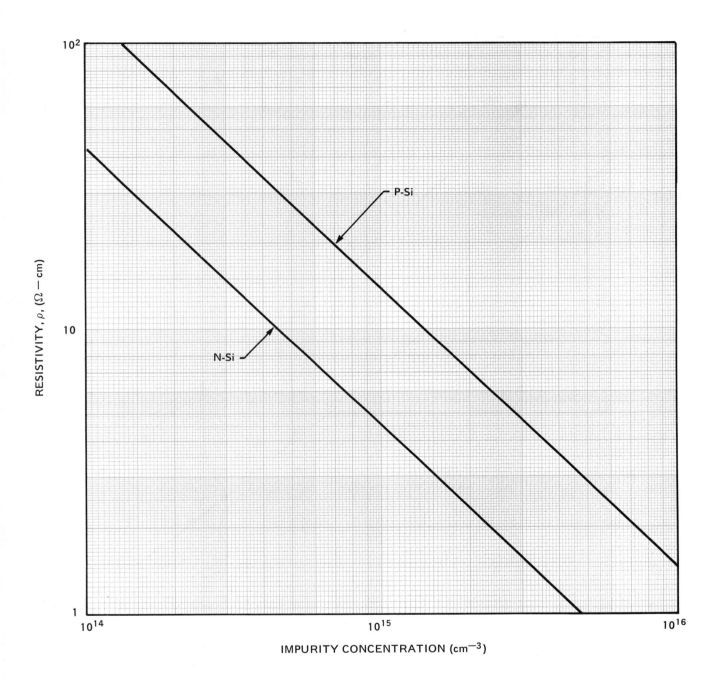

Figure 2-14 — Resistivity Versus Impurity Concentration at 300K

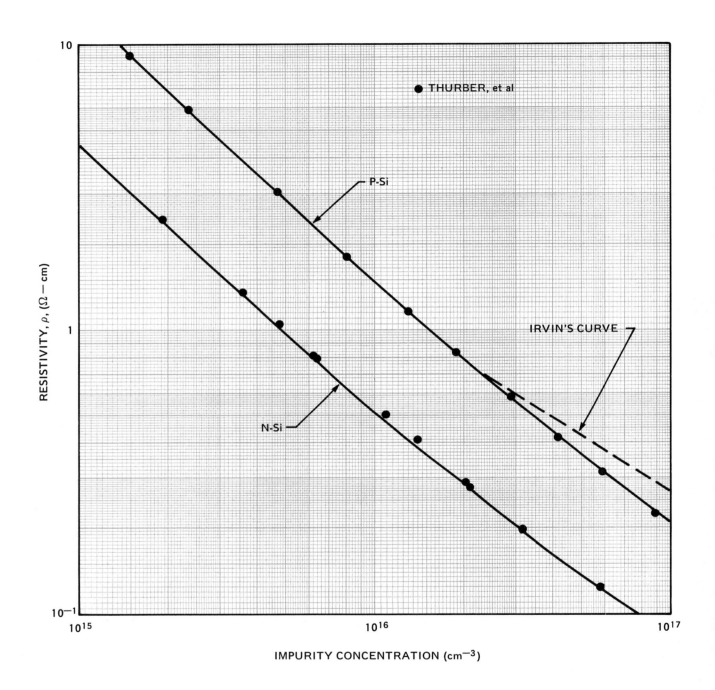

Figure 2-15 — Resistivity Versus Impurity Concentration at 300K

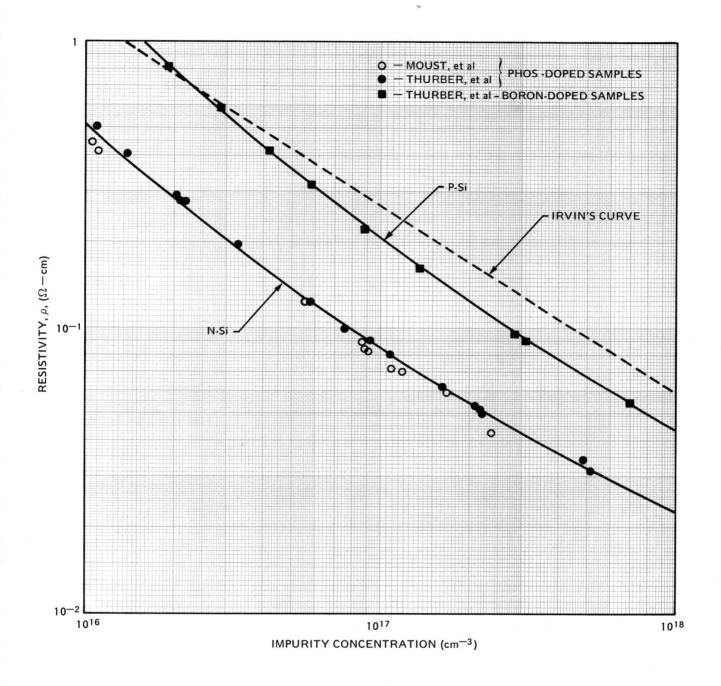

Figure 2-16 — Resistivity Versus Impurity Concentration at 300K

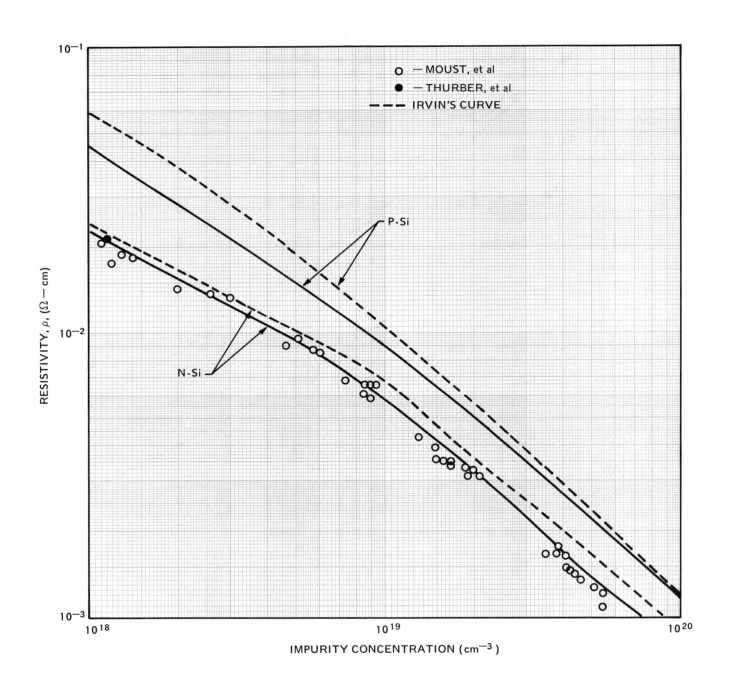

Figure 2-17 — Resistivity Versus Impurity Concentration at 300K

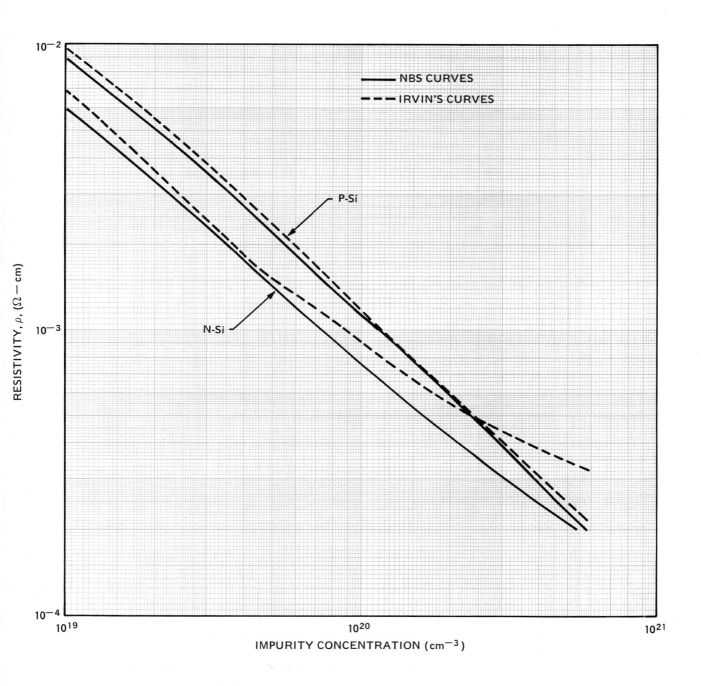

Figure 2-18 — Resistivity Versus Impurity Concentration

Phosphorus-Doped Silicon [16]

Resistivity-to-concentration conversion at 23°C

$$N = \frac{1}{q\rho} \; 10^{(A_1)},$$

$$A_1 = \frac{a_0 + a_1 x + a_2 x^2 + a_3 x^3}{1 + b_1 x + b_2 x^2 + b_3 x^3},$$

where

q = electronic charge, 1.6021×10^{-19}

ρ = resistivity (ohm-cm)

N = carrier concentration = impurity concentration (when $N \lesssim 10^{20}$) (cm^{-3})

$x = \log_{10}(\rho)$

$a_0 = -3.1083 \pm 0.0038$

$a_1 = -3.2626 \pm 0.0952$

$a_2 = -1.2196 \pm 0.0341$

$a_3 = -0.13923 \pm 0.00468$

$b_1 = 1.0265 \pm 0.0318$

$b_2 = 0.38755 \pm 0.0109$

$b_3 = 0.041833 \pm 0.00168.$

Tables 2-2 and 2-3 show resistivity to dopant-density conversion for boron- and phosphorus-doped wafers. Tables 2-4 and 2-5 show dopant-density to resistivity conversion for boron- and phosphorus-doped wafers. Composite curves are shown in Figure 2-19.

High Concentration Considerations

The effect of impurity ionization is presented in this paragraph. At impurity concentrations greater than 10^{20} cm^{-3}, the electron and hole mobilities and impurity ionizations depend not only on impurity concentrations but also on the sample thermal history, crystallographic defects, impurity-vacancy complex formations, and impurity precipitations. Hence, in this concentration range, the experimental data shows wider spread than one would like to have for curve fittings. The reader is cautioned in using either Irvin's curves or the NBS curves for impurity concentrations beyond ~3×10^{20} cm^{-3} for both p- and n-type silicon that the curves represent an extrapolation in this range. Two examples from arsenic- and phosphorus-diffused samples are given in Figures 2-20 and 2-21.[18] [20] The electron concentrations differ significantly from the phosphorus or arsenic concentrations at concentrations greater than $\cong 10^{20}$ cm^{-3}. If the diffusion conditions are comparable to those data given in references 17 and 19, the use of these curves is preferred to the NBS curve for n-type silicon at concentrations $>10^{20}$ cm^{-3}. For detailed diffusion conditions, the readers should refer to the original publications. For arsenic-diffused layers, Figure 2-20 shows the resistivity versus electron concentration only. It has been observed that in the high arsenic doping range the resistivity versus arsenic concentration relationship depends on the diffusion source (chemical diffusion source, ion implantation, etc.) and diffusion conditions.[20] [21] The resistivity versus arsenic concentration cannot be presented in simple relationships and it is not so presented.

TABLE 2-2
RESISTIVITY TO DOPANT DENSITY CONVERSION AT 23°C FOR BORON-DOPED SILICON

RESISTIVITY (OHM-CM)	DOPANT DENSITY (/CM**3)										
	0.	0.1	0.2	0.3	0.4	0.5	0.6	0.7	0.8	0.9	
1.0E-04	1.2E 21	1.1E 21	1.0E 21	9.3E 20	8.6E 20	8.1E 20	7.6E 20	7.1E 20	6.7E 20	6.4E 20	BORON
2.0E-04	6.0E 20	5.8E 20	5.5E 20	5.2E 20	5.0E 20	4.8E 20	4.6E 20	4.5E 20	4.3E 20	4.2E 20	BORON
3.0E-04	4.0E 20	3.9E 20	3.8E 20	3.6E 20	3.5E 20	3.4E 20	3.3E 20	3.2E 20	3.2E 20	3.1E 20	BORON
4.0E-04	3.0E 20	2.9E 20	2.9E 20	2.8E 20	2.7E 20	2.7E 20	2.6E 20	2.5E 20	2.5E 20	2.4E 20	BORON
5.0E-04	2.4E 20	2.3E 20	2.3E 20	2.3E 20	2.2E 20	2.2E 20	2.1E 20	2.1E 20	2.1E 20	2.0E 20	BORON
6.0E-04	2.0E 20	2.0E 20	1.9E 20	1.9E 20	1.9E 20	1.8E 20	1.8E 20	1.8E 20	1.7E 20	1.7E 20	BORON
7.0E-04	1.7E 20	1.7E 20	1.6E 20	1.6E 20	1.6E 20	1.6E 20	1.6E 20	1.5E 20	1.5E 20	1.5E 20	BORON
8.0E-04	1.5E 20	1.5E 20	1.4E 20	1.4E 20	1.4E 20	1.4E 20	1.4E 20	1.4E 20	1.3E 20	1.3E 20	BORON
9.0E-04	1.31E 20	1.29E 20	1.28E 20	1.27E 20	1.25E 20	1.24E 20	1.22E 20	1.21E 20	1.20E 20	1.19E 20	BORON
1.0E-03	1.17E 20	1.06E 20	9.70E 19	8.92E 19	8.25E 19	7.67E 19	7.17E 19	6.72E 19	6.32E 19	5.97E 19	BORON
2.0E-03	5.65E 19	5.36E 19	5.09E 19	4.85E 19	4.63E 19	4.43E 19	4.24E 19	4.07E 19	3.91E 19	3.76E 19	BORON
3.0E-03	3.62E 19	3.49E 19	3.37E 19	3.25E 19	3.15E 19	3.04E 19	2.95E 19	2.86E 19	2.77E 19	2.69E 19	BORON
4.0E-03	2.61E 19	2.54E 19	2.47E 19	2.40E 19	2.34E 19	2.28E 19	2.22E 19	2.17E 19	2.11E 19	2.06E 19	BORON
5.0E-03	2.01E 19	1.97E 19	1.92E 19	1.88E 19	1.84E 19	1.80E 19	1.76E 19	1.72E 19	1.69E 19	1.65E 19	BORON
6.0E-03	1.62E 19	1.59E 19	1.56E 19	1.53E 19	1.50E 19	1.47E 19	1.44E 19	1.41E 19	1.39E 19	1.36E 19	BORON
7.0E-03	1.34E 19	1.32E 19	1.29E 19	1.27E 19	1.25E 19	1.23E 19	1.21E 19	1.19E 19	1.17E 19	1.15E 19	BORON
8.0E-03	1.13E 19	1.11E 19	1.10E 19	1.08E 19	1.06E 19	1.05E 19	1.03E 19	1.02E 19	1.00E 19	9.88E 18	BORON
9.0E-03	9.74E 18	9.60E 18	9.47E 18	9.33E 18	9.20E 18	9.08E 18	8.95E 18	8.83E 18	8.72E 18	8.60E 18	BORON
1.0E-02	8.49E 18	7.48E 18	6.65E 18	5.97E 18	5.39E 18	4.89E 18	4.47E 18	4.10E 18	3.78E 18	3.49E 18	BORON
2.0E-02	3.24E 18	3.02E 18	2.82E 18	2.64E 18	2.48E 18	2.33E 18	2.20E 18	2.07E 18	1.96E 18	1.86E 18	BORON
3.0E-02	1.77E 18	1.68E 18	1.60E 18	1.53E 18	1.46E 18	1.40E 18	1.34E 18	1.28E 18	1.23E 18	1.18E 18	BORON
4.0E-02	1.14E 18	1.09E 18	1.05E 18	1.02E 18	9.80E 17	9.46E 17	9.14E 17	8.84E 17	8.56E 17	8.29E 17	BORON
5.0E-02	8.03E 17	7.79E 17	7.55E 17	7.33E 17	7.12E 17	6.92E 17	6.73E 17	6.55E 17	6.37E 17	6.21E 17	BORON
6.0E-02	6.05E 17	5.89E 17	5.75E 17	5.61E 17	5.47E 17	5.34E 17	5.22E 17	5.10E 17	4.98E 17	4.87E 17	BORON
7.0E-02	4.76E 17	4.66E 17	4.56E 17	4.47E 17	4.37E 17	4.29E 17	4.20E 17	4.12E 17	4.03E 17	3.96E 17	BORON
8.0E-02	3.88E 17	3.81E 17	3.74E 17	3.67E 17	3.60E 17	3.54E 17	3.48E 17	3.42E 17	3.36E 17	3.30E 17	BORON
9.0E-02	3.24E 17	3.19E 17	3.14E 17	3.09E 17	3.04E 17	2.99E 17	2.94E 17	2.90E 17	2.85E 17	2.81E 17	BORON
1.0E-01	2.77E 17	2.40E 17	2.11E 17	1.88E 17	1.69E 17	1.53E 17	1.40E 17	1.28E 17	1.18E 17	1.10E 17	BORON
2.0E-01	1.02E 17	9.59E 16	9.01E 16	8.49E 16	8.02E 16	7.60E 16	7.22E 16	6.88E 16	6.56E 16	6.27E 16	BORON
3.0E-01	6.00E 16	5.76E 16	5.53E 16	5.32E 16	5.13E 16	4.94E 16	4.77E 16	4.61E 16	4.46E 16	4.32E 16	BORON
4.0E-01	4.19E 16	4.07E 16	3.95E 16	3.84E 16	3.73E 16	3.63E 16	3.54E 16	3.45E 16	3.36E 16	3.28E 16	BORON
5.0E-01	3.21E 16	3.13E 16	3.06E 16	2.99E 16	2.93E 16	2.87E 16	2.81E 16	2.75E 16	2.69E 16	2.64E 16	BORON
6.0E-01	2.59E 16	2.54E 16	2.49E 16	2.45E 16	2.40E 16	2.36E 16	2.32E 16	2.28E 16	2.24E 16	2.21E 16	BORON
7.0E-01	2.17E 16	2.14E 16	2.10E 16	2.07E 16	2.04E 16	2.01E 16	1.98E 16	1.95E 16	1.92E 16	1.89E 16	BORON
8.0E-01	1.87E 16	1.84E 16	1.82E 16	1.79E 16	1.77E 16	1.75E 16	1.72E 16	1.70E 16	1.68E 16	1.66E 16	BORON
9.0E-01	1.64E 16	1.62E 16	1.60E 16	1.58E 16	1.56E 16	1.54E 16	1.53E 16	1.51E 16	1.49E 16	1.47E 16	BORON
1.0E 00	1.46E 16	1.32E 16	1.20E 16	1.10E 16	1.01E 16	9.41E 15	8.79E 15	8.24E 15	7.76E 15	7.33E 15	BORON
2.0E 00	6.95E 15	6.60E 15	6.29E 15	6.01E 15	5.75E 15	5.51E 15	5.29E 15	5.08E 15	4.90E 15	4.72E 15	BORON
3.0E 00	4.56E 15	4.41E 15	4.27E 15	4.13E 15	4.01E 15	3.89E 15	3.78E 15	3.68E 15	3.58E 15	3.48E 15	BORON
4.0E 00	3.39E 15	3.31E 15	3.23E 15	3.15E 15	3.08E 15	3.01E 15	2.94E 15	2.88E 15	2.82E 15	2.76E 15	BORON
5.0E 00	2.70E 15	2.65E 15	2.60E 15	2.55E 15	2.50E 15	2.45E 15	2.41E 15	2.37E 15	2.32E 15	2.28E 15	BORON
6.0E 00	2.25E 15	2.21E 15	2.17E 15	2.14E 15	2.10E 15	2.07E 15	2.04E 15	2.01E 15	1.98E 15	1.95E 15	BORON
7.0E 00	1.92E 15	1.89E 15	1.87E 15	1.84E 15	1.82E 15	1.79E 15	1.77E 15	1.74E 15	1.72E 15	1.70E 15	BORON
8.0E 00	1.68E 15	1.66E 15	1.64E 15	1.62E 15	1.60E 15	1.58E 15	1.56E 15	1.54E 15	1.52E 15	1.51E 15	BORON
9.0E 00	1.49E 15	1.47E 15	1.46E 15	1.44E 15	1.43E 15	1.41E 15	1.40E 15	1.38E 15	1.37E 15	1.35E 15	BORON
1.0E 01	1.34E 15	1.22E 15	1.11E 15	1.03E 15	9.55E 14	8.91E 14	8.35E 14	7.85E 14	7.42E 14	7.02E 14	BORON
2.0E 01	6.67E 14	6.35E 14	6.06E 14	5.80E 14	5.56E 14	5.33E 14	5.13E 14	4.94E 14	4.76E 14	4.60E 14	BORON
3.0E 01	4.44E 14	4.30E 14	4.16E 14	4.04E 14	3.92E 14	3.81E 14	3.70E 14	3.60E 14	3.50E 14	3.41E 14	BORON
4.0E 01	3.33E 14	3.25E 14	3.17E 14	3.10E 14	3.03E 14	2.96E 14	2.89E 14	2.83E 14	2.77E 14	2.72E 14	BORON
5.0E 01	2.66E 14	2.61E 14	2.56E 14	2.51E 14	2.46E 14	2.42E 14	2.38E 14	2.34E 14	2.29E 14	2.26E 14	BORON

TABLE 2-2 (Contd)

RESISTIVITY TO DOPANT DENSITY CONVERSION AT 23°C FOR BORON-DOPED SILICON

RESISTIVITY (OHM-CM)	DOPANT DENSITY (/CM**3)										
	0.	0.1	0.2	0.3	0.4	0.5	0.6	0.7	0.8	0.9	
6.0E 01	2.22E 14	2.18E 14	2.15E 14	2.11E 14	2.08E 14	2.05E 14	2.02E 14	1.99E 14	1.96E 14	1.93E 14	BORON
7.0E 01	1.90E 14	1.87E 14	1.85E 14	1.82E 14	1.80E 14	1.77E 14	1.75E 14	1.73E 14	1.71E 14	1.68E 14	BORON
8.0E 01	1.66E 14	1.64E 14	1.62E 14	1.60E 14	1.58E 14	1.57E 14	1.55E 14	1.53E 14	1.51E 14	1.49E 14	BORON
9.0E 01	1.48E 14	1.46E 14	1.45E 14	1.43E 14	1.42E 14	1.40E 14	1.39E 14	1.37E 14	1.36E 14	1.34E 14	BORON
1.0E 02	6.6E 13	6.3E 13	6.0E 13	5.8E 13	5.5E 13	5.3E 13	5.1E 13	4.9E 13	4.7E 13	4.6E 13	BORON
2.0E 02	4.4E 13	4.3E 13	4.2E 13	4.0E 13	3.9E 13	3.8E 13	3.7E 13	3.6E 13	3.5E 13	3.4E 13	BORON
3.0E 02	3.3E 13	3.2E 13	3.2E 13	3.1E 13	3.0E 13	3.0E 13	2.9E 13	2.8E 13	2.8E 13	2.7E 13	BORON
4.0E 02	2.7E 13	2.6E 13	2.6E 13	2.5E 13	2.5E 13	2.4E 13	2.4E 13	2.3E 13	2.3E 13	2.3E 13	BORON
5.0E 02	2.2E 13	2.2E 13	2.1E 13	2.1E 13	2.1E 13	2.0E 13	2.0E 13	2.0E 13	2.0E 13	1.9E 13	BORON
6.0E 02	1.9E 13	1.9E 13	1.8E 13	1.8E 13	1.8E 13	1.8E 13	1.7E 13	1.7E 13	1.7E 13	1.7E 13	BORON
7.0E 02	1.7E 13	1.6E 13	1.6E 13	1.6E 13	1.6E 13	1.6E 13	1.5E 13	1.5E 13	1.5E 13	1.5E 13	BORON
8.0E 02	1.5E 13	1.5E 13	1.4E 13	1.4E 13	1.4E 13	1.4E 13	1.4E 13	1.4E 13	1.4E 13	1.3E 13	BORON
9.0E 02	1.3E 13	1.2E 13	1.1E 13	1.0E 13	9.5E 12	8.9E 12	8.3E 12	7.8E 12	7.4E 12	7.0E 12	BORON
1.0E 03	6.6E 12	6.3E 12	6.0E 12	5.8E 12	5.5E 12	5.3E 12	5.1E 12	4.9E 12	4.7E 12	4.6E 12	BORON
2.0E 03	4.4E 12	4.3E 12	4.2E 12	4.0E 12	3.9E 12	3.8E 12	3.7E 12	3.6E 12	3.5E 12	3.4E 12	BORON
3.0E 03	3.3E 12	3.2E 12	3.2E 12	3.1E 12	3.0E 12	3.0E 12	2.9E 12	2.8E 12	2.8E 12	2.7E 12	BORON
4.0E 03	2.7E 12	2.6E 12	2.6E 12	2.5E 12	2.5E 12	2.4E 12	2.4E 12	2.3E 12	2.3E 12	2.3E 12	BORON
5.0E 03	2.2E 12	2.2E 12	2.1E 12	2.1E 12	2.1E 12	2.0E 12	2.0E 12	2.0E 12	2.0E 12	1.9E 12	BORON
6.0E 03	1.9E 12	1.9E 12	1.8E 12	1.8E 12	1.8E 12	1.8E 12	1.7E 12	1.7E 12	1.7E 12	1.7E 12	BORON
7.0E 03	1.7E 12	1.6E 12	1.6E 12	1.6E 12	1.6E 12	1.6E 12	1.5E 12	1.5E 12	1.5E 12	1.5E 12	BORON
8.0E 03	1.5E 12	1.5E 12	1.4E 12	1.4E 12	1.4E 12	1.4E 12	1.4E 12	1.4E 12	1.4E 12	1.3E 12	BORON
9.0E 03	1.3E 12	1.2E 12	1.1E 12	1.0E 12	9.5E 11	8.9E 11	8.3E 11	7.8E 11	7.4E 11	7.0E 11	BORON
1.0E 04	6.6E 11	6.3E 11	6.0E 11	5.8E 11	5.5E 11	5.3E 11	5.1E 11	4.9E 11	4.7E 11	4.6E 11	BORON
2.0E 04	4.4E 11	4.3E 11	4.2E 11	4.0E 11	3.9E 11	3.8E 11	3.7E 11	3.6E 11	3.5E 11	3.4E 11	BORON
3.0E 04	3.3E 11	3.2E 11	3.2E 11	3.1E 11	3.0E 11	3.0E 11	2.9E 11	2.8E 11	2.8E 11	2.7E 11	BORON
4.0E 04	2.7E 11	2.6E 11	2.6E 11	2.5E 11	2.5E 11	2.4E 11	2.4E 11	2.3E 11	2.3E 11	2.3E 11	BORON
5.0E 04	2.2E 11	2.2E 11	2.1E 11	2.1E 11	2.1E 11	2.0E 11	2.0E 11	2.0E 11	2.0E 11	1.9E 11	BORON
6.0E 04	1.9E 11	1.9E 11	1.8E 11	1.8E 11	1.8E 11	1.8E 11	1.7E 11	1.7E 11	1.7E 11	1.7E 11	BORON
7.0E 04	1.7E 11	1.6E 11	1.6E 11	1.6E 11	1.6E 11	1.6E 11	1.5E 11	1.5E 11	1.5E 11	1.5E 11	BORON
8.0E 04	1.5E 11	1.5E 11	1.4E 11	1.4E 11	1.4E 11	1.4E 11	1.4E 11	1.4E 11	1.4E 11	1.3E 11	BORON
9.0E 04	1.3E 11	1.2E 11	1.1E 11	1.0E 11	9.5E 10	8.9E 10	8.3E 10	7.8E 10	7.4E 10	7.0E 10	BORON

TABLE 2-3

RESISTIVITY TO DOPANT DENSITY CONVERSION AT 23°C FOR PHOSPHORUS-DOPED SILICON

RESISTIVITY (OHM-CM)	DOPANT DENSITY (/CM**3)										
	0.	0.1	0.2	0.3	0.4	0.5	0.6	0.7	0.8	0.9	
1.0E-04	1.6E 21	1.4E 21	1.2E 21	1.1E 21	9.4E 20	8.5E 20	7.7E 20	7.0E 20	6.4E 20	5.9E 20	PHOS
2.0E-04	5.5E 20	5.1E 20	4.8E 20	4.5E 20	4.3E 20	4.0E 20	3.8E 20	3.6E 20	3.5E 20	3.3E 20	FHCS
3.0E-04	3.15E 20	3.02E 20	2.89E 20	2.78E 20	2.67E 20	2.58E 20	2.48E 20	2.40E 20	2.32E 20	2.24E 20	FHCS
4.0E-04	2.17E 20	2.10E 20	2.04E 20	1.98E 20	1.93E 20	1.87E 20	1.82E 20	1.78E 20	1.73E 20	1.69E 20	PHOS
5.0E-04	1.65E 20	1.61E 20	1.57E 20	1.53E 20	1.50E 20	1.47E 20	1.44E 20	1.41E 20	1.38E 20	1.35E 20	PHOS
6.0E-04	1.32E 20	1.30E 20	1.27E 20	1.25E 20	1.23E 20	1.20E 20	1.18E 20	1.16E 20	1.14E 20	1.12E 20	PHOS
7.0E-04	1.10E 20	1.09E 20	1.07E 20	1.05E 20	1.04E 20	1.02E 20	1.00E 20	9.90E 19	9.75E 19	9.61E 19	FHOS
8.0E-04	9.48E 19	9.34E 19	9.21E 19	9.09E 19	8.97E 19	8.85E 19	8.73E 19	8.62E 19	8.51E 19	8.40E 19	PHOS
9.0E-04	8.30E 19	8.19E 19	8.09E 19	8.00E 19	7.90E 19	7.81E 19	7.72E 19	7.63E 19	7.54E 19	7.46E 19	FHOS
1.0E-03	7.38E 19	6.64E 19	6.04E 19	5.53E 19	5.11E 19	4.74E 19	4.42E 19	4.14E 19	3.90E 19	3.68E 19	PHOS
2.0E-03	3.48E 19	3.30E 19	3.14E 19	2.99E 19	2.86E 19	2.73E 19	2.62E 19	2.51E 19	2.41E 19	2.32E 19	FHOS
3.0E-03	2.24E 19	2.16E 19	2.08E 19	2.01E 19	1.94E 19	1.88E 19	1.82E 19	1.76E 19	1.71E 19	1.66E 19	PHOS
4.0E-03	1.61E 19	1.56E 19	1.52E 19	1.47E 19	1.43E 19	1.40E 19	1.36E 19	1.32E 19	1.29E 19	1.26E 19	PHOS
5.0E-03	1.22E 19	1.19E 19	1.16E 19	1.14E 19	1.11E 19	1.08E 19	1.06E 19	1.03E 19	1.01E 19	9.87E 18	PHOS
6.0E-03	9.65E 18	9.44E 18	9.24E 18	9.04E 18	8.84E 18	8.66E 18	8.47E 18	8.30E 18	8.13E 18	7.96E 18	PHOS
7.0E-03	7.80E 18	7.65E 18	7.49E 18	7.35E 18	7.20E 18	7.06E 18	6.93E 18	6.80E 18	6.67E 18	6.54E 18	FHOS
8.0E-03	6.42E 18	6.30E 18	6.19E 18	6.07E 18	5.96E 18	5.86E 18	5.75E 18	5.65E 18	5.55E 18	5.46E 18	PHOS
9.0E-03	5.36E 18	5.27E 18	5.18E 18	5.09E 18	5.01E 18	4.92E 18	4.84E 18	4.76E 18	4.68E 18	4.61E 18	PHOS
1.0E-02	4.53E 18	3.87E 18	3.34E 18	2.91E 18	2.55E 18	2.25E 18	2.00E 18	1.78E 18	1.60E 18	1.45E 18	PHOS
2.0E-02	1.31E 18	1.20E 18	1.09E 18	1.00E 18	9.26E 17	8.56E 17	7.94E 17	7.39E 17	6.89E 17	6.44E 17	PHOS
3.0E-02	6.04E 17	5.68E 17	5.35E 17	5.05E 17	4.77E 17	4.52E 17	4.29E 17	4.08E 17	3.88E 17	3.70E 17	PHOS
4.0E-02	3.53E 17	3.38E 17	3.23E 17	3.10E 17	2.98E 17	2.86E 17	2.75E 17	2.65E 17	2.55E 17	2.46E 17	FHOS
5.0E-02	2.37E 17	2.29E 17	2.22E 17	2.15E 17	2.08E 17	2.02E 17	1.95E 17	1.90E 17	1.84E 17	1.79E 17	PHOS
6.0E-02	1.74E 17	1.69E 17	1.65E 17	1.61E 17	1.56E 17	1.53E 17	1.49E 17	1.45E 17	1.42E 17	1.38E 17	FHOS
7.0E-02	1.35E 17	1.32E 17	1.29E 17	1.27E 17	1.24E 17	1.21E 17	1.19E 17	1.16E 17	1.14E 17	1.12E 17	PHOS
8.0E-02	1.10E 17	1.08E 17	1.06E 17	1.04E 17	1.02E 17	9.99E 16	9.81E 16	9.64E 16	9.48E 16	9.32E 16	FHOS
9.0E-02	9.16E 16	9.01E 16	8.87E 16	8.72E 16	8.59E 16	8.45E 16	8.32E 16	8.20E 16	8.07E 16	7.95E 16	PHOS
1.0E-01	7.84E 16	6.83E 16	6.04E 16	5.40E 16	4.88E 16	4.45E 16	4.09E 16	3.78E 16	3.51E 16	3.27E 16	PHOS
2.0E-01	3.07E 16	2.88E 16	2.72E 16	2.58E 16	2.45E 16	2.33E 16	2.22E 16	2.12E 16	2.03E 16	1.95E 16	PHOS
3.0E-01	1.87E 16	1.80E 16	1.73E 16	1.67E 16	1.61E 16	1.56E 16	1.51E 16	1.46E 16	1.42E 16	1.38E 16	PHOS
4.0E-01	1.34E 16	1.30E 16	1.26E 16	1.23E 16	1.20E 16	1.17E 16	1.14E 16	1.11E 16	1.09E 16	1.06E 16	PHOS
5.0E-01	1.04E 16	1.02E 16	9.94E 15	9.73E 15	9.53E 15	9.34E 15	9.15E 15	8.97E 15	8.80E 15	8.64E 15	FHOS
6.0E-01	8.48E 15	8.32E 15	8.18E 15	8.03E 15	7.89E 15	7.76E 15	7.63E 15	7.51E 15	7.39E 15	7.27E 15	PHOS
7.0E-01	7.16E 15	7.05E 15	6.94E 15	6.83E 15	6.73E 15	6.64E 15	6.54E 15	6.45E 15	6.36E 15	6.27E 15	FHOS
8.0E-01	6.19E 15	6.10E 15	6.02E 15	5.95E 15	5.87E 15	5.79E 15	5.72E 15	5.65E 15	5.58E 15	5.51E 15	PHOS
9.0E-01	5.45E 15	5.38E 15	5.32E 15	5.26E 15	5.20E 15	5.14E 15	5.08E 15	5.03E 15	4.97E 15	4.92E 15	PHOS
1.0E 00	4.86E 15	4.39E 15	4.00E 15	3.68E 15	3.40E 15	3.16E 15	2.95E 15	2.77E 15	2.61E 15	2.47E 15	PHOS
2.0E 00	2.34E 15	2.22E 15	2.11E 15	2.02E 15	1.93E 15	1.85E 15	1.78E 15	1.71E 15	1.65E 15	1.59E 15	PHOS
3.0E 00	1.53E 15	1.48E 15	1.43E 15	1.39E 15	1.35E 15	1.31E 15	1.27E 15	1.23E 15	1.20E 15	1.17E 15	PHOS
4.0E 00	1.14E 15	1.11E 15	1.08E 15	1.06E 15	1.03E 15	1.01E 15	9.86E 14	9.64E 14	9.43E 14	9.24E 14	FHOS
5.0E 00	9.05E 14	8.86E 14	8.69E 14	8.52E 14	8.36E 14	8.20E 14	8.05E 14	7.91E 14	7.77E 14	7.63E 14	PHOS
6.0E 00	7.50E 14	7.37E 14	7.25E 14	7.13E 14	7.02E 14	6.91E 14	6.80E 14	6.70E 14	6.60E 14	6.50E 14	FHOS
7.0E 00	6.40E 14	6.31E 14	6.22E 14	6.13E 14	6.05E 14	5.97E 14	5.89E 14	5.81E 14	5.73E 14	5.66E 14	PHOS
8.0E 00	5.58E 14	5.51E 14	5.45E 14	5.38E 14	5.31E 14	5.25E 14	5.19E 14	5.13E 14	5.07E 14	5.01E 14	PHOS
9.0E 00	4.95E 14	4.89E 14	4.84E 14	4.79E 14	4.74E 14	4.68E 14	4.63E 14	4.59E 14	4.54E 14	4.49E 14	PHOS
1.0E 01	4.45E 14	4.03E 14	3.69E 14	3.40E 14	3.15E 14	2.94E 14	2.75E 14	2.59E 14	2.44E 14	2.31E 14	PHOS
2.0E 01	2.19E 14	2.09E 14	1.99E 14	1.90E 14	1.82E 14	1.75E 14	1.68E 14	1.62E 14	1.56E 14	1.50E 14	PHOS
3.0E 01	1.45E 14	1.40E 14	1.36E 14	1.32E 14	1.28E 14	1.24E 14	1.21E 14	1.17E 14	1.14E 14	1.11E 14	FHOS
4.0E 01	1.08E 14	1.06E 14	1.03E 14	1.01E 14	9.83E 13	9.61E 13	9.40E 13	9.20E 13	9.00E 13	8.81E 13	PHOS
5.0E 01	8.64E 13	8.46E 13	8.30E 13	8.14E 13	7.99E 13	7.84E 13	7.70E 13	7.56E 13	7.43E 13	7.30E 13	FHOS

TABLE 2-3 (Contd)

RESISTIVITY TO DOPANT DENSITY CONVERSION AT 23°C FOR PHOSPHORUS-DOPED SILICON

RESISTIVITY (OHM-CM)	DOPANT DENSITY (/CM**3)										
	0.	0.1	0.2	0.3	0.4	0.5	0.6	0.7	0.8	0.9	
6.0E 01	7.18E 13	7.06E 13	6.94E 13	6.83E 13	6.72E 13	6.62E 13	6.51E 13	6.42E 13	6.32E 13	6.23E 13	PHOS
7.0E 01	6.14E 13	6.05E 13	5.96E 13	5.88E 13	5.80E 13	5.72E 13	5.65E 13	5.57E 13	5.50E 13	5.43E 13	PHOS
8.0E 01	5.36E 13	5.29E 13	5.23E 13	5.16E 13	5.10E 13	5.04E 13	4.98E 13	4.92E 13	4.87E 13	4.81E 13	FHCS
9.0E 01	4.76E 13	4.70E 13	4.65E 13	4.60E 13	4.55E 13	4.50E 13	4.45E 13	4.41E 13	4.36E 13	4.32E 13	PHOS
1.0E 02	4.27E 13	3.88E 13	3.55E 13	3.28E 13	3.04E 13	2.83E 13	2.65E 13	2.50E 13	2.36E 13	2.23E 13	PHOS
2.0E 02	2.1E 13	2.0E 13	1.9E 13	1.8E 13	1.8E 13	1.7E 13	1.6E 13	1.6E 13	1.5E 13	1.5E 13	PHOS
3.0E 02	1.4E 13	1.4E 13	1.3E 13	1.3E 13	1.2E 13	1.2E 13	1.2E 13	1.1E 13	1.1E 13	1.1E 13	FHCS
4.0E 02	1.0E 13	1.0E 13	1.0E 13	9.8E 12	9.5E 12	9.3E 12	9.1E 12	8.9E 12	8.7E 12	8.5E 12	PHOS
5.0E 02	8.4E 12	8.2E 12	8.0E 12	7.9E 12	7.7E 12	7.6E 12	7.5E 12	7.3E 12	7.2E 12	7.1E 12	FHCS
6.0E 02	7.0E 12	6.8E 12	6.7E 12	6.6E 12	6.5E 12	6.4E 12	6.3E 12	6.2E 12	6.1E 12	6.0E 12	PHOS
7.0E 02	6.0E 12	5.9E 12	5.8E 12	5.7E 12	5.6E 12	5.6E 12	5.5E 12	5.4E 12	5.3E 12	5.3E 12	PHOS
8.0E 02	5.2E 12	5.1E 12	5.1E 12	5.0E 12	5.0E 12	4.9E 12	4.8E 12	4.8E 12	4.7E 12	4.7E 12	PHOS
9.0E 02	4.6E 12	4.6E 12	4.5E 12	4.5E 12	4.4E 12	4.4E 12	4.3E 12	4.3E 12	4.2E 12	4.2E 12	PHOS
1.0E 03	4.2E 12	3.8E 12	3.5E 12	3.2E 12	3.0E 12	2.8E 12	2.6E 12	2.4E 12	2.3E 12	2.2E 12	PHOS
2.0E 03	2.1E 12	2.0E 12	1.9E 12	1.8E 12	1.7E 12	1.6E 12	1.6E 12	1.5E 12	1.5E 12	1.4E 12	PHOS
3.0E 03	1.4E 12	1.3E 12	1.3E 12	1.2E 12	1.2E 12	1.2E 12	1.1E 12	1.1E 12	1.1E 12	1.0E 12	PHOS
4.0E 03	1.0E 12	1.0E 12	1.0E 12	9.5E 11	9.3E 11	9.1E 11	8.9E 11	8.7E 11	8.5E 11	8.3E 11	PHOS
5.0E 03	8.2E 11	8.0E 11	7.8E 11	7.7E 11	7.5E 11	7.4E 11	7.3E 11	7.1E 11	7.0E 11	6.9E 11	FHCS
6.0E 03	6.8E 11	6.7E 11	6.6E 11	6.5E 11	6.4E 11	6.3E 11	6.2E 11	6.1E 11	6.0E 11	5.9E 11	PHOS
7.0E 03	5.8E 11	5.7E 11	5.6E 11	5.6E 11	5.5E 11	5.4E 11	5.3E 11	5.3E 11	5.2E 11	5.1E 11	PHOS
8.0E 03	5.1E 11	5.0E 11	4.9E 11	4.9E 11	4.8E 11	4.8E 11	4.7E 11	4.7E 11	4.6E 11	4.6E 11	FHCS
9.0E 03	4.5E 11	4.5E 11	4.4E 11	4.4E 11	4.3E 11	4.3E 11	4.2E 11	4.2E 11	4.1E 11	4.1E 11	PHOS
1.0E 04	4.0E 11	3.7E 11	3.4E 11	3.1E 11	2.9E 11	2.7E 11	2.5E 11	2.4E 11	2.2E 11	2.2E 11	PHOS
2.0E 04	2.0E 11	1.9E 11	1.8E 11	1.8E 11	1.7E 11	1.6E 11	1.5E 11	1.5E 11	1.4E 11	1.4E 11	PHOS
3.0E 04	1.3E 11	1.3E 11	1.3E 11	1.2E 11	1.2E 11	1.1E 11	1.1E 11	1.1E 11	1.1E 11	1.0E 11	PHOS
4.0E 04	1.0E 11	9.7E 10	9.5E 10	9.3E 10	9.1E 10	8.9E 10	8.7E 10	8.5E 10	8.3E 10	8.1E 10	FHCS
5.0E 04	8.0E 10	8.0E 10	7.7E 10	7.5E 10	7.4E 10	7.2E 10	7.1E 10	7.0E 10	6.9E 10	6.7E 10	PHOS
6.0E 04	6.6E 10	6.5E 10	6.4E 10	6.3E 10	6.2E 10	6.1E 10	6.0E 10	5.9E 10	5.8E 10	5.8E 10	PHOS
7.0E 04	5.7E 10	5.6E 10	5.5E 10	5.4E 10	5.4E 10	5.3E 10	5.2E 10	5.2E 10	5.1E 10	5.0E 10	PHOS
8.0E 04	5.0E 10	4.9E 10	4.8E 10	4.8E 10	4.7E 10	4.7E 10	4.6E 10	4.6E 10	4.5E 10	4.5E 10	PHOS
9.0E 04	4.4E 10	4.4E 10	4.3E 10	4.3E 10	4.2E 10	4.2E 10	4.1E 10	4.1E 10	4.0E 10	4.0E 10	FHCS

TABLE 2-4
DOPANT DENSITY TO RESISTIVITY CONVERSION AT 23°C FOR BORON-DOPED SILICON

DOPANT DENSITY (/CM**3)	RESISTIVITY (OHM-CM)										
	0.	0.1	0.2	0.3	0.4	0.5	0.6	0.7	0.8	0.9	
1.0E 12	1.3E 04	1.2E 04	1.1E 04	1.0E 04	9.3E 03	8.7E 03	8.2E 03	7.7E 03	7.2E 03	6.9E 03	BORON
2.0E 12	6.5E 03	6.2E 03	5.9E 03	5.7E 03	5.4E 03	5.2E 03	5.0E 03	4.8E 03	4.7E 03	4.5E 03	BORON
3.0E 12	4.3E 03	4.2E 03	4.1E 03	4.0E 03	3.8E 03	3.7E 03	3.6E 03	3.5E 03	3.4E 03	3.3E 03	BORON
4.0E 12	3.3E 03	3.2E 03	3.1E 03	3.0E 03	3.0E 03	2.9E 03	2.8E 03	2.8E 03	2.7E 03	2.7E 03	BORCN
5.0E 12	2.6E 03	2.6E 03	2.5E 03	2.5E 03	2.4E 03	2.4E 03	2.3E 03	2.3E 03	2.3E 03	2.2E 03	BORON
6.0E 12	2.2E 03	2.1E 03	2.1E 03	2.1E 03	2.0E 03	2.0E 03	2.0E 03	1.9E 03	1.9E 03	1.9E 03	BCRCN
7.0E 12	1.9E 03	1.8E 03	1.8E 03	1.8E 03	1.8E 03	1.7E 03	1.7E 03	1.7E 03	1.7E 03	1.7E 03	BORON
8.0E 12	1.6E 03	1.6E 03	1.6E 03	1.6E 03	1.6E 03	1.5E 03	1.5E 03	1.5E 03	1.5E 03	1.5E 03	BORON
9.0E 12	1.5E 03	1.4E 03	1.4E 03	1.4E 03	1.4E 03	1.4E 03	1.5E 03	1.3E 03	1.3E 03	1.3E 03	BCRCN
1.0E 13	1.3E 03	1.2E 03	1.1E 03	1.0E 03	9.3E 02	8.7E 02	8.2E 02	7.7E 02	7.3E 02	6.9E 02	BORON
2.0E 13	6.5E 02	6.2E 02	5.9E 02	5.7E 02	5.4E 02	5.2E 02	5.0E 02	4.8E 02	4.7E 02	4.5E 02	BORON
3.0E 13	4.4E 02	4.2E 02	4.1E 02	4.0E 02	3.8E 02	3.7E 02	3.6E 02	3.5E 02	3.4E 02	3.4E 02	BORON
4.0E 13	3.3E 02	3.2E 02	3.1E 02	3.0E 02	3.0E 02	2.9E 02	2.8E 02	2.8E 02	2.7E 02	2.7E 02	BORON
5.0E 13	2.6E 02	2.6E 02	2.5E 02	2.5E 02	2.4E 02	2.4E 02	2.3E 02	2.3E 02	2.3E 02	2.2E 02	BORON
6.0E 13	2.2E 02	2.1E 02	2.1E 02	2.1E 02	2.0E 02	2.0E 02	2.0E 02	2.0E 02	1.9E 02	1.9E 02	BCRCN
7.0E 13	1.9E 02	1.8E 02	1.8E 02	1.8E 02	1.8E 02	1.7E 02	1.7E 02	1.7E 02	1.7E 02	1.7E 02	BORON
8.0E 13	1.6E 02	1.6E 02	1.6E 02	1.6E 02	1.6E 02	1.5E 02	1.5E 02	1.5E 02	1.5E 02	1.5E 02	BORON
9.0E 13	1.5E 02	1.4E 02	1.4E 02	1.4E 02	1.4E 02	1.4E 02	1.4E 02	1.3E 02	1.3E 02	1.3E 02	BORON
1.0E 14	1.3E 02	1.3E 02	1.1E 02	1.0E 02	9.4E 01	8.7E 01	8.2E 01	7.7E 01	7.3E 01	6.9E 01	BCRCN
2.0E 14	6.56E 01	6.25E 01	5.97E 01	5.71E 01	5.47E 01	5.26E 01	5.05E 01	4.87E 01	4.69E 01	4.53E 01	BORON
3.0E 14	4.38E 01	4.24E 01	4.11E 01	3.99E 01	3.87E 01	3.76E 01	3.66E 01	3.56E 01	3.47E 01	3.38E 01	BCRCN
4.0E 14	3.29E 01	3.21E 01	3.14E 01	3.07E 01	3.00E 01	2.93E 01	2.87E 01	2.81E 01	2.75E 01	2.69E 01	BORON
5.0E 14	2.64E 01	2.59E 01	2.54E 01	2.49E 01	2.45E 01	2.40E 01	2.36E 01	2.32E 01	2.28E 01	2.24E 01	BORON
6.0E 14	2.20E 01	2.17E 01	2.13E 01	2.10E 01	2.07E 01	2.04E 01	2.01E 01	1.98E 01	1.95E 01	1.92E 01	BCRCN
7.0E 14	1.89E 01	1.87E 01	1.84E 01	1.81E 01	1.79E 01	1.77E 01	1.74E 01	1.72E 01	1.70E 01	1.68E 01	BORON
8.0E 14	1.66E 01	1.64E 01	1.62E 01	1.60E 01	1.58E 01	1.56E 01	1.54E 01	1.53E 01	1.51E 01	1.49E 01	BORON
9.0E 14	1.48E 01	1.46E 01	1.44E 01	1.43E 01	1.41E 01	1.40E 01	1.38E 01	1.37E 01	1.36E 01	1.34E 01	BORON
1.0E 15	1.33E 01	1.21E 01	1.11E 01	1.03E 01	9.55E 00	8.93E 00	8.38E 00	7.90E 00	7.47E 00	7.08E 00	BORON
2.0E 15	6.74E 00	6.42E 00	6.14E 00	5.88E 00	5.64E 00	5.42E 00	5.22E 00	5.03E 00	4.85E 00	4.69E 00	BORON
3.0E 15	4.54E 00	4.40E 00	4.26E 00	4.14E 00	4.02E 00	3.91E 00	3.81E 00	3.71E 00	3.61E 00	3.52E 00	BORON
4.0E 15	3.44E 00	3.36E 00	3.28E 00	3.21E 00	3.14E 00	3.07E 00	3.01E 00	2.95E 00	2.89E 00	2.83E 00	BCRCN
5.0E 15	2.78E 00	2.72E 00	2.67E 00	2.63E 00	2.58E 00	2.53E 00	2.49E 00	2.45E 00	2.41E 00	2.37E 00	BORON
6.0E 15	2.33E 00	2.30E 00	2.26E 00	2.23E 00	2.19E 00	2.16E 00	2.13E 00	2.10E 00	2.07E 00	2.04E 00	BORON
7.0E 15	2.02E 00	1.99E 00	1.96E 00	1.96E 00	1.91E 00	1.89E 00	1.87E 00	1.84E 00	1.82E 00	1.80E 00	BORON
8.0E 15	1.78E 00	1.76E 00	1.74E 00	1.72E 00	1.70E 00	1.68E 00	1.66E 00	1.64E 00	1.63E 00	1.61E 00	BORON
9.0E 15	1.59E 00	1.58E 00	1.56E 00	1.54E 00	1.53E 00	1.51E 00	1.50E 00	1.48E 00	1.47E 00	1.46E 00	BORON
1.0E 16	1.44E 00	1.32E 00	1.24E 00	1.13E 00	1.06E 00	9.93E-01	9.37E-01	8.87E-01	8.42E-01	8.03E-01	BCRCN
2.0E 16	7.67E-01	7.34E-01	7.04E-01	6.77E-01	6.52E-01	6.29E-01	6.08E-01	5.88E-01	5.70E-01	5.53E-01	BORON
3.0E 16	5.37E-01	5.22E-01	5.08E-01	4.95E-01	4.82E-01	4.70E-01	4.59E-01	4.49E-01	4.39E-01	4.29E-01	BORON
4.0E 16	4.20E-01	4.12E-01	4.03E-01	3.95E-01	3.88E-01	3.81E-01	3.74E-01	3.67E-01	3.61E-01	3.55E-01	BORON
5.0E 16	3.49E-01	3.43E-01	3.38E-01	3.33E-01	3.28E-01	3.23E-01	3.18E-01	3.14E-01	3.09E-01	3.05E-01	BORON
6.0E 16	3.01E-01	2.97E-01	2.93E-01	2.89E-01	2.86E-01	2.82E-01	2.79E-01	2.76E-01	2.72E-01	2.69E-01	BORON
7.0E 16	2.66E-01	2.63E-01	2.60E-01	2.58E-01	2.55E-01	2.52E-01	2.50E-01	2.47E-01	2.44E-01	2.42E-01	BORON
8.0E 16	2.40E-01	2.37E-01	2.35E-01	2.33E-01	2.31E-01	2.29E-01	2.27E-01	2.25E-01	2.23E-01	2.21E-01	BORON
9.0E 16	2.19E-01	2.17E-01	2.15E-01	2.13E-01	2.12E-01	2.10E-01	2.08E-01	2.07E-01	2.05E-01	2.04E-01	BCRCN
1.0E 17	2.02E-01	1.88E-01	1.76E-01	1.66E-01	1.58E-01	1.50E-01	1.43E-01	1.37E-01	1.32E-01	1.27E-01	BORON
2.0E 17	1.22E-01	1.18E-01	1.15E-01	1.11E-01	1.08E-01	1.05E-01	1.02E-01	9.99E-02	9.75E-02	9.53E-02	BORON
3.0E 17	9.32E-02	9.12E-02	8.93E-02	8.76E-02	8.59E-02	8.43E-02	8.27E-02	8.13E-02	7.99E-02	7.86E-02	BORON
4.0E 17	7.73E-02	7.61E-02	7.49E-02	7.38E-02	7.27E-02	7.17E-02	7.07E-02	6.98E-02	6.88E-02	6.79E-02	BORON
5.0E 17	6.71E-02	6.63E-02	6.54E-02	6.47E-02	6.39E-02	6.32E-02	6.25E-02	6.18E-02	6.11E-02	6.05E-02	BORON

TABLE 2-4 (Contd)

DOPANT DENSITY TO RESISTIVITY CONVERSION AT 23°C FOR BORON-DOPED SILICON

DOPANT DENSITY (/CM**3)	RESISTIVITY (OHM-CM)										
	0.	0.1	0.2	0.3	0.4	0.5	0.6	0.7	0.8	0.9	
6.0E 17	5.98E-02	5.92E-02	5.86E-02	5.80E-02	5.75E-02	5.69E-02	5.64E-02	5.58E-02	5.53E-02	5.48E-02	BORON
7.0E 17	5.43E-02	5.39E-02	5.34E-02	5.29E-02	5.25E-02	5.21E-02	5.16E-02	5.12E-02	5.08E-02	5.04E-02	BORON
8.0E 17	5.00E-02	4.96E-02	4.92E-02	4.89E-02	4.85E-02	4.82E-02	4.78E-02	4.75E-02	4.71E-02	4.68E-02	BORON
9.0E 17	4.65E-02	4.62E-02	4.58E-02	4.55E-02	4.52E-02	4.49E-02	4.47E-02	4.44E-02	4.41E-02	4.38E-02	BORON
1.0E 18	4.35E-02	4.10E-02	3.88E-02	3.69E-02	3.53E-02	3.38E-02	3.24E-02	3.12E-02	3.00E-02	2.90E-02	BORON
2.0E 18	2.81E-02	2.72E-02	2.64E-02	2.56E-02	2.49E-02	2.42E-02	2.36E-02	2.30E-02	2.25E-02	2.19E-02	BORON
3.0E 18	2.14E-02	2.10E-02	2.05E-02	2.01E-02	1.97E-02	1.93E-02	1.89E-02	1.86E-02	1.82E-02	1.79E-02	BORON
4.0E 18	1.76E-02	1.73E-02	1.70E-02	1.67E-02	1.64E-02	1.62E-02	1.59E-02	1.57E-02	1.54E-02	1.52E-02	BORON
5.0E 18	1.50E-02	1.48E-02	1.46E-02	1.44E-02	1.42E-02	1.40E-02	1.38E-02	1.36E-02	1.35E-02	1.33E-02	BORON
6.0E 18	1.31E-02	1.30E-02	1.28E-02	1.27E-02	1.25E-02	1.24E-02	1.22E-02	1.21E-02	1.19E-02	1.18E-02	BORON
7.0E 18	1.17E-02	1.16E-02	1.14E-02	1.13E-02	1.12E-02	1.11E-02	1.10E-02	1.09E-02	1.08E-02	1.07E-02	BORON
8.0E 18	1.06E-02	1.05E-02	1.04E-02	1.03E-02	1.02E-02	1.01E-02	9.98E-03	9.89E-03	9.80E-03	9.72E-03	BORON
9.0E 18	9.63E-03	9.55E-03	9.47E-03	9.39E-03	9.31E-03	9.23E-03	9.16E-03	9.08E-03	9.01E-03	8.94E-03	BORON
1.0E 19	8.87E-03	8.22E-03	7.67E-03	7.18E-03	6.76E-03	6.39E-03	6.05E-03	5.75E-03	5.48E-03	5.24E-03	BORON
2.0E 19	5.01E-03	4.81E-03	4.62E-03	4.45E-03	4.29E-03	4.14E-03	4.00E-03	3.87E-03	3.75E-03	3.63E-03	BORON
3.0E 19	3.53E-03	3.43E-03	3.33E-03	3.24E-03	3.16E-03	3.08E-03	3.00E-03	2.93E-03	2.86E-03	2.79E-03	BORON
4.0E 19	2.73E-03	2.67E-03	2.61E-03	2.56E-03	2.50E-03	2.45E-03	2.40E-03	2.36E-03	2.31E-03	2.27E-03	BORON
5.0E 19	2.23E-03	2.19E-03	2.15E-03	2.11E-03	2.08E-03	2.04E-03	2.01E-03	1.98E-03	1.94E-03	1.91E-03	BORON
6.0E 19	1.88E-03	1.86E-03	1.83E-03	1.80E-03	1.77E-03	1.75E-03	1.72E-03	1.70E-03	1.68E-03	1.66E-03	BORON
7.0E 19	1.63E-03	1.61E-03	1.59E-03	1.57E-03	1.55E-03	1.53E-03	1.51E-03	1.49E-03	1.48E-03	1.46E-03	BORON
8.0E 19	1.44E-03	1.43E-03	1.41E-03	1.39E-03	1.38E-03	1.36E-03	1.35E-03	1.33E-03	1.32E-03	1.30E-03	BORON
9.0E 19	1.29E-03	1.28E-03	1.26E-03	1.25E-03	1.24E-03	1.23E-03	1.21E-03	1.20E-03	1.19E-03	1.18E-03	BORON
1.0E 20	1.17E-03	1.07E-03	9.83E-04	9.11E-04	8.49E-04	7.94E-04	7.47E-04	7.04E-04	6.67E-04	6.33E-04	BORON
2.0E 20	6.0E-04	5.7E-04	5.5E-04	5.3E-04	5.0E-04	4.9E-04	4.7E-04	4.5E-04	4.3E-04	4.2E-04	BORON
3.0E 20	4.1E-04	3.9E-04	3.8E-04	3.7E-04	3.6E-04	3.5E-04	3.4E-04	3.3E-04	3.2E-04	3.1E-04	BORON
4.0E 20	3.1E-04	3.0E-04	2.9E-04	2.9E-04	2.8E-04	2.7E-04	2.7E-04	2.6E-04	2.6E-04	2.5E-04	BORON
5.0E 20	2.5E-04	2.4E-04	2.4E-04	2.3E-04	2.3E-04	2.2E-04	2.2E-04	2.2E-04	2.1E-04	2.1E-04	BORON
6.0E 20	2.1E-04	2.0E-04	2.0E-04	2.0E-04	1.9E-04	1.9E-04	1.9E-04	1.8E-04	1.8E-04	1.8E-04	BORON
7.0E 20	1.8E-04	1.7E-04	1.7E-04	1.7E-04	1.7E-04	1.7E-04	1.6E-04	1.6E-04	1.6E-04	1.6E-04	BORON
8.0E 20	1.6E-04	1.5E-04	1.5E-04	1.5E-04	1.5E-04	1.5E-04	1.4E-04	1.4E-04	1.4E-04	1.4E-04	BORON
9.0E 20	1.4E-04	1.4E-04	1.4E-04	1.3E-04	1.3E-04	1.3E-04	1.3E-04	1.3E-04	1.3E-04	1.3E-04	BORON

TABLE 2-5
DOPANT DENSITY TO RESISTIVITY CONVERSION AT 23°C FOR PHOSPHORUS-DOPED SILICON

DOPANT DENSITY (/CM**3)	RESISTIVITY (OHM-CM)										
	0.	0.1	0.2	0.3	0.4	0.5	0.6	0.7	0.8	0.9	
1.0E 12	4.3E 03	3.9E 03	3.6E 03	3.3E 03	3.0E 03	2.8E 03	2.7E 03	2.5E 03	2.4E 03	2.2E 03	PHOS
2.0E 12	2.1E 03	2.0E 03	1.9E 03	1.9E 03	1.8E 03	1.7E 03	1.6E 03	1.6E 03	1.5E 03	1.5E 03	PHOS
3.0E 12	1.4E 03	1.4E 03	1.3E 03	1.3E 03	1.3E 03	1.2E 03	1.2E 03	1.1E 03	1.1E 03	1.1E 03	PHOS
4.0E 12	1.0E 03	1.0E 03	1.0E 03	9.9E 02	9.7E 02	9.4E 02	9.2E 02	9.0E 02	8.9E 02	8.7E 02	PHOS
5.0E 12	8.5E 02	8.3E 02	8.2E 02	8.0E 02	7.9E 02	7.7E 02	7.6E 02	7.5E 02	7.3E 02	7.2E 02	PHOS
6.0E 12	7.1E 02	7.0E 02	6.9E 02	6.7E 02	6.6E 02	6.5E 02	6.4E 02	6.3E 02	6.3E 02	6.2E 02	PHOS
7.0E 12	6.1E 02	6.0E 02	5.9E 02	5.8E 02	5.7E 02	5.7E 02	5.6E 02	5.5E 02	5.4E 02	5.4E 02	PHOS
8.0E 12	5.3E 02	5.2E 02	5.2E 02	5.1E 02	5.1E 02	5.0E 02	4.9E 02	4.9E 02	4.8E 02	4.8E 02	PHOS
9.0E 12	4.7E 02	4.7E 02	4.6E 02	4.6E 02	4.5E 02	4.5E 02	4.4E 02	4.4E 02	4.3E 02	4.3E 02	PHOS
1.0E 13	4.3E 02	3.9E 02	3.5E 02	3.3E 02	3.0E 02	2.8E 02	2.7E 02	2.5E 02	2.4E 02	2.2E 02	PHOS
2.0E 13	2.1E 02	2.0E 02	1.9E 02	1.8E 02	1.8E 02	1.7E 02	1.6E 02	1.6E 02	1.5E 02	1.5E 02	PHOS
3.0E 13	1.42E 02	1.37E 02	1.33E 02	1.29E 02	1.25E 02	1.22E 02	1.18E 02	1.15E 02	1.12E 02	1.09E 02	PHOS
4.0E 13	1.07E 02	1.04E 02	1.02E 02	9.92E 01	9.69E 01	9.48E 01	9.27E 01	9.08E 01	8.89E 01	8.71E 01	PHOS
5.0E 13	8.53E 01	8.37E 01	8.21E 01	8.05E 01	7.90E 01	7.76E 01	7.62E 01	7.49E 01	7.36E 01	7.24E 01	PHOS
6.0E 13	7.12E 01	7.00E 01	6.89E 01	6.78E 01	6.68E 01	6.57E 01	6.48E 01	6.38E 01	6.29E 01	6.20E 01	PHOS
7.0E 13	6.11E 01	6.02E 01	5.94E 01	5.86E 01	5.78E 01	5.70E 01	5.63E 01	5.56E 01	5.49E 01	5.42E 01	PHOS
8.0E 13	5.35E 01	5.28E 01	5.22E 01	5.16E 01	5.10E 01	5.04E 01	4.98E 01	4.92E 01	4.87E 01	4.81E 01	PHOS
9.0E 13	4.76E 01	4.71E 01	4.66E 01	4.61E 01	4.56E 01	4.51E 01	4.46E 01	4.42E 01	4.37E 01	4.33E 01	PHOS
1.0E 14	4.29E 01	3.90E 01	3.58E 01	3.31E 01	3.07E 01	2.87E 01	2.69E 01	2.54E 01	2.40E 01	2.27E 01	PHOS
2.0E 14	2.16E 01	2.06E 01	1.97E 01	1.88E 01	1.80E 01	1.73E 01	1.67E 01	1.61E 01	1.55E 01	1.50E 01	PHOS
3.0E 14	1.45E 01	1.40E 01	1.36E 01	1.32E 01	1.28E 01	1.25E 01	1.21E 01	1.18E 01	1.15E 01	1.12E 01	PHOS
4.0E 14	1.09E 01	1.07E 01	1.04E 01	1.02E 01	9.96E 00	9.74E 00	9.53E 00	9.33E 00	9.14E 00	8.96E 00	PHOS
5.0E 14	8.79E 00	8.62E 00	8.46E 00	8.30E 00	8.15E 00	8.01E 00	7.87E 00	7.73E 00	7.60E 00	7.48E 00	PHOS
6.0E 14	7.36E 00	7.24E 00	7.12E 00	7.01E 00	6.91E 00	6.80E 00	6.70E 00	6.61E 00	6.51E 00	6.42E 00	PHOS
7.0E 14	6.33E 00	6.24E 00	6.16E 00	6.08E 00	6.00E 00	5.92E 00	5.85E 00	5.77E 00	5.70E 00	5.63E 00	PHOS
8.0E 14	5.56E 00	5.50E 00	5.43E 00	5.37E 00	5.30E 00	5.24E 00	5.19E 00	5.13E 00	5.07E 00	5.02E 00	PHOS
9.0E 14	4.96E 00	4.91E 00	4.86E 00	4.81E 00	4.76E 00	4.71E 00	4.66E 00	4.62E 00	4.57E 00	4.53E 00	PHOS
1.0E 15	4.48E 00	4.09E 00	3.76E 00	3.48E 00	3.24E 00	3.03E 00	2.85E 00	2.69E 00	2.55E 00	2.42E 00	PHOS
2.0E 15	2.31E 00	2.20E 00	2.11E 00	2.02E 00	1.94E 00	1.87E 00	1.80E 00	1.74E 00	1.68E 00	1.62E 00	PHOS
3.0E 15	1.57E 00	1.53E 00	1.48E 00	1.44E 00	1.40E 00	1.36E 00	1.33E 00	1.29E 00	1.26E 00	1.23E 00	PHOS
4.0E 15	1.20E 00	1.18E 00	1.15E 00	1.12E 00	1.10E 00	1.08E 00	1.06E 00	1.04E 00	1.02E 00	9.97E-01	PHOS
5.0E 15	9.78E-01	9.61E-01	9.44E-01	9.27E-01	9.12E-01	8.96E-01	8.82E-01	8.68E-01	8.54E-01	8.41E-01	PHOS
6.0E 15	8.28E-01	8.16E-01	8.04E-01	7.92E-01	7.81E-01	7.70E-01	7.59E-01	7.49E-01	7.39E-01	7.29E-01	PHOS
7.0E 15	7.20E-01	7.11E-01	7.02E-01	6.93E-01	6.85E-01	6.76E-01	6.68E-01	6.60E-01	6.53E-01	6.45E-01	PHOS
8.0E 15	6.38E-01	6.31E-01	6.24E-01	6.17E-01	6.11E-01	6.04E-01	5.98E-01	5.92E-01	5.86E-01	5.80E-01	PHOS
9.0E 15	5.74E-01	5.69E-01	5.63E-01	5.58E-01	5.52E-01	5.47E-01	5.42E-01	5.37E-01	5.32E-01	5.28E-01	PHOS
1.0E 16	5.23E-01	4.81E-01	4.45E-01	4.15E-01	3.89E-01	3.67E-01	3.47E-01	3.30E-01	3.14E-01	3.00E-01	PHOS
2.0E 16	2.87E-01	2.76E-01	2.65E-01	2.56E-01	2.47E-01	2.39E-01	2.31E-01	2.24E-01	2.17E-01	2.11E-01	PHOS
3.0E 16	2.06E-01	2.00E-01	1.95E-01	1.90E-01	1.86E-01	1.82E-01	1.78E-01	1.74E-01	1.70E-01	1.67E-01	PHOS
4.0E 16	1.64E-01	1.61E-01	1.58E-01	1.55E-01	1.52E-01	1.49E-01	1.47E-01	1.44E-01	1.42E-01	1.40E-01	PHOS
5.0E 16	1.38E-01	1.36E-01	1.34E-01	1.32E-01	1.30E-01	1.28E-01	1.27E-01	1.25E-01	1.23E-01	1.22E-01	PHOS
6.0E 16	1.20E-01	1.19E-01	1.17E-01	1.16E-01	1.15E-01	1.13E-01	1.12E-01	1.11E-01	1.10E-01	1.09E-01	PHOS
7.0E 16	1.08E-01	1.06E-01	1.05E-01	1.04E-01	1.03E-01	1.02E-01	1.01E-01	1.00E-01	9.95E-02	9.87E-02	PHOS
8.0E 16	9.78E-02	9.69E-02	9.61E-02	9.53E-02	9.45E-02	9.37E-02	9.30E-02	9.22E-02	9.15E-02	9.08E-02	PHOS
9.0E 16	9.01E-02	8.94E-02	8.87E-02	8.81E-02	8.74E-02	8.68E-02	8.62E-02	8.56E-02	8.50E-02	8.44E-02	PHOS
1.0E 17	8.38E-02	7.86E-02	7.43E-02	7.05E-02	6.72E-02	6.44E-02	6.19E-02	5.96E-02	5.76E-02	5.57E-02	PHOS
2.0E 17	5.41E-02	5.25E-02	5.11E-02	4.98E-02	4.87E-02	4.75E-02	4.65E-02	4.55E-02	4.46E-02	4.38E-02	PHOS
3.0E 17	4.22E-02	4.15E-02	4.15E-02	4.08E-02	4.02E-02	3.96E-02	3.90E-02	3.84E-02	3.79E-02	3.74E-02	PHOS
4.0E 17	3.69E-02	3.65E-02	3.60E-02	3.56E-02	3.52E-02	3.48E-02	3.44E-02	3.40E-02	3.37E-02	3.33E-02	PHOS
5.0E 17	3.30E-02	3.27E-02	3.23E-02	3.20E-02	3.17E-02	3.15E-02	3.12E-02	3.09E-02	3.07E-02	3.04E-02	PHOS

TABLE 2-5 (Contd)

DOPANT DENSITY TO RESISTIVITY CONVERSION AT 23°C FOR PHOSPHORUS-DOPED SILICON

DOPANT DENSITY (/CM**3)	RESISTIVITY (OHM-CM)										
	0.	0.1	0.2	0.3	0.4	0.5	0.6	0.7	0.8	0.9	
6.0E 17	3.01E-02	2.99E-02	2.97E-02	2.94E-02	2.92E-02	2.90E-02	2.88E-02	2.86E-02	2.84E-02	2.82E-02	PHOS
7.0E 17	2.80E-02	2.78E-02	2.76E-02	2.74E-02	2.72E-02	2.71E-02	2.69E-02	2.67E-02	2.66E-02	2.64E-02	PHOS
8.0E 17	2.62E-02	2.61E-02	2.59E-02	2.58E-02	2.56E-02	2.55E-02	2.53E-02	2.52E-02	2.51E-02	2.49E-02	PHOS
9.0E 17	2.48E-02	2.47E-02	2.45E-02	2.44E-02	2.43E-02	2.41E-02	2.40E-02	2.39E-02	2.38E-02	2.37E-02	PHOS
1.0E 18	2.36E-02	2.25E-02	2.16E-02	2.07E-02	2.00E-02	1.93E-02	1.87E-02	1.81E-02	1.76E-02	1.71E-02	PHOS
2.0E 18	1.66E-02	1.62E-02	1.58E-02	1.54E-02	1.51E-02	1.47E-02	1.44E-02	1.41E-02	1.38E-02	1.35E-02	PHOS
3.0E 18	1.32E-02	1.30E-02	1.27E-02	1.25E-02	1.23E-02	1.21E-02	1.19E-02	1.17E-02	1.15E-02	1.13E-02	PHOS
4.0E 18	1.11E-02	1.09E-02	1.08E-02	1.06E-02	1.04E-02	1.03E-02	1.01E-02	1.00E-02	9.86E-03	9.73E-03	PHOS
5.0E 18	9.60E-03	9.47E-03	9.34E-03	9.22E-03	9.11E-03	8.99E-03	8.88E-03	8.77E-03	8.66E-03	8.56E-03	PHOS
6.0E 18	8.46E-03	8.36E-03	8.27E-03	8.17E-03	8.08E-03	7.99E-03	7.90E-03	7.82E-03	7.74E-03	7.65E-03	PHOS
7.0E 18	7.57E-03	7.50E-03	7.42E-03	7.34E-03	7.27E-03	7.20E-03	7.13E-03	7.06E-03	6.99E-03	6.92E-03	PHOS
8.0E 18	6.86E-03	6.79E-03	6.73E-03	6.67E-03	6.61E-03	6.55E-03	6.49E-03	6.44E-03	6.38E-03	6.32E-03	PHOS
9.0E 18	6.27E-03	6.22E-03	6.16E-03	6.11E-03	6.06E-03	6.01E-03	5.96E-03	5.91E-03	5.87E-03	5.82E-03	PHOS
1.0E 19	5.78E-03	5.35E-03	4.99E-03	4.68E-03	4.40E-03	4.15E-03	3.93E-03	3.73E-03	3.56E-03	3.39E-03	PHOS
2.0E 19	3.25E-03	3.11E-03	2.99E-03	2.87E-03	2.77E-03	2.67E-03	2.58E-03	2.49E-03	2.41E-03	2.34E-03	PHOS
3.0E 19	2.27E-03	2.20E-03	2.14E-03	2.08E-03	2.03E-03	1.98E-03	1.93E-03	1.88E-03	1.84E-03	1.79E-03	PHOS
4.0E 19	1.75E-03	1.71E-03	1.68E-03	1.64E-03	1.61E-03	1.58E-03	1.54E-03	1.51E-03	1.49E-03	1.46E-03	PHOS
5.0E 19	1.43E-03	1.41E-03	1.38E-03	1.36E-03	1.34E-03	1.31E-03	1.29E-03	1.27E-03	1.25E-03	1.23E-03	PHOS
6.0E 19	1.21E-03	1.20E-03	1.18E-03	1.16E-03	1.15E-03	1.13E-03	1.11E-03	1.10E-03	1.09E-03	1.07E-03	PHOS
7.0E 19	1.06E-03	1.04E-03	1.03E-03	1.02E-03	1.01E-03	9.94E-04	9.82E-04	9.71E-04	9.60E-04	9.49E-04	PHOS
8.0E 19	9.38E-04	9.28E-04	9.18E-04	9.08E-04	8.98E-04	8.89E-04	8.80E-04	8.71E-04	8.62E-04	8.53E-04	PHOS
9.0E 19	8.45E-04	8.37E-04	8.29E-04	8.21E-04	8.13E-04	8.06E-04	7.98E-04	7.91E-04	7.84E-04	7.77E-04	PHOS
1.0E 20	7.70E-04	7.08E-04	6.57E-04	6.13E-04	5.76E-04	5.43E-04	5.15E-04	4.89E-04	4.67E-04	4.47E-04	PHOS
2.0E 20	4.29E-04	4.12E-04	3.97E-04	3.84E-04	3.71E-04	3.60E-04	3.49E-04	3.39E-04	3.30E-04	3.21E-04	PHOS
3.0E 20	3.1E-04	3.1E-04	3.0E-04	2.9E-04	2.9E-04	2.8E-04	2.8E-04	2.7E-04	2.7E-04	2.6E-04	PHOS
4.0E 20	2.6E-04	2.5E-04	2.5E-04	2.4E-04	2.4E-04	2.4E-04	2.3E-04	2.3E-04	2.3E-04	2.0E-04	PHOS
5.0E 20	2.2E-04	2.2E-04	2.2E-04	2.1E-04	2.1E-04	2.1E-04	2.1E-04	2.1E-04	2.1E-04	2.0E-04	PHOS
6.0E 20	2.0E-04	2.0E-04	2.0E-04	1.9E-04	1.9E-04	1.9E-04	1.9E-04	1.9E-04	1.9E-04	1.9E-04	PHOS
7.0E 20	1.8E-04	1.8E-04	1.8E-04	1.8E-04	1.8E-04	1.8E-04	1.8E-04	1.8E-04	1.8E-04	1.7E-04	PHOS
8.0E 20	1.7E-04	1.7E-04	1.7E-04	1.7E-04	1.7E-04	1.7E-04	1.7E-04	1.7E-04	1.7E-04	1.7E-04	PHOS
9.0E 20	1.7E-04	1.7E-04	1.6E-04	1.6E-04	1.6E-04	1.6E-04	1.6E-04	1.6E-04	1.6E-04	1.6E-04	PHOS

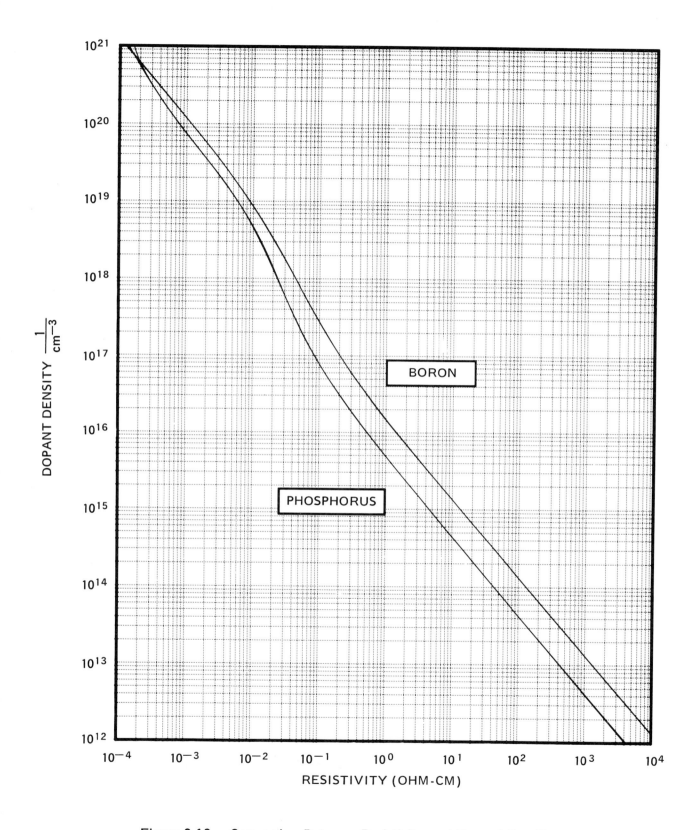

Figure 2-19 — Conversion Between Resistivity and Dopant Density

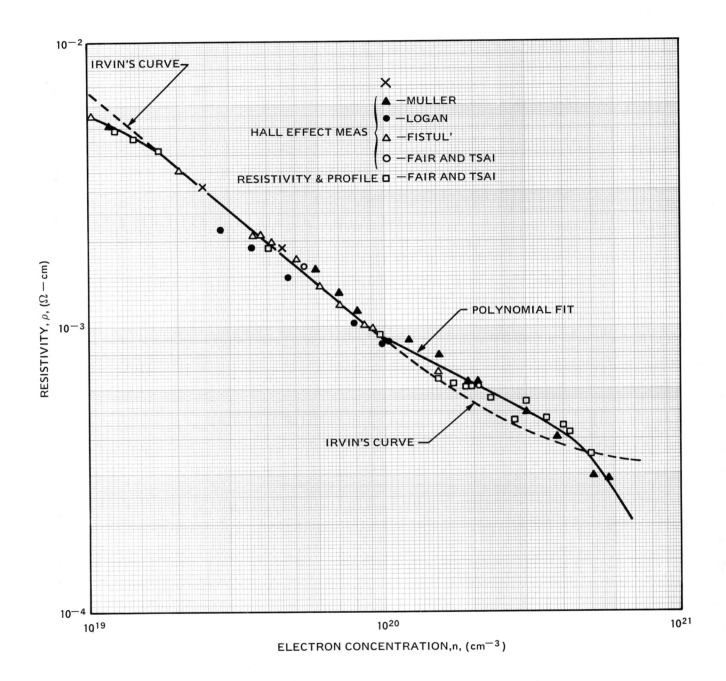

Figure 2-20 — Resistivity Versus Electron Concentration for Arsenic-Doped Silicon

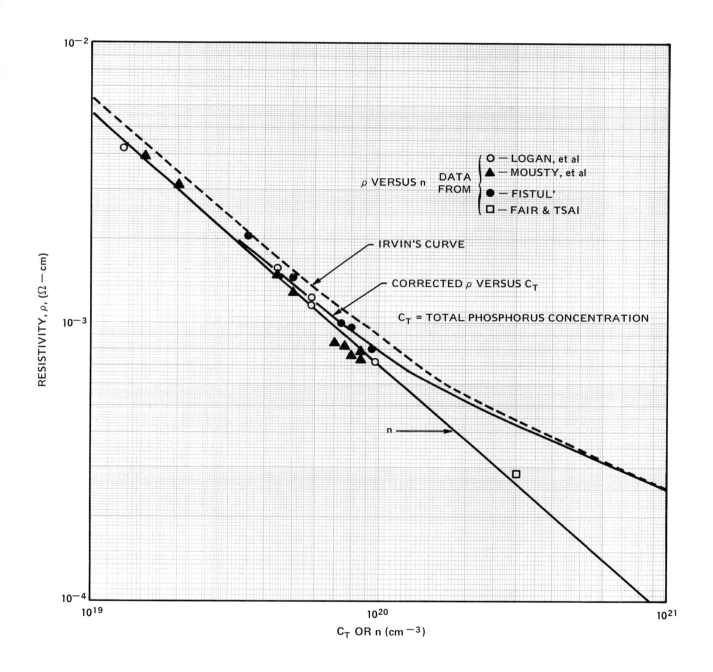

Figure 2-21 — Resistivity Versus Phosphorus and Electron Concentration in Silicon

The equation for conversion from electron concentration to resistivity for arsenic is given below.

$$\log \rho = \sum_{i=0}^{10} a_i (\log n)^i \qquad \text{for } 10^{19} \leqslant n \leqslant 7 \times 10^{20},$$

where

n = electron concentration, cm^{-3}

ρ = resistivity, ohm-cm

a_0 = -6.633667×10^3

a_1 = 7.682531×10^2

a_2 = -2.577373×10

a_3 = 0.9658177

a_4 = -5.643443×10^{-2}

a_5 = -8.008543×10^{-4}

a_6 = 9.055838×10^{-5}

a_7 = -1.776701×10^{-6}

a_8 = 1.953279×10^{-7}

a_9 = -5.754599×10^{-9}

a_{10} = $-1.316567 \times 10^{-11}$.

Mobility Versus Concentration

The revised curves for the hole and electron mobilities versus impurity concentration are given in Figures 2-22 and 2-23. The equations which approximate the experimental data are given. Different studies showed slight variations in the constants for the curve fittings.[13] [14] [15] [22] [23] However, all of them agree within a 20 percent range. The curve fittings by Antoniadis et al, for hole mobility and Baccarani and Ostoja for electron mobility are given in Figures 2-22 and 2-23, respectively. The NBS formula for mobility versus phosphorus doping is given in the following[16] :

$$\log_{10} (\mu/\mu_o) = \frac{a_0 + a_1 x + a_2 x^2 + a_3 x^3}{1 + b_1 x + b_2 x^2 + b_3 x^3},$$

where

μ_o = $1 \dfrac{cm^2}{V \bullet s}$

x = $\log_{10} (n/n_o)$

n_o = $10^{16} \ cm^{-3}$.

At 23°C the constants are

a_0 = 3.0746 ± 0.0025

a_1 = -2.2679 ± 0.0076

a_2 = 0.62998 ± 0.00245

$$a_3 = -0.061285 \pm 0.00087$$
$$b_1 = -0.70017 \pm 0.00290$$
$$b_2 = 0.19839 \pm 0.00113$$
$$b_3 = -0.020150 \pm 0.00041$$

Although the NBS formula is more complicated than the one shown in Figure 2-23, we recommend its use for consistency and as an industry standard . However, a similar curve fitting has not been developed for the hole mobility versus boron concentration.

Resistivity Versus Concentration for Deep-Level Impurities

In infrared detectors, silicon is often doped with elements which have energy levels deep within the band gap. The resistivity versus impurity concentration curves are therefore modified. Sclar calculated the relationships considering the fractional ionization of these impurities at 300K.[24] His calculation used the mobility versus concentration curves of Sze and Irvin.[12] With the corrections for p-type silicon taken into consideration, the resistivity versus concentration curves are given in Figures 2-24 and 2-25 for p-type and n-type dopants, respectively. The reader is referred to the original paper by Sclar for detailed explanations of these curves.

Since few experimental data are available, one should use these curves with caution. Data on indium- and gallium-doped samples are summarized by W. R. Thurber.[25] Work in this area is in progress. Linares and Li[26] measured resistivity versus concentration for boron-, gallium- and indium-doped silicon for concentrations in the range of 10^{16} to 10^{18} cm^{-3}. They also presented a theoretical model to calculate the resistivity versus concentration for silicon doped with these elements. For boron-doped samples, the results agreed with Thurber's and for gallium-doped samples, the results agreed with Wolfstirn's data.[27] The indium-doped samples were within the range of data presented by Schroder et al.[28]

From Sclar's calculation, shown in Figure 2-24, the boron curve (Si:B) deviates from that of Thurber's (the NBS curve) by more than 10 percent over the concentration range from 5×10^{16} to 10^{18} cm^{-3}. Sclar used the mobility-derived curve for the boron concentration versus resistivity as corrections to Irvin's curve in the 10^{16} to 10^{18} cm^{-3} range. Because the reason for this discrepancy has not been identified, we recommend that the reader use the NBS curve for boron-doped silicon (Figures 2-12 through 2-18).

Figure 2-26 shows the calculated degree of ionization versus impurity concentration at 300K for various dopant elements.[24]

2.6   Resistivity and Mobility Versus Temperature for Germanium and Silicon

Both carrier mobility and impurity ionization are functions of temperature. Hence, the resistivities of germanium and silicon are complex functions of temperature. Prior to 1960, only two papers had been published giving the experimental data for antimony-doped germanium and arsenic-doped silicon. These are shown in Figures 2-27 and 2-28, respectively. Recently, Linares and Li [26] measured the resistivity change as a function of temperature for boron-, gallium-, and indium-doped silicon in the temperature range from 100°C to 400°C. Their results are given in Figures 2-29, 2-30, and 2-31, respectively. Figures 2-32 and 2-33 show the hole and electron mobilities as functions of temperature for p-type and n-type silicon, respectively.

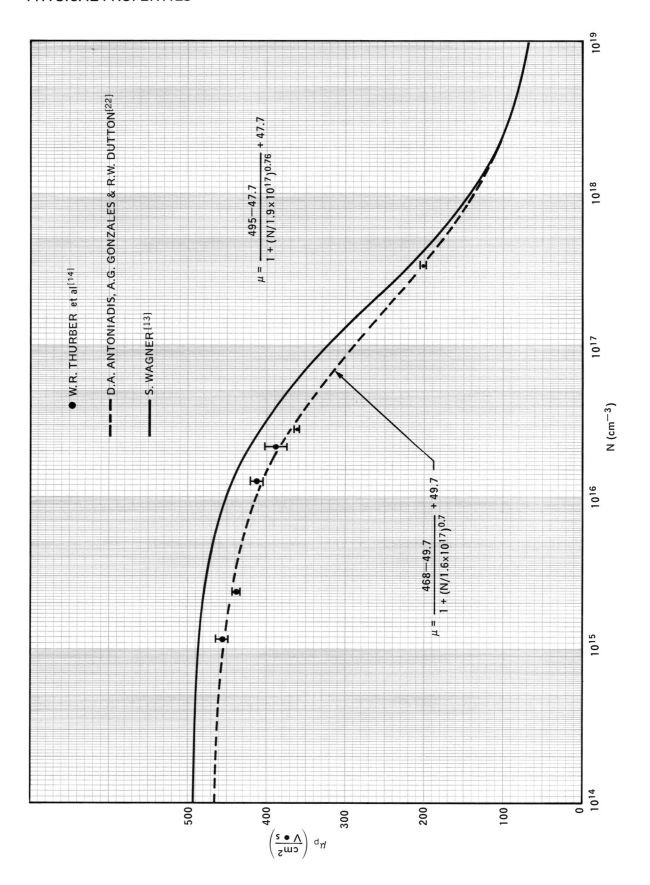

Figure 2-22 — Hole Mobility Versus Impurity Concentration in Boron-Doped Silicon

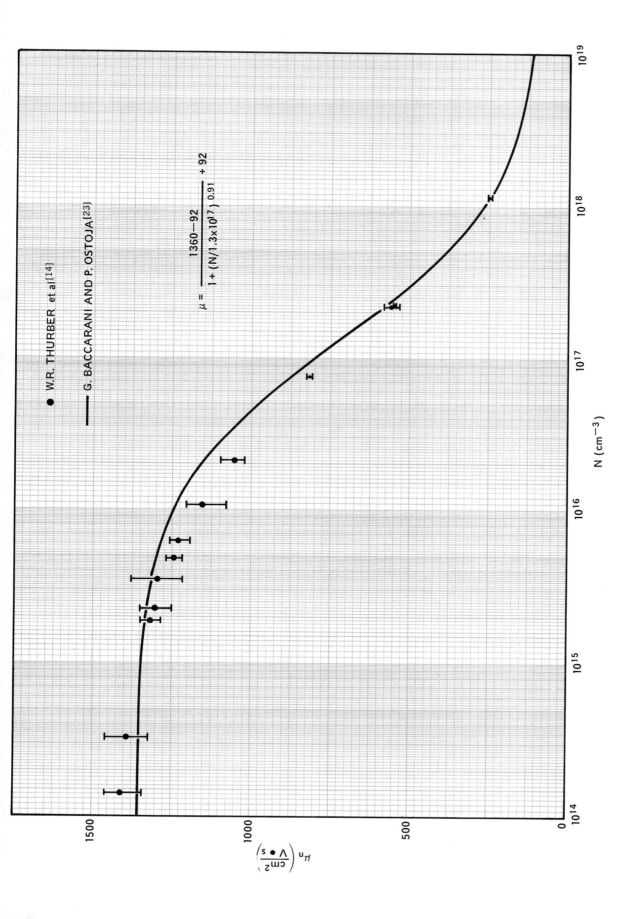

Figure 2-23 — Electron Mobility Versus Impurity Concentration in Phosphorus-Doped Silicon

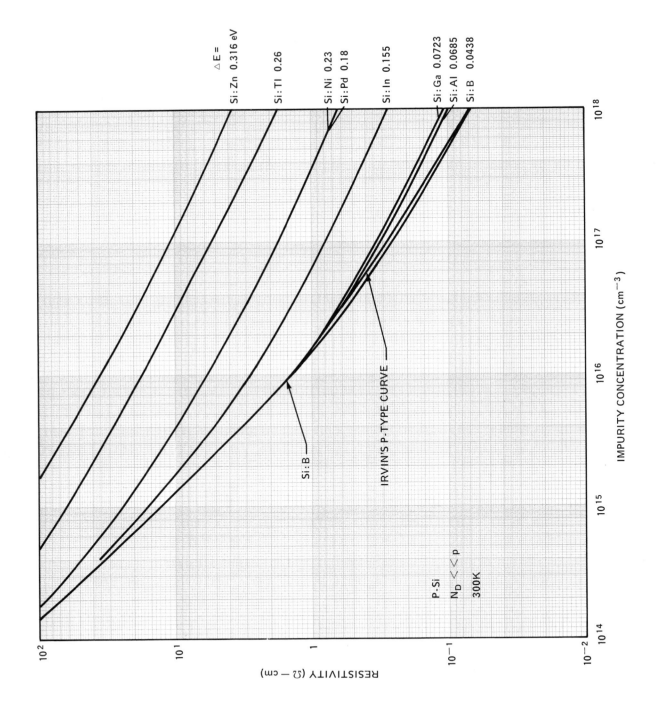

Figure 2-24 — Resistivity Versus Impurity Concentration[24] (© 1977 IEEE)

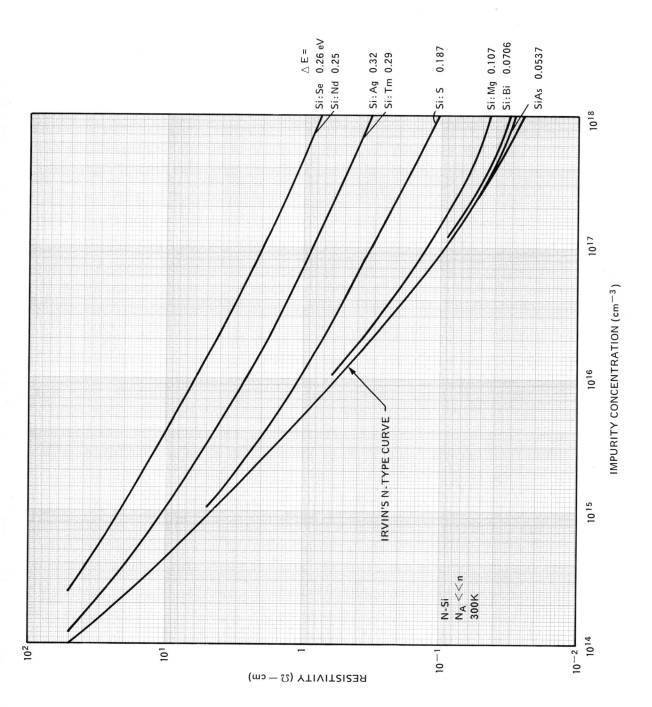

Figure 2-25 — Resistivity Versus Impurity Concentration[24] (© 1977 IEEE)

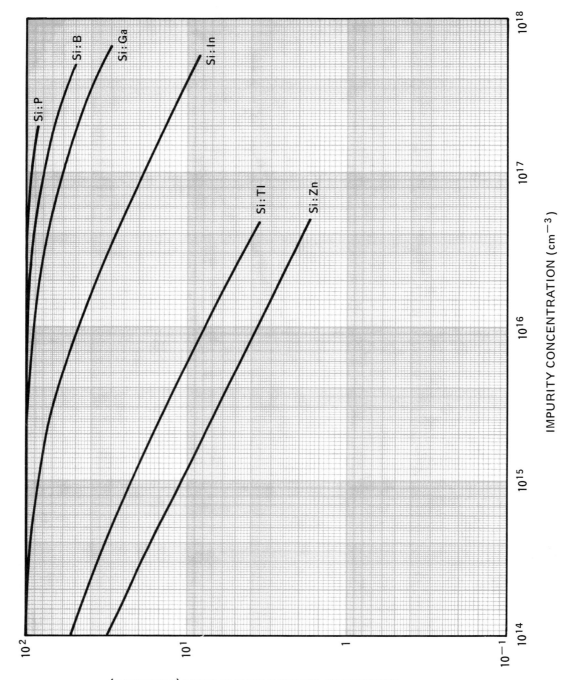

Figure 2-26 — Degree of Ionization Versus Impurity Concentration at 300K[24] (© 1977 IEEE)

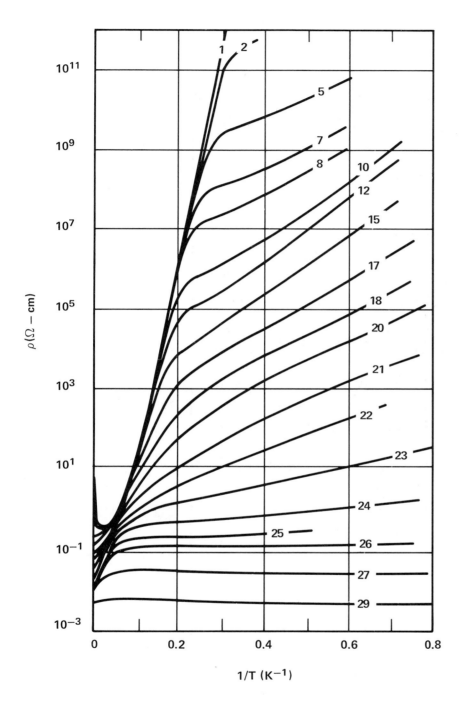

Specimen	Donor Concentration (cm^{-3})
1	5.3×10^{14}
2	9.3×10^{14}
5	1.6×10^{15}
7	2.3×10^{15}
8	3.0×10^{15}
10	5.2×10^{15}
12	8.5×10^{15}
15	1.3×10^{16}
17	2.4×10^{16}
18	3.5×10^{16}
20	5.5×10^{16}
21	5.5×10^{16}
22	6.4×10^{16}
23	7.4×10^{16}
24	8.4×10^{16}
25	1.2×10^{17}
26	1.3×10^{17}
27	2.7×10^{17}
29	9.5×10^{17}

Figure 2-27 — Resistivity Versus Temperature for Ge, Sb-Doped[29]
(Reprinted by permission of John Wiley & Sons, Inc.)

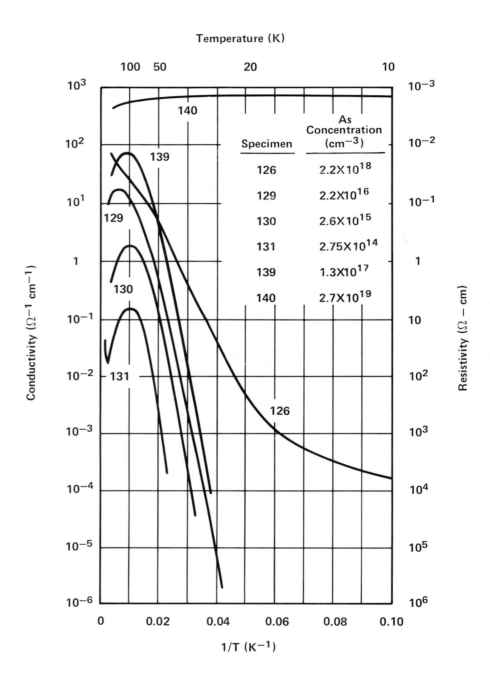

Figure 2-28 -- Electrical Conductivity Versus Temperature for Si, As-Doped[5]
(Reprinted with permission from the American Physical Society)

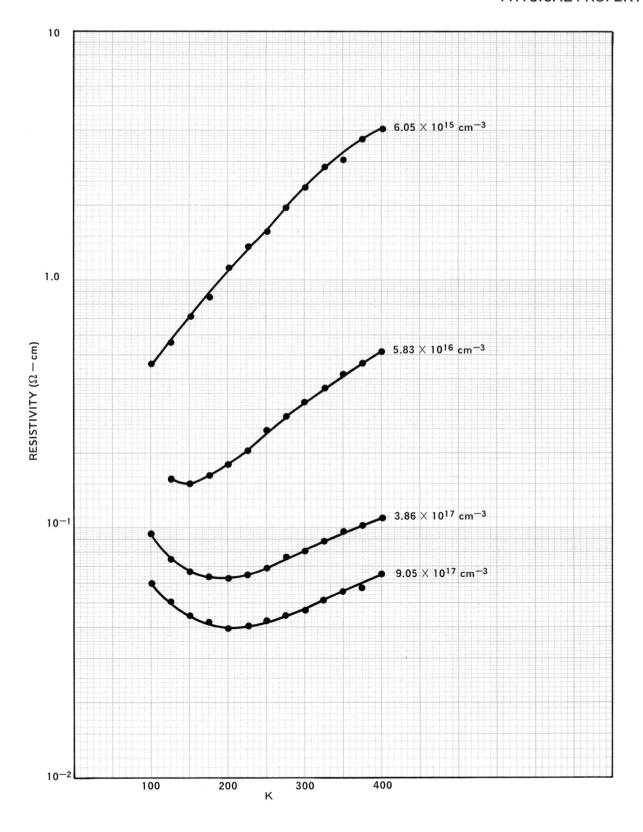

Figure 2-29 — Resistivity Versus Temperature for Boron-Doped Silicon[26]

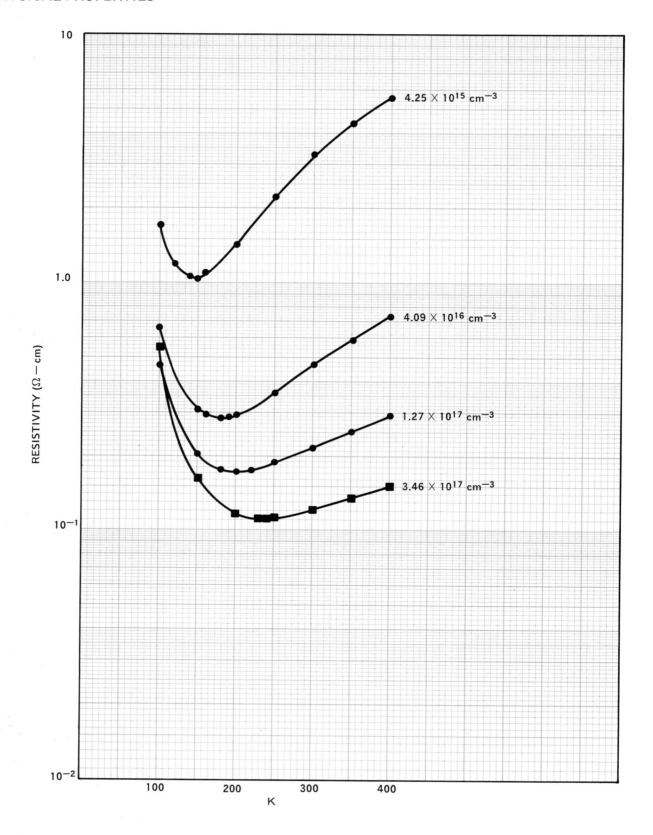

Figure 2-30 — Resistivity Versus Temperature for Gallium-Doped Silicon

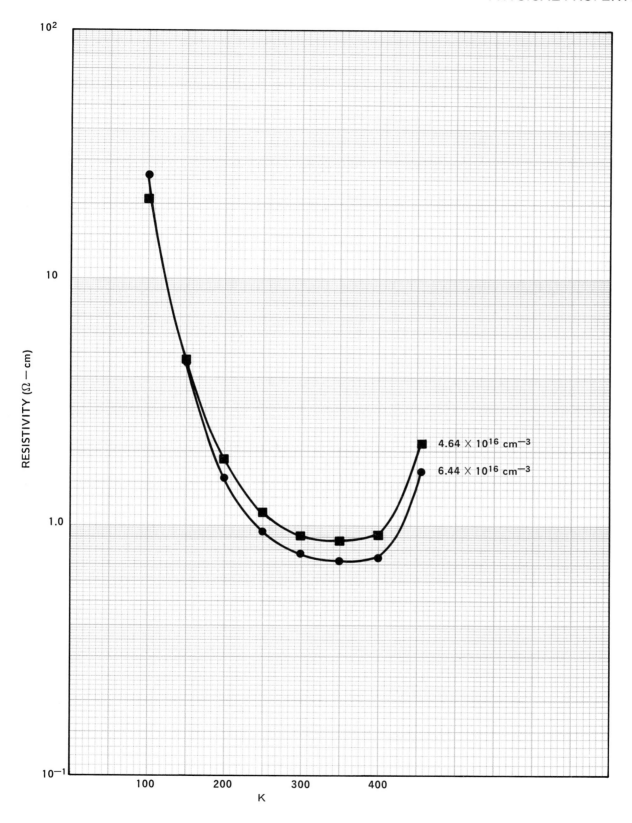

Figure 2-31 — Resistivity Versus Temperature for Indium-Doped Silicon[26]

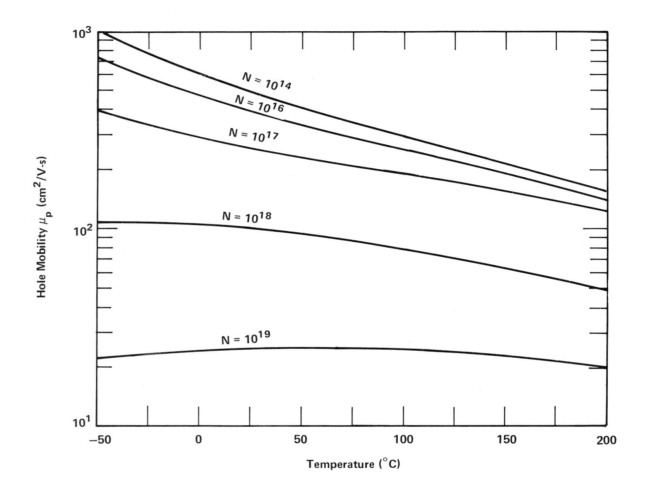

Figure 2-32 — Temperature Dependence of Hole Mobility in Si[30]

© 1957 IRE (now IEEE)

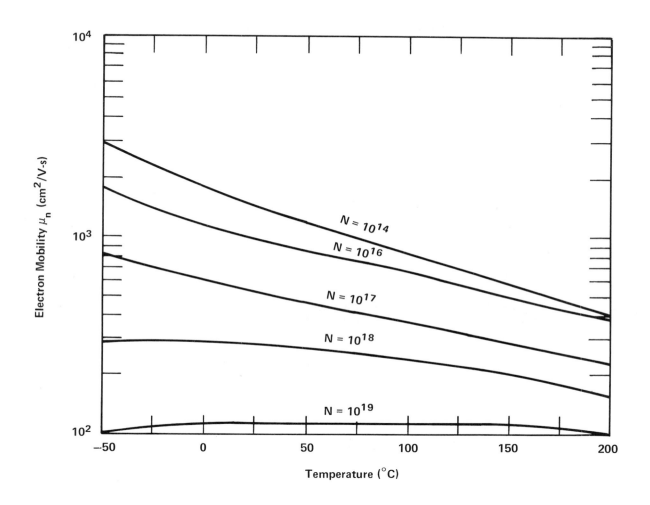

Figure 2-33 — Temperature Dependence of Electron Mobility in Si[30]
© 1957 IRE (now IEEE)

2.7   Drift Velocity

The drift velocities of both electrons and holes are functions of the electric field and show saturation at high electric fields. The relationships for hole and electron drift velocities versus electric field in silicon at 300K are shown in Figure 2-34. The temperature effects are given in the table in Figure 2-34.

2.8   Thermal Expansion Coefficient

The linear thermal expansion coefficient of single and polycrystalline silicon as a function of temperature from -150°C to 850°C is shown in Figure 2-35. The equation for a between T = 120 — 1500 K is:[42]

$$a(T) = 3.725 \times 10^{-6} \left\{ 1 - \exp\left[-0.00588\,(T-24)\right] \right\} + 5.548 \times 10^{-10} T \ (K^{-1})$$

2.9   Optical Properties

The optical properties of silicon, silicon dioxide, and silicon nitride are shown in Figures 2-36 through 2-39.

The emissivity correction, the brightness temperature, and the spectral emissivity as functions of true temperature for germanium and silicon are shown in Figures 2-40 through 2-42, respectively. These curves are of value when using an infrared optical pyrometer for measuring the temperature of germanium or silicon above 1000°C. This method is often used to determine the growth temperature for epitaxial film growth.

2.10   Thermal Properties

The thermal conductivity of silicon is given in Figures 2-43 and 2-44 for differently doped samples. Related information is given in Tables 2-6 and 2-8. Table 2-7 gives the thermal diffusivity of silicon at different temperatures.

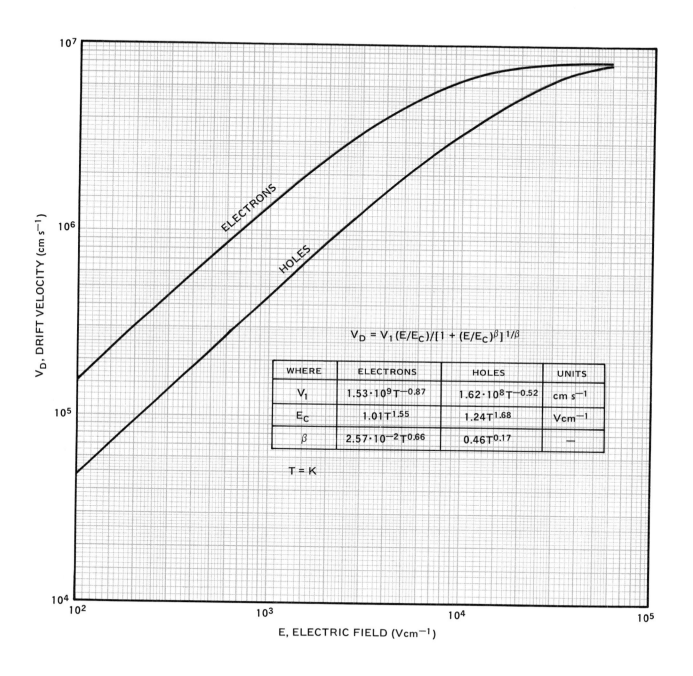

Figure 2-34 — Carrier Velocity Versus Field Strength for Si, T = 300K[31]
(Reprinted with permission from Pergamon Press, Ltd.)

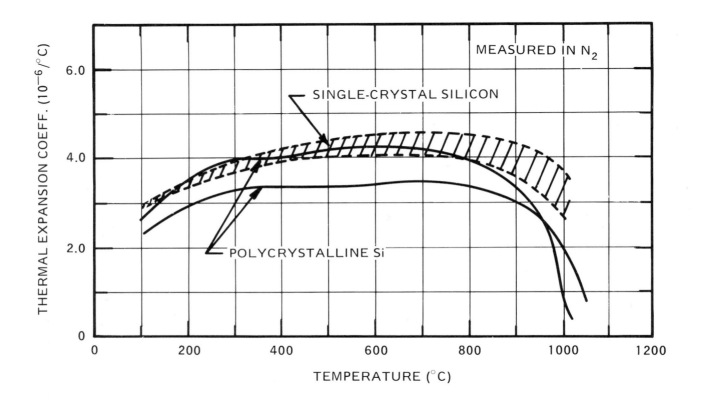

Figure 2-35 — Thermal Expansion Coefficient of Polycrystalline-Silicon Layer and Single Crystal Silicon at High Temperature[32] (Reprinted by permission of the publisher, The Electrochemical Society, Inc.)

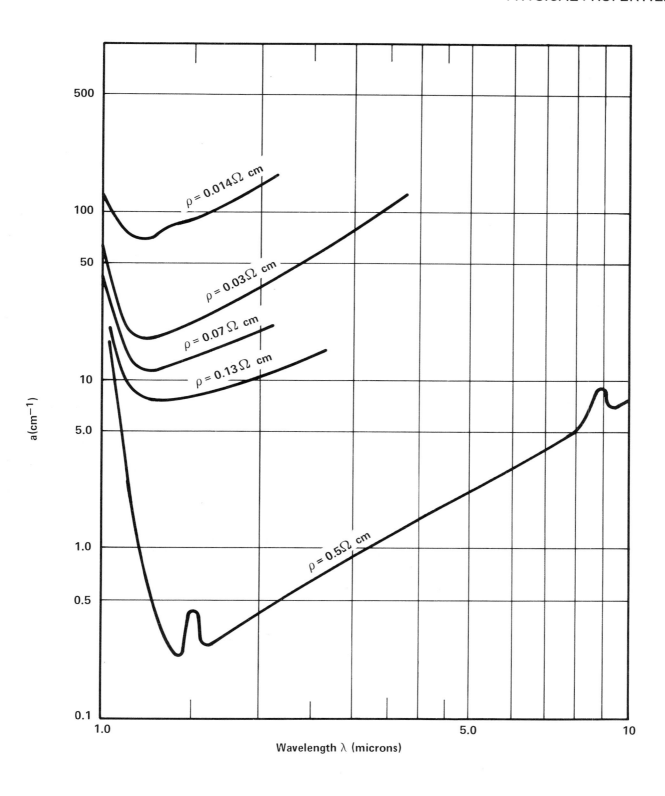

Figure 2-36 — Absorption Coefficient of p- Type Si as a Function of Wavelength[33]

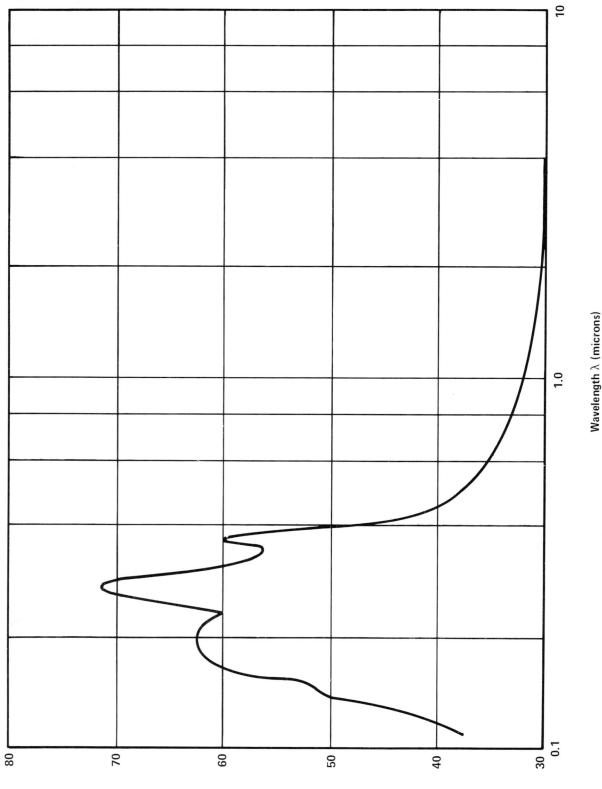

Figure 2-37 — Reflection Coefficient Versus Wavelength for Pure Si[34]

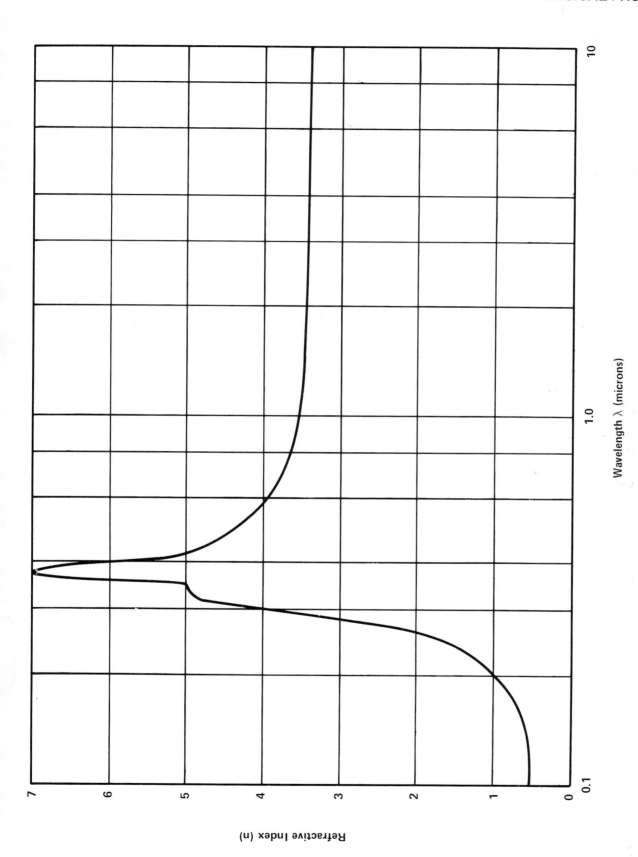

Figure 2-38 — Refractive Index of Si at Room Temperature[35]

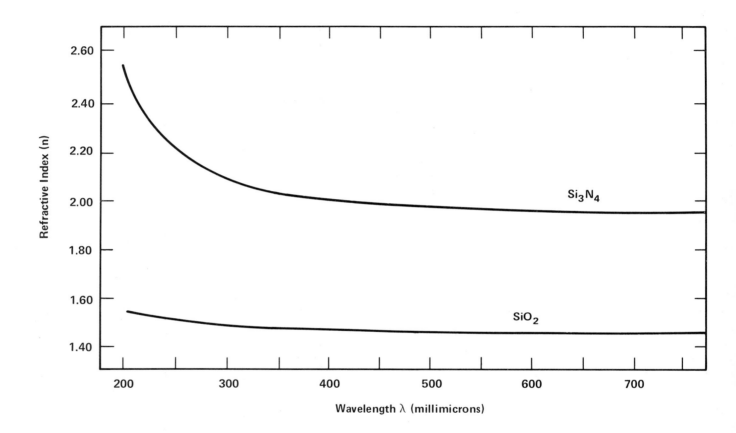

Figure 2-39 — Refractive Index Versus Wavelength for SiO_2 and Si_3N_4 [36]
(Reprinted with permission from Pergamon Press, Ltd.)

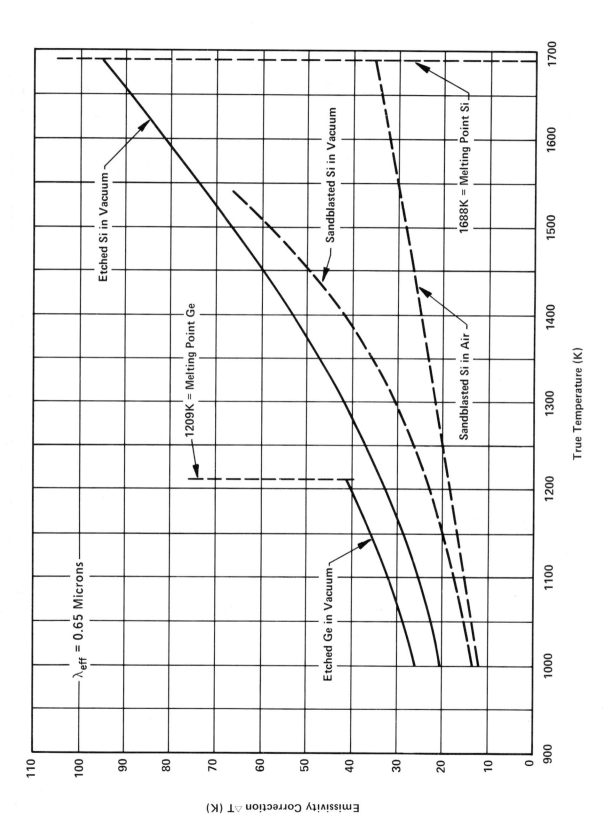

Figure 2-40 — Emissivity Correction $\triangle$ T for Si and Ge[37]

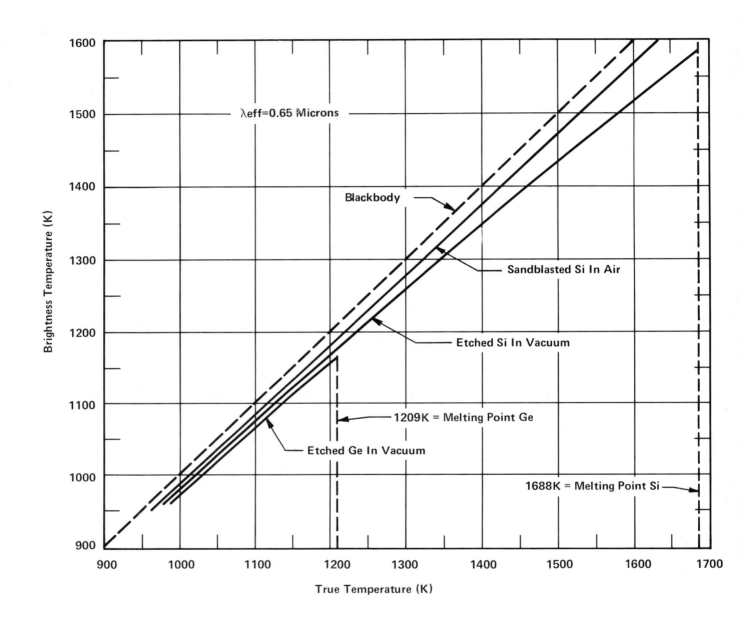

Figure 2-41 — Brightness Temperature Versus True Temperature for Si and Ge[37]

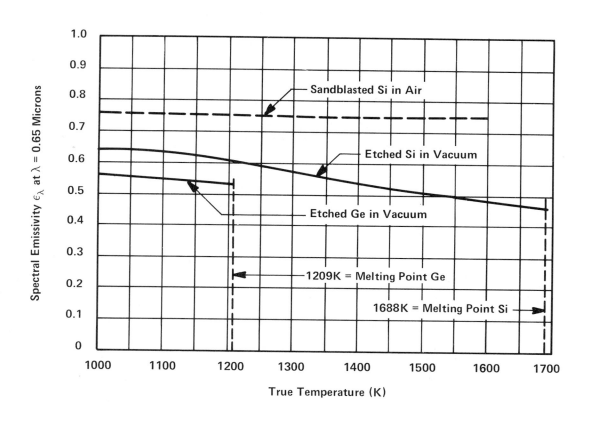

Figure 2-42 — Spectral Emissivity Versus True Temperature for Ge and Si[37]

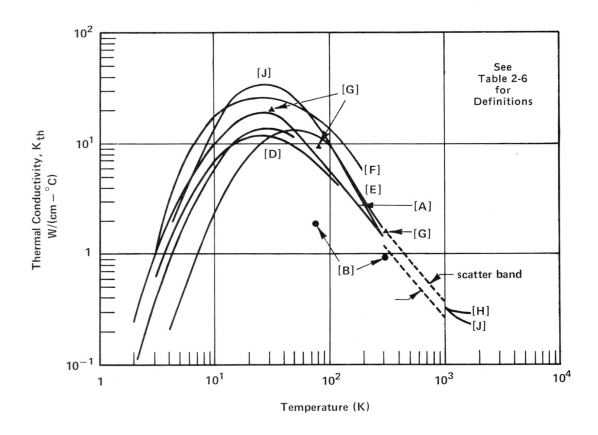

Figure 2-43 — Thermal Conductivity of Silicon [38]

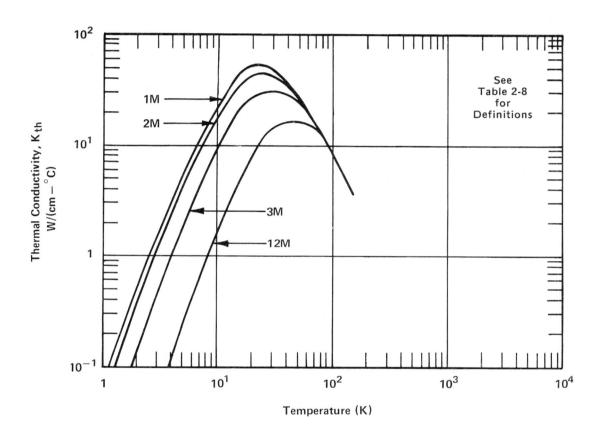

Figure 2-44 — Thermal Conductivity of B-Doped Si[38]

TABLE 2-6

APPROXIMATE PHYSICAL PROPERTIES OF Si SAMPLES IN FIGURE 2-43

CURVE	IMPURITY	IMPURITY CONTENT (cm^{-3})	CROSS SECTION $(mm \times mm)$
A	no added impurity	5×10^{14}	2.15 x 1.98
B	As	2.5×10^{19}	unknown
C	Au	10^{15}	2.36 x 2.21
D	unknown	7×10^{14}	1.75 x 1.5
E	B	5×10^{14}	3 x 3
F	B	2×10^{12}	3 x 3
G	unknown	4.8×10^{14}	unknown
J	p-Type	7×10^{12}	unknown

TABLE 2-7

THERMAL DIFFUSIVITY OF Si[38]

TEMPERATURE (K)	D_{th} (cm^2/s)
400	0.52
600	0.29
800	0.19
1000	0.14
1200	0.12

TABLE 2-8

THERMAL CONDUCTIVITY OF DOPED SILICON
Properties of Samples Containing Impurities in Figure 2-44 [41]

SAMPLE	MAJORITY IMPURITY[a] (cm^{-3})	MINORITY IMPURITY[a] (cm^{-3})	K_{th} MAX[b] (W/cm • °C)	K_{th} (3K) (W/cm • °C)
1M	1.0×10^{13}B	3.7×10^{11}P	48 (22K)	1.45
2M	4.2×10^{14}B	1.6×10^{13}P	43 (25K)	1.04
3M	4.0×10^{15}B	4.3×10^{13}P	33.3 (27K)	0.418
12M	4.0×10^{16}B	$\leqslant 1.0 \times 10^{15}$P	18 (37K)	0.056
T1	4×10^{15}Al(c)	(c)	28.6 (31K)	0.72
T2	3×10^{15}Ga(c)	(c)	31 (29K)	0.75
T3	1.9×10^{15}Sb	6.8×10^{14}(d)	49 (22K)	1.55
T4	3.5×10^{15}In(e)	4×10^{14}P(e)	21.3 (37K)	1.03
2M–S	4.2×10^{14}B	1.6×10^{13}P	38 (26K)	0.71
9M	1.0×10^{15}B	3.9×10^{14}P	43 (25K)	0.89
4M	3.8×10^{13}P	2.3×10^{13}B	51 (22K)	1.57
6M	1.1×10^{15}P	1.7×10^{13}B	51 (22K)	1.48
18M	2.5×10^{15}As	$\sim 1 \times 10^{13}$B	46 (22K)	1.48

(a) The chemical symbol of the impurity is listed next to its value of concentration.

(b) The temperature at which the thermal conductivity attains its maximum value is listed in the parentheses.

(c) The Hall coefficient against 1/T data for these samples had two slopes in the region below 200K. This indicates the presence of a second type of group III impurity, probably boron at an estimated concentration of $\sim 10^{14}$ cm^{-3}. The concentration of group V impurities is difficult to estimate.

(d) Boron or arsenic.

(e) The Hall coefficient against 1/T data exhibited three slopes in the region below 200K. In addition to the impurities listed, there was present an estimated 4×10^{14}B cm^{-3} and 2×10^{14}Ga cm^{-3}.

Recent data on carrier mobility versus carrier concentration at high concentrations of arsenic, phosphorus, and boron, showed decreasing mobility values as the concentrations of the dopants increased above 10^{20} atoms/cm^3.[39] The expression of mobility as a function of carrier concentration is given as

$$\mu_n = \mu_o + \frac{\mu_{max} - \mu_o}{1 + \left(\dfrac{n}{C_r}\right)^a} - \frac{\mu_i}{1 + \left(\dfrac{C_s}{n}\right)^\beta} \; .$$

The constants for arsenic-, phosphorus-, and boron- doped silicon are given in Table 2-9 for carrier concentrations ranging from 10^{13} to 5×10^{21} cm^{-3}. Some data for greater than 10^{20} atoms/cm^3 were obtained with laser-annealed samples. Figure 2-45 shows the electron mobility versus carrier concentration for phosphorus-doped silicon in the concentration range from 10^{19} to 6×10^{21} cm^{-3}. Figure 2-46 shows the carrier concentration versus resistivity for arsenic-, phosphorus-, and boron-doped silicon at concentrations greater than 10^{18} cm^{-3}. The dotted lines in Figure 2-46 are extrapolations. This data should be used as a supplement to the data given in Figures 2-17 through 2-23.

TABLE 2-9

CARRIER MOBILITY CONSTANTS FOR EXPERIMENTAL DATA FITTING[39]

	ARSENIC	PHOSPHORUS	BORON
μ_o	52.2	68.5	44.9
μ_{max}	1417	1414	470.5
μ_1	43.4	56.1	29.0
C_r	9.68×10^{16}	9.20×10^{16}	2.23×10^{17}
C_s	3.43×10^{20}	3.41×10^{20}	6.10×10^{20}
a	0.680	0.711	0.719
β	2.00	1.98	2.00
p_c	—	—	9.23×10^{16}

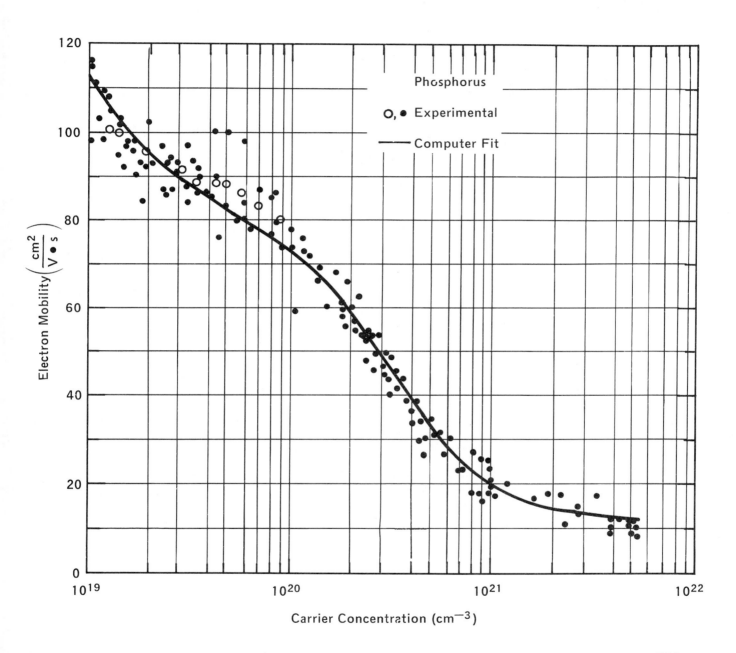

Figure 2-45 — Electron Mobility Versus Carrier Concentration for Phosphorus-Doped Silicon[39]
(© 1980 IEEE)

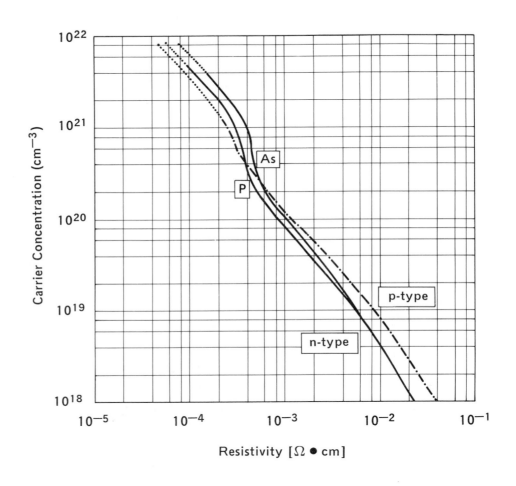

Figure 2-46 — Carrier Concentration Versus Resistivity for Arsenic-, Phosphorus-, and Boron-Doped Silicon[39] (© 1980 IEEE)

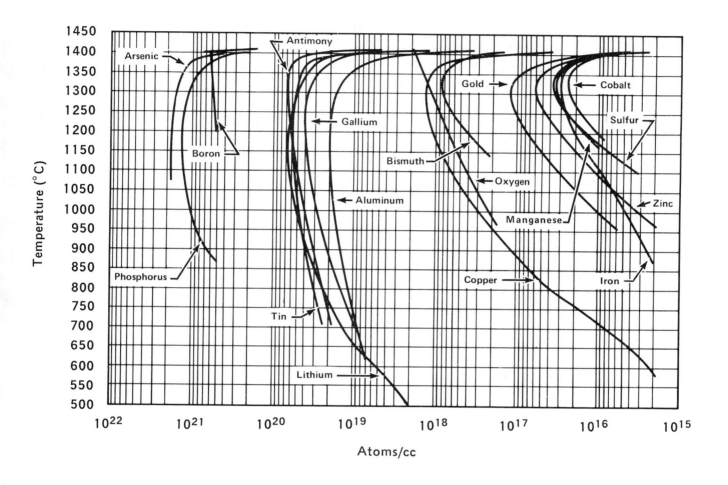

Figure 2-46 — Solid Solubilities of Impurity Elements in Silicon[40]

REFERENCES

1. *Semiconductors.* ed. N. B. Hannay, New York: Reinhold, 1959, pp. 329-333.

2. J. C. Hensel, H. Hasegawa, and M. Nakayama, "Cyclotron Resonance in Uniaxially Stressed Silicon," *Physical Review,* Vol. 138, No. 1A (April 5, 1965), p. A232.

3. H. D. Barber, "Effective Mass and Intrinsic Concentration in Silicon," *Solid-State Electronics,* Vol. 10 (1967), p. 1039.

4. F. J. Morin and J. P. Maita, "Conductivity and Hall Effect in the Intrinsic Range of Germanium," *Physical Review,* Vol. 94, No. 6 (June 15, 1954), p. 1525.

5. F. J. Morin and J. P. Maita, "Electrical Properties of Silicon Containing Arsenic and Boron," *Physical Review,* Vol. 96, No. 1 (October 1, 1954), p. 28.

6. E. H. Putley and W. H. Mitchell, "The Electrical Conductivity and Hall Effect of Silicon," *Proc. Phys. Soc.,* Vol. 72 (1958) p. 193.

7. A. Herlet, *Z. angew. Phys.,* Vol. 9, No. 4 (1957), p. 155.

8. C. D. Thurmond, "The Standard Thermodynamic Functions for the Formation of Electrons and Holes in Ge, Si, GaAs, and GaP," *J. Electrochem. Soc.: Solid-State Science and Technology,* Vol. 122, No. 8 (August 1975), p. 1133.

9. A. K. Jonscher, *Principles of Semiconductor Device Operation.* New York: John Wiley & Sons, 1960, p. 9.

10. G. L. Miller and D. A. H. Robinson, private communication.

11. J. C. Irvin, "Resistivity of Bulk Silicon and of Diffused Layers in Silicon," B. S. T. J., *41,* No. 2 (March 1962), p. 387.

12. S. M. Sze and J. C. Irvin, "Resistivity, Mobility and Impurity Levels in GaAs, Ge, and Si at 300K," *Solid-State Electron.,* Vol. 11 (1968), p. 599.

13. S. Wagner, "Diffusion of Boron from Shallow Ion Implants in Silicon," *J. Electrochem. Soc.: Solid-State Science and Technology,* Vol. 119 (November 1972), p. 1570.

14. W. R. Thurber et al, "Reevaluation of Irvin's Curves," *NBS Special Publication 400-29,* April 1977 (National Bureau of Standards), pp. 15-17.

15. W. R. Thurber et al, "Nuclear-Track Technique," *NBS Special Publication 400-36,* July 1978 (National Bureau of Standards), pp. 36-37.

16. W. R. Thurber, R. L. Mattis, Y. M. Liu, and J. J. Filliben, "Resistivity-Dopant Density Relationship for Phosphorus-Doped Silicon," *J. Electrochem. Soc.: Solid-State Science and Technology,* Vol. 127, No. 8 (August 1980), p. 1807.

17. L. Esaki and Y. Miyahara, "A New Device Using the Tunneling Process in Narrow p-n Junctions," *Solid-State Electron.,* Vol. 1 (1960), p. 13.

18. R. B. Fair and J. C. C. Tsai, "A Quantitative Model for the Diffusion of Phosphorus in Silicon and the Emitter Dip Effect," *J. Electrochem. Soc.: Solid-State Science and Technology,* Vol. 124, No. 7 (July 1977), pp. 1107-1108.

19. F. Mousty, P. Ostoja, and L. Passari, "Relationship Between Resistivity and Phosphorus Concentration in Silicon," *J. Appl. Phys.,* Vol. 45, No. 10 (October 1974), P. 4576.

20. R. B. Fair and J. C. C. Tsai, "The Diffusion of Ion-Implanted Arsenic in Silicon," *J. Electrochem. Soc.: Solid-State Science and Technology,* Vol. 122, No. 12 (December 1975), p. 1689.

21. R. B. Fair and G. R. Weber, "Effect of Complex Formation on Diffusion of Arsenic in Silicon," *J. Appl. Phys.,* Vol. 44, No. 1 (January 1973), p. 273.

22. D. A. Antoniadis, A. G. Gonzalez, and R. W. Dutton, "Boron in Near-Intrinsic <100> and <111> Silicon Under Inert and Oxidizing Ambients – Diffusion and Segregation," *J. Electrochem. Soc.: Solid-State Science and Technology,* Vol. 125, No. 5 (May 1978), p. 813.

23. G. Baccarani and P. Ostoja, "Electron Mobility Empirically Related to the Phosphorus Concentration in Silicon," *Solid-State Electronics,* Vol. 18 (1975), p. 579.

24. N. Sclar, "Resistivity and Deep Impurity Levels in Silicon at 300K," *IEEE Trans. on Electron Dev.,* Vol. ED-24, No. 6 (June 1977), pp. 709-712.

25. W.R. Thurber, "Resistivity-Dopant Density Evaluation," *Semiconductor Technology Program Process Briefs,* NBSIR 79-1591-5 (National Bureau of Standards), pp. 5-6.

26. L. C. Linares and S. S. Li, unpublished work.

27. K. B. Wolfstirn, "Hole and Electron Mobilities in Doped Silicon From Radiochemical and Conductivity Measurements," *J. Phys. Chem. Solids,* Vol. 16 (1960), p. 279.

28. D. K. Schroder, T. T. Braggins, and H. M. Hobgood, "The Doping Concentrations of Indium-Doped Silicon Measured by Hall, C-V, and Junction–Breakdown Techniques," *J. Appl. Phys.,* Vol. 49, No. 10 (October 1978), p. 5256.

29. C. A. Hogarth, "Germanium," *Materials Used in Semiconductor Devices.* ed. C. A. Hogarth, New York: John Wiley & Sons, 1965, p. 13.

30. W. W. Gartner, "Temperature Dependence of Junction Transistor Parameters," *Proceedings of the IRE,* Vol. 45, No. 5 (May 1957), p. 667.

31. C. Canall, G. Ottaviani, and A. A. Quaranta, "Drift Velocity of Electrons and Holes and Associated Anisotropic Effects in Silicon," *J. Phys. Chem. Solids,* Vol. 32 (1971), p. 1719.

32. T. Suzuki, A. Mimura, and T. Ogawa, "The Deformation of Polycrystalline-Silicon Deposited on Oxide-Covered Single Crystal Silicon Substrates," *J. Electrochem. Soc.: Solid-State Science and Technology,* Vol. 124, (November 1977), p. 1778.

33. H. Y. Fan and M. Becker, "Infrared Optical Properties of Silicon and Germanium," *Semi-Conducting Materials.* ed. H. K. Henisch, London: Butterworths Scientific Publications, 1951, p. 136.

34. H. R. Philipp and E. A. Taft, "Optical Constants in the Region 1 to 10 ev," *Phys. Rev.,* Vol. 120, No. (October 1, 1960), p. 37.

35. C. D. Salzberg and J. J. Villa, "Infrared Refractive Indexes of Silicon, Germanium, and Modified Selenium Glass," *J. Opt. Soc. of Am.,* Vol. 47, No. 3 (March 1957), p. 244.

36. F. Reizman and W. Van Gelder, "Optical Thickness Measurements of SiO_2-Si_3N_4 Films on Silicon," *Solid-State Electron.,* Vol. 10 (1967), p. 625.

37. F. G. Allen, private communication.

38. "Physical/Electrical Properties of Silicon," *Integrated Silicon Device Technology,* Vol. 5, Ad-605-558, under Contract AF 33 (657) — 10340, Research Triangle Institute, July 1964, pp. 98-106.

39. G. Masetti, M. Severi, and S. Solmi, "Modeling of Carrier Mobility Against Carrier Concentration in Arsenic-, Phosphorus-, and Boron-Doped Silicon," *IEEE Trans. on Electron Dev.,* Vol. ED-30, No. 7 (July 1983), pp. 764-765.

40. F. A. Trumbore, "Solid Solubilities of Impurity Elements in Germanium and Silicon," B. S. T. J., *39,* No. 1 (January 1960), p. 210.

41. M. G. Holland and L. J. Neuringer, "The Effect of Impurities on the Lattice Thermal Conductivity of Silicon," *Proceedings of the International Conference on the Physics of Semiconductors Held at Exeter,* July 1962, p. 475.

42. Y. Okada and Y. Tokumaru, "Precise Determination of Lattice and Thermal Expansion Coefficient of Silicon Between 300 and 1500 K," *J. Appl. Phys.,* Vol. 56, No. 2 (July 1984), p. 314.

3. BASIC EXPRESSIONS

INTRODUCTION

This section presents a listing of the fundamental relationships (basic expressions) that are pertinent to semiconductor devices, as well as some discussion of the pn junction and the Gummel-Poon equations. The material is organized from the most general to the more specific. Derived expressions are presented, whenever possible, in the order they would occur in the derivation process. This allows the user to examine the assumptions underlying an expression by tracing its development in reverse order.

The first sub-section contains a listing of the more commonly used symbols. Presentation of the basic expressions starts in subsection 3.2 with the general field relationships. Subsections 3.3 to 3.5 follow with the charge transport, charge carrier material interaction, and the hole-electron interaction relationships. Thermal relationships are provided in subsection 3.6 in conjunction with a table of the electrical-thermal duals. Subsections 3.7 to 3.9 pertain to diode current flow, capacitance, and avalanche breakdown.

A table of useful constants is provided in Table 3-17.

3.1 Symbols

Symbols used most commonly in Section 3 are presented in this subsection. Organizationly, the symbols are divided into the three following categories:

      (a) Operators (Table 3-1).

      (b) Vectors (Table 3-2).

      (c) Scalars (Table 3-3).

Descriptions of the symbols are presented in alphabetical order.

Some of the scalars, in general, may be vectors or tensors; but, for the purpose of this text, the greater generality was assumed to be of little use.

3.2 Field Relationships

The field equations which describe the classical laws of electromagnetics are shown in Table 3-4 in their various forms with their associated names and definitions. In addition, the special cases of electrostatic and magnetostatic approximation are also presented as Tables 3-5 and 3-6. The semiconductor charge density expression and the definition of charge neutrality are included in the electrostatic category.

3.3 Charge Transport

This subsection presents the relationships associated with the charge transport phenomena: current continuity, current flow, and isothermal current flow.

Table 3-7, Current Continuity, defines the total current, J_T, and introduces the general recombination/ generation function U. Current flow expressions are presented in Table 3-8, including drift, diffusion, and thermal currents. Quasi-Fermi potentials are introduced in Table 3-9, in conjunction with the definition of equilibrium.

TABLE 3-1

OPERATORS

DESCRIPTION	SYMBOL	UNIT
Closed region integral	$\oint$	
Cross product	$\times$	
Exponential	exp	
Implication	$\Rightarrow$	
Inner product	$\bullet$	
Integral	$\int$	
Spatial derivative	∇	(cm^{-1})
Temporal derivative	$\partial/\partial t$	(s^{-1})

TABLE 3-2

VECTORS

DESCRIPTION	SYMBOL	UNIT
Conduction current density	$\overline{J}_c$	(A/cm^2)
Displacement current density	$\overline{J}_D$	(A/cm^2)
Electric field	$\overline{E}$	(V/cm)
Force vector	$\overline{F}$	$(C \bullet V/cm)$
Hole and electron current densities	$\overline{J}_p, \overline{J}_n$	(A/cm^2)
Magnetic field	$\overline{B}$	$(Weber/cm^2)$
Thermal energy flow	$\overline{P}$	(W/cm^2)
Total current density	$\overline{J}_T$	(A/cm^2)

TABLE 3-3

SCALARS

DESCRIPTION	SYMBOL	UNIT
Absolute temperature	T	(K)
Acceptor and donor concentration	N_a, N_d	(cm^{-3})
Auger charge carrier recombination/generation	R_A	$(cm^{-3}s^{-1})$
Boltzmann constant	k	(J/K)
Charge density	ρ	(C/cm^3)
Conductivity for holes and electrons	σ_p, σ_n	$(mhos \bullet cm^{-1})$
Current	I	(A)
Density of recombination sites	N_t	(cm^{-3})
Diffusion coefficient for holes and electrons	D_p, D_n	(cm^2/s)
Electron-hole coupling	U	$(cm^{-3}s^{-1})$
Energy gap	E_g	(eV)
Energy gap at 0K	E_{g0}	(eV)
Fermi energy	E_F	(eV)
Generation by impact ionization	G	$(cm^{-3}s^{-1})$
Hole and electron concentration	p, n	(cm^{-3})
Hole and electron lifetime	τ_p, τ_n	(s)
Hole and electron quasi-Fermi potentials	ϕ_p, ϕ_n	(eV)
Intrinsic carrier concentration	n_i	(cm^{-3})
Intrinsic Fermi energy	E_i	(eV)
Ionized acceptor and donor concentration	$N_a{}^-, N_d{}^+$	(cm^{-3})
Mobility of holes and electrons	μ_p, μ_n	$(cm^2/V \bullet s)$
Permeability	μ	(h/cm)
Resistivity	ρ	$(ohms \bullet cm)$
Schockley-Hall-Reed charge carrier recombination/ generation	R	$(cm^{-3}s^{-1})$

TABLE 3-3 (Contd)
SCALARS

DESCRIPTION	SYMBOL	UNIT
Specific heat	C_p	$(J/g \bullet K)$
Thermal capacitance	C_{TH}	(J/K)
Thermal conductivity	K	$(W/cm \bullet K)$
Thermal diffusion coefficient for holes and electrons	D_P^T	$(cm^2/s \bullet K)$
Thermal energy density	Q_E	(J/cm^3)
Thermal power source	P_s	(W)
Thermal resistance	R_{TH}	(K/W)
Thermal velocity	v_{TH}	(cm/s)
Thermal voltage	V_T	$(volts)$
Total charge	Q_T	(C)
Total permittivity	ϵ	(F/cm)
Total permittivity of Si	ϵ_s	(F/cm)
Unit charge	q	(C)
Valence and conduction band energy levels	E_v, E_c	(eV)

TABLE 3-4

ELECTROMAGNETIC RELATIONSHIPS

DESCRIPTION	RELATIONSHIP
Charge and electric field	$\nabla \bullet (\epsilon \overline{E}) = \rho$
Gauss' law	$\int \nabla \bullet \overline{E} dV = \oint \overline{E} \bullet \overline{dS} = \dfrac{1}{\epsilon} \int \rho \, dV = \dfrac{Q_T}{\epsilon}$
Electric field for a point charge	$E_r = \dfrac{Q_T}{4\pi\epsilon r^2}$
Magnetic and electric field	$\nabla \times \overline{E} = -\partial \overline{B}/\partial t$
Magnetic field	$\nabla \bullet B = 0$
Magnetic field and current	$\nabla \times \left(\dfrac{B}{\mu}\right) = \overline{J}_c + \dfrac{\partial \epsilon \overline{E}}{\partial t}$
Electromagnetic force	$\overline{F} = q\,(\overline{E} + \overline{v} \times \overline{B})$

TABLE 3-5

ELECTROSTATIC APPROXIMATION, $\partial \overline{B}/\partial t = 0$

DESCRIPTION	RELATIONSHIP
Scalar potential	$\overline{E} = -\nabla \psi$
Kirchoff's voltage law	$\int \nabla \times \overline{E} \bullet \overline{dS} = \oint \overline{E} \bullet \overline{d\ell} = 0$
Poisson's equation	$\nabla^2 \psi = -\rho/\epsilon$
Semiconductor charge density	$\rho = q(p - n + N_d^+ - N_a^-)$
Charge neutrality	$p - n + N_d^+ - N_a^- = 0$
Coulomb's law	$F = \dfrac{q\,Q_T}{4\pi\epsilon r^2}$

TABLE 3-6

MAGNETOSTATIC APPROXIMATION, $\partial(\epsilon\overline{E})/\partial t = 0$

DESCRIPTION	RELATIONSHIP
Ampère's law	$\int \nabla \times \overline{B} \bullet \overline{dS} = \oint \overline{B} \bullet \overline{d\ell} = \mu I_c$
Kirchoff's current law	$\int \nabla \bullet \overline{J_c} \, dV = \oint \overline{J_c} \bullet \overline{dS}$
Faraday's law	$\oint \overline{E} \bullet \overline{dI} = -\dfrac{\partial}{\partial t} \oint \overline{B} \bullet \overline{dS}$
Vector potential	$\overline{B} = \nabla \times \overline{A}$
Kirchoff's voltage law (applies to inductive circuits)	$\int \nabla \times \left(\overline{E} - \dfrac{\partial \overline{A}}{\partial t} \right) dS = \oint \left(\overline{E} - \dfrac{\partial \overline{A}}{\partial t} \right) \bullet \overline{d\ell}$

TABLE 3-7

CURRENT CONTINUITY, $\nabla \bullet \nabla \times \overline{B} \equiv 0$

DESCRIPTION	RELATIONSHIP
Continuity equation	$\nabla \bullet \overline{J_T} = \nabla \bullet \left(\overline{J_c} + \dfrac{\partial \epsilon \overline{E}}{\partial t} \right)$ $\nabla \bullet \overline{J_c} + \dfrac{\partial \rho}{\partial t} = 0$
Conduction current	$\overline{J_c} = \overline{J_n} + \overline{J_p}$
Displacement current	$\overline{J_d} = \dfrac{\partial(\epsilon \overline{E})}{\partial t}$
Total current	$\overline{J_T} = \overline{J_c} + \overline{J_d}$
Kirchoff's current law (applies to capacitive circuits)	$\int \nabla \bullet \overline{J_T} dV = \oint \overline{J_T} \bullet \overline{dS} = 0$
Coupled continuity equation	$\nabla \bullet \overline{J_n} + \nabla \bullet \overline{J_p} - q \dfrac{\partial n}{\partial t} + q \dfrac{\partial p}{\partial t} = 0$
Electron continuity equation	$\nabla \bullet \overline{J_n} = q \left(\dfrac{\partial n}{\partial t} + U \right)$
Hole continuity equation	$\nabla \bullet \overline{J_p} = -q \left(\dfrac{\partial p}{\partial t} + U \right)$
Current coupling recombination/generation rate	$U = \Sigma \, (R + G)$

TABLE 3-8

CURRENT FLOW

DESCRIPTION	RELATIONSHIP
Electron drift, diffusion and thermal currents	$\overline{J}_n = q\mu_n n\overline{E} + qD_n\nabla n + qnD_n^T\nabla T$
Hole drift, diffusion and thermal currents	$\overline{J}_p = q\mu_p p\overline{E} - qD_p\nabla p - qpD_p^T\nabla T$
Einstein approximations	$kT + E_v < E_F < E_c - kT$
Electrons	$D_n = \dfrac{kT}{q}\mu_n \quad D_n^T = \dfrac{k}{2q}\mu_n$
Holes	$D_p = \dfrac{kT}{q}\mu_p \quad D_p^T = \dfrac{k}{2q}\mu_p$
Thermal voltage	$V_T \equiv \dfrac{kT}{q}$

TABLE 3-9

ISOTHERMAL CURRENT FLOW, $\nabla T = 0$

DESCRIPTION	RELATIONSHIP
Electron current	$\overline{J}_n = -q\mu_n(n\nabla\psi - V_T\nabla n)$
	$= -q\mu_n n\nabla\phi_n$
Electron quasi-Fermi potential	$\phi_n \equiv \psi - V_T \ln(n/n_i)$
	$n = n_i \exp\left(\dfrac{\psi - \phi_n}{V_T}\right)$
Equilibrium (electron)	$\overline{J}_n = 0 \quad \nabla\psi = V_T\dfrac{\nabla n}{n}, \nabla\phi_n = 0$
Hole current	$\overline{J}_p = -q\mu_p(p\nabla\psi + V_T\nabla p)$
	$= -q\mu_p p\nabla\phi_p$
Hole quasi-Fermi potential	$\phi_p \equiv \psi + V_T \ln(p/n_i)$
	$p = n_i \exp\left(\dfrac{\phi_p - \psi}{V_T}\right)$
Equilibrium (hole)	$\overline{J}_p = 0 \quad \nabla\psi = -V_T\dfrac{\nabla p}{p}, \nabla\phi_p = 0$
Intrinsic carrier concentration (See Section 2.1)	$n_i = n_{io} T^{3/2}\exp(-E_g/2kT)$

3.4 Charge Carrier Material Interaction

Expressions for conductivity, resistivity, and mobility are presented in Table 3-10. A function which describes mobility as a function of both the net doping concentration and the electric field is given. The scattering limited velocity obtained for the large electric field limit in this expression differs from that given in Figure 2-34. The two values are presented in Table 3-16. In addition, expressions for Hall mobility are also presented.

3.5 Hole-Electron Interaction

Hole-electron interactions through recombination/generation expressions are presented in Tables 3-12, 3-13, and 3-14. The material is subdivided into bulk and surface mechanisms as well as generation and recombination phenomena. Figure 3-1 illustrates lifetime as a function of resistivity.

3.6 Thermal Relationships

This subsection covers thermal relationships (Table 3-15) with special emphasis on semiconductors. The expressions are associated with their electrical duals. Table 3-16 provides the electrical and thermal duals. Table 3-17 is a collection of useful constants.

3.7 The pn Junction

The pn junction is covered extensively in current literature. Elementary treatments can be found in most texts, for example, S. M. Sze.[1] This subsection is a review of the basic equations and boundary conditions for the pn junction depletion region. As such, it draws heavily from a review by J. R. Hauser.[2]

Equilibrium Model

Consider the equilibrium pn junction of Figure 3-2. The equilibrium carrier densities, assuming Maxwell-Boltzmann statistics, are described by the displacement of the Fermi energy (E_f) from the intrinsic level of energy (E_i) as follows:

$$n = n_i \, \exp \left(\frac{E_f - E_i}{kT} \right) = n_i \, \exp - \left(\frac{q(\phi - \psi)}{kT} \right) \qquad 3.7.1$$

and

$$p = n_i \, \exp - \left(\frac{E_f - E_i}{kT} \right) = n_i \, \exp \left(\frac{q(\phi - \psi)}{kT} \right), \qquad 3.7.2$$

and the np product is everywhere equal to n_i^2.

The intrinsic level of energy (E_i) is defined as

$$E_i \equiv \frac{1}{2} \, [E_c + E_v + kT \, \ln (N_v/N_c)], \qquad 3.7.3$$

where N_v and N_c are the state densities in the valence and conduction bands, respectively. For Si, E_i differs little from the forbidden energy band center.

TABLE 3-10

CHARGE CARRIER MATERIAL INTERACTION

DESCRIPTION	RELATIONSHIP
Conductivity/Ohm's law (See Table 3-16.)	$\sigma = \dfrac{J_c}{E} = q\,(\mu_n n + \mu_p p)$
Resistivity (See Section 2.5.)	$\rho = \dfrac{1}{\sigma} = \dfrac{1}{q(\mu_n n + \mu_p p)}$
n-type material	$\sigma \cong q\mu_n n$
p-type material	$\sigma \cong q\mu_p p$
Mobility (See Section 2.6.) [3]	$\left(\dfrac{\mu_o}{\mu}\right)^2 = 2 + \dfrac{\lvert N_d + N_a \rvert}{\dfrac{\lvert N_d + N_a \rvert}{S} + N} + \dfrac{(E/A)^2}{E/A + F} + \left(\dfrac{E}{B}\right)^2$ (See Table 3-11)
Carrier velocity	
Electrons	$\overline{v}_n = \mu_n \,\overline{E}$
Holes	$\overline{v}_p = \mu_p \,\overline{E}$
Scattering limited velocity (See Section 2.6.)	$(E \gg B)$ $v_{SL} = \mu E \cong B\mu_o$
Hall mobility	$\mu_H = \lvert R_H\, \sigma \rvert$
Hall coefficient [1]	$R_H = \dfrac{r[p - (\mu_n/\mu_p)^2 n]}{q[p + (\mu_n/\mu_p)n]^2}$
n-type material	$R_H \cong -\dfrac{r}{qn}$
p-type material	$R_H \cong \dfrac{r}{qp}$
Phonon scattering	$r = 1.18$
Ionized impurity scattering	$r = 1.93$

TABLE 3-11

MOBILITY PARAMETERS FOR Si (300K)

	$\mu_0\left(\frac{cm^2}{V\cdot s}\right)$	$S(cm^{-3})$	N (V/cm)	A	F	B (V/cm)
μ_n	1400	350	3×10^{16}	3500	8.8	7400
μ_p	480	8.1	4×10^{16}	6100	1.6	25000

TABLE 3-12

BULK RECOMBINATION MECHANISM ($pn > n_i^2$)

DESCRIPTION	RELATIONSHIP
Shockley-Hall-Reed (SHR) [1] (See Figure 3-1.)	$R_{SHR} = \dfrac{pn - n_i^2}{\tau_{n0}\left[p + n_i \exp\left(\dfrac{E_i - E_t}{kT}\right)\right] + \tau_{p0}\left[n + n_i \exp\left(\dfrac{E_t - E_i}{kT}\right)\right]}$
SHR with a single recombination center at midband (See Figure 3-1.)	$R_{SHR} = \dfrac{pn - n_i^2}{\tau_{n0}(p + n_i) + \tau_{p0}(n + n_i)}$
Hole lifetime (See Figure 3-1.)	$\tau_{p0} = \dfrac{1}{\sigma_p\, V_{TH}\, N_t}$
Electron lifetime (See Figure 3-1.)	$\tau_{n0} = \dfrac{1}{\sigma_n\, V_{TH}\, N_t}$
n-type material approximation	$(n \gg n_i \gg p,\ p_{n0} = n_i^2/n_n)$ $R_{SHR} \cong \dfrac{p_n - p_{n0}}{\tau_{p0}}$
p-type material approximation (See Figure 3-1.)	$(p \gg n_i \gg n,\ n_{n0} = n_i^2/p_p)$ $R_{SHR} \cong \dfrac{n_p - n_{p0}}{\tau_{n0}}$
Thermal equilibrium	$pn = n_i^2$
Auger recombination	$R_A = (A_n n + A_p p)\,(pn - n_i^2)$

TABLE 3-13

BULK GENERATION MECHANISM ($pn < n_i^2$)

DESCRIPTION	RELATIONSHIP
Thermal SHR	$$G_{SHR} = \frac{pn - n_i^2}{\tau_{n0}(p + p_{n0}) + \tau_{p0}(n + n_{p0})}$$
Impact Ionization[1] (See Figure 3-7)	$$G = \alpha_n \frac{|J_n|}{q} + \alpha_p \frac{|J_p|}{q}$$ $$\alpha = Ae^{-b/E}$$

TABLE 3-14

SURFACE CARRIER RECOMBINATION

DESCRIPTION	RELATIONSHIP
SHR Surface	$$U^s = \frac{p_s n_s - n_i^2}{\tau_{n0}\left[p_s + n_i \exp\left(\frac{E_i - E_{st}}{kT}\right)\right] + \tau_{n0}\left[n_s + n_i \exp\left(\frac{E_{st} - E_i}{kT}\right)\right]}$$
n-type material	$$U_n^s = S_n(p - p_{n0})$$
Surface recombination velocity (electron)	$$S_n = S_0 \frac{N_D}{n_s + p_s + 2n_i}$$
p-type material	$$U_p^s = S_p(n - n_{p0})$$
Surface recombination velocity (hole)	$$S_p = S_0 \frac{N_A}{n_s + p_s + 2n_i}$$
Surface velocity	$$S_0 = \sigma v_{TH} N_{st}$$

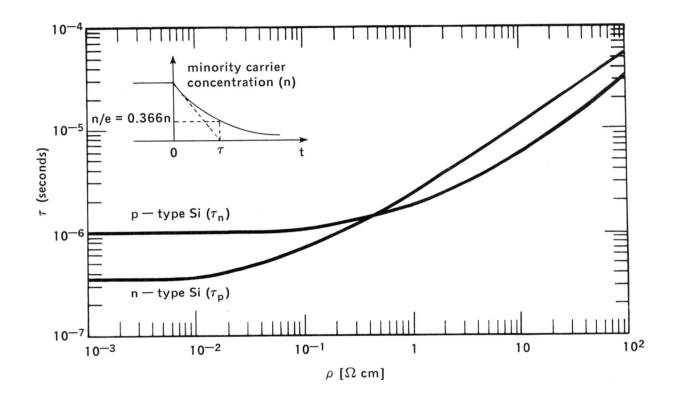

Figure 3-1 — Minority Carrier Lifetime (τ) Versus Resistivity (ρ) [4] (Reprinted with permission from Pergamon Press, Ltd.)

TABLE 3-15
THERMAL RELATIONSHIPS

DESCRIPTION	RELATIONSHIP
Energy density flow (Ohm's law dual) (See Table 3-16.)	$\overline{P} = -K \nabla T$
Thermal conductivity of Si (See Figure 2-43.)	$K = \dfrac{K_o}{T^{4/3}}$
Conservation of energy (continuity equation dual)	$\nabla \cdot \overline{P} = -\dfrac{\partial Q_E}{\partial t}$
Energy source and storage	$Q_E = -\int P_S\, dt + \rho C_p T$
Energy density generation $p_s > 0$: absorption $p_s < 0$:	$P_s = \overline{J} \cdot \overline{E} + E_g\,(R - G)$
Heat flow equation	$\nabla \cdot K \nabla T = -P_s + \delta\, C_p \dfrac{\partial T}{\partial t}$
Thermal dual of KVL	$\int \nabla T \cdot \overline{d\ell} = 0$
Thermal dual of KCL for $\partial Q/\partial T = 0$	$\int \overline{P} \cdot \overline{dS} = 0$
Thermal dual of resistance	$R_{TH} = \dfrac{1}{K}\,\dfrac{L}{A}$
Thermal dual of capacitance	$C_{TH} = \delta C_p Vol$

TABLE 3-16
ELECTRICAL AND THERMAL DUAL QUANTITIES

ELECTRICAL			THERMAL		
Current density	$\overline{J}$	(A/cm^2)	Power density	$\overline{P}$	(w/cm^2)
Voltage	V	(volts)	Temperature	T	(K)
Electric field	E	(volts/cm)	Temperature gradient	∇T	(K/cm)
Conductivity	σ	(mhos•cm)	Thermal conductivity	K	(W/cm•K)
Charge concentration	ρ	(C/cm^3)	Energy density	Q_E	(J/cm^3)
Resistance	R	(ohms)	Thermal resistance	R_{TH}	(K/W)
Capacitance	C	(farads)	Thermal capacitance	C_{TH}	(J/K)
Voltage source	V_S	(volts)	Temperature source	T_S	(K)
Current source	I_S	(amperes)	Power source	P_{ST}	(W)

TABLE 3-17

USEFUL CONSTANTS

DESCRIPTION	QUANTITY
Auger electron recombination for Si	$A_n = 1.4 \times 10^{-31}$ cm^{6}s^{-1}
Auger hole recombination for Si	$A_p = 9.9 \times 10^{-32}$ cm^{6}s^{-1}
Boltzmann constant	$k = 8.62 \times 10^{-5}$ eV/K
Density of Si	$\delta = 2.33$ g/cm^3
Electron lifetime n-type for Si	$\tau_{n0} = 1.2 \times 10^{-6}$ s
Energy gap at 0 K	$E_{g0} = 1.17$ eV
Hole lifetime p-type for Si	$\tau_{p0} = 3.5 \times 10^{-7}$ s
Intrinsic carrier concentration at T = 300 K for Si	$n_i = 1.38 \times 10^{10}$ cm^{-3}
Permeability	$\mu_0 = 4\pi \times 10^{-9}$ H/cm
Scattering limited velocity of electrons	$V_{s\ell} = 10^7$ cm/s
Scattering limited velocity of holes	$V_{s\ell} = 1.2 \times 10^7$ cm/s (See Table 3-10)
	$= 3.4 \times 10^6$ (See Figure 2-34)
Specific heat of Si	$C_p = 0.7$ J/g $\bullet$ K
Thermal conductivity at T = 300 K of Si	$K = 1.5$ W/cm $\bullet$ K
Thermal conductivity constant of Si	$K_0 = 3110$ W $\bullet$ K$^{1/3}$/cm
Thermal voltage at T = 300 K	$V_T = 0.02586$ volts
Total permittivity of Si	$\epsilon_s = 1.04 \times 10^{-12}$ F/cm
Unit charge	$q = 1.6 \times 10^{-19}$ C

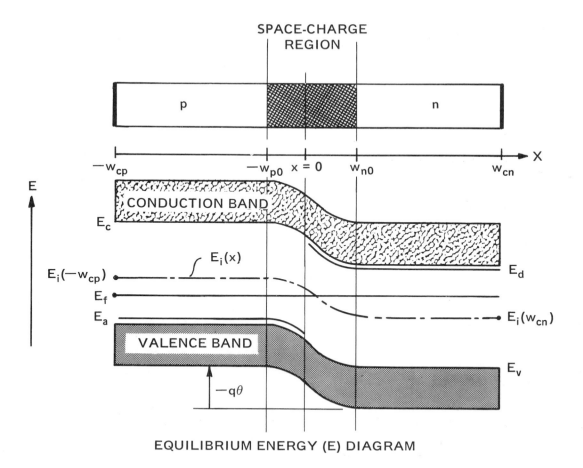

EQUILIBRIUM ENERGY (E) DIAGRAM

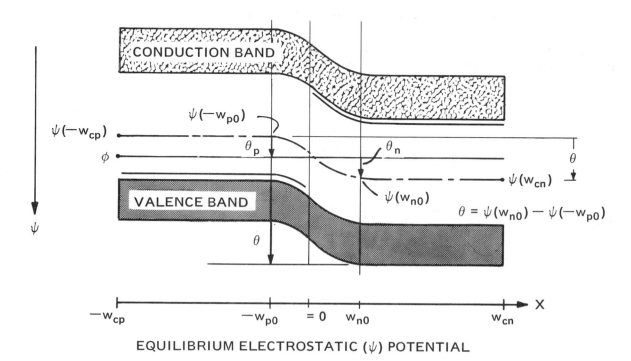

EQUILIBRIUM ELECTROSTATIC (ψ) POTENTIAL

Figure 3-2 — Equilibrium Diagrams for pn Junction Showing Built-In Voltage, θ

In Equations 3.7.1 and 3.7.2, the electrostatic potential (ψ) and the Fermi potential (ϕ) are defined as

$$\psi \equiv -E_i/q,$$ 3.7.4

and

$$\phi \equiv -E_f/q.$$ 3.7.5

The built-in voltage (θ) of the junction is defined as the relative displacement of the bands between the p and n regions (Figure 3-2):

$$\theta \equiv \psi(w_{n0}) - \psi(-w_{p0}).$$ 3.7.6

Substituting Equations 3.7.1 and 3.7.2 into 3.7.6 yields

$$\theta = V_T \ln \frac{p_0(-w_{p0})}{n_i} + V_T \ln \frac{n_0(w_{n0})}{n_i} ,$$ 3.7.7

where $p_0(-w_{p0})$ and $n_0(w_{n0})$ are the majority carrier equilibrium concentrations at the edges of the space-charge (depletion) region and $V_T \equiv kT/q$.

Equation 3.7.7 can be rewritten as

$$\theta = V_T \ln \frac{n_0(w_{n0})p_0(-w_{p0})}{n_i^2} .$$ 3.7.8

Equation 3.7.8 can be expressed in terms of doping concentration at the depletion edges as

$$\theta = V_T \ln \frac{N_d(w_{n0})N_a(-w_{p0})}{n_i^2} ,$$ 3.7.9

which is the useful expression for totally ionized doping levels significantly greater than n_i.

The relationship between carrier concentrations across the depletion region ($-w_{p0} < x < w_{n0}$) can be determined from a consideration of detailed current flow. The governing current expressions are

$$J_n = q\mu_n nE + qD_n \frac{\partial n}{\partial x} ,$$ 3.7.10

and

$$J_p = q\mu_p pE - qD_p \frac{\partial p}{\partial x} .$$ 3.7.11

In equilibrium, $J_n = J_p = 0$. Using the Einstein relationship ($D/\mu = \frac{kT}{q}$), Equations 3.7.10 and 3.7.11 can be simplified as

$$\frac{\partial \psi}{\partial x} = \frac{V_T}{n} \frac{\partial n}{\partial x} ,$$ 3.7.12

and

$$\frac{\partial \psi}{\partial x} = -\frac{V_T}{p} \frac{\partial p}{\partial x} .$$ 3.7.13

Integrating Equations 3.7.12 and 3.7.13 from $-w_{p0}$ to w_{n0} and substituting the expression for the built-in voltage (θ) from Equation 3.7.6 gives

$$n_0(w_{n0}) = n_0(-w_{p0}) \exp\left(\frac{\theta}{V_T}\right),$$

3.7.14

and

$$p_0(w_{n0}) = p_0(-w_{p0}) \exp\left(-\frac{\theta}{V_T}\right).$$

3.7.15

Nonequilibrium Model

For the nonequilibrium situation, quasi-Fermi levels are defined that express carrier concentration as

$$n = n_i \exp\left(\frac{E_{fn} - E_i}{kT}\right) = n_i \exp\left[-\frac{q(\phi_n - \psi)}{kT}\right],$$

3.7.16

and

$$p = n_i \exp\left(-\frac{E_{fp} - E_i}{kT}\right) = n_i \exp\left[\frac{q(\phi_p - \psi)}{kT}\right],$$

3.7.17

where E_{fn} and E_{fp} are the quasi-Fermi levels for electrons and holes respectively, and ϕ_n and ϕ_p are the corresponding quasi-Fermi potentials

$$\phi_n = -E_{fn}/q, \text{ and } \phi_p = -E_{fp}/q.$$

3.7.18

Note that the nonequilibrium np product

$$np = n_i^2 \exp\left(\frac{E_{fn} - E_{fp}}{kT}\right) = n_i^2 \exp-\left[\frac{q(\phi_n - \phi_p)}{kT}\right]$$

3.7.19

is exponentially different from n_i^2 by the difference between the quasi-equilibrium Fermi energies. It is important to note that it is the differences in Fermi potentials that are the quantities measured by an external voltmeter.

Figure 3-3 shows the pn junction with a forward bias voltage, V_a, applied. This voltage appears between the quasi-Fermi potentials of the n and p regions. If ohmic contacts at w_{nc} and $-w_{pc}$ are assumed, then

$$V_a = \phi_n(w_{nc}) - \phi_p(-w_{pc})$$

3.7.20

$$= [\psi_0(w_{nc}) - \psi_0(-w_{pc})] - [\psi(w_{nc}) - \psi(-w_{pc})].$$

The effect of V_a on the band bending can be determined through the current flow equations

$$J_n = -q\mu_n n \frac{\partial \phi_n}{\partial x},$$

3.7.21

and

$$J_p = q\mu_p p \frac{\partial \phi_p}{\partial x}.$$

3.7.22

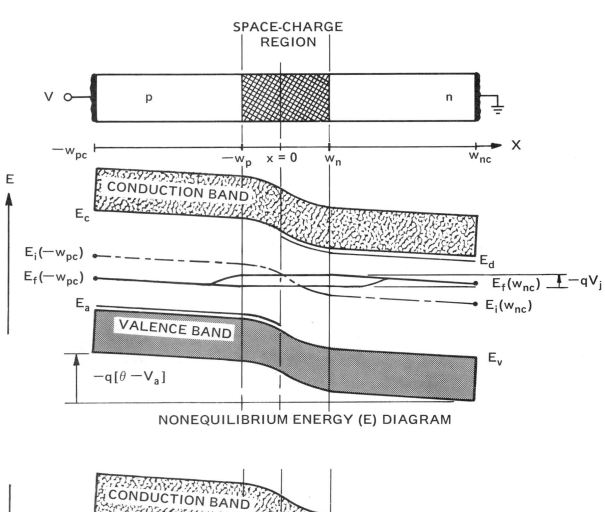

NONEQUILIBRIUM ENERGY (E) DIAGRAM

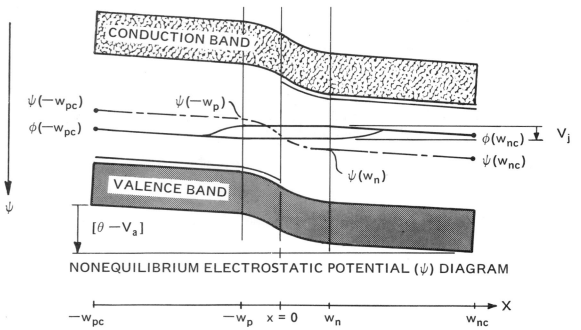

NONEQUILIBRIUM ELECTROSTATIC POTENTIAL (ψ) DIAGRAM

Figure 3-3 — Nonequilibrium Diagrams for pn Junctions Showing Applied Voltage, V_a, and Junction Voltage, V_j [5] (Reprinted with permission from Pergamon Press, Ltd.)

Integrating Equations 3.7.21 and 3.7.22 from the appropriate depletion region edge to the metal contact gives

$$\phi_n(w_{nc}) - \phi_n(w_n) = -\int_{w_n}^{w_{nc}} \frac{J_n}{q\mu_n n} dx ,$$ 3.7.23

and

$$\phi_p(-w_p) - \phi_p(-w_{pc}) = -\int_{-w_{pc}}^{-w_p} \frac{J_p}{q\mu_p p} dx .$$ 3.7.24

Addition of Equations 3.7.23 and 3.7.24 shows that V_a reduces the depletion region equilibrium band bending by an amount, $\phi_p(-w_p) - \phi_n(w_n)$, such that

$$\phi_p(-w_p) - \phi_n(w_n) = V_a - \int_{-w_{pc}}^{-w_p} \frac{J_p}{q\mu_p p} dx - \int_{w_n}^{w_{nc}} \frac{J_n}{q\mu_n n} dx .$$ 3.7.25

Since

$$\int_{-w_{pc}}^{-w_p} \frac{J_p}{q\mu_p p} dx + \int_{w_n}^{w_{nc}} \frac{J_n}{q\mu_n n} dx \leq I (R_p + R_n),$$ 3.7.26

the correction to V_a is less than or equal to the ohmic drops due to the bulk resistances, R_p and R_n, and the total current, I.[2]   This inequality is a result of the neglect of the depletion region resistance and minority current flow in the bulk regions.

The change in potential drop from equilibrium across the depletion region is defined as

$$V_j \equiv \theta - \psi(w_n) + \psi(w_p),$$ 3.7.27

and is related to the applied potential as follows:

$$V_a = V_j + IR_{total}.$$ 3.7.28

Under low injection conditions ($n_n \cong n_{n0} \cong N_D$, $p_p \cong p_{p0} \cong N_A$),

$$V_j \cong \phi_p(-w_p) - \phi_n(w_n).$$ 3.7.29

Very often, it is adequate to assume that $V_a = V_j$.

The potential difference, $\theta - V_j$, controls the nonequilibrium carrier concentrations at the depletion layer edges. This can be shown from consideration of the integral solutions[2] of Equations 3.7.10 and 3.7.11 as follows:

$$n(w_n) = \exp\left[\frac{\psi(w_n)}{V_T}\right]\left\{n(-w_p) \exp\left[\frac{-\psi(-w_p)}{V_T}\right] + \int_{-w_p}^{w_n} \frac{J_n(\xi)}{qD_n} \exp\left[\frac{-\psi(\xi)}{V_T}\right]d\xi\right\} ,$$ 3.7.30

and

$$p(w_n) = \exp\left[\frac{\psi(w_n)}{V_T}\right]\left\{p(-w_p) \exp\left[\frac{\psi(-w_p)}{V_T}\right] + \int_{-w_p}^{w_n} \frac{J_p(\xi)}{qD_p} \exp\left[\frac{\psi(\xi)}{V_T}\right]d\xi\right\}.$$ 3.7.31

Assuming that the integral current terms of Equations 3.7.30 and 3.7.31 can be neglected, resulting in a very good approximation according to Hauser[2] , then the following applies:*

$$n(w_n) = n(-w_p) \exp\left[\frac{(\theta - V_j)}{V_T}\right], \qquad\qquad\qquad 3.7.32$$

and

$$p(w_n) = p(-w_p) \exp\left[\frac{(\theta - V_j)}{V_T}\right]. \qquad\qquad\qquad 3.7.33$$

The influence of the detailed doping profile on these expressions can be seen by rewriting Equations 3.7.32 and 3.7.33 in terms of the equilibrium concentrations (as given by Equation 3.7.9), as

$$n(-w_p) = n_0 (-w_{p0}) \frac{n(w_n)}{n_0(w_{n0})} \exp\left(\frac{V_j}{V_T}\right), \qquad\qquad 3.7.34$$

and

$$p(w_n) = p_0(w_{n0}) \frac{p(-w_p)}{p_0(-w_{p0})} \exp\left(\frac{V_j}{V_T}\right). \qquad\qquad 3.7.35$$

If the doping density does not vary too much in the vicinity of the depletion edge of the junction, it is convenient to assume the following:

$$p_0(w_{n0}) \cong p_0(w_n) = p_{n0},$$

$$\qquad\qquad\qquad\qquad\qquad\qquad\qquad\qquad\qquad\qquad 3.7.36$$

$$n_0(w_{n0}) \cong n_0(w_n) = n_{n0},$$

and

$$p_0(-w_{p0}) \cong p_0(-w_p) = p_{p0},$$

$$\qquad\qquad\qquad\qquad\qquad\qquad\qquad\qquad\qquad\qquad 3.7.37$$

$$n_0(-w_{p0}) \cong n_0(-w_p) = n_{p0}.$$

(Normally, doping variations are neglected.)   If, in addition, low-level injection is assumed, the majority carrier densities are effectively constant and equal to the doping. Equations 3.7.32 and 3.7.33 can be expressed as

$$n(-w_p) = n_{p0} \exp\left(\frac{V_j}{V_T}\right), \qquad\qquad\qquad 3.7.38$$

and

$$p(w_n) = p_{n0} \exp\left(\frac{V_j}{V_T}\right). \qquad\qquad\qquad 3.7.39$$

---

*Equations 3.7.32 and 3.7.33 can also be obtained directly from the solution of Equations 3.7.12 and 3.7.13.

These important equations that express the minority carrier concentrations at the edges of the depletion layer, are commonly referred to as the "Law of the Junction." Furthermore, the excess carrier concentrations at the depletion edges are

$$\Delta_n(-w_p) = n(-w_p) - n_{p0} = n_{p0}\left[\exp\left(\frac{V_j}{V_T}\right) - 1\right],\qquad\qquad 3.7.40$$

and

$$\Delta_p(w_n) = p(w_n) - p_{n0} = p_{n0}\left[\exp\left(\frac{V_j}{V_T}\right) - 1\right].\qquad\qquad 3.7.41$$

If low-level injection cannot be assumed, J.R. Mathews[6] provides a more exact formulation for the excess minority carriers at the depletion region edges. That is, the variations in the majority carrier concentration are retained as

$$n(w_n) = n_{n0} + \Delta_n(w_n),\qquad\qquad 3.7.42$$

and

$$p(-w_p) = p_{p0} + \Delta_p(-w_p).\qquad\qquad 3.7.43$$

These expressions, in conjunction with Equations 3.7.32 and 3.7.33 and Equations 3.7.14 and 3.7.15, result in the following expressions for the excess carrier concentration at the depletion edges:

$$\Delta_n(-w_p) = n_{p0}\left[\exp\left(\frac{V_j}{V_T}\right)-1\right]\left[1+\frac{p_{p0}}{n_{n0}}\exp\left(\frac{V_j-\theta}{V_T}\right)\right]/\left\{1-\exp\left[2\left(\frac{V_j-\theta}{V_T}\right)\right]\right\},\quad 3.7.44$$

and

$$\Delta_p(w_n) = p_{n0}\left[\exp\left(\frac{V_j}{V_T}\right)-1\right]\left[1+\frac{n_{n0}}{p_{p0}}\exp\left(\frac{V_j-\theta}{V_T}\right)\right]/\left\{1-\exp\left[2\left(\frac{V_j-\theta}{V_T}\right)\right]\right\}.\quad 3.7.45$$

In this formulation, charge neutrality in the bulk regions ($\Delta n = \Delta p$) has been assumed.

Equations 3.7.44 and 3.7.45 can further be simplified by assuming a symmetrical junction (i.e. $p_{po} = n_{n0}$), to the following

$$\Delta n(-w_p) = n_{p0}\left[\exp\left(\frac{V_j}{V_t}\right)-1\right]/\left[1-\exp\left(\frac{V_j-\theta}{V_t}\right)\right],\qquad\qquad 3.7.46$$

and

$$\Delta p(w_n) = p_{n0}\left[\exp\left(\frac{V_j}{V_T}\right)-1\right]/\left[1-\exp\left(\frac{V_j-\theta}{V_t}\right)\right].\qquad\qquad 3.7.47$$

Note that the denominators of Equations 3.7.44 and 3.7.45 are significant only under large forward biases — a condition that implies narrow depletion widths. Since most p←n junctions are nearly symmetrical near their centers, the symmetrical approximation is not very restrictive, and Equations 3.7.46 and 3.7.47 are usually adequate.

Current-Voltage Characteristics

The diffusion model for the pn diode is based on the following simplifying assumptions:

(a) Minority carriers move primarily by diffusion.

(b) The excess densities of carriers at the edges of the depletion region are given by the Law of the Junction. (See Equations 3.7.38 and 3.7.39.)

(c) Low-level injection is assumed.

(d) Ohmic contacts to the diode couple the conduction electrons of the bonding wire to the majority carrier of the contacted semiconductor crystal.

(e) All the applied voltage is dropped across the junction.

Figure 3-4 outlines the diffusion model current calculations. The resultant low injection diffusion current density is

$$J_D = J_1 \left[\exp \frac{V_j}{V_T} - 1 \right], \qquad\qquad 3.7.48$$

where

$$J_1 = \frac{q D_p p_{n0}}{L_p} + \frac{q D_n n_{p0}}{L_n} . \qquad\qquad 3.7.49$$

In the familiar diode relationship, J_1 is usually referred to as the reverse saturation current. We will refer to it as the ideal current coefficient.

For high injection conditions, a more accurate expression is

$$J_D = J_1 \left[\exp \frac{V_j}{V_T} - 1 \right] / \left[1 - \exp \frac{V_j - \theta}{V_T} \right]. \qquad\qquad 3.7.50$$

This expression is necessary when both the current flow and depletion charge in a pn junction are calculated simultaneously. If the denominator of equation 3.7.50 is neglected, the internal junction voltage, V_j, can exceed the built-in voltage, θ, (which is physically impossible).

Furthermore, if V_j exceeds θ, then Equations 3.8.4 and 3.8.5 are physically impossible results.

The denominator of Equation 3.7.50 guarantees that the junction voltage, V_j, will not exceed the built-in voltage, θ. Physically, this means that the depletion region approaches and ideal short as the junction voltage approaches the built-in voltage. For a real diode, the applied voltage, V_a, to the diode, is not limited because all of the applied external potential above the built-in potential occurs across the series resistance, R_{total}, of the diode, as presented in Equation 3.7.28. Consequently, as the junction potential approaches the built-in voltage, the characteristic behavior of a diode becomes more ohmic.

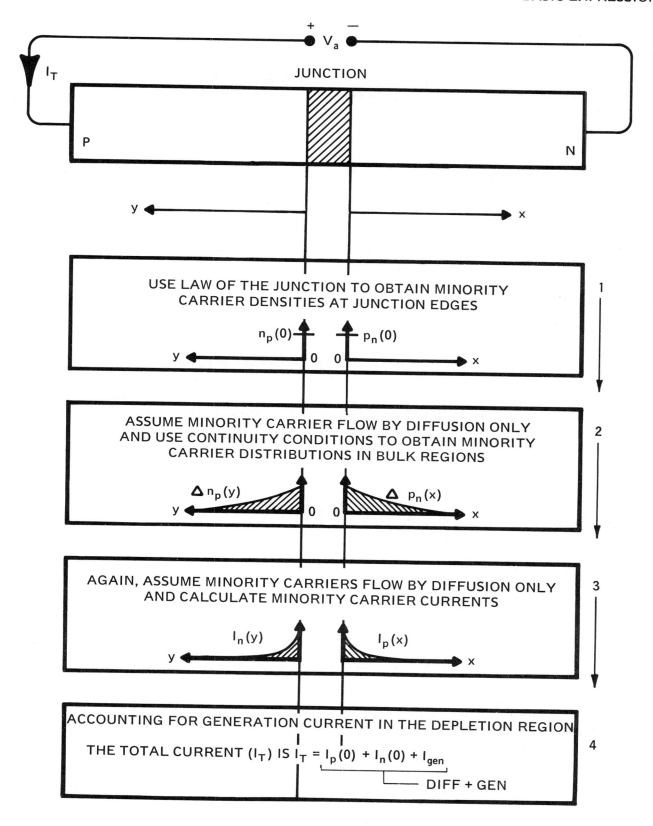

Figure 3-4 — Analysis of pn Diode Model

Recombination Within the Space-Charge Region

Equation 3.7.46 was derived assuming no recombination or generation within the space-charge region. Space-charge currents can be modeled by adding to the diffusion current, J_D, a term which accounts for generation or recombination within the space-charge region, J_R, as

$$J_R = q\sigma V_{th} N_t \, w \, \frac{n_i \, [\exp \, (\, |V_j| /V_T) - 1]}{2 \, [\exp \, (\, |V_j| /2V_T) + 1]}.$$

3.7.51

Again, w is the space-charge width. Equation 3.7.51 gives the maximum generation current for a single-level trap of density N_t at band center. The capture cross-section is δ, and V_{th} is the thermal velocity. For forward bias ($V_a \gg kT/q$), space-charge current is the result of recombining in the space-charge region as

$$J_R = q \, \frac{1}{2} \, \frac{n_i w}{\tau_0} \, \exp \, (\, |V_j| /2V_t),$$

3.7.52

where

$$\tau_0 = \frac{1}{\sigma v_{th} N_t}$$

3.7.53

is an equivalent lifetime.

For reverse bias ($V_a \ll -\frac{kT}{q}$), the space-charge current is the result of generation in the space-charge region,

$$J_R = -q \, \frac{n_i w}{2\tau_0}$$

3.7.54

The space-charge current can thus be written

$$J_R = J_2 \, (\exp \, \frac{V_j}{2V_T} - 1),$$

where

$$J_2 = \frac{q n_i w}{2\tau_0}.$$

Reverse-Bias Space-Charge Region Properties

Detailed properties of the pn junction depend on the doping profile of the diode. Two doping profiles are typically assumed: the step-junction and the linearly-graded junction. In both these models, the depletion-layer approximation is assumed; that is, the free carrier concentration is negligible in the space-charge region. Furthermore, the free carrier concentration changes abruptly to zero at the space-charge edge.

The Step-Junction Approximation

The simplest approximation to the charge density in the space-charge region assumes constant doping on either side of the metallurgical ($N_A = N_D$) junction. This gives rise to constant charge densities on either side of the metallurgical junction. The charge distribution for the step-junction approximation is shown in Figure 3-5. The calculations of electric field intensity electrostatic potential and capacitance for this charge distribution are outlined in Figure 3-5. Ohmic (IR) voltage drops in the neutral regions are assumed negligible.

The Linearly Graded Junction

A second useful approximation to a practical doping configuration is the linearly-graded junction. The properties of a linearly-graded junction are given in Figure 3-6. These calculations parallel those of the step-junction.

3.8 Depletion-Layer Capacitance

A differential form of the depletion-layer capacitance, shown below, has important applications in doping evaluations. (See J. L. Moll[7] for derivation.)

$$\frac{dC}{dV} = \frac{C^3}{A^2 2q\epsilon} \left[\frac{1}{N_d(w_n)} + \frac{1}{N_a(w_p)} \right] .$$

3.8.1

A generalized solution of this equation, Equation 3.8.2, gives the general relationship for the capacitance for arbitrary doping density. (Note again that $V_j > 0$ corresponds to forward bias, where the p-region is defined to be positive with respect to the n-region).

$$\int_\infty^C \frac{dC}{C^3} = \frac{1}{A^2} \int_\theta^{V_a} \frac{dV}{q\epsilon} \left[\frac{1}{N_d(w_n)} + \frac{1}{N_a(w_p)} \right] .$$

3.8.2

For the step-junction, where N_d and N_a are constants, Equation 3.8.2 becomes

$$\frac{1}{2C^2} = \frac{(\theta - V_j)}{q\epsilon A^2} \left(\frac{1}{N_d} + \frac{1}{N_a} \right) .$$

3.8.3

For the asymmetric step-junction, where $N_a \gg N_d$, the capacitance is

$$C = \left(\frac{q\epsilon N_d}{2} \right)^{1/2} \frac{1}{(\theta - V_j)^{1/2}} .$$

3.8.4

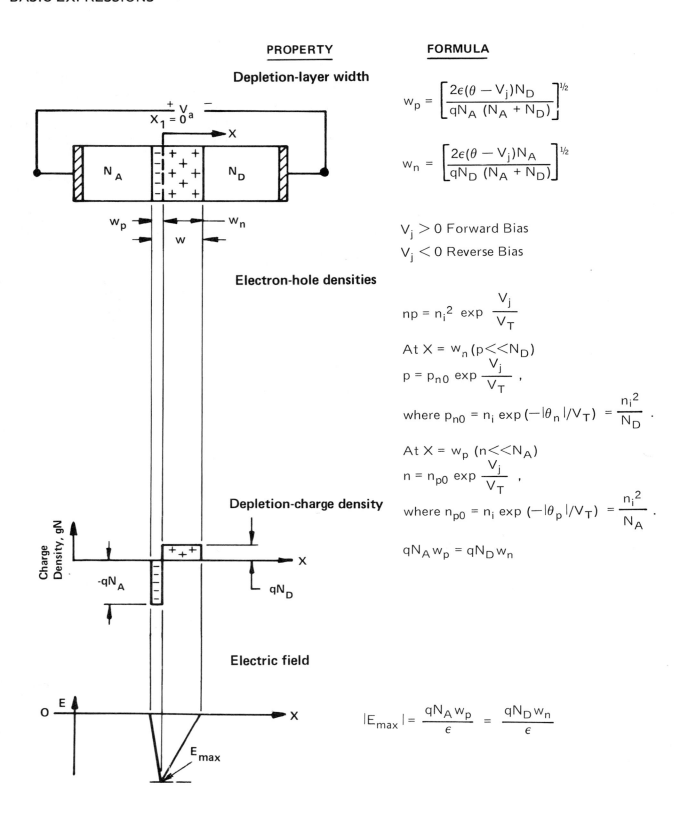

PROPERTY

FORMULA

Depletion-layer width

$$w_p = \left[\frac{2\epsilon(\theta - V_j)N_D}{qN_A(N_A + N_D)}\right]^{1/2}$$

$$w_n = \left[\frac{2\epsilon(\theta - V_j)N_A}{qN_D(N_A + N_D)}\right]^{1/2}$$

$V_j > 0$ Forward Bias

$V_j < 0$ Reverse Bias

Electron-hole densities

$$np = n_i^2 \ \exp \ \frac{V_j}{V_T}$$

At $X = w_n \ (p \ll N_D)$

$$p = p_{n0} \ \exp \ \frac{V_j}{V_T} \ ,$$

where $p_{n0} = n_i \ \exp \ (-|\theta_n|/V_T) \ = \ \frac{n_i^2}{N_D} \ .$

At $X = w_p \ (n \ll N_A)$

$$n = n_{p0} \ \exp \ \frac{V_j}{V_T} \ ,$$

where $n_{p0} = n_i \ \exp \ (-|\theta_p|/V_T) \ = \ \frac{n_i^2}{N_A} \ .$

Depletion-charge density

$$qN_A w_p = qN_D w_n$$

Electric field

$$|E_{max}| = \frac{qN_A w_p}{\epsilon} \ = \ \frac{qN_D w_n}{\epsilon}$$

Figure 3-5 — Relationships for Step-Junction Diode Depletion Layer
(Sheet 1 of 2)

PROPERTY	FORMULA

Potential distribution

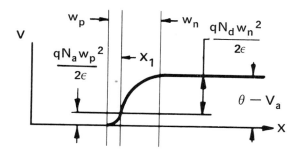

Built-in voltage:

$$\theta = \frac{kT}{q} \ln \frac{N_A N_D}{n_i^2}$$

$$= |\theta_p| + |\theta_n|$$

Capacitance

$$C = \frac{\epsilon A}{w_n + w_p} = \frac{A\left[\dfrac{\epsilon q N_A N_D}{2(N_A + N_D)}\right]^{1/2}}{[\theta - V_j]^{1/2}}$$

Breakdown voltage (one-sided)

$$V_B = E_m \ w/2 = \frac{\epsilon E_m^2}{2q} \ \frac{1}{N_B}$$

$$= V_{0A} \ [N_B(cm^{-3})/10^{16}]^{-0.75}$$

$$= 60(E_g/1.1)^{3/2}(N_B/10^{16})^{-0.75}$$

where:

$$V_{0A} = 25V \text{ for Ge}$$

$$= 60V \text{ for Si}$$

$$= 135V \text{ for GaP}$$

$$= 320V \text{ for GaAs}$$

Figure 3-5 — Relationships for Step-Junction Diode Depletion Layer
(Sheet 2 of 2)

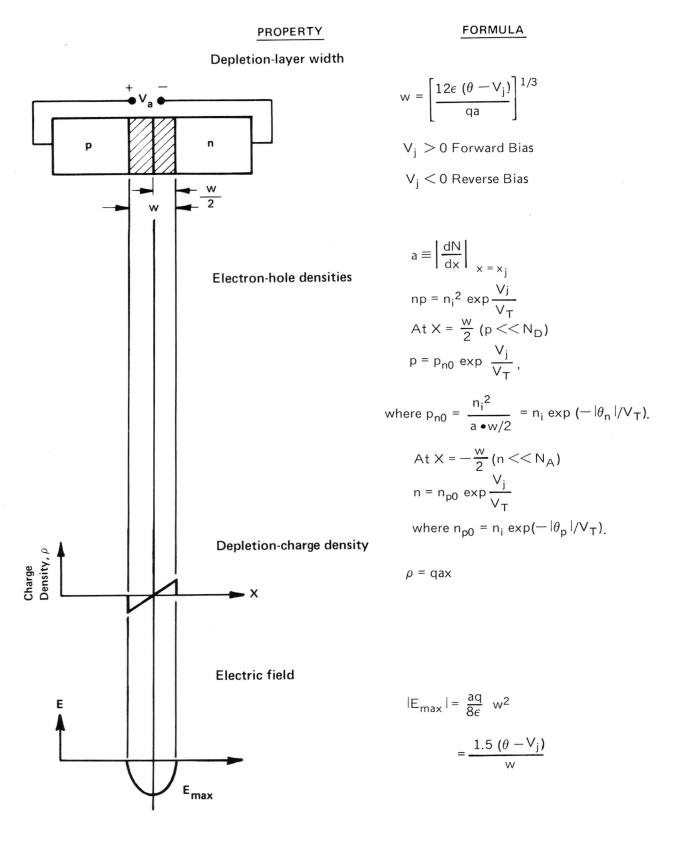

PROPERTY | FORMULA

Depletion-layer width

$$w = \left[\frac{12\epsilon\,(\theta - V_j)}{qa}\right]^{1/3}$$

$V_j > 0$ Forward Bias

$V_j < 0$ Reverse Bias

Electron-hole densities

$$a \equiv \left|\frac{dN}{dx}\right|_{x = x_j}$$

$$np = n_i^2 \exp\frac{V_j}{V_T}$$

At $X = \frac{w}{2}$ $(p \ll N_D)$

$$p = p_{n0} \exp\frac{V_j}{V_T},$$

where $p_{n0} = \dfrac{n_i^2}{a \cdot w/2} = n_i \exp(-|\theta_n|/V_T).$

At $X = -\frac{w}{2}$ $(n \ll N_A)$

$$n = n_{p0} \exp\frac{V_j}{V_T}$$

where $n_{p0} = n_i \exp(-|\theta_p|/V_T).$

Depletion-charge density

$$\rho = qax$$

Electric field

$$|E_{max}| = \frac{aq}{8\epsilon}\,w^2$$

$$= \frac{1.5\,(\theta - V_j)}{w}$$

Figure 3-6 — Relationships for Linear-Graded Junction (Sheet 1 of 2)

PROPERTY	FORMULA

Potential distribution

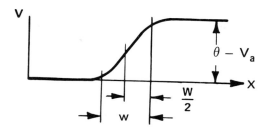

Built-in voltage:

$$\theta = 2V_T \ln \frac{aw_0}{2n_i}$$

See Figure 11-8 for θ vs. a for Si

Capacitance

$$C = \frac{\epsilon A}{w} = 0.436 \ (aq)^{1/3} \epsilon^{2/3} A \ (\theta - V_j)^{-1/3}$$

Breakdown voltage

$$V_B = V_{0G} \ [a(cm^{-4})/10^{21}]^{-0.4}$$

where:

$$V_{0G} \quad = \quad 18V \text{ for Ge}$$
$$= \quad 40V \text{ for Si}$$
$$= \quad 80V \text{ for GaP}$$
$$= \quad 145V \text{ for GaAs}$$

$$V_B = 60(E_g/1.1)^{6/5} (a/3 \times 10^{20})^{-2/5},$$

where E_g = bandgap in eV.

Figure 3-6 — Relationships for Linear-Graded Junction (Sheet 2 of 2)

Note that the depletion-layer capacitance varies inversely as the square root of the total junction voltage. The capacitance from Equation 3.8.4 is plotted in Figures 11-3 and 11-4 of Section 11.

For the linearly-graded junction, the capacitance is

$$C = \left(\frac{\epsilon^2 q}{12}\right)^{1/3} \left(\frac{a}{\theta - V_a}\right)^{1/3}. \qquad\qquad 3.8.5$$

Here, depletion-layer capacitance varies inversely as the cube root of the total junction voltage. The capacitance (and space-charge width) from Equation 3.8.5 is plotted against $(\theta - V_j)$ in Fig. 11-9 of Section 11.

3.9 Avalanche Breakdown

Given a reverse-biased junction, consider the effect of the electric field in the depletion region on the kinetic energy of a hole. As the hole drifts through this space-charge region, it is accelerated by the electric field. The kinetic energy of the hole increases until it collides with the semiconductor lattice. As the junction voltage is increased, however, there exists a critical field, E_C, above which the kinetic energy of the hole is sufficient to shatter silicon-to-silicon bonds, leading to the formation of additional electron hole pairs. The junction current increases very rapidly with further increase in voltage, and the junction is in the avalanche breakdown regime.

In actuality, the avalanche breakdown voltage is not simply determined by a critical field. It also depends on how the field is distributed across the space-charge region, as this will effect how much energy the hole will acquire between collisions. The analytical solution for breakdown thus must take geometric effects into consideration, in conjunction with the ionization coefficients for the appropriate ionizing particle.

The equations governing the behavior of a p n junction reverse-biased in avalanche [8] are the secondary ionization rate equation, Poisson's equation, and the carrier continuity equations for holes and electrons. The effect of carrier diffusion is usually neglected with respect to the drift transport in the high electric field. The boundary conditions are determined by the total current density and the minority carrier saturation currents at the edges of the depletion layer. The total generation rate, G, of hole-electron pairs is obtained from the ionization rates as follows:

$$G = \alpha_p v_p \, p + \alpha_n v_n \, n, \qquad\qquad 3.9.1$$

where

$$\alpha_p, \alpha_n = \text{hole, electron ionization rate}$$

$$v_p, v_n = \text{hole, electron velocity.}$$

The continuity equations for holes and electrons for steady state (dc) are:

$$\frac{\partial p}{\partial t} = -v_p \frac{\partial p}{\partial x} + G = 0, \qquad\qquad 3.9.2$$

and

$$\frac{\partial n}{\partial t} = v_n \frac{\partial n}{\partial x} + G = 0. \qquad\qquad 3.9.3$$

Consider the equation for the hole current as a function of distance, x, for n^+pp^+ diode. Combining Equations 3.9.1 and 3.9.2 and including the electronic charge, q, the differential equation for the hole current is

$$q\, v_p\, \frac{dp}{dx} = \frac{dJ_p}{dx} = qG = \alpha_p\, J_p + \alpha_n\, J_n \ . \qquad\qquad 3.9.4$$

For the dc case, $\nabla \cdot J = 0$, and the total current (J) is independent of x. This allows J_n to be eliminated from Equation 3.9.4 by substituting

$$J_n = J - J_p. \qquad\qquad 3.9.5$$

Inserting Equation 3.9.5 into Equation 3.9.4 and dividing by the total current we get the following differential equation for the hole current fraction:

$$\frac{dr}{dx} - (\alpha_p - \alpha_n)r = \alpha_n, \qquad\qquad 3.9.6$$

where

$$r = J_p/J.$$

The boundary conditions to be used in solving this equation are

$$r = r_1 = J_{ps}/J \quad \text{at} \quad x = x_1, \qquad\qquad 3.9.7$$

and

$$r = r_2 = (J - J_{ns})/J \quad \text{at} \quad x = x_2, \qquad\qquad 3.9.8$$

where

$$x_1 = \text{n-type edge of the depletion layer,}$$

$$x_2 = \text{p-type edge of the depletion layer,}$$

and

$$J_{ps} \text{ and } J_{ns} = \text{hole and electron saturation current densities, respectively.}$$

The integrating factor for this first order differential equation is

$$\exp \int_{x_1}^{x} p(\xi)d\xi = \exp - \int_{x_1}^{x} (\alpha_p - \alpha_n)d\xi \ . \qquad\qquad 3.9.9$$

The solution of Equation 3.9.6 may be written formally using the integrating factor and the boundary conditions, as

$$r(x) = \exp \int_{x_1}^{x} (\alpha_p - \alpha_n)d\xi \ [r_1 + \int_{x_1}^{x} \alpha_n \exp \int_{x_1}^{x} (\alpha_p - \alpha_n)d\xi \ dx] \ . \qquad\qquad 3.9.10$$

Equation 3.9.10 expresses the hole current fraction as a function of position when the hole and electron ionization rates are known functions of position. However, α_p and α_n are dependent on the electric field, and the boundary condition, r_2, has not yet been matched. For a given bias current density, J, it is necessary to determine the position of the depletion layer edges x_1 and x_2 so that this boundary condition may be met.

The dependence of the hole and electron ionization ratio upon electric field is expressible in the form

$$\alpha = Ae^{-(b/E)^m} .$$

3.9.11

The four ionization coefficients for Si and Ge are given in the following table. See also Figure 3-7[9].

TABLE 3-18

PARAMETERS OF IONIZATION RATE
$\alpha = Ae^{-(b/E)^m}$, E in V/cm

SEMICONDUCTOR	ELECTRONS		HOLES		m	REF
	A (cm^{-1})	b (V/cm)	A (cm^{-1})	b (V/cm)		
Ge	1.55×10^7	1.56×10^6	1.0×10^7	1.28×10^6	1	11
	4.9×10^5	7.9×10^5	2.15×10^5	7.1×10^5	1	10
Si	3.8×10^6	1.75×10^6	2.25×10^7	3.26×10^6	1	11
	1.4×10^6	1.3×10^6	5.8×10^5	1.6×10^6	1	10

C.N. Dunn[11] considers the temperature dependence of ionization rates and saturated drift velocity in silicon. The above data on Si reflects room temperature fitting of his data.

The ionization rates must be expressible as functions of position through the dependence on electric field. The electric field to be used in solving the equation is a solution of Poisson's equation

$$\frac{dE}{dX} = \frac{q}{\epsilon}(N_D - N_A + p - n),$$

3.9.12

with the boundary conditions

$$E(x_1) = E(x_2) = 0,$$

3.9.13

where

N_D = donor doping density

N_A = acceptor doping density.

Equation 3.9.10 is a nonlinear integral equation involving the electric field and the carrier densities. Computer solution[11] for a one-sided step-junction gives the breakdown voltage as

$$V_{BD} = E_c(w/2) = \frac{\epsilon E_c^2}{2q} (N_B)^{-1}.$$

3.9.14

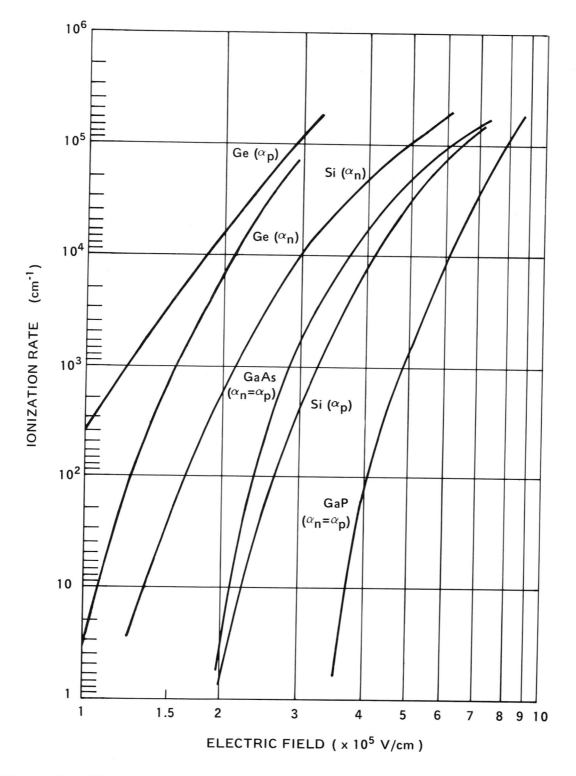

Figure 3-7 — Measured Ionization Coefficient for Avalanche Multiplication Versus Electric Field for Ge, Si, GaAs, and GaP. [9] (Reprinted by permission of John Wiley & Sons, Inc.)

BASIC EXPRESSIONS

An approximate universal expression for the step-junction is

$$V_{BD} \cong 60 \, (Eg/1.1)^{3/2} \, (N_B/10^{16})^{-3/4},$$

3.9.15

where Eg is the bandgap energy in eV and N_B is the background doping (cm^{-3}) on the lightly doped side of the junction.

For the linearly graded junction [10]

$$V_{BD} = \frac{2E_c w}{3} = \frac{4E_c^{3/2}}{3} \left(\frac{2\epsilon}{q}\right)^{1/2} (a)^{-1/2},$$

3.9.16

where a is the impurity gradient. The approximate universal expression in this case is

$$V_{BD} \cong 60 \, (Eg/1.1)^{6/5} \, (a/3 \bullet 10^{20})^{-2/5}.$$

3.9.17

For Si, Equation 3.9.17 is

$$V_{BD} \, (Si) \cong 40 \, (a/10^{21})^{-0.4}.$$

3.9.18

Extensive curves for Si and Ge depletion-layer width and maximum electric field at breakdown and breakdown voltage are included in Section 11. This data also considers curvature effects.

Again, the equations discussed thus far are summarized in Figures 3-5 and 3-6 for step-junction and linearly-graded junctions, respectively.

REFERENCES

1. S. M. Sze, *Physics of Semiconductor Devices.* New York: John Wiley & Sons, 1981.

2. J. R. Hauser, "Boundary Conditions at p-n Junctions," *Solid-State Electron.,* Vol. 14 (1971), pp. 133-139.

3. D. L. Scharfetter and H. K. Gummel, "Large-Signal Analysis of a Silicon Read Diode Oscillator," *IEEE Trans. on Electron Dev.,* Vol. ED-16, No. 1 (January 1969), pp. 64-77.

4. H. F. Wolf, *Silicon Semiconductor Data.* New York: Pergamon Press, 1969, p. 501.

5. A. Nussbaum, "Boundary Conditions for the Space-Charge Region of a p-n Junction," *Solid-State Electron.,* Vol. 12 (1969), pp. 177-183.

6. J. R. Mathews, private communication.

7. J. L. Moll, *Physics of Semiconductors.* New York: McGraw-Hill, 1964, p. 125.

8. D. R. Decker, "The Effect of Lattice Temperature on the Ionization Rates and Saturated Velocities in Germanium," Dissertation for PhD in EE, Lehigh University, 1970.

9. S. M. Sze, *Physics of Semiconductor Devices.* New York: John Wiley & Sons, 1969, p. 60.

10. S. M. Sze and G. Gibbons, "Avalanche Breakdown Voltages of Abrupt and Linearly Graded p-n Junctions in Ge, Si, GaAs and GaP," *Appl. Phys. Lett.,* Vol. 8, No. 5 (March 1, 1966), pp. 111-113.

11. D. R. Decker and C. N. Dunn, private communication.

4. MEASUREMENTS

4.1   Resistivity Measurements

The most direct approach to measuring the resistivity of semiconductor materials is to pass a current through the sample and to measure the voltage drop across a known distance. This is most readily achieved with a four-point resistivity probe with appropriate current supplies and voltage measuring circuitry. Similar apparatus can also be used to measure other characteristics, such as mobility.

Four-Point Probe Method

Figure 4-1 illustrates the use of a collinear four-probe array for resistivity measurements. Four correction factors are applied, as necessary, to obtain accurate measurements. They are

$$F_{SP} = \text{probe spacing correction factor}$$

$$F\left(\frac{W}{S}\right) = \text{thickness correction factor}$$

$$F_2 = \text{geometrical correction factor}$$

$$C = \text{correction factor for conducting skin of finite thickness (wraparound diffusion).}$$

Resistivity, then, is defined as $R_S = \frac{V}{I} \times \text{correction factor}$.

The probe spacing correction factor, F_{SP}, is defined as

$$F_{SP} = 1 + 1.082 \left(1 - \frac{S_2}{S} \right),$$

where

$$S = 1/3 \, (S_1 + S_2 + S_3).$$

Thus, $R_S = \frac{V}{I} \times F_{SP}$.

The thickness correction factor, $F\left(\frac{W}{S}\right)$, is obtained from the table below.

$\left(\frac{W}{S}\right)$	$F\left(\frac{W}{S}\right)$
0.5	0.997
0.6	0.992
0.7	0.982
0.8	0.966
0.9	0.944
1.0	0.921

Thus, $R_S = \frac{V}{I} \times F\left(\frac{W}{S}\right)$.

Figure 4-2 illustrates the use of the geometrical correction factor, F_2, and Figure 4-3 illustrates the use of correction factor C.

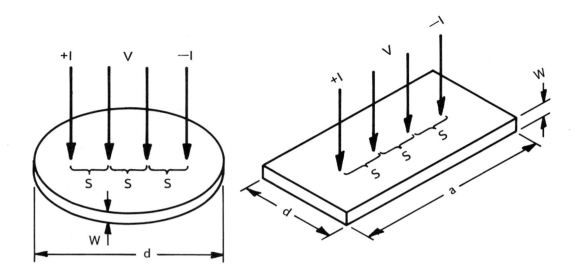

APPARATUS: collinear four-probe array

APPLICATION: silicon slices: diameter >16 mm
                                 thickness <1.6 mm

PARAMETERS: I = current applied through outer probes

                V = voltage measured across inner probes

                d = diameter of sample (in rectangular case, dimensions
                     perpendicular to probe direction)

                a = sample dimension parallel to probe direction

                W = thickness of sample

                S = probe spacing

Figure 4-1 — Four-Point Probe Measurement of Resistivity

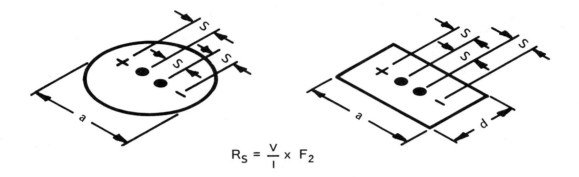

$$R_S = \frac{V}{I} \times F_2$$

$\frac{d}{s}$	CIRCLE	$\frac{a}{d} = 1$	$\frac{a}{d} = 2$	$\frac{a}{d} = 3$	$\frac{a}{d} \geqslant 4$
1.00				0.9988	0.9994
1.25				1.2467	1.2248
1.50			1.4788	1.4893	1.4893
1.75			1.7196	1.7238	1.7238
2.00			1.9454	1.9475	1.9475
2.50			2.3532	2.3541	2.3541
3.00	2.2662	2.4575	2.7000	2.7005	2.7005
4.00	2.9289	3.1137	3.2246	3.2248	3.2248
5.00	3.3625	3.5098	3.5749	3.5750	3.5750
7.50	3.9273	4.0095	4.0361	4.0362	4.0362
10.00	4.1716	4.2209	4.2357	4.2357	4.2357
15.00	4.3646	4.3882	4.3947	4.3947	4.3947
20.00	4.4364	4.4516	4.4553	4.4553	4.4553
40.00	4.5076	4.5120	4.5129	4.5129	4.5129
inf	4.5324	4.5324	4.5324	4.5324	4.5324

Figure 4-2 — Geometrical Correction Factor F_2 (for Sample Thickness or Junction Depth $< \frac{S}{2}$)

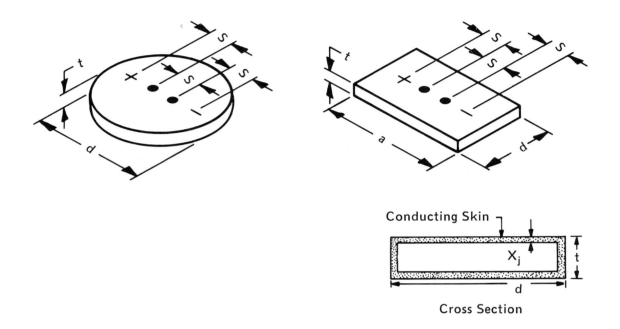

$$R_S = \frac{V}{I} \times C; \quad X_j < \frac{t}{2}$$

$\dfrac{d + t}{s}$	CIRCLE	$\dfrac{a + t}{d + t} = 1$	$\dfrac{a + t}{d + t} = 2$	$\dfrac{a + t}{d + t} = 3$	$\dfrac{a + t}{d + t} \geqslant 4$
1.00	Note:			1.9976	1.9497
1.25	factor also			2.3741	2.3550
1.50	independent		2.9575	2.7113	2.7010
1.75	of probe		3.1596	2.0053	2.9887
2.00	location		3.3381	3.2295	3.2248
2.50	on surface		3.6408	3.5778	3.5751
3.00	4.5324	4.9124	3.8543	3.8127	3.8109
4.00	4.5324	4.6477	4.1118	4.0899	4.0888
5.00	4.5324	4.5790	4.2504	4.2362	4.2356
7.50	4.5324	4.5415	4.4008	4.3946	4.3943
10.00	4.5324	4.5353	4.4571	4.4536	4.4535
15.00	4.5324	4.5329	4.4985	4.4969	4.4969
20.00	4.5324	4.5326	4.5132	4.5124	4.5124
40.00	4.5324	4.5325	4.5275	4.5273	4.5273
inf	4.5324	4.5324	4.5324	4.5324	4.5324

Figure 4-3 — Correction Factor C (for Conducting Skin of Depth X_j on a Sample of Thickness t Assuming Diffusion From all Surfaces)

Van der Pauw Method[1]

This section describes a general method for measuring the resistivity of irregularly shaped samples using a nonlinear four-probe array. Figure 4-4 shows a suggested circuit diagram and probe arrangement for such measurements. For irregularly shaped samples, a geometric correction factor, $F(Q)$, must be applied to the measured value to give the true resistivity, as

$$\rho = \frac{\pi d}{\ln 2} \frac{(R_{12,34} + R_{23,41})}{2} F(Q).$$

The correction factor, $F(Q)$, is taken from Figure 4-5 as a function of Q, which is measured as follows:

$$Q = \left[\frac{V_1(+I)}{V_S(+I),} + \frac{V_1(-I)}{V_S(-I)} \right] \div \left[\frac{V_2(+I)}{V_S(+I)} + \frac{V_2(-I)}{V_S(-I)} \right] = \frac{R_{12,34}}{R_{23,41}},$$

where V_1, V_2, and V_s are defined according to Figure 4-4. Figure 4-6 shows typical geometries used for Van der Pauw-type measurements. Other correction factors to be applied are described in Four-Point Probe Method above. For a more detailed discussion of the Van der Pauw method, see reference 1.

Temperature Corrections

If the temperature of the sample structure is different from that of the resistivity reference standard, a temperature correction must be applied. The graphs in Figures 4-7 and 4-8 show the temperature factors (C_r) for all methods of resistivity measurement for n- and p-type silicon. Corrections are applied as follows:

$$\rho = \rho_{unc} [1 + C_r (T - T_{ref})]^{-1},$$

where

ρ = corrected resistivity

ρ_{unc} = uncorrected resistivity

C_r = temperature coefficient appropriate to specimen

T = temperature of measurement ($^\circ$C)

T_{ref} = temperature of reference standard ($^\circ$C).

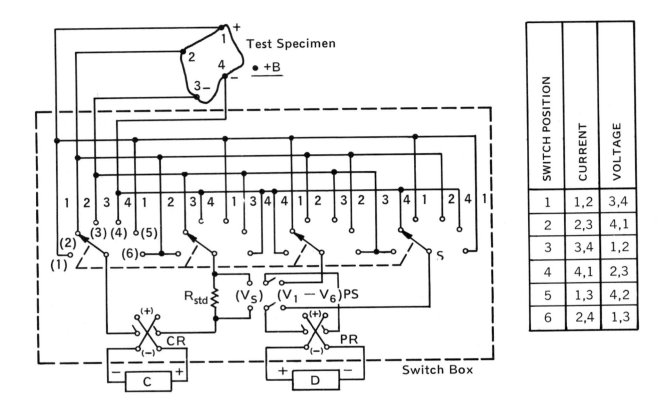

SWITCH POSITION	CURRENT	VOLTAGE
1	1,2	3,4
2	2,3	4,1
3	3,4	1,2
4	4,1	2,3
5	1,3	4,2
6	2,4	1,3

C = constant current supply

CR = current reversing switch

D = potentiometer-galvanometer system (or electrometer)

PR = potential reversing switch

PS = potential selector switch

R_{std} = standard resistor

S = contact selector switch

Figure 4-4 — General Four-Probe Resistivity Measurements — Van der Pauw Method[2]
(Copyright, ASTM, 1916 Race Street, Philadelphia, PA. 19103. Reprinted, with permission.)

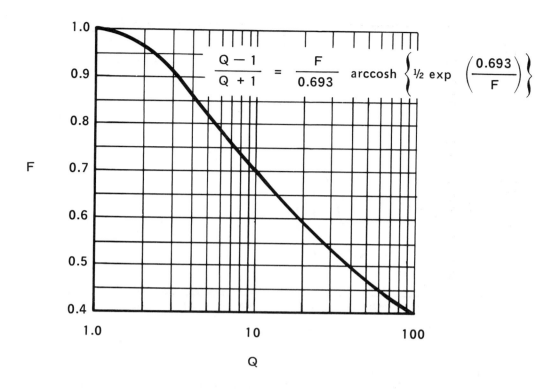

$$\frac{Q-1}{Q+1} = \frac{F}{0.693} \; arccosh \left\{ \tfrac{1}{2} \exp \left(\frac{0.693}{F} \right) \right\}$$

Figure 4-5 — The Factor F Plotted as a Function of Q[2]
(Copyright, ASTM, 1916 Race Street, Philadelphia, PA. 19103.
Reprinted, with permission.)

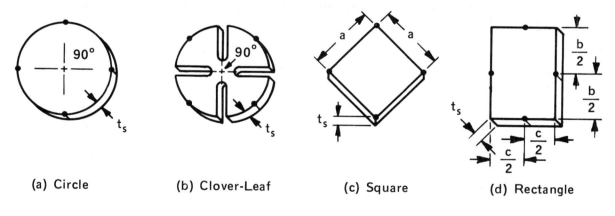

(a) Circle   (b) Clover-Leaf   (c) Square   (d) Rectangle

Note: Contact positions are indicated schematically by the small dots.

Figure 4-6 — Typical Symmetrical Lamellar Specimens[2]
(Copyright, ASTM, 1916 Race Street, Philadelphia, PA. 19103.
Reprinted, with permission.)

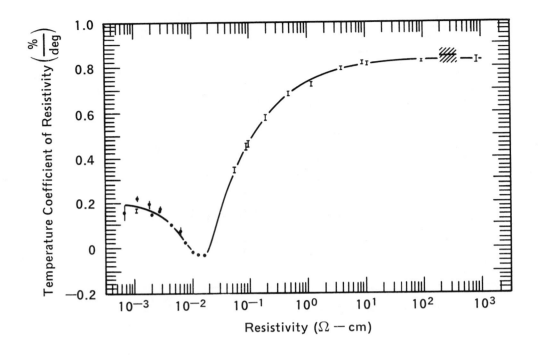

Figure 4-7 — Temperature Coefficient (C_r) as a Function of Specimen Resistivity (ρ_{AV}) for n-Type Silicon[3] (Copyright, ASTM, 1916 Race Street, Philadelphia, PA. 19103. Reprinted, with permission.)

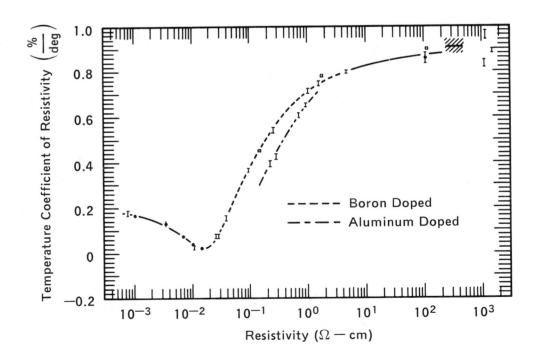

Figure 4-8 — Temperature Coefficient (C_r) as a Function of Specimen Resistivity (ρ_{AV}) for p-Type Silicon[3] (Copyright, ASTM, 1916 Race Street, Philadelphia, PA. 19103. Reprinted, with permission.)

4.2 Carrier Concentration Measurements

While carrier concentration can be inferred from resistivity data by comparison with tabulated values of N versus ρ (see Section 2.5 of this manual), several preferred methods are available which give a more direct measurement of N.

Plasma Minimum

A rough estimate of carrier concentration can be made by infrared (IR) reflectance measurements. The reflectance curve is characterized by a minimum caused by plasma absorption due to the free carriers. Empirical data relating this minimum to carrier concentration for n- and p-type substrates are shown in Figures 4-9 and 4-10.[4]

MOS Capacitance-Voltage (C-V) Technique

Techniques that utilize the change in measured capacitance as a function of applied voltage are especially useful in the measurement of carrier concentration of thin films, such as epitaxial silicon layers. A technique particularly suited to MOS technology uses a metal-oxide-semiconductor structure fabricated on the sample surface for the measurement contact. (See Figure 4-11.) The technique is relatively nondestructive, as the test structure can be stripped, leaving the substrate intact.

Data are obtained by generating a voltage ramp across the structure in Figure 4-11 and measuring the capacitance-versus-voltage response of the system. (See Figure 4-12.) Details of the relation between capacitance and voltage as a function of carrier concentration are described in detail in Section 14 of this manual and will not be discussed here. For purposes of measuring carrier concentration, the following formulae are required:

$$C_{measured} = \frac{C_{OX} \times C_{SC}}{C_{OX} + C_{SC}} .$$

$$C_{OX} = C_{max}/A, \quad \text{(oxide capacitance)}$$

where

$$A = \text{area of capacitor structure.}$$

$$C_{measured} = C_{min}/A. \quad \text{(measured capacitance)}$$

$$C_{SC} = \frac{C_{max}/A}{(C_{max}/C_{min}) - 1} .$$

Carrier concentration is obtained from the following relationship which was obtained by polynomial regression[7] of published data.[8]

$$N = 10^m,$$

where

$$m = 34.5888 + 3.4076\, L + 0.2018\, L^2 + 0.010514\, L^3,$$

and

$$L = \log(C_{SC}).$$

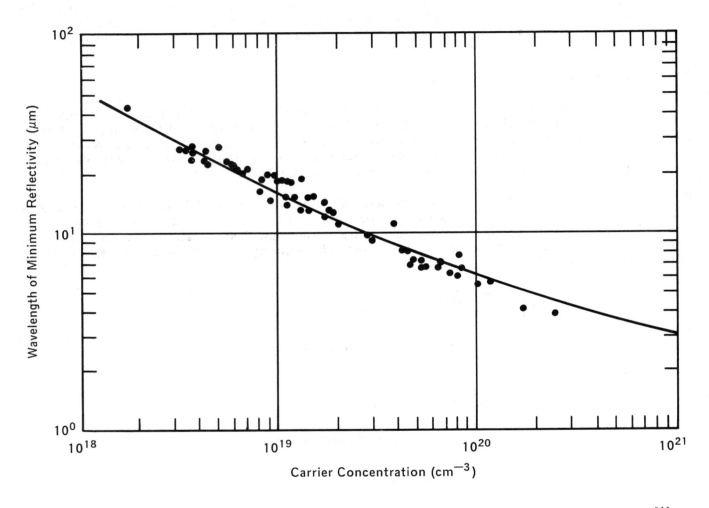

Figure 4-9 — Wavelength of Plasma Minimum Versus Carrier Concentration for n-Type Si[4]
(Reprinted with permission from Solid State Technology, Technical Publishing, a company of Dun & Bradstreet.)

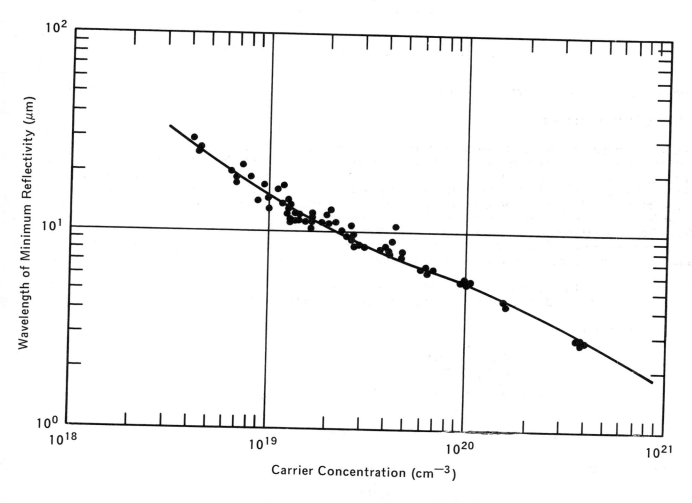

Figure 4-10 — Wavelength of Plasma Minimum Versus Carrier Concentration for p-Type Si[4]
(Reprinted with permission from Solid State Technology, Technical Publications, a company of Dun & Bradstreet.)

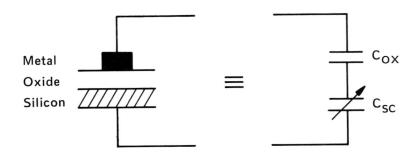

Figure 4-11 — Circuit Representation of MOS Capacitor

Figure 4-13 is a graph of net carrier concentration versus C_{min}/C_{max} for several values of oxide thickness.

Diode C-V Measurements

Diode C-V measurements that can be performed on several different structures are particularly useful for carrier concentration measurements in epitaxial layers and other samples where depth profiling is desired. Mesa and planar junction diodes [Figure 4-14 (a) and (b)] are formed by diffusion of impurities of conductivity type opposite to that of the material under test, forming a junction. In the case of mesa diodes, the material surrounding the diode is etched away to a depth equal to the junction depth. This eliminates the need for certain corrections to the calculation of carrier concentration. In the case of the mercury contact Schottky diode shown in Figure 4-14 (c), an apparatus is used that forms a mercury contact of known area on the sample surface through a mylar orifice. This technique is especially useful for product material testing, since it is nondestructive. The measurement technique that follows is similar for the three diodes types. [9]

For the index, i = 1 to n, measure C_i (capacitance in pF) versus V_i (voltage in volts) across the diode, varying V_i in increments such that $|C_{i+1} - C_i|$ changes by approximately 5 percent.

For i = 1 to n − 3, calculate

$$S_i = \frac{\ln\left(\dfrac{V_{i+3} + 0.6}{V_i + 0.6}\right)}{\ln\left(C_i/C_{i+3}\right)} \; ;$$

$$W_i = 10404\,(A/C_i), \qquad \text{(depletion width)}$$

where A = diode area;

$$W_i' = \left[\frac{(W_{i+3})^{S_i} - (W_i)^{S_i}}{S_i\,(W_{i+3} - W_i)}\right]^{(1/S_i - 1)} ; \qquad \text{(average depth)}$$

$$N_i = (6.493 \times 10^{14}) \times \left[\frac{V_{i+3} - V_i}{W_i'\,(W_{i+3} - W_i)}\right]; \qquad \text{(carrier concentration)}$$

and

$$W_{ti}' = W_i' + W_j, \qquad \text{(true depth)}$$

where

$$W_j = \text{junction depth.}$$

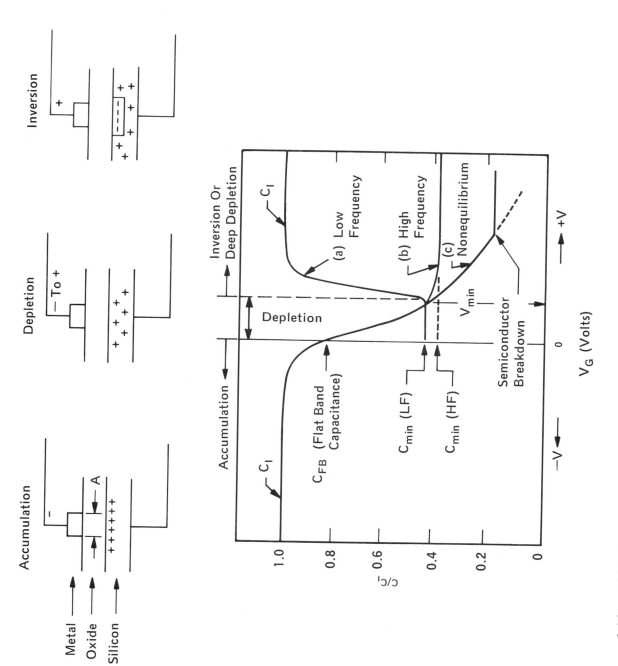

Figure 4-12 — MOS Capacitance-Voltage Curves. (a) Low Frequency, (b) High Frequency, (c) Nonequilibrium Case.[5] [6]

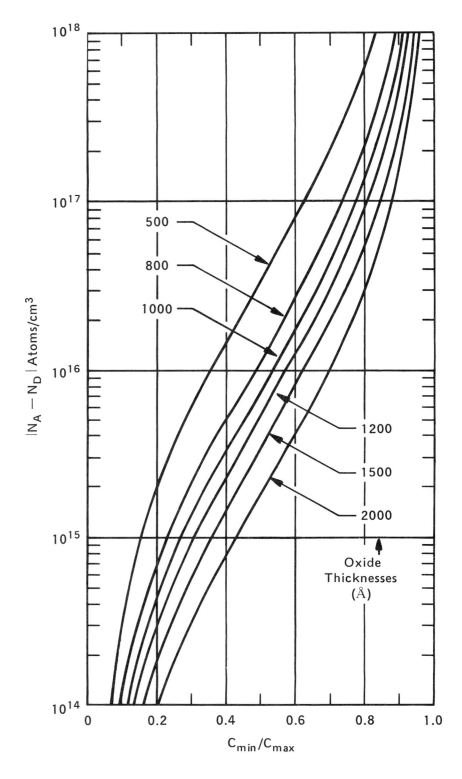

Figure 4-13 -- Net Impurity Concentration, $|N_A - N_D|$ Versus C_{min}/C_{max} for SiO_2 on Si

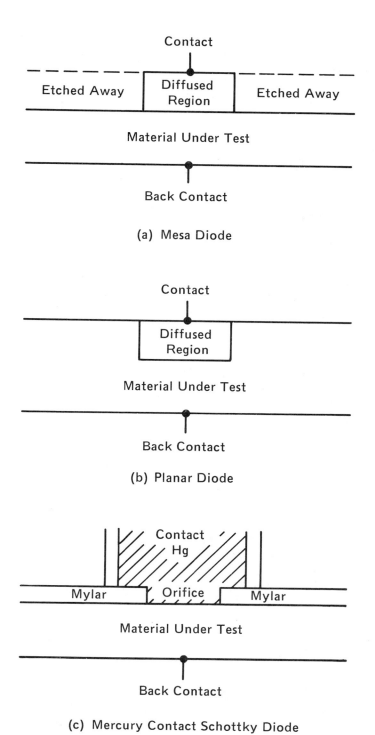

Figure 4-14 — Diode Formation for C-V Carrier Concentration Measurement

For convenience, arrange the calculations in tabular form on a data sheet as shown in Figure 4-15.

The following correction factor must be applied to C-V measurements for planar diodes. First, calculate

$$X_0 = 1.0404 \left(\frac{A}{C_i} \right). \qquad \text{(depletion width)}$$

$$C_{pi} = \frac{5.7931 \sqrt{A}}{\ln \left[\left(1 + \dfrac{X_0}{X_j} \right) \middle/ \left(1 + \dfrac{4X_0}{(3.5449\sqrt{A} + 4X_j)} \right) \right]} \qquad \text{(peripheral capacitance)}$$

$$C_{ci} = C_i - C_{pi}. \qquad \text{(corrected capacitance)}$$

Repeat the calculations, substituting C_{ci} for C_i. This results in new estimates for X_0, C_{pi}, and C_{ci}. Continue until old and new X_0 differ by less than 0.01 percent. Use C_{ci} instead of C_i and X_0 instead of W_i in performing the main calculation of carrier capacitance.

4.3 Epitaxial Layer Thickness Measurements

There are several methods for measuring epitaxial layer thickness, a crucial parameter in many bipolar and MOS applications. Some methods are destructive; others are nondestructive. Most methods rely on the fact that the epitaxial layer is characterized by properties (one is carrier concentration) that differ from those of the substrate. The bevel-and-stain method utilizes differential staining of substrate and epitaxial layer to delineate the boundary between them. The stacking fault method relies on the known morphology of stacking fault growth. The infrared reflectance method depends upon reflectance from the epilayer-substrate interface caused by a difference in optical properties.

Bevel-and-Stain Method[10]

This destructive method is generally applicable to many types of depth measurements, such as epitaxial thickness, junction depth, and buried layer depth. The sample to be measured is beveled at a shallow angle (typically less than 5°) and stained, and the width of the exposed surface layer is measured by a video micrometer or from a photograph taken with a fringe camera. The layer thickness is then calculated from the geometrical relationship between the exposed surface width of the layer and the bevel angle, as shown in Figure 4-16, or from the fringe count. When the resistivity of the substrate differs significantly from that of the epilayer, the apparent junction may be shifted due to redistribution of mobile charge at the interface.

Stacking Fault Method[11]

This nondestructive technique utilizes surface measurements of the side lengths of visible stacking faults to infer epitaxial thickness from the known morphology of stacking fault growth as a function of surface orientation. Assuming that the stacking fault nucleates at the epilayer-substrate interface and that the sides grow at a fixed angle relative to the surface normal, the surface attains dimensions geometrically related to the epitaxial thickness. These relationships are illustrated in Figure 4-17.

i	MEASURED		CALCULATED			CORRECTIONS		RESULTS	
	V_i (VOLTS)	C_i (pF)	S_i	W_i	W_i'	x_0	C_{ci}	W_{ti}' (um)	N_i (cm^{-3})

Figure 4-15 — Data Sheet for Calculation of Carrier Concentration From C-V Measurements

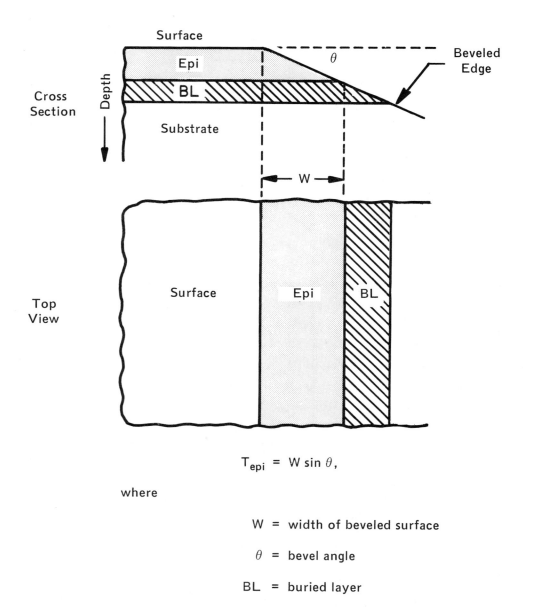

$$T_{epi} = W \sin \theta,$$

where

W = width of beveled surface

θ = bevel angle

BL = buried layer

Figure 4-16 — Bevel-and-Stain Method for Measuring Epitaxial Layer Thickness

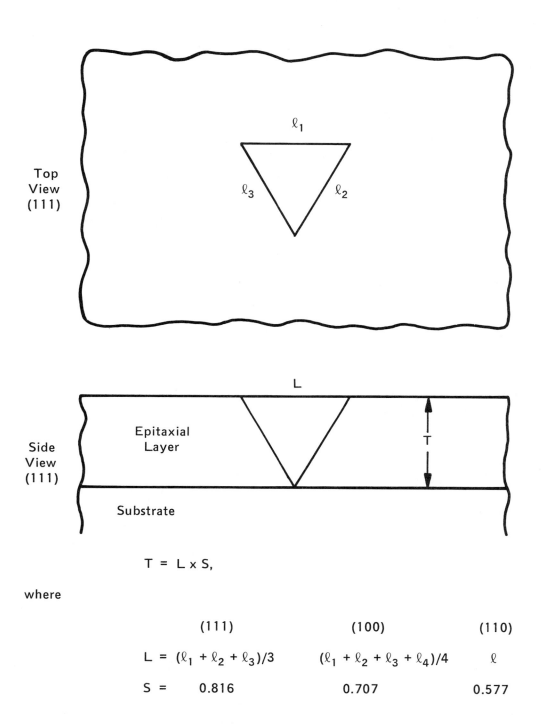

$$T = L \times S,$$

where

	(111)	(100)	(110)
L =	$(\ell_1 + \ell_2 + \ell_3)/3$	$(\ell_1 + \ell_2 + \ell_3 + \ell_4)/4$	ℓ
S =	0.816	0.707	0.577

Figure 4-17 — Stacking Fault Method for Measuring Epitaxial Layer Thickness

Infrared Reflectance Method

This nondestructive technique takes advantage of the differences in optical properties between the epitaxial layer and the underlying substrate. Incident radiation is reflected from both the epitaxial surface and the epilayer-substrate interface. (See Figure 4-18.) Interference between the two outgoing wavefronts causes fringes in the reflectance-versus-frequency spectrum. Figure 4-19 shows a typical reflectance spectrum for a 22.5 μm epitaxial layer on a substrate for which $N_D = 1.5 \times 10^{19}$ cm^{-3}. The position and spacing of the fringes can be used to calculate the epitaxial thickness according to the formula in Figure 4-18, provided accurate phase shifts (ϕ_2) for the reflectance at the epilayer-substrate interface are known. Tables 4-1 and 4-2 contain values of ϕ_2 for a range of wavelengths and impurity concentrations taken from ASTM F-95.[12]

For very thin layers or layers on very thin buried layer structures, these phase shift values are crucial.[13]

Interferometric Method

An alternate IR reflectance method currently in use in commercial epitaxial thickness instruments is illustrated in Figure 4-20. These instruments are manufactured by Digilab, Incorporated, Cambridge Massachusetts, Nicolet Instrument Corporation, Madison, Wisconsin, and IBM Instruments, Danbury Connecticut. Here, an interferometer is used to directly measure the interference properties of the epilayer-substrate system. When the two path lengths of the split beam are equal (solid line), the beam is nearly coherent, and a large signal is detected. When the movable mirror is moved a distance equal to the optical path length of radiation within the epitaxial layer, a second coherence condition is reached, causing the "side bursts" seen in the lower part of Figure 4-20 and in Figure 4-21, which is an "interferogram" of the same 22.5 μm layer shown in Figure 4-19. The thickness can be calculated from the mirror displacement at which the side bursts occur by employing the relationship

$$T_{epi} = \frac{x}{2\,n_{epi}\cos\theta},$$

where x is the mirror displacement relative to the zero path difference condition, n_{epi} is the index of refraction of the epitaxial layer, and θ is the angle of incidence of the beam. This relationship is not strictly correct, as phase shifts in the reflected beam alter the shape and position of the side burst peak. These phase shifts are not easy to include in the analysis, because the detector sees radiation over a broad range of frequencies simultaneously, and, thus, sees the effects of a range of phase shifts. (See Tables 4-1 and 4-2.) In practice, an empirical relationship is established between side burst position and layer thickness. For this reason, measurements made using these instruments yield results which are dependent (through ϕ_2) on the substrate carrier concentration and should be carefully calibrated. This is particularly important for very thin ($\sim 1\ \mu$m) layers.

4.4 Pattern Shift and Pattern Distortion Measurements

When an epitaxial silicon layer is deposited on a patterned silicon substrate, features are replicated on the epitaxial surface that are not in exact registration with the underlying substrate. Figure 4-22 illustrates this effect and provides definitions for the shift of a feature relative to the substrate (Pattern Shift) and for the change in linear dimensions of a feature (Pattern Distortion) after epitaxial growth. A photograph of a beveled sample is shown in Figure 4-23. These effects can lead to misregistration of subsequent masking steps and to shorting of normally isolated areas. A full description of these effects is discussed in reference 14.

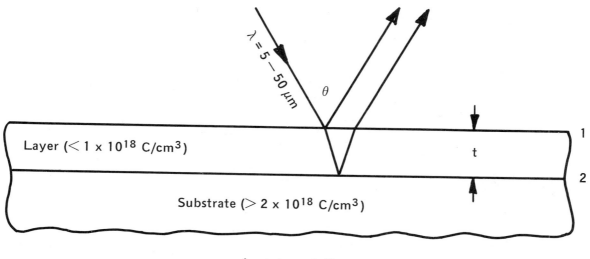

$$t = \frac{(m + \phi_1 - \phi_2)\lambda_i}{\sqrt{n^2 - \sin^2 \theta}}$$

t = epitaxial layer thickness

m = order of max or min in interference spectra

ϕ_1 = phase shift at air-epilayer interface (2π)

ϕ_2 = phase shift at epilayer-substrate interface (a function of substrate concentration and wavelength)

θ = angle of incidence

λ_i = wavelength of ith extrema in spectrum

n = index of refraction of silicon $\sim$3.42

      (a)   range = 0.5 to 200 μm

      (b)   area sampled = 0.25 cm^2

      (c)   precision = $\pm$ 1 percent

Figure 4-18 — IR Reflectance Method for Measuring Epitaxial Layer Thickness

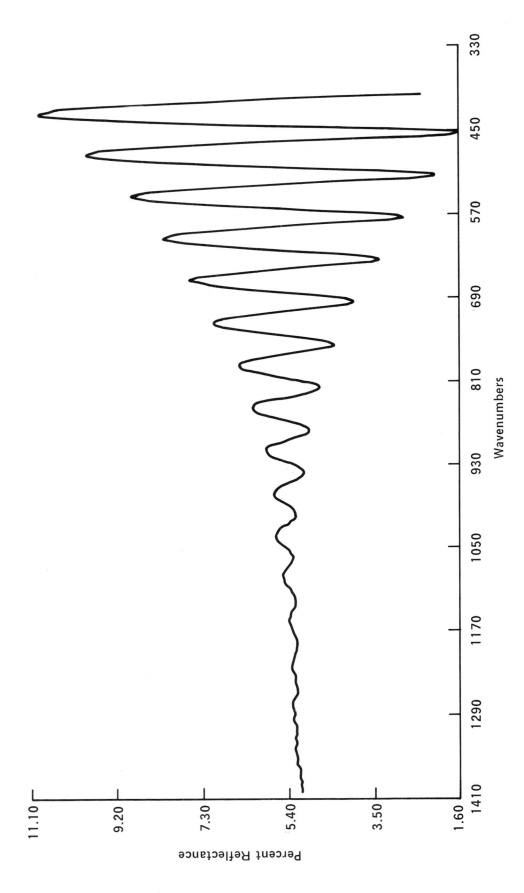

Figure 4-19 — Typical Reflectance Spectrum of Epitaxial Layer on a Reflective Substrate

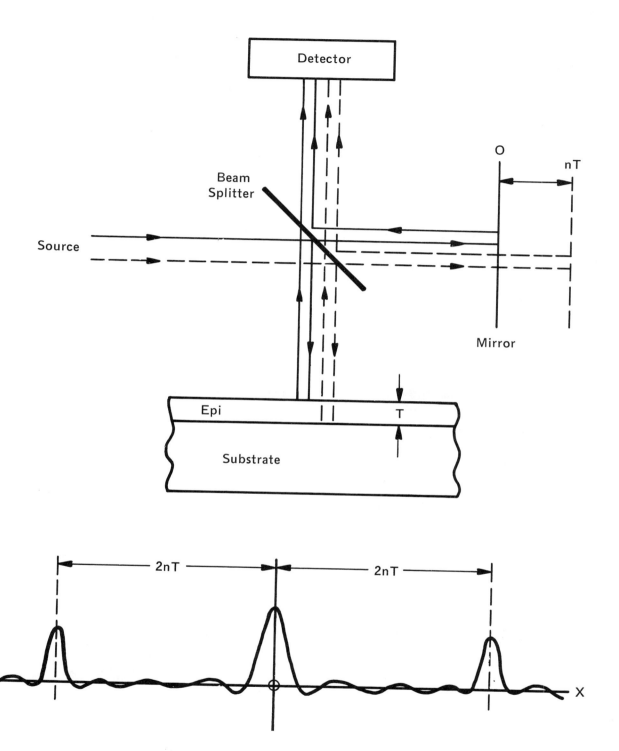

Figure 4-20 — Interferometric Method for Measuring Epitaxial Layer Thickness

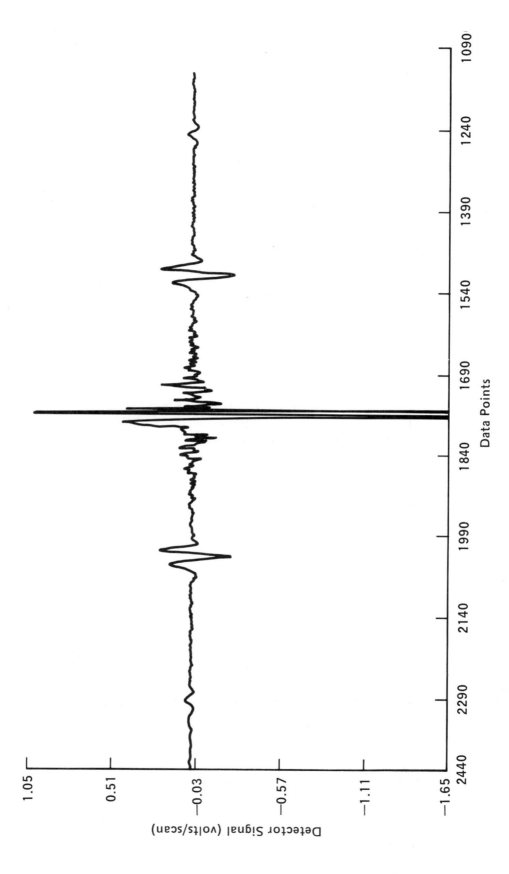

Figure 4-21 — Typical Interferogram for Epitaxial Thickness Measurement

TABLE 4-1

PHASE SHIFTS ($\phi/2\pi$) FOR n-TYPE SILICON[12]

WAVE-LENGTH, μm	RESISTIVITY, $\Omega \cdot$cm							
	0.001	0.002	0.003	0.004	0.005	0.006	0.007	0.008
2	0.033	0.029	0.028	0.027	0.027	0.026	0.025	0.024
4	0.061	0.050	0.047	0.046	0.045	0.043	0.041	0.039
6	0.105	0.072	0.064	0.062	0.060	0.057	0.055	0.052
8	0.182	0.099	0.083	0.078	0.075	0.071	0.067	0.064
10	0.247	0.137	0.105	0.095	0.090	0.084	0.079	0.075
12	0.289	0.183	0.132	0.115	0.106	0.098	0.091	0.084
14	0.318	0.225	0.164	0.137	0.124	0.113	0.104	0.097
16	0.339	0.258	0.197	0.163	0.144	0.129	0.117	0.109
18	0.355	0.283	0.226	0.189	0.166	0.146	0.131	0.121
20	0.368	0.303	0.251	0.214	0.188	0.165	0.147	0.134
22	0.378	0.319	0.272	0.236	0.209	0.183	0.163	0.148
24	0.387	0.333	0.289	0.255	0.229	0.202	0.179	0.162
26	0.394	0.344	0.303	0.272	0.246	0.219	0.196	0.177
28	0.401	0.353	0.316	0.286	0.261	0.235	0.211	0.191
30	0.406	0.362	0.326	0.298	0.275	0.250	0.226	0.206
32	0.411	0.369	0.336	0.309	0.287	0.263	0.240	0.219
34	0.415	0.375	0.344	0.319	0.297	0.274	0.252	0.232
36	0.419	0.381	0.351	0.327	0.307	0.285	0.263	0.243
38	0.422	0.386	0.357	0.335	0.315	0.294	0.273	0.254
40	0.425	0.391	0.363	0.341	0.323	0.302	0.283	0.264

WAVE-LENGTH, μm	RESISTIVITY, $\Omega \cdot$cm							
	0.009	0.010	0.012	0.014	0.016	0.018	0.020	
2	0.023	0.022	0.020	0.019	0.017	0.016	0.021	
4	0.038	0.036	0.034	0.031	0.029	0.027	0.025	
6	0.050	0.048	0.044	0.042	0.039	0.036	0.033	
8	0.061	0.059	0.054	0.051	0.047	0.043	0.040	
10	0.071	0.069	0.063	0.059	0.055	0.051	0.047	
12	0.081	0.078	0.072	0.067	0.062	0.057	0.053	
14	0.092	0.087	0.080	0.074	0.069	0.064	0.059	
16	0.102	0.097	0.088	0.082	0.075	0.070	0.065	
18	0.113	0.107	0.096	0.089	0.082	0.076	0.070	
20	0.124	0.117	0.105	0.096	0.088	0.081	0.075	
22	0.136	0.127	0.113	0.104	0.095	0.087	0.081	
24	0.148	0.138	0.122	0.111	0.101	0.093	0.086	
26	0.161	0.150	0.131	0.119	0.108	0.099	0.091	
28	0.175	0.161	0.141	0.127	0.115	0.104	0.096	
30	0.188	0.173	0.150	0.135	0.121	0.110	0.101	
32	0.201	0.185	0.160	0.143	0.128	0.116	0.106	
34	0.213	0.197	0.170	0.151	0.135	0.122	0.112	
36	0.225	0.209	0.180	0.160	0.143	0.129	0.117	
38	0.236	0.220	0.191	0.167	0.150	0.135	0.123	
40	0.246	0.230	0.200	0.178	0.158	0.141	0.128	

TABLE 4-2

PHASE SHIFTS ($\phi/2\pi$) FOR p-TYPE SILICON[12]

WAVE-LENGTH, μm	RESISTIVITY, $\Omega \cdot$cm							
	0.001	0.0015	0.002	0.003	0.004	0.005	0.006	0.007
2	0.036	0.034	0.033	0.033	0.033	0.034	0.034	0.033
4	0.067	0.060	0.057	0.055	0.055	0.055	0.055	0.054
6	0.119	0.091	0.082	0.076	0.074	0.073	0.072	0.071
8	0.200	0.140	0.114	0.099	0.094	0.091	0.089	0.086
10	0.261	0.199	0.158	0.127	0.115	0.110	0.105	0.102
12	0.300	0.247	0.205	0.160	0.140	0.130	0.123	0.117
14	0.327	0.282	0.244	0.194	0.167	0.152	0.141	0.133
16	0.346	0.307	0.274	0.226	0.195	0.175	0.161	0.151
18	0.361	0.327	0.297	0.253	0.221	0.198	0.182	0.168
20	0.373	0.342	0.315	0.274	0.243	0.220	0.202	0.186
22	0.383	0.354	0.330	0.292	0.263	0.240	0.220	0.204
24	0.391	0.365	0.342	0.307	0.279	0.257	0.238	0.220
26	0.398	0.374	0.352	0.320	0.294	0.272	0.253	0.236
28	0.404	0.381	0.361	0.331	0.306	0.285	0.267	0.250
30	0.409	0.387	0.369	0.340	0.316	0.297	0.279	0.262
32	0.414	0.393	0.376	0.348	0.326	0.307	0.290	0.273
34	0.418	0.398	0.381	0.355	0.334	0.316	0.299	0.284
36	0.421	0.403	0.387	0.362	0.341	0.324	0.308	0.293
38	0.425	0.407	0.391	0.368	0.348	0.331	0.316	0.301
40	0.428	0.410	0.396	0.373	0.354	0.338	0.323	0.309

WAVE-LENGTH, μm	RESISTIVITY, $\Omega \cdot$cm							
	0.008	0.009	0.010	0.012	0.014	0.016	0.018	0.020
2	0.032	0.031	0.030	0.028	0.027	0.025	0.024	0.024
4	0.052	0.050	0.049	0.045	0.043	0.040	0.038	0.037
6	0.068	0.066	0.064	0.059	0.056	0.053	0.050	0.049
8	0.083	0.080	0.077	0.072	0.067	0.064	0.060	0.059
10	0.097	0.093	0.089	0.083	0.078	0.073	0.070	0.068
12	0.111	0.106	0.101	0.094	0.088	0.083	0.078	0.076
14	0.126	0.119	0.113	0.104	0.097	0.091	0.087	0.084
16	0.141	0.132	0.126	0.115	0.106	0.100	0.094	0.091
18	0.157	0.146	0.138	0.125	0.116	0.108	0.102	0.099
20	0.173	0.160	0.151	0.136	0.125	0.117	0.100	0.106
22	0.188	0.175	0.164	0.147	0.134	0.125	0.117	0.113
24	0.204	0.189	0.177	0.158	0.144	0.133	0.125	0.120
26	0.219	0.203	0.190	0.169	0.153	0.142	0.132	0.127
28	0.233	0.217	0.203	0.180	0.163	0.150	0.140	0.134
30	0.245	0.229	0.215	0.191	0.173	0.159	0.148	0.141
32	0.257	0.241	0.227	0.202	0.182	0.167	0.155	0.148
34	0.268	0.252	0.238	0.213	0.192	0.176	0.163	0.155
36	0.277	0.262	0.248	0.223	0.201	0.185	0.171	0.162
38	0.286	0.271	0.258	0.232	0.211	0.193	0.178	0.169
40	0.294	0.280	0.266	0.241	0.219	0.201	0.186	0.176

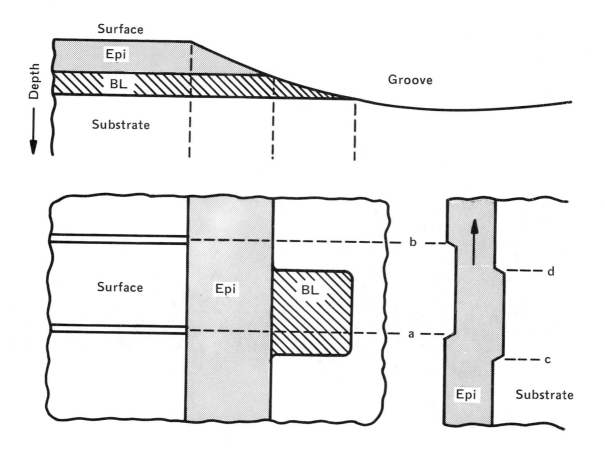

Shift and Distortion Measurements

$$\text{Pattern Shift} = \frac{(a - c) + (b - d)}{2}$$

$$\text{Pattern Distortion} = \overline{ab} - \overline{cd}$$

Figure 4-22 — Definition of Epitaxial Pattern Shift and Pattern Distortion[14]
(Reprinted with permission from Solid State Technology, Technical Publishing,
a company of Dun & Bradstreet.)

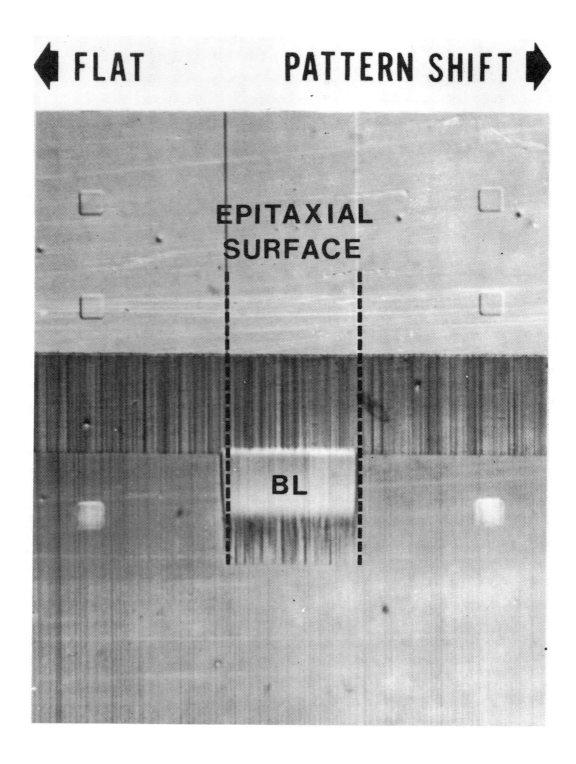

Figure 4-23 — Photograph of Beveled Sample Showing Buried-Layer (BL) Region and Indicating Direction of Pattern Shift[14] (Reprinted with permission from Solid State Technology, Technical Publishing, a company of Dun & Bradstreet.)

4.5   Carbon and Oxygen Measurements[15] [16] [17]

The levels of substitutional carbon and interstitial oxygen present in a silicon sample can be determined by infrared absorbance spectroscopy. A spectrum of percent transmittance is obtained from a double-side, polished slug of the sample material of known thickness, either in a dual-beam IR spectrometer or Fourier Transform infrared (FTIR) spectrometer. The concentration of carbon or oxygen can then be calculated from IR absorbance due to the presence of the impurities.

Oxygen

Interstitial oxygen present in the silicon lattice causes absorption of IR radiation at frequencies and band-widths shown in Table 4-3. Two methods can be used to establish the baseline absorbance:

(a)   Air reference method: In this method, the percent transmittance relative to air is measured and an appropriate baseline (I_0) drawn at the peak location. (See Figure 4-24). Absorbance is then calculated according to the relationship

$$a = \frac{1}{x} \ln (I_0/I),$$

where

x = sample thickness   (cm)

I = percent transmitted intensity at peak absorption

I_0 = baseline intensity at peak absorption.

The value of a is then adjusted by $0.4 cm^{-1}$ for 300K measurements and by $0.2 cm^{-1}$ for 77K measurements to account for the shape of the background (baseline). The concentration of interstitial oxygen is then calculated using a and the formula shown in Table 4-3.

(b)   Difference method: This method utilizes a reference sample placed in the reference beam of a dual-beam spectrometer. This reference is usually a polished slug of float-zone silicon with negligible oxygen or carbon content. Following standard procedures, the absorbance of the test sample relative to that of the reference is measured, and the oxygen content is calculated as in (a) above. In this case, however, no adjustment is made to a, as the baseline is flat and well-characterized.

Carbon

Substitutional carbon present in the silicon lattice can be measured similarly to oxygen. Values of peak locations, bandwidths, and conversion factors are given in Table 4-3. In the case of carbon, strong interference by silicon phonon bands (Figure 4-24) necessitates the use of the difference method for this measurement.

TABLE 4-3

PARAMETERS USED IN CARBON AND OXYGEN MEASUREMENTS

TEMP	PEAK LOCATION (cm)	FWHM (cm)	AIR REF ADJ	PPMA	ATOMS/cm^3	DETECTION LIMITS (PPMA)
Oxygen						
300K	1105.0	32	0.4	$4.9a$	$2.45 \times 10^{17} a$	
77K	1127.6	10	0.2	$1.9a$	$0.95 \times 10^{17} a$	0.050
Carbon						
300K	607.2	6	—	$2.0a$	$1.0 \times 10^{17} a$	0.20
77K	607.5	3	—	$0.9a$	$4.5 \times 10^{16} a$	0.10

Use of Fourier Transform Infrared (FTIR) Method

Fourier Transform infrared instruments can be used to greatly simplify the calculations necessary to measure both carbon and oxygen in silicon. Since these instruments utilize a computer to perform the Fourier Transform required to produce a frequency spectrum from the interference data, they offer enhanced data manipulation capabilities. By calculating a for all frequencies, an "absorbance spectrum" can be produced, as shown in Figure 4-25. A similar absorbance spectrum for the float-zone silicon reference can be obtained and subtracted, producing a "difference spectrum", also shown in Figure 4-25. This amounts to calculating the absorbance for every point in a normal dual-beam dispersive spectrum using the difference method outlined previously. This "difference spectrum" will have a baseline at zero absorbance, and the peak value for either carbon or oxygen will correspond to the absorbance for each element. This value is then adjusted for sample thickness to produce a (cm^{-1}), and is multiplied by the conversion factor to yield ppma or atoms/cm^3.

When the test sample and reference sample are of different thicknesses, the reference spectrum can be scaled up or down to the sample thickness before subtraction of absorbance spectra. It is possible to use this technique to determine the thickness of the sample by dividing the reference thickness by the scaling factor. This method also assures that a good baseline is obtained by observing the cancellation of silicon phonon bands after subtraction.

4.6   Oxide Film Color Chart

Multiple internal reflections within thin films of oxide grown on silicon substrates result in the appearance of color when substrates are viewed under particular conditions. Table 4-4 is a chart of the thickness of thermally grown oxide films when observed perpendicularly under daylight fluorescent lighting. It can be useful in visually verifying the accuracy and uniformity of oxidation processes.

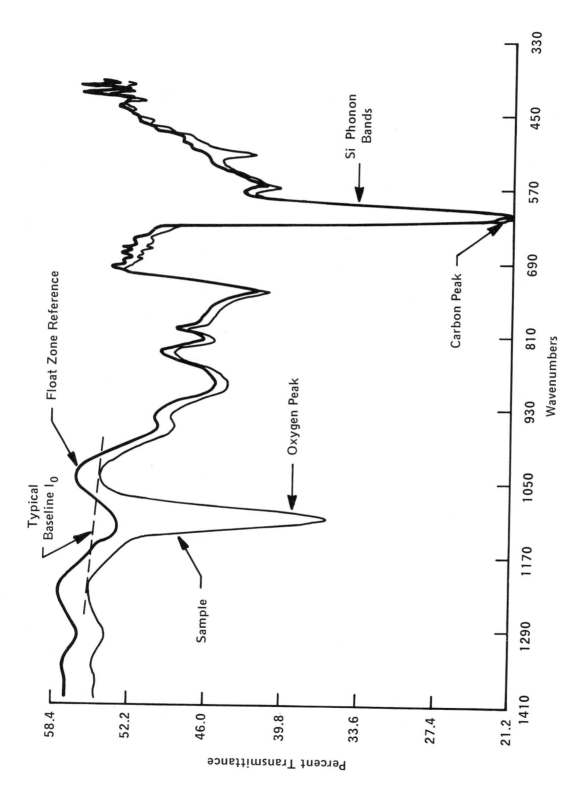

Figure 4-24 — Transmittance Spectra for Carbon and Oxygen Measurements

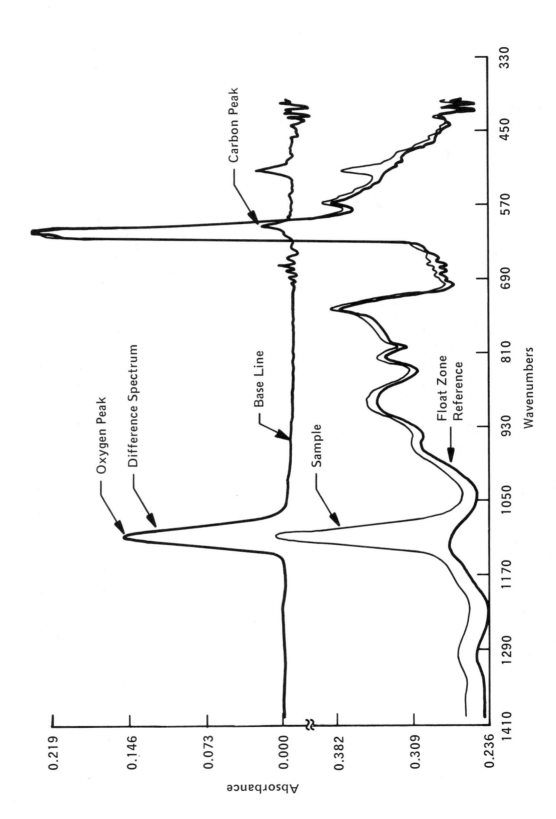

Figure 4-25 — Absorbance Spectra for Carbon and Oxygen Measurements

TABLE 4-4

COLOR CHART FOR THERMALLY GROWN OXIDE FILMS OBSERVED
PERPENDICULARLY UNDER DAYLIGHT FLUORESCENT LIGHTING[18]

FILM THICKNESS (MICRONS)	ORDER (5450 A)	COLOR AND COMMENTS
0.05_0		Tan
0.07_5		Brown
0.10_0		Dark violet to red-violet
0.12_5		Royal blue
0.15_0		Light blue to metallic blue
0.17_5	I	Metallic to very light yellow-green
0.20_0		Light gold to yellow; slightly metallic
0.22_5		Gold with slight yellow-orange
0.25_0		Orange to melon
0.27_5		Red-violet
0.30_0		Blue to violet-blue
0.31_0		Blue
0.32_5		Blue to blue-green
0.34_5		Light green
0.35_0		Green to yellow-green
0.36_5	II	Yellow-green
0.37_5		Green-yellow
0.39_0		Yellow
0.41_2		Light orange
0.42_6		Carnation pink
0.44_3		Violet-red
0.46_5		Red-violet
0.47_6		Violet

TABLE 4-4 (Contd)

COLOR CHART FOR THERMALLY GROWN OXIDE FILMS OBSERVED
PERPENDICULARLY UNDER DAYLIGHT FLUORESCENT LIGHTING[18]

FILM THICKNESS (MICRONS)	ORDER (5450 A)	COLOR AND COMMENTS
0.48_0		Blue-violet
0.49_3		Blue
0.50_2		Blue-green
0.52_0		Green (broad)
0.54_0		Yellow-green
0.56_0	III	Green-yellow
0.57_4		Yellow to "yellowish" (not yellow but is in position where yellow is to be expected; at times it appears to be light creamy grey or metallic.)
0.58_5		Light-orange or yellow to pink borderline
0.60_0		Carnation pink
0.63_0		Violet-red
0.68_0		"Bluish" (not blue but borderline between violet and blue-green; it appears more like a mixture between violet-red and blue-green, and overall looks greyish.)
0.72	IV	Blue-green to green (quite broad)
0.77		"Yellowish"
0.80		Orange (rather broad for orange)
0.82		Salmon
0.85		Dull, light red-violet
0.86		Violet
0.87		Blue-violet
0.89		Blue

TABLE 4-4 (Contd)

COLOR CHART FOR THERMALLY GROWN OXIDE FILMS OBSERVED PERPENDICULARLY UNDER DAYLIGHT FLUORESCENT LIGHTING[18]

FILM THICKNESS (MICRONS)	ORDER (5450 A)	COLOR AND COMMENTS
0.92	V	Blue-green
0.95		Dull yellow-green
0.97		Yellow to "yellowish"
0.99		Orange
1.00		Carnation pink
1.02		Violet-red
1.05		Red-violet
1.06		Violet
1.07		Blue-violet
1.10		Green
1.11		Yellow-green
1.12	VI	Green
1.18		Violet
1.19		Red-violet
1.21		Violet-red
1.24		Carnation pink to salmon
1.25		Orange
1.28		"Yellowish"
1.32	VII	Sky blue to green-blue
1.40		Orange
1.45		Violet
1.46		Blue-violet
1.50	VIII	Blue
1.54		Dull yellow-green

REFERENCES

1. L. J. van der Pauw, "A Method of Measuring Specific Resistivity and Hall Effect of Discs of Arbitrary Shape," *Phillips Res. Repts.,* Vol. 13, No. 1 (February 1958), pp. 1-9.

2. *Annual Book of ASTM Standards, 1981,* Part 43, Standard F-76.

3. *Ibid.,* Part 43, Standard F-43.

4. P. A. Schumann, Jr., "Plasma Resonance Calibration Curves for Silicon, Germanium, and Gallium Arsenide," *Solid State Tech.,* Vol. 13, No. 1 (January 1970), p. 50.

5. A. S. Grove et al, "Investigation of Thermally Oxidized Silicon Surfaces Using Metal-Oxide Semiconductor Structures," *Solid-State Electron.,* Vol. 8, (1965), pp. 145-163.

6. S. M. Sze, *Physics of Semiconductor Devices.* New York: John Wiley & Sons, 1969, p. 436.

7. L. P. Adda, unpublished work.

8. C. T. Sah, T. L. Chin, and R. F. Pierret, Technical Report No. 2, Solid State Electronics Laboratory, Electrical Engineering Research Laboratory, University of Illinois.

9. *Annual Book of ASTM Standards, 1981,* Part 43, Standard F-419.

10. *Ibid.,* Part 43, Standard F-110.

11. *Ibid.,* Part 43, Standard F-143.

12. *Ibid.,* Part 43, Standard F-95.

13. B. Senitzky and S. P. Weeks, "Infrared Reflectance Spectra of Thin Epitaxial Silicon Layers," *J. Appl. Phys.,* Vol. 52, No. 8 (August 1981), pp. 5308-5313.

14. S. P. Weeks, "Pattern Shift and Pattern Distortion During CVD Epitaxy on (111) and (100) Silicon," *Solid State Tech.,* Vol. 24 (November 1981), pp. 111-117.

15. *Annual Book of ASTM Standards, 1981,* Part 43, Standards F-120, F-121, and F-123.

16. B. Pajot, "Characterization of Oxygen in Silicon by Infrared Absorption," *Analusis,* Vol. 5, No. 7, (1977), pp. 293-303.

17. D. W. Vidrine, "Room Temperature Carbon and Oxygen Determination in Single-Crystal Silicon," *Anal. Chem.,* Vol. 52, No. 1 (January 1980), pp. 92-96.

18. W. A. Pliskin and E. E. Conrad, "Nondestructive Determination of Thickness and Refractive Index of Transparent Films," IBM Journal, January 1964, p. 48.

5. CHEMICAL RECIPES

General

Various chemicals have been used in silicon device technology for fabrication of the devices or evaluation of the device or control wafers. Because chemical activity varies with silicon resistivity, impurity doping concentration, and processing steps, many modifications of basic chemical mixtures exist. Because it is difficult to anticipate individual needs, only a selection of chemical formulations for various purposes are given in this section.

Table 5-1 shows several compositions for polishing, junction staining, V-groove etching, and removing particulates. The most commonly used silicon etchants consist of mixtures of HF and HNO_3, with H_2O or CH_3COOH as diluents. Robbins and Schwartz[1] thoroughly investigated the $HF-HNO_3-H_2O$ system. Holmes included a list of various etchants for many semiconductor materials in his book.[2] Many techniques for p-n junction delineations have been published with varying success. The use of HF with strong illumination is by far the simplest and the most useful method of junction staining. (No. 6 in Table 5-1.) Occasionally, there are junctions which will not stain satisfactorily no matter what the method used. Interest in dielectric isolation using deep slots in silicon has been growing. In this process, deep trenches are etched in silicon with the use of KOH solutions. The etch composition shown as No. 9 in Table 5-1 gives more consistent etch features for high resistivity floating-zone silicon than the other etchants which are in the referenced literature. Additional details may be found in references [3][4][5]. When silicon wafers are exposed to a very high concentration diffusion source (such as P_2O_5, BBr_3, or BN) in an ambient with insufficient oxygen present, a stain film is often formed. This film is a suboxide of silicon which cannot be etched off in HF. When using chemical polishing wafers with high boron doping ($>10^{19} cm^{-3}$), a stain film consisting of a suboxide of silicon is also formed. The stain film may affect the electrical contact to the wafer. A simple removal mixture is listed as No. 11 in Table 5-1. It should be noted that when using this mixture a thin layer of silicon will be removed.

Dislocation etchants are useful for device and process diagnosis. Etches described by Dash [6][7] and Sirtl[8] have been commonly used for a long time. However, Sirtl's etch does not delineate defects in (100) planes with consistent results. Secco[9] published a mixture for etching (100) silicon. This was improved by Schimmel.[10] Table 5-2 shows a few variations of Secco's etch by Schimmel. The chemical etchants listed in Table 5-2 will reveal the following crystal defects in <100> or <111> silicon:

   (a) Dislocations

      (1) slip

      (2) misfit.

   (b) Stacking faults

      (1) bulk

            } oxidation induced

      (2) surface

      (3) epitaxial.

   (c) Surface damages

      (1) scratches

      (2) residual saw damages.

   (d) Swirl.

   (e) Resistivity striations.

Pictures of the various etch pit configurations can be found in references[10] through [15]. Another etchant which is sometimes quoted in the relevant literature is called the "Wright etch."[16]

Anodization of silicon has not been widely used in device manufacturing. However, it is often used for diffusion profile study or removing silicon at room temperature with good control of the rate of removal. Two mixtures for anodic oxidation of silicon are shown below. Both solutions remove approximately 5Å/volt. The minimum applied voltage is close to 30~40 volts, depending on the resistivity of the material.

    (a) 900 ml ethylene glycol

        4 gm KNO_3

                    [17]

        100 ml H_2O

    (b) 1 gal tetrahydrofurfuryl alcohol

                      [18]

        57 gm KNO_2.

Table 5-3 gives the etch compositions of BHF (buffered HF) and p-etch. The etch rates for thermal SiO_2 are also given. The p-etch can be used to remove high concentration diffused SiO_2 layers (such as boron and phosphorus after the predeposition diffusions). A 15:1 (H_2O:HF) is used for this purpose. The removal rate of thermal oxide is slower for p-etch than the 15:1 (H_2O:HF) mixture specified in Table 5-4. Table 5-3 also shows a method for removing P-glass and B-glass. Owing to the rather involved procedure, this method is not widely used.

Table 5-4 shows etch rates for the various oxides of silicon, and Table 5-5 shows the etch rates for SiO_2 (thermal) after ion implantation. In fabricating silicon integrated circuits (SIC) with beam leads, the circuit chips are separated by etching. Table 5-6 shows the etch rates of CP-4, CP-6, and CP-8 mixtures on completed SIC.

Table 5-7 shows the etch rates for Si, SiO_2, Si_3N_4, and metal films in a Freon plasma. Since the etch rates are functions of gas mixtures, reactor geometry, and substrates, the table gives a few examples for possible applications.

Table 5-8 gives the etchants for evaporated metal films. These are collections by J. A. Wenger and A. K. Sinha. Detailed references are not given here for the purpose of saving space. Certain preferred etchants are recommended for defining fine features by photoresist masking technique. These are indicated with an asterisk (*) in front of the etchant composition.

TABLE 5-1

CHEMICAL ETCHANTS FOR SILICON

NO.	NAME	PURPOSE	VOLUME COMPOSITION	REFERENCE
1	CP-4A	Polishing or lapping saw damage removal	3 HF 5 HNO_3 3 CH_3COOH	19
2		Polishing	1 HF 4 HNO_3 2 CH_3COOH	20
3		Slow polishing	1 HF 6 HNO_3 1 CH_3COOH	21
4		Removing shallow arsenic ion-implanted layer using PR-102 or IC-28 photoresist	1 HF 100 HNO_3 99 H_2O	22
5	Iodine etch	Polishing	5 HF 10 HNO_3 11 CH_3COOH 0.3gm I_2/250 ml solution	23
6		Junction staining	HF or HF + 0.1% HNO_3	24,25
7		n^+n delineation	40 ml HF (49%) 20 ml HNO_3 (70%) 100 ml H_2O 2 gm Ag NO_3	26
8		n^+n or p^+p delineation	100 ml HF (40%) 1 ml HNO_3 (70%) under electric pulse	27
9	Anisotropic etch	Groove etching at 80-82°C	23.4% KOH (by weight) 13.3% normal propyl alcohol 63.3% H_2O	28
10	Copper displacement plating solution	Cleaning and removing particulate from mechanically lapped surface	50 ml HF 950 ml H_2O 55 gm $CuSo_4$ 5 H_2O	29
11	Stain film removal	Removing stains from a high concentration diffusion in the reducing ambient at high temperature or removing stains from chemical etching of low resistivity crystal	2 HF 1 $KMnO_4$ (0.038M)	15

TABLE 5-2

CHEMICAL ETCHANTS FOR SILICON DEFECTS

<100> or <111> Si TEST SAMPLE	VOLUME COMPOSITION			APPROXIMATE ETCH TIME (MINUTES)
	49% HF	0.75M CrO_3	H_2O	
Ingot or substrate				
(n or p) >0.2 Ω — cm	2	1		5
p-type <0.2 Ω — cm	2	1	1.5	5
n-type <0.2 Ω — cm	2	1	1.5	10
Oxidized wafers, discrete devices, or ICs* (HF strip oxide before defect etching)	2	1		1 to 2

* For heavily doped, p-type diffused areas it may be necessary to use 1.5 parts H_2O dilution.

TABLE 5-3

ETCHANTS FOR SiO_2 AND DOPED OXIDES

ETCHANT	ETCH RATE	REF
BHF (buffered HF) 34.6% (wt) NH_4F/6.8% HF/58.6% H_2O	1000Å/min for thermal oxide	30
p-etch 15 HF (49%)/10 HNO_3 (70%)/300 H_2O		
1. Thermal SiO_2	2Å/s	30
2. P_2O_5 + SiO_2 from P_2O_5 diffusion	570Å/s	30
3. Furnace-grown pyrolitic oxide at 675°C	13Å/s	31
4. Boron-doped oxide deposited at 300°C and annealed at 1050°C in argon		32
14 molar % of B_2O_3	3Å/s	
34 molar % of B_2O_3	50Å/s	
Phosphorus-and boron-doped SiO_2 1. Five minutes in concentrated HNO_3 at 95 ± 2°C 2. Three minutes in flowing DI water 3. Five minutes in 0.1M ethylenedinitrilotetraacetic solution at 95 ± 2°C (pH must be adjusted to 8.2 ± 0.3 with concentrated NH_4OH) 4. Three minutes (minimum) rinse in flowing DI water 5. Repeat steps 1 through 4 for boron-rich glasses. This process will reduce 2000Å of boron-rich glasses to a residual film of <20Å. The EDTA solution is made up as a 0.1 molar aqueous solution. The pH is adjusted to 9.5 at 25°C with concentrated ammonium hydroxide.		33

TABLE 5-4

ETCH RATES FOR VARIOUSLY TREATED OXIDES[34]

MATERIAL	DOPING, TEMP TREATMENT OR GROWTH CONDITIONS	DEPTH OF DOPING (Å)	ETCH RATE (Å/MIN) 15:1 H₂O:HF	ETCH RATE (Å/MIN) BHF	Å REMOVED CRUD REMOVAL (1)	Å REMOVED P-GLASS (2)
Thermal Oxide SiO₂	**1050° 98° H₂O - O₂** Undoped	—	160	1100	—	—
B Doped	10 min BN @ 1100°	700	—	—	1350	—
	30 min BN @ 1140°	1200	1000	300	1750	—
	60 min BN @ 1140°	1800	1000	—	2400	—
B Doped	30 min BN @ 890°	—	—	—	1100	—
	75 min BN @ 890°	—	—	—	1225	—
	90 min BN @ 890°	—	—	—	1240	—
P Doped	40 min POCL₃ @ 1000°	1600	~20000	~9000	—	1900
	50 min POCL₃ @ 1000°	1800	~20000	~9000	—	2150
	55 min POCL₃ @ 1000°	1900	~20000	~9000	—	2250
	60 min POCL₃ @ 1000°	2000	~20000	~9000	—	2300
P Doped	30 min POCL₃ @ 1040°	1800	~20000	~9000	—	2000
	40 min POCL₃ @ 1040°	2100	~20000	~9000	—	2300
	60 min POCL₃ @ 1040°	2600	~20000	~9000	—	2800
P Doped	30 min POCL₃ @ 1100°	2600	~20000	~9000	—	—
Deposited Oxide	**650° Ethyl Orthosilicate**					
Oxide	Undoped-Undensified	—	1960	5000		—
P Doped on Deposited Oxide	30 min POCL₃ @ 1100°C	3000	~10000	~9000	—	—
Deposited Oxide and Densified	30 min @ 1100°	—	260	1650	—	—
	100 min @ 1200°	—	—	1000	—	—

Notes:
1. Crud removal—1 min 15:1 H₂O:HF; rinse; 5 minutes 85°C Nitric; rinse; 4 min 15:1 H₂O:HF; rinse; dry.
2. P-Glass—30s 15:1 H₂O:HF; rinse; dry.

TABLE 5-5

ETCH RATES FOR ION-IMPLANTED SiO$_2$ [35]

ELEMENT	ENERGY (keV)	DOSE (cm^{-2} x 10^{15})	ETCHANT	ETCH RATE (Å/s)	FAST-ETCHING REGION (Å)
B	50	0.7-0.2	BHF	50	3300
B	50	5	15:1	15	3300
P	50	5	15:1	17-20	1600
As	50	1-10	15:1	18	1100

TABLE 5-6

ETCH RATES OF CP-X ETCHANTS [36] *

ETCH NAME \ MATERIAL	Si$_3$N$_4$	PERMALLOY	Si	THERMAL SiO$_2$	B SiO$_2$	P SiO$_2$	Ti
CP-4	—	13.7 μm/min	34.8 μm/min	2750Å/min	—	—	
CP-6	—	18.0 μm/min	27.7 μm/min	2260Å/min	2200Å/min	—	3570Å/min
CP-8	—	17.8 μm/min	7.4 μm/min	900Å/min	725Å/min	—	725Å/min

CHEMICAL COMPOSITION OF ETCHES

CP-4 — 5:3:3   HNO$_3$:HF:CH$_3$COOH

CP-6 — 5:2:3   HNO$_3$:HF:CH$_3$COOH + I$_2$

CP-8 — 5:1:2:2 HNO$_3$:HF:CH$_3$COOH + I$_2$

*The etch rates presented here are the result of a compilation from many sources.

TABLE 5-7

FLUORINE GAS PLASMA ETCH RATES

MATERIAL		RELATIVE ETCH RATE	MEASURED ETCH RATE ($\pm$ 10%)
SiO_2			
	Dry thermal (1295°C)	0.75 — 1.0	50Å/min
	Wet thermal (1150°C-steam)	1.0	100Å/min
	Emitter (1080°C-P rich)	3.0	
	Deposited (undoped) 350°C	2.5	
	400°C	2.0	
	500°C	1.5	
	Deposited (p-doped, 5-7%, 400°C)	4.0	
Si_3N_4			
	"Stoichiometric" (n = 2.05)	3.0	300 Å/min
	O-rich (n = 1.85)	2.0 — 3.0	
	Si-rich (n = 2.20)	5.0 — 10	
Si			
	Single [1]	20-40	2000Å/min
	Polycrystalline [1], [2]	30 — 50	
	Amorphous	30 — 50	
Metal Films	Ta_2O_5, Ta_2N	10	
	W, Ta, Mo, Ti	5.0 — 10	
	$TaSi_2 WSi_2$, $TiSi_{21}$, $MoSi_2$	10 — 15	
	$PtSi_2$, Pt [3]		~ 0
	Cu, Pt, Cr, Cr_2O_3 [3]		~ 0
	GaAs, GaP, InP [3]		~ 0

Notes:
1. N-type doped materials etch more rapidly.
2. Etch rate dependent on grain size.
3. Metal flourides are involatile and are, therefore, difficult to etch in flourine based plasmas.

TABLE 5-8

METAL AND METAL FILM ETCHANTS [37]

METAL	COMPOSITION (VOLUME RATIO)		COMMENTS
Ag		$1\ NH_4OH/1\ H_2O_2$	
Al	*(1)	$5\ HNO_3/5\ CH_3COOH/85\ H_3PO_4/5\ H_2O$ [38] at 40-45°C	Rinse in DI H_2O followed by 1 HCl/3 H_2O to remove phosphorus.
	(2)	$1\ NaOH/5\ H_2O$	Not recommended for production use when Na contamination is of concern.
Au	*(1)	1/4 lb potassium iodate (KI)/65 gm I/100 ml H_2O	1000~1200Å/s
	(2)	$3\ HNO_3/1\ HCl$	
Bi		$1\ HCl/10\ H_2O$	
Co	*(1)	$30\ HNO_3/10\ H_2SO_4/50\ CH_3COOH/10\ H_3PO_4$ [38]	
	(2)	$1\ HNO_3/1\ H_2O$	
	(3)	$3\ HCl/1\ H_2O_2$	
Cr	(1)	$3\ HCl/1\ H_2O$	
	(2)	1 HCl/1 glycerin	
	(3)	$8\ KMnO_4/4\ NaOH/100\ H_2O$ by weight	
Cu	*(1)	$5\ HNO_3/1\ H_2O$	
	(2)	Ammonium persulphate	
	(3)	6 formic acid/1 H_2O_2/3 H_2O	
Fe		$1\ HCl/1\ H_2O$	
Hf		$1\ HF/1\ H_2O_2/20\ H_2O$	
In		$1\ HNO_3/3\ HCl$ at 40°C	
Ir		$1\ HNO_3/3\ HCl$ at 40°C	
Mg		$1\ NaOH/10\ H_2O$ by wt followed by $1\ CrO_3/5\ H_2O$ by wt	

* Preferred etchant for delineation with photoresist masking.

TABLE 5-8 (Contd)

METAL AND METAL FILM ETCHANTS

METAL	COMPOSITION (VOLUME RATIO)		COMMENTS
Mo	*(1)	85 H_3PO_3/5CH_3COOH/1 HNO_3/5 H_2O followed by 1 HCl/3 H_2O	@ 40°C
	(2)	1 H_2SO_4/4 HNO_3 (dry)	A slow etch for patterning use
	(3)	1 H_2SO_4/1 HNO_3/2 H_2O	A fast etch for nonpatterning use
Nb		1 HNO_3/1 HF	
Ni	*(1)	Ferric chloride 40%	
	(2)	1 HNO_3/5 HCl	
	(3)	150 CH_3COOH/50 HNO_3/3 HCl	At 70°C
NiCr	(1)	1 HNO_3/1 HCl/3 H_2O	
	(2)	3 $FeCl_3$/20 HCl (37%)	On Ta thin film circuits
Pb		4 CH_3COOH/4 H_2O_2/10 H_2O	
Pd		1 HNO_3/3 HCl/4 H_2O	At 95°C
Pt	(1)	1 HNO_3/3 HCl/4 H_2O	At 95°C
	(2)	1 HNO_3/8 HCl	Age 1 hour, etch at 70°C
Rh		62% hydrobromic	At 100°C
Ru		Anode in HCl or H_2SO_4 (A.C.)	
Sb		1 HNO_3/1 HCl/1 H_2O	
Sn	*(1)	1 HCl/4 H_2O	
	(2)	1 HF/1 HCl	
Ta		4 HNO_3/4 HF/10 H_2O	
$TaSi_2$		4 HNO_3/6 BHF (7:1 dilution) [38]	

* Preferred etchant for delineation with photoresist masking.

TABLE 5-8 (Contd)

METAL AND METAL FILM ETCHANTS

METAL	COMPOSITION (VOLUME RATIO)	COMMENTS
Ti	(1) 1 H_2SO_4/1 H_2O	At 155-120°C in ultra-sonic
	(2) 1 HF/30 H_2SO_4/69 H_2O	At 70°C
	(3) 1 H_2O_2/2 EDTA	At 65°C
V	1 HNO_3/1 HF/1 H_2O	
W	*(1) (0.25M) KH_2PO_4/(0.24M) KOH/(0.1M) $K_3Fe(CN)_6$	
	(2) 1 HNO_3/1 HF	
	(3) 1 H_2O_2/2 EDTA	
Zr	1 HNO_3/1 HF/50 H_2O	

*Preferred etchant for delineation with photoresist masking.

REFERENCES

1. H. Robbins and B. Schwartz, "Chemical Etching of Silicon," *J. Electrochem Soc.: Solid State Science and Technology,* Vol. 106, No. 6 (June 1959), p. 505 and Vol. 107, No. 2 (February 1960), p. 108.

2. *The Electrochemistry of Semiconductors.* ed. P.J. Holmes, New York: Academy Press, 1962, pp. 367-375.

3. K.E. Bean and W.R. Runyan, "Dielectric Isolation: Comprehensive, Current and Future," *J. Electrochem Soc.: Solid State Science and Technolgoy,* Vol. 124, No. 1 (January 1977), p. 5C.

4. E. Bassous, "Fabrication of Novel Three-Dimensional Microstructures by the Anisotropic Etching of (100) and (110) Silicon," *IEEE Trans. on Electron. Dev.,* Vol. ED-25 (October 1978), p. 1178.

5. D.L. Kendall, "Vertical Etching of Silicon at Very High Aspect Ratios," *Ann. Rev. Mater. Sci.,* Vol. 9 (1979), p. 373.

6. W.C. Dash, "Copper Precipitation on Dislocations in Silicon," *J. Appl. Phys.,* Vol. 27, No. 10 (October 1956), p. 1193.

7. W.C. Dash, "Evidence of Dislocation Jogs in Deformed Silicon," *J. Appl. Phys.,* Vol. 29, No. 4 (April 1958), p. 705.

8. E. Sirtl and A. Adler, *Z. Mettalkd.,* Vol. 52, No. 8 (August 1961), p. 529.

9. F. Secco d'Aragona, "Dislocation Etch for (100) Planes in Silicon," *J. Electrochem. Soc.: Solid State Science and Technology,* Vol. 119, No. 7 (July 1972), p. 948.

10. D.G. Schimmel, "Defect Etch for <100> Silicon Ingot Evaluation," *J. Electrochem Soc.: Solid State Science and Technology,* Vol. 126, No. 3 (March 1979), p. 479.

11. *Annual Book of ASTM STANDARDS, 1979,* Part 43, ANSI/ASTM F47-70, ANSI/ASTM F154-76, and ANSI/ASTM F416-77.

12. S.F. Moyer and D.G. Schimmel, private communication.

13. D.G. Schimmel, private communication.

14. G.A. Rozgonyi and T.E. Seidel, private communication.

15. D.G. Schimmel and M.J. Elkind, "An Examination of the Chemical Staining of Silicon," *J. Electrochem. Soc.: Solid-State Science and Technology,* Vol. 125, No. 1 (January 1978), p. 152.

16. M.W. Jenkins, "A New Preferential Etch for Defects in Silicon Crystals," *J. Electrochem Soc.: Solid-State Science and Technology,* Vol. 124, No. 5 (May 1977), p. 757.

17. K.M. Busen and R. Linzey, Removal of Thin Layers of n-Type Silicon by Anodic Oxidation," *Trans. Am. Inst. Min. Met. Pet. Engr.,* Vol. 236 (March 1966), p. 306.

18. P.F. Schmidt and A.E. Owen, "Anodic Oxide Films for Device Fabrication in Silicon," *J. Electrochem. Soc.: Solid State Science and Technology,* Vol. 111, No. 6 (June 1964), p. 682.

19. P.J. Holmes, "The Use of Etchants in Assessment of Semiconductor Crystal Properties," *Proc. Inst. Elec. Engrs.,* Vol. 106B, Suppl. 15 (May 1959), p. 861.

20. R.A. Clapper, private communication.

21. J.C.C. Tsai, private communication.

22. K.H. Cho, private communication.

23. P. Wang, "Etching of Germanium and Silicon," *Sylv. Tech.* Vol. 11, No. 2 (April 1958), p. 50.

24. C.S. Fuller and J.A. Ditzenberger, "Diffusion of Donor and Acceptor Elements in Silicon," *J. Appl. Phys.,* Vol. 27, No. 5 (May 1956), p. 544.

25. P.J. Whoriskey, "Two Chemical Stains for Marking p-n Junctions in Silicon," *J. Appl. Phys.,* Vol. 29, No. 5 (May 1958), p. 867.

26. I. Berman, "N+N Delineation in Silicon," *J. Electrochem. Soc.: Solid State Science and Technology,* Vol. 109, No. 10 (October 1962), p. 1002.

27. B.A. Joyce, "A Pulse Staining Method for Delineating $n-n^+$ and $p-p^+$ Junctions in Silicon," *Solid-State Electron.,* Vol. 5 (March 1962), p. 102.

28. R.C. Kragness and H.A. Waggener, "Precision Etching of Semiconductors," U.S. Patent 3,765,969, issued October 16, 1973.

29. V.C. Garbarini, D.L. Klein, and R. Lieberman, private communication.

30. W.A. Pliskin and R.P. Gnall, "Evidence for Oxidation Growth at the Oxide-Silicon Interface From Controlled Etch Studies," *J. Electrochem. Soc.: Solid State Science and Technology,* Vol. 111, No. 7 (July 1964), p. 872.

31. W.A. Pliskin and H.S. Lehman, "Structural Evaluation of Silicon Oxide Films," *J. Electrochem. Soc.: Solid State Science and Technology,* Vol. 112, No. 10 (October 1965), p. 1013.

32. A.H. El-Hoshy, "Measurement of P-Etch Rates for Boron-Doped Glass Films," *J. Electrochem Soc.: Solid-State Science and Technology,* Vol. 117, No. 12 (December 1970), p. 1583.

33. F.J. Biondi, Request for Patent Study, March 14, 1969; subsequently U.S. Patent 3,669,775, issued June 13, 1972.

34. J.W. Brossman, private communication.

35. J.C.C. Tsai, private communication.

36. G.K. Herb and H.W. Wivell, private communication.

37. J.A. Wenger, private communication.

38. A.K. Sinha, private communication.

6. DIFFUSION

List of Symbols 6-2

6.1   Commonly Used Expressions 6-3

Erf and Erfc Functions 6-3

Normalized Gaussian and Erfc Curves (Figure 6-1) 6-4

Erfc, Composite, and Gaussian Curves (Figure 6-2) 6-5

6.2   Diffusion Equations 6-6

Low Impurity Concentration Diffusion Equations 6-6

    Constant Surface Concentration 6-6

    Fixed Total Impurity - Single-Sided Gaussian Distribution 6-7

    Fixed Total Impurity - Limited Source of Width h 6-7

    Diffusion From a Uniformly Doped Oxide Source 6-8

    Redistribution Diffusion 6-9

    Outdiffusion From a Buried, Uniformly Doped Layer 6-10

    Diffusion of Ion-Implanted Impurities During Oxidation 6-10

High Impurity Concentration Diffusion Equations 6-12

    Arsenic Diffusion in Silicon. 6-12

    Sheet Resistance Versus Junction Depth for Implanted
    As Layers Diffused in N_2 or O_2 (Figure 6-3) 6-14

    Time, Temperature, and As Dose Dependence of Sheet,
    Resistance (Diffusions in N_2 or O_2)(Figure 6-4) 6-15

    Time To Complete Activation as a Function of Arsenic Dose
    and Diffusion Temperature (Figure 6-5) 6-16

    Boron Diffusion in Silicon. 6-13

    Diffusivity of Boron in Dry O_2 and Nonoxidizing Ambient (Figure 6-6) 6-17

    Segregation Coefficients of Boron at the Si/SiO_2 Interface (Figure 6-7) 6-18

    Phosphorus Diffusion in Silicon 6-19

    Surface Concentration Versus Average Resistivity for
    Phosphorus-Diffused Silicon (Figure 6-8) 6-20

    Sheet Resistance Versus Time for Phosphorus Diffusion
    Into Silicon in Low O_2 Ambient (Figure 6-9) 6-22

    Sheet Resistance Versus Junction Depth for Implanted
    Phosphorus Layers Diffused in N_2 or O_2 (Figure 6-10) 6-24

    Sheet Resistance Versus Time, Temperature, and Dose for
    Phosphorus-Implanted, Diffused Layers in Silicon (Figure 6-11) 6-25

    Critical Time to Dislocation Array Formation for Phosphorus
    Diffusion in Low O_2 Ambients (Figure 6-12) 6-26

LIST OF SYMBOLS

Symbol	Definition
C	Concentration
C_0 or C_s	Surface concentration
C_1 or C_{SiO_2}	Concentration in SiO$_2$
C_2 or C_{Si}	Concentration in Si
C_{MAX}	Maximum concentration
C_{sub}	Substrate concentration
D	Diffusivity or diffusion coefficient
D_i	Diffusivity at low concentration
D_0	Preexpoential factor for diffusivity vs. temperature expressions
E	Activation energy
h	Distance
J	Diffusion current
k_l	The linear oxidation rate constant
k_p	The parabolic oxidation rate constant
m	Impurity segregation coefficient at Si — SiO$_2$ interface
n_i	Intrinsic electron concentration
p	Hole concentration
Q	Total impurity concentration
Q_A	Electrically active impurity concentration
Q_T	Total impurity concentration
R_P	Ion projected range
ΔR_p	Projected standard deviation of Gaussian distribution of ion implant profile
R_s	Sheet resistance
t	Time
x or X	Distance
x or X_j	Junction depth
z or Z	Normalized variable

6.1   Commonly Used Expressions

Erf and Erfc Functions

Error Function (erf) and Complementary Error Function (erfc)

$$erf\left(\frac{x}{2\sqrt{Dt}}\right) = \frac{1}{\sqrt{\pi Dt}} \int_{o}^{x} exp\left(-\frac{x^2}{4Dt}\right) dx.$$

When $Z = \frac{x}{2\sqrt{Dt}}$, the general expressions for erf (z) and erfc (z) are:

$$erf(z) = \frac{2}{\sqrt{\pi}} \int_{o}^{z} exp(-y^2)\, dy.$$

$$erfc(z) = 1 - erf\,(z) = \frac{2}{\sqrt{\pi}} \int_{z}^{\infty} exp(-y^2)\, dy \qquad \text{for } z \geqslant 0.$$

$$\frac{d}{dz}\,[erf(z)] = \frac{2}{\sqrt{\pi}}\, exp\,(-z^2).$$

$$\frac{d^2}{dz^2}\,[erf(z)] = -\frac{4}{\sqrt{\pi}}\, z\, exp\,(-z^2).$$

$$ierfc(z) \equiv \int_{z}^{\infty} erfc(z)\, dz = \frac{1}{\sqrt{\pi}}\, exp\,(-z^2) - z\, erfc(z).$$

$$i^2 erfc(z) \equiv \int_{z}^{\infty}\int_{z}^{\infty} erfc(z)\, dz\, dz = 1/4\,[(1 + 2z^2)\, erfc(z) - \frac{2}{\sqrt{\pi}}\, z\, exp\,(-z^2)\,].$$

where:

$$erf(-z) = -erf\,(z) \qquad\qquad erf(0) = 0.0$$

$$\frac{2}{\sqrt{\pi}} = 1.12838 \qquad\qquad erf(\infty) = 1.0.$$

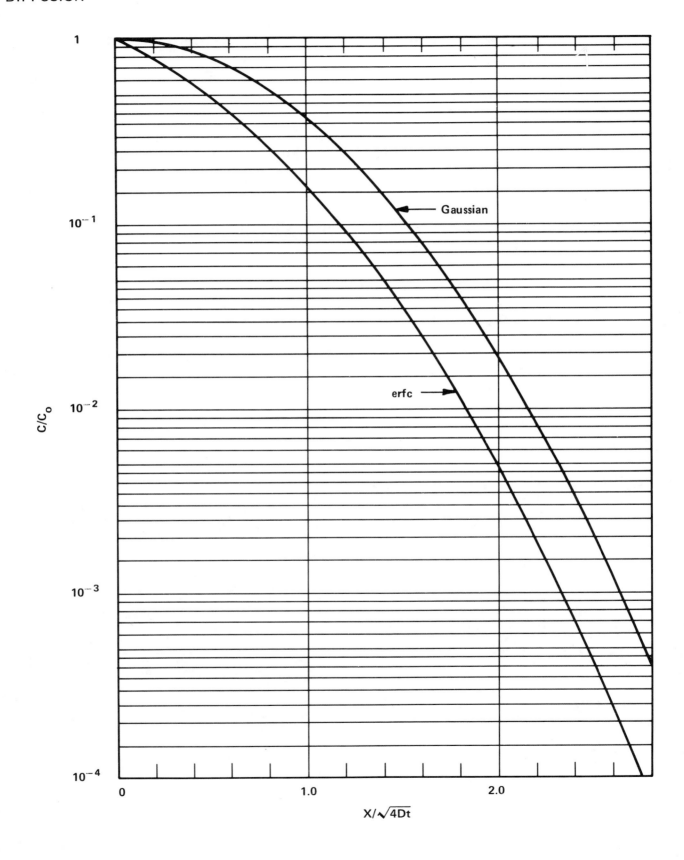

Figure 6-1 — Normalized Gaussian and Erfc Curves

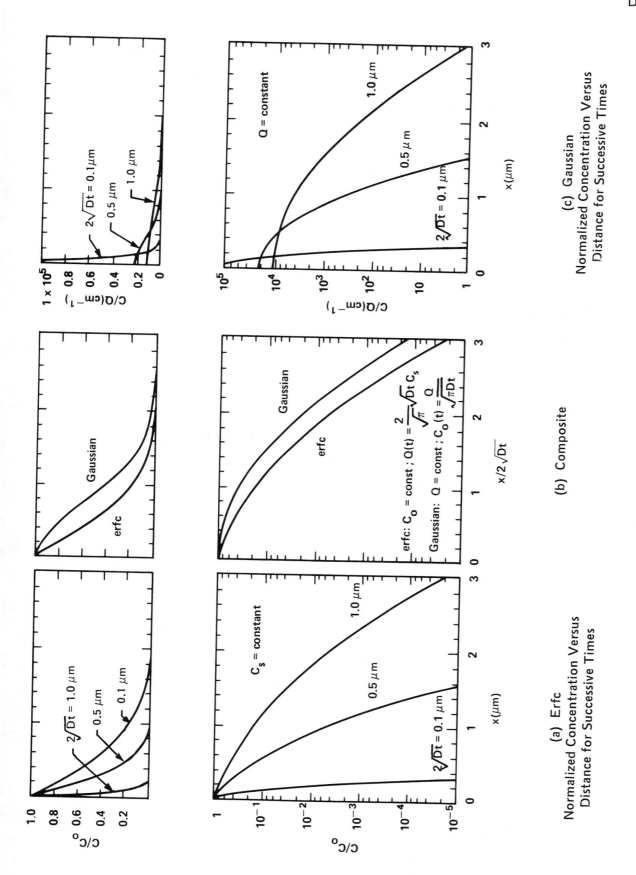

(a) Erfc
Normalized Concentration Versus
Distance for Successive Times

(b) Composite

(c) Gaussian
Normalized Concentration Versus
Distance for Successive Times

Figure 6-2 — Erfc, Composite, and Gaussian Curves

6.2   Diffusion Equations

Low Impurity Concentration Diffusion Equations

When the impurity concentration is below n_i at the diffusion temperature, the impurity diffusivity is a constant for a given temperature.

Constant Surface Concentration

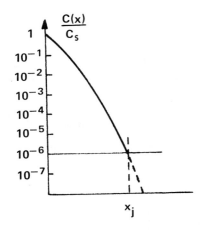

Semilog Plot

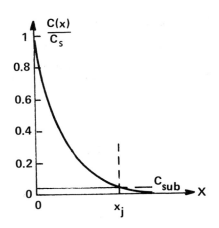

Linear Plot

$$C(x,t) = C_s \, \text{erfc}\left(\frac{x}{2\sqrt{Dt}}\right) - C_{sub}.$$

$$x_j = 2\sqrt{Dt} \, \text{erfc}^{-1}\left(\frac{C_{sub}}{C_s}\right).$$

The rate of impurity flow through the surface is

$$J = -D \left.\frac{\partial C}{\partial x}\right|_{x=0} C_s \, [D/(\pi t)]^{1/2}.$$

The total amount of impurity in the diffused layer is

$$Q(t) = \frac{2}{\sqrt{\pi}} \sqrt{Dt} \, C_s.$$

Fixed Total Impurity - Single-Sided Gaussian Distribution

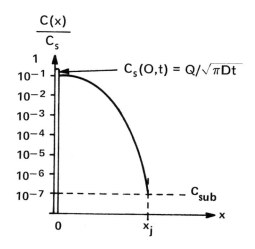

Semilog Plot

$$C(x,t) = \frac{Q}{\sqrt{\pi Dt}} \exp \left(-\frac{x^2}{4Dt} \right) - C_{sub},$$

where Q is the total quantity of impurity per cm^2.

$$C(0,t) = \frac{Q}{\sqrt{\pi Dt}} \; .$$

$$X_j = [4Dt \; \ell n \; (Q/C_{sub} \sqrt{\pi Dt})]^{1/2} .$$

Fixed Total Impurity - Limited Source of Width h

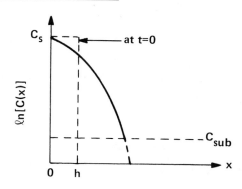

$$C(x,t) = \frac{C_s}{2} \left[\text{erfc} \left(\frac{x-h}{2\sqrt{DT}} \right) - \text{erfc} \left(\frac{x+h}{2\sqrt{Dt}} \right) \right] - C_{sub} .$$

The surface concentration is

$$C(0,t) = C_s \left[1 - \text{erfc} \left(\frac{h}{2\sqrt{Dt}} \right) \right] .$$

$$Q = C_s h.$$

When $x < \dfrac{Dt}{h}$, $t > \dfrac{h^2}{D}$, the profile is the Gaussian distribution.

Diffusion From a Uniformly Doped Oxide Source

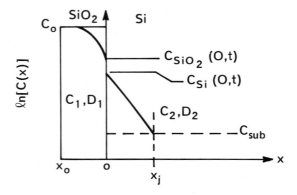

For relatively short diffusion times such that

$$X_o > 4\sqrt{D_1 t},$$

$$C_2(x,t) \simeq \frac{C_o (D_1/D_2)^{1/2}}{(1+k)} \; \text{erfc} \left(\frac{x}{2\sqrt{D_2 t}} \right) ,$$

where:

C_o = impurity concentration in the oxide at the beginning of the diffusion

C_1 = impurity concentration in the oxide

C_2 = impurity concentration in silicon

D_1 = diffusion coefficient in the oxide

D_2 = diffusion coefficient in silicon

$k = \dfrac{1}{m} \left(\dfrac{D_1}{D_2} \right)^{1/2}$

m = impurity segregation coefficient at the oxide-silicon interface.

For long diffusion times or thin oxide layers, the following approximations may be used:

$$C_2(0,t) = \frac{C_o R m}{m + R} \left[1 - 2 \text{ erfc } (\phi_o) + \text{erfc } (2\phi_o) \right]$$

$$C_2(x,t) = \frac{C_o R m}{m + R} \left[\text{erfc } (\phi) - 2 \text{ erfc } (\phi_o + \phi) + \text{erfc } (2\phi_o + \phi) \right],$$

where:

$$m = \frac{C_2(0, t)}{C_1(0, t)}$$

$$R = (D_1/D_2)^{1/2}$$

$$\phi_o = x_o/2(D_1 t)^{0.5}$$

$$\phi = x/2(D_2 t)^{0.5}.$$

Redistribution Diffusion [1]

When redistribution diffusion length, $(D_2 t_2)^{1/2}$, is not large in comparison with predeposition diffusion length, $(D_1 t_1)^{1/2}$, the diffusion profile is not a Gaussian distribution. For certain simple cases, exact solutions are available.

A. Assume that the diffusion profile for predeposition diffusion is represented by an erfc. The redistribution diffusion is at a different temperature and for a different length of time in a nonoxidizing ambient. We can assume that the surface atoms are depleted, $C_2(0,t_2) = 0$.

The concentration profile is given by

$$C_2(x,t_2) = C_{so} \left[\text{erfc} \left(\frac{x}{2\sqrt{D_1 t_1 + D_2 t_2}} \right) - \text{erfc} \left(\frac{x}{2\sqrt{D_2 t_2}} \right) \right].$$

B. Assume the same conditions as in A, except that there is a capping oxide present on the surface during the redistribution diffusion. The concentration of impurity in silicon at the oxide-silicon interface is determined by the segregation coefficient at the diffusion temperature.

$$C_2(x,t_2) = C_{so} \text{ erfc } \left(\frac{x}{2\sqrt{D_1 t_1 + D_2 t_2}} \right) + [C_2(0,t_2) - C_{so}] \text{ erfc } \left(\frac{x}{2\sqrt{D_2 t_2}} \right)$$

$$C_2(0,t_2) = m C_{so},$$

where:

D_1 = diffusion coefficient at prediffusion temperature

D_2 = diffusion coefficient at drive-in diffusion temperature

C_{so} = surface concentration during prediffusion

C_2 = impurity concentration in silicon during redistribution diffusion

t_1 = prediffusion time

t_2 = redistribution-diffusion time

m = segregation coefficient.

Outdiffusion From a Buried, Uniformly Doped Layer

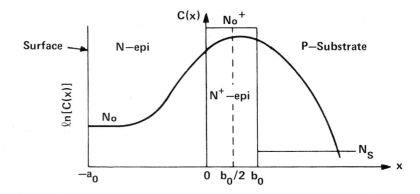

$$C_1(x,t) = \frac{N_o^+}{2} \, \text{erfc}\left(\frac{2a_o+x}{2\sqrt{Dt}}\right) + \frac{N_o^+}{2} \, \text{erfc}\left(-\frac{x}{2\sqrt{Dt}}\right) - \frac{N_o^+}{2} \, \text{erfc}\left(\frac{b_o-x}{2\sqrt{Dt}}\right)$$

$$- \frac{N_o^+}{2} \, \text{erfc}\left(\frac{2a_o-b_o+x}{2\sqrt{Dt}}\right) + N_o \, ; \, -a_o < x > b_o,$$

where:

N_o^+ = concentration of the buried layer

N_o = concentration of the uniform-surface layer($\simeq$ epi layer doping)

a_o = thickness of the surface layer

b_o = thickness of the N_o^+ buried layer

D = diffusion coefficient for the N_o^+ dopant.

Diffusion of Ion-Implanted Impurities During Oxidation[2]

For an implanted impurity profile given by

$$C(x) = C_{max} \, \exp \left(\frac{-(x-R_p)^2}{2\Delta R_p^2}\right),$$

$C(x,t)$ following oxidation as measured from the Si/SiO$_2$ interface is

$$C(x,t) = -\frac{k' - m_s}{2}\left[\frac{\pi k_p}{D(1+k_p/4k_\ell^2\, t)}\right]^{1/2} C_s \;\exp\left(\frac{m_s x_o}{2\sqrt{Dt}}\right)^2 \mathrm{erfc}\left(\frac{x+m_s x_o}{2\sqrt{Dt}}\right)$$

where

$$C_s = \frac{\dfrac{C_{max}}{4\sqrt{Dt}}\,\Omega(m_s x_o,\, t)}{1 + \dfrac{k' m_s}{2}\left[\dfrac{\pi\, k_p}{D(1+k_p/4k_\ell^2 t)}\right]^{1/2}\exp\left[\dfrac{m_s x_o}{2\sqrt{Dt}}\right]^2 \mathrm{erfc}\left[\dfrac{m_s x_o}{2\sqrt{Dt}}\right]}$$

$$\Omega(m_s x_o,t) = E^{1/2}\exp\left[-(A-\frac{B^2}{4C})\right]\frac{\pi^{1/2}}{2}\left[1 + \mathrm{erf}\left(\frac{m_s x_o - B/2}{E^{1/2}}\right)\right]$$

$$A = \frac{(m_s x_o - R_p)^2}{2\Delta R_p^{\,2}}$$

$$B = \frac{4Dt\,(m_s x_o - R_p)}{\Delta R_p^{\,2} + 2Dt}$$

$$E = \frac{4Dt\,\Delta R_p^{\,2}}{\Delta R_p^{\,2} + 2Dt}$$

$$x_o = \frac{k_p}{2k_\ell}\left\{\left[\frac{4(t+\tau_i)}{k_p}k_\ell^2 + 1\right]^{1/2} - 1\right\}$$

$(k')^{-1} = m$, impurity segregation coefficient

$m_s\quad = 0.44$

$k_p\quad$ = parabolic oxidation rate constant

$k_\ell\quad$ = linear oxidation rate constant

$\tau_i\quad$ = initial oxide thickness time parameter

$R_p\quad$ = projected range of implanted impurity

ΔR_p = straggle of implanted impurity.

High Impurity Concentration Diffusion Equations

Arsenic Diffusion in Silicon

For $C_s \gg n_i$, the diffusivity of arsenic is

$$D(C) = 2 \frac{D_i C}{n_i}.$$

An approximate solution for the impurity profile is [3]

$$C = C_s (1.00 - 0.87Y - 0.45Y^2),$$

where:

$$Y = x \left(\frac{8C_s}{n_i} D_i t \right)^{-1/2}$$

D_i = diffusion coefficient at low impurity concentration, $(C < n_i)$,

$\quad\quad$ = $22.9 \exp (-4.10eV/kT)$ (cm^2/sec)

n_i = intrinsic electron concentration at the diffusion temperature, cm^{-3}

C_s = surface concentration, cm^{-3}

For constant surface concentration,

$$X_j = 2.3 \left(\frac{C_s D_i t}{n_i} \right)^{1/2}.$$

$$C_s = \frac{1.56 \times 10^{17}}{R_s X_j}.$$

For ion-implanted arsenic layers that are diffused, [4]

$$X_j = 2 \left(\frac{Q D_i t}{n_i} \right)^{1/3}.$$

$$C_s = 0.91 \left(\frac{Q^2 n_i}{D_i t} \right)^{1/3}.$$

$$Q = 0.55 C_s X_J.$$

$$C_s = \frac{6.26 \times 10^{15}}{(R_s X_j)^{3/2}}.$$

$$R_s = \frac{1.76 \times 10^{10}}{Q^{7/9}} \left(\frac{n_i}{D_i t} \right)^{1/9}.$$

The sheet resistance of a diffused implanted arsenic layer versus junction depth is shown in Figure 6-3 for various arsenic implant doses. [4] The times and temperatures required to achieve a given sheet resistance for a given dose are shown in Figure 6-4. The times required to completely activate an implanted arsenic layer at some temperature are shown in Figure 6-5.

Boron Diffusion in Silicon

The diffusivity of boron is

$$D(C) = \frac{D_i p}{n_i}.$$

An approximate solution for the impurity profile for $C \gg n_i$ ($C = p$) is [5]

$$C = C_s (1 - Y^{2/3}),$$

Where Y is dimensionless and

$$X_j = 2.45 \left(\frac{C_s D_i t}{n_i} \right)^{1/2} \text{ at } C_B \simeq 10^{18}.$$

$$C_s = \frac{2.78 \times 10^{17}}{R_s X_j}.$$

For an implanted, diffused boron layer,

$$Q = 0.4 \, C_s X_j \ (X_j \text{ at } C_B \simeq 10^{18}).$$

$$C_s = 0.53 \left(\frac{Q^2 n_i}{D_i t} \right)^{1/3}.$$

$$D_i = 3.17 \exp \left(\frac{-3.59 eV}{kT} \right) \ (cm^2/sec).$$

The diffusivity in (100) silicon is enhanced for boron diffusion in oxidizing ambients. The effective diffusivity for dry O_2 oxidation is shown in Figure 6-6. [6] The segregation coefficients of boron in various oxidizing ambients are shown in Figure 6-7. [7]

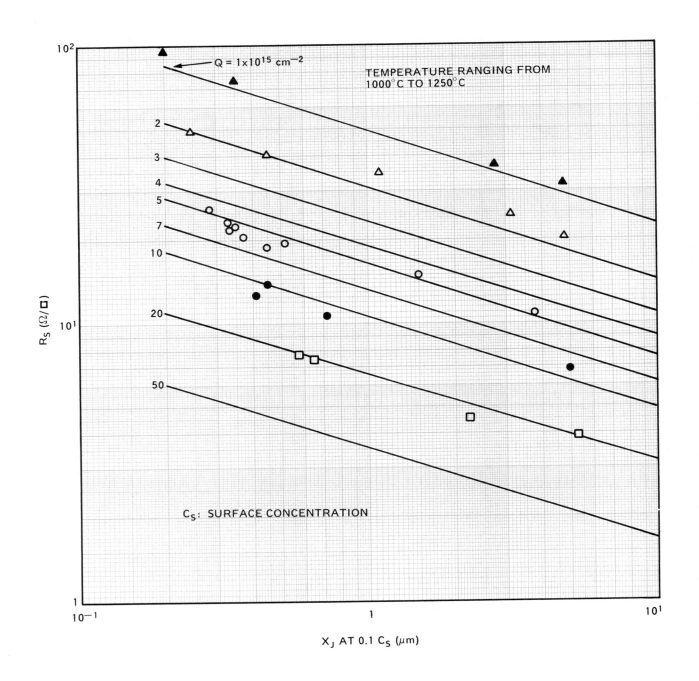

Figure 6-3—Sheet Resistance Versus Junction Depth for Implanted As Layers Diffused in N_2 or O_2

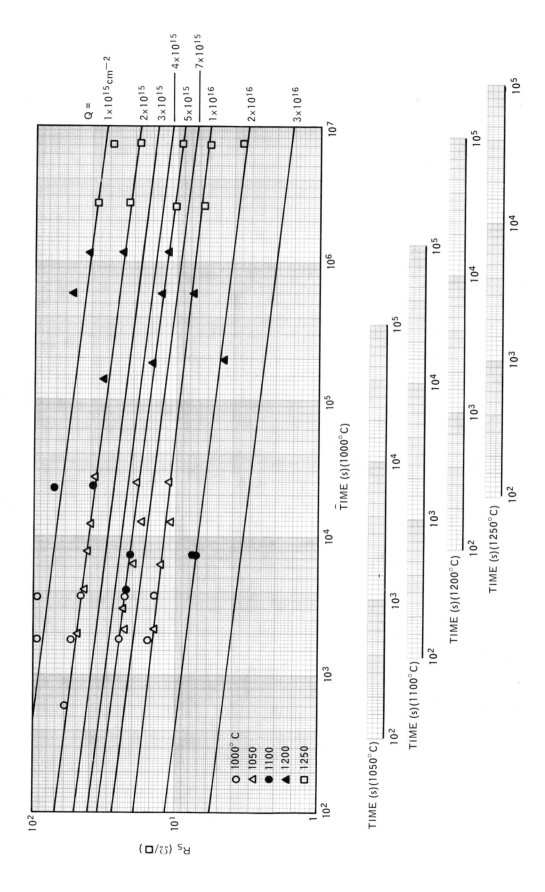

Figure 6-4 — Time, Temperature, and As Dose Dependence of Sheet Resistance (Diffusions in N_2 or O_2)

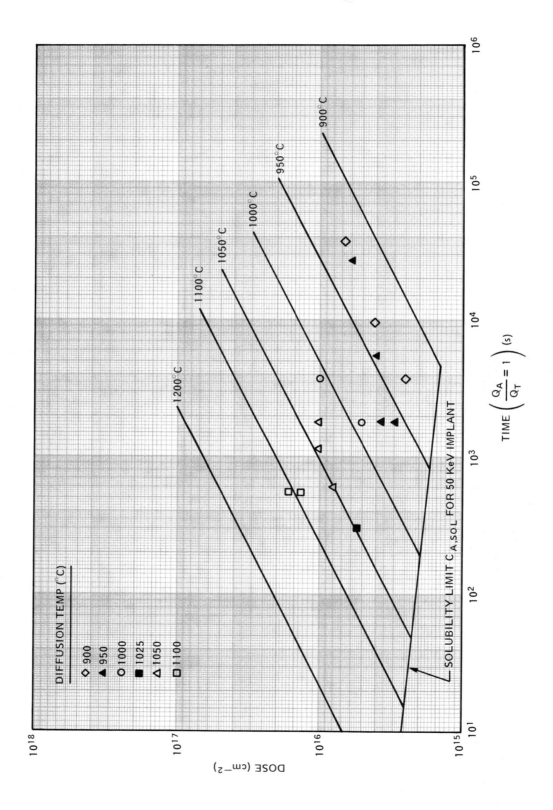

Figure 6-5 — Time To Complete Activation as a Function of Arsenic Dose and Diffusion Temperature

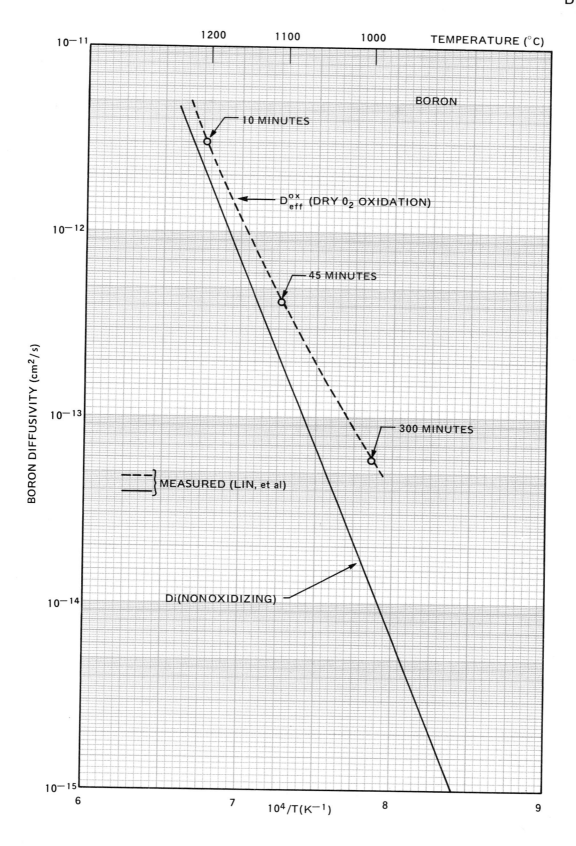

Figure 6-6 — Diffusivity of Boron in Dry O_2 and Nonoxidizing Ambient for (100) Silicon[6]

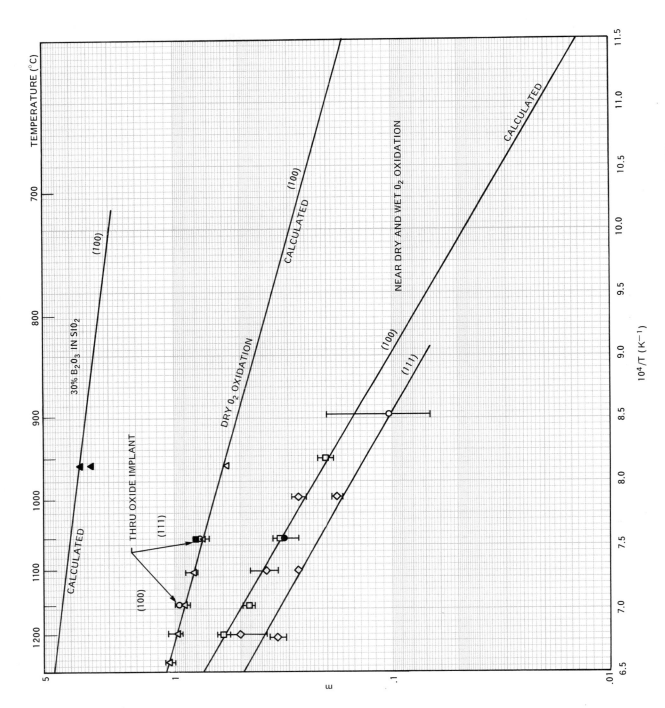

Figure 6-7 — Segregation Coefficients of Boron at the Si/SiO$_2$ Interface[7]
(Reprinted by permission of the Electrochemical Society, Inc.)

Phosphorus Diffusion in Silicon

High concentration phosphorus diffusion in silicon is complicated by mulitiple-species diffusion mechanisms.[8] Thus, no simple solution exists to the diffusion equation. However, some useful curves exist to characterize phosphorus diffusions. In Figure 6-8, the total phosphorus surface concentration and the free electron surface concentration are plotted for several temperatures versus the average resistivity of the diffused layer.[9]   In Figure 6-9, sheet resistance versus diffusion time for constant surface concentration is plotted for junction depths of 0.1μm to 5.0μm.

The free electron surface concentration for ion-implanted phosphorus after diffusion is

$$n_s = \frac{7.5 \times 10^{25}}{T^{4.03}} \left(\frac{Q}{\sqrt{t}} \right)^{5/11} \quad (cm^{-3})$$

or

$$n_s = 4000 \left(\frac{Q}{X_j} \right)^{5/6}$$

with T in $^\circ$C, X_j in cm, t in s, and Q in cm^{-2}.

Curves of sheet resistance versus junction depth for a range of doses are shown in Figure 6-10.[9] The times and temperatures required to achieve a required sheet resistance for a given dose are shown in Figure 6-11.

If the integrated concentration of phosphorus exceeds a certain value, sufficient elastic energy can be stored in the diffused layer to create a network of misfit dislocations. It is important to generate such a network for efficient gettering. The times and temperatures required to generate misfit dislocation arrays (MDA) for constant surface concentration diffusions are shown in Figure 6-12.[10]

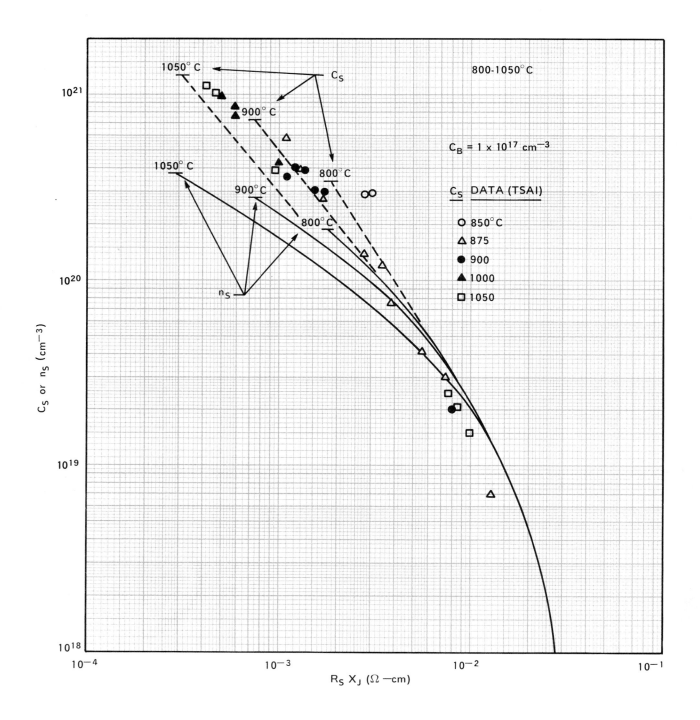

Figure 6-8a — Surface Concentration Versus Average Resistivity for Phosphorus-Diffused Silicon[9]
(Reprinted by permission of the Electrochemical Society, Inc.)

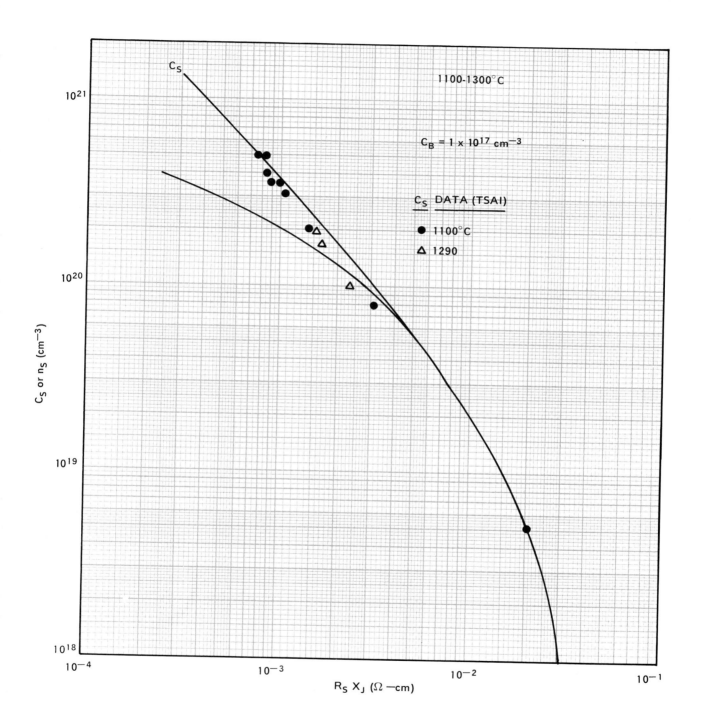

Figure 6-8b — Surface Concentration Versus Average Resistivity for Phosphorus-Diffused Silicon[9]
(Reprinted by permission of The Electrochemical Society, Inc.)

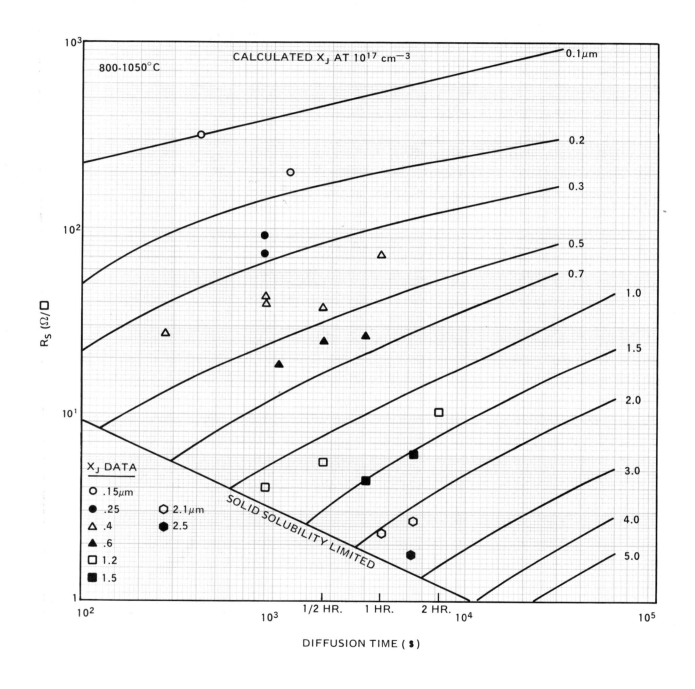

Figure 6-9a — Sheet Resistance Versus Time for Phosphorus Diffusion Into Silicon in Low O_2 Ambient

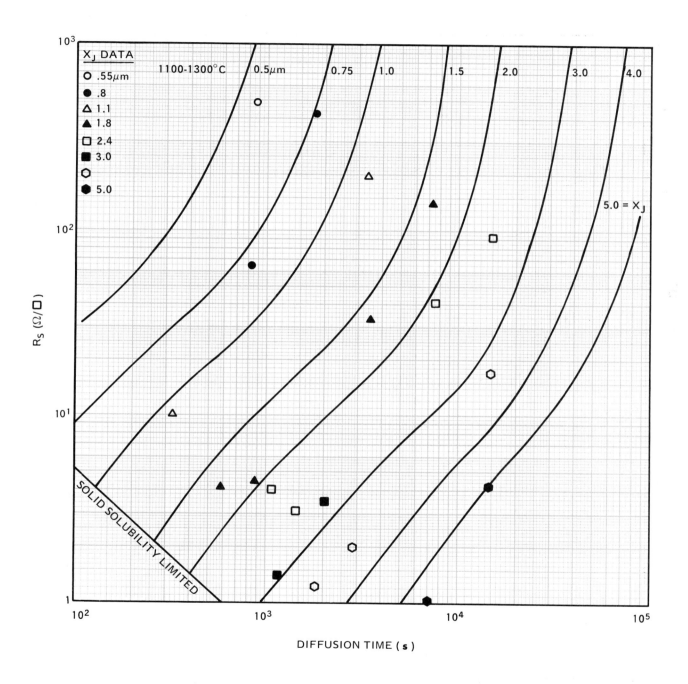

Figure 6-9b — Sheet Resistance Versus Time for Phosphorus Diffusion Into Silicon in Low O_2 Ambient

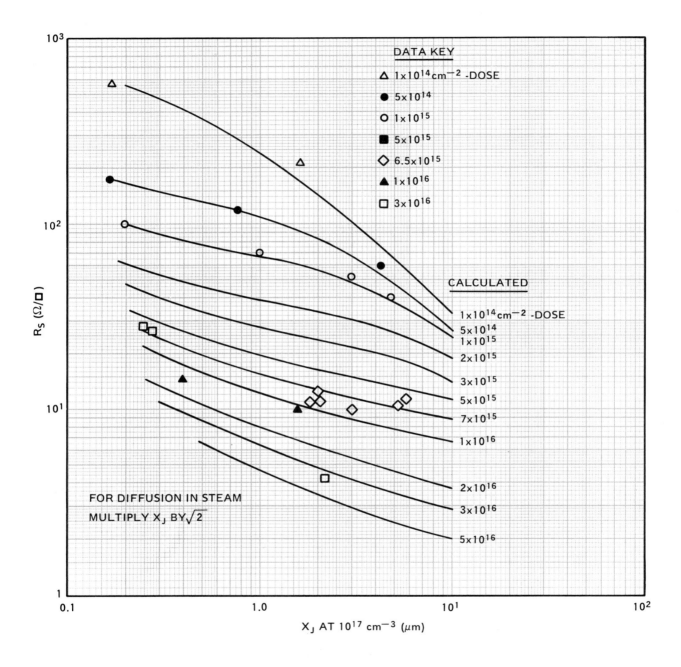

Figure 6-10 — Sheet Resistance Versus Junction Depth for Implanted Phosphorus Layers Diffused in N_2 or O_2 [9] (Reprinted by permission of The Electrochemical Society, Inc.)

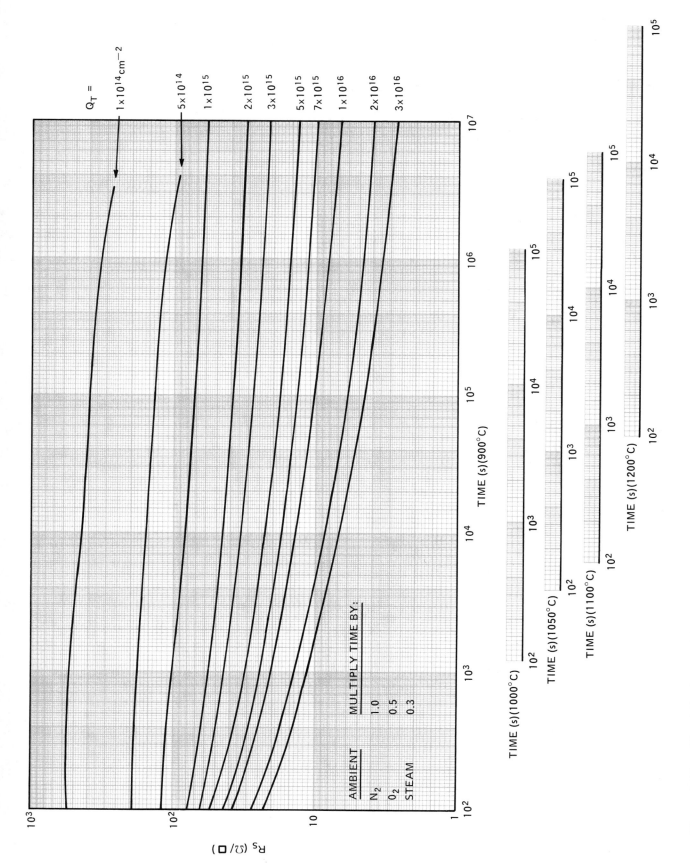

Figure 6-11 — Sheet Resistance Versus Time, Temperature, and Dose for Phosphorus-Implanted, Diffused Layers in (100) Silicon

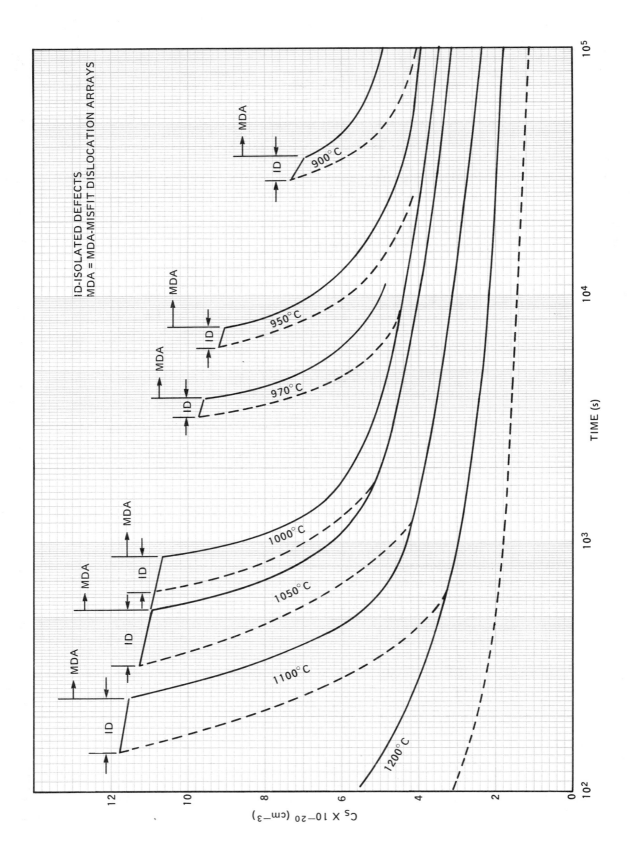

Figure 6-12 — Critical Time to Dislocation Array Formation for Phosphorus Diffusion in Low O₂ Ambients

Maximum Carrier Concentration

For shallow junctions, the diffusion temperatures are low and sometimes without intentional diffusion, except an annealing of ion-implanted dopants. In these cases, the electrical activities are not 100 percent. Figures 6-13 and 6-14 show the maximum carrier concentrations for arsenic- and boron- implanted silicon as a function of the annealing temperatures. The annealing ambient is often nitrogen or argon for temperatures below 1150°C, and a mixture of nitrogen and a small percentage of oxygen for temperatures above 1150°C.

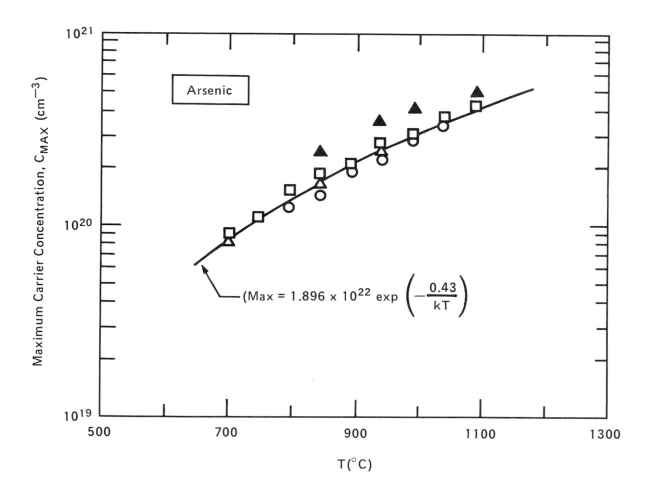

Figure 6-13 — Maximum Carrier Concentration of Arsenic-Implanted Silicon Versus Annealing Diffusion Temperature[11] (Reprinted by permission of the Electrochemical Society, Inc.)

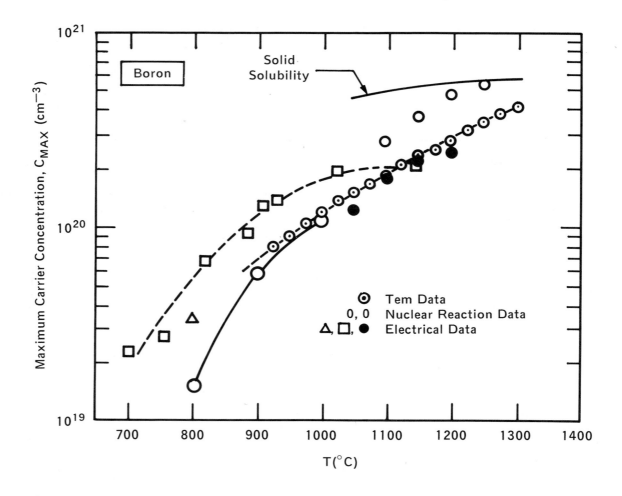

Figure 6-14 —Maximum Carrier Concentration of Boron-Implanted Silicon Versus Annealing
Diffusion Temperature[12] (Reprinted with permission from Springer- Verlag, Inc.)

6.3   Diffusivities in Silicon

Figure 6-15 shows the diffusivities in silicon calculated by Kendall and DeVries.[13]   These calculations are derived from data published prior to 1969, and should be used as an overview estimation of the orders of magnitude of the elements shown.

Table 6-1 shows selected values of the intrinsic diffusion coefficients, called the intrinsic diffusivities, in silicon. The intrinsic diffusivities represent the values determined from diffusion from low concentration sources when the diffusivity is independent of the dopant concentrations of the diffusants and the substrate dopant. When the local impurity concentration is below the intrinsic carrier concentration, n_i, at the diffusion temperature, the intrinsic diffusivity can be applied.

In Table 6-1, most of the values come from two sources — namely, the collections of Kendall and DeVries,[13] and Fair.[14]   Fair proposed a multi-charge state impurity-defect interaction mechanism to explain the diffusion phenomenon of group III elements, group V elements, and silicon, in silicon. Based on this theory, the intrinsic diffusivities can be dominated by interactions with neutral vacancies, V^x, singly-charged acceptor vacancies, V^-, doubly-charged acceptor vacancies, V^{-2}, and singly-charged donor vacancies, V^+. The corresponding diffusivities are labeled as D_i^x, D_i^-, D_i^{-2} and D_i^+, respectively. The values of these D_i's, as shown in Table 6-1 are well fitted to the experimental data. The dominating interactions are also given in the table. The expressions of $(D_i^x + D_i^+)$ for Al, Ga, and In are given to represent the measured diffusivities. For phosphorus, the D_i^x is dominating at low concentrations. In Table 6-1, CD means that the concentration dependence was not taken into account, but the data were for concentrations less than n_i.

Table 6-2 shows the diffusivities and solubilities of fast diffusants in silicon. Most of these elements reduce carrier lifetime and cause leakage currents, and are therefore undesirable. Also shown, are gaseous elements and miscellaneous elements. As word of caution, note that the consistencies of the diffusivity measurements of the fast diffusants, such as Cu, Au, Pt, Ag, etc., were not good. Many factors affect the distribution and diffusion rate of these elements. One must be very careful in using the data given in Table 6-2.

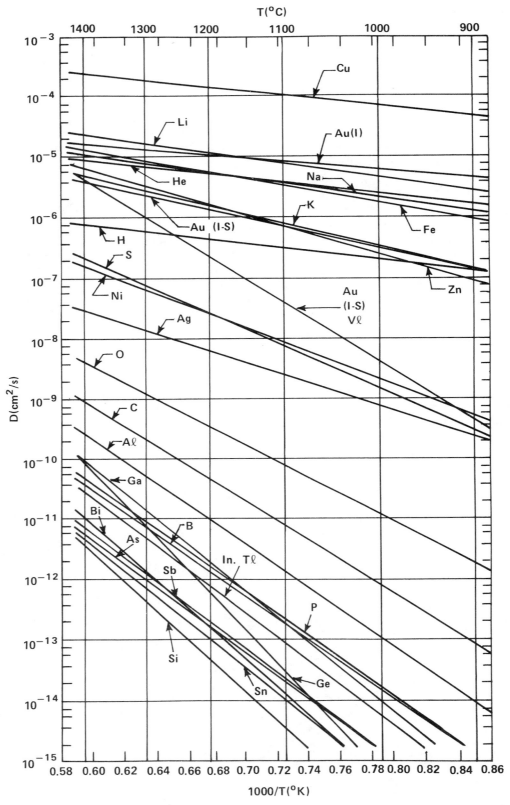

Figure 6-15 —Diffusion Coefficients in Intrinsic Silicon[13]
(Originally presented at the Spring 1969 meeting of the
Electrochemical Society, Inc. held in New York, N.Y.)

TABLE 6-1
SELECTED VALUES OF INTRINSIC DIFFUSIVITIES IN SILICON

Element	Ref.	$\dfrac{D_o}{cm^2/s}$	E (eV)	Remarks
Al	13	0.5	3.0	CD
	14	1350	4.1	$D_i^x + D_i^+$
B	14	0.76	3.46	D_i^+
Ga	13	225	4.12	CD
	14	13.1	3.7	$D_i^x + D_i^+$
In, Tl	13	16.5	3.9	CD
	14	269	4,19	$D_i^x + D_i^+$
As	14	12	4.05	D_i^-
	15	24	4.08	Chemical Source
	16	22.9	4.1	Ion Implant Predeposition
	17	60	4.2	Isoconcentration diffusion experiment
N	14	0.05	3.65	D_i^x
P	14	3.85	3.66	D_i^x
	14	4.44	4.0	D_i^-
	14	44.2	4.37	D_i^{-2}
Sb	13	12.9	3.98	$2 \times 10^{19} - 5 \times 10^{19}$
	14	15	4.08	D_i^-
Bi	14	396	4.12	D_i^-
C	13	1.9	3.1	
Ge	13	6.26×10^5	5.28	Ge powder
Sn	13	32	4.25	Tracer measurement
Si	14	0.015	3.89	D_i^x
	14	16	4.54	D_i^-
	14	1180	5.09	D_i^+

NOTE: More recent data on diffusion and solubility of transition metals in silicon can be found in a review paper ''Transition Metals in Silicon'', authored by E. R. Weber and published in *Applied Physics,* Vol. A30, No. 1 (January 1983), pp. 1-22.

TABLE 6-2

DIFFUSIVITY AND SOLUBILITIES OF THE FAST DIFFUSANT IN SILICON

ELEMENT	REF	$\dfrac{D_o}{cm^2/s}$	E (eV)	SOLUBILITY cm^{-3}
Li (25 − 1350°C)	18	2.3×10^{-3} to 9.4×10^{-4}	0.63 to 0.78	max. 7×10^{19} 1200°C
Na (800 − 1100°C)	18	1.6×10^{-3}	0.76	10^{18} to 9×10^{18} (600 − 1200°C)
K (800 − 1100°C)	18	1.1×10^{-3}	0.76	9×10^{17} to 7×10^{18} (600 − 1200°C)
Cu (800 − 1100°C)	18	4×10^{-2}	1.0	5×10^{15} to 3×10^{18} (600 − 1300°C)
$(Cu)_i$ (300 − 700°C)	18	4.7×10^{-3}	0.43	
Ag (1100 − 1350°C)	18	2×10^{-3}	1.6	6.5×10^{15} to 2×10^{17} (1200 − 1350°C)
Au (800 − 1200°C) $(Au)_i$ $(Au)_s$ (700 − 1300°C)	18	1.1×10^{-3} 2.4×10^{-4} 2.8×10^{-3}	1.12 0.39 2.04	5×10^{14} to 5×10^{16} (900 − 1300°C)
Pt (800 − 1000°C)	19	1.5×10^2 to 1.7×10^2	2.22 to 2.15	4×10^{16} to 5×10^{17} (800 − 1000°C)
Fe (1100 − 1250°C)	18	6.2×10^{-3}	0.87	10^{13} to 5×10^{16} 900 − 1300°C)
Ni (450 − 800°C)	18	0.1	1.9	6×10^{18} (1200 − 1300°C)
Cr (1100 − 1250°C)	20	0.01	1.0	2×10^{13} to 2.5×10^{15} (900 − 1280°C)
Co (900 − 1200°C)	21	9.2×10^4	2.8	max. 2.5×10^{16} (1300°C)
O_2 (700 − 1240°C)	22	7×10^{-2}	2.44	1.5×10^{17} to 2×10^{18} (1000 − 1400°C)
H_2	18	9.4×10^{-3}	0.48	$2.4 \times 10^{21} \exp \dfrac{-1.86}{kT}$ at 1 atm.
He	13	0.11	1.26	
S	13	0.92	2.2	
Zn	13	0.1	1.4	

6.4 Diffusivities in SiO$_2$

Table 6-3 shows the diffusivities in SiO$_2$ for the common dopant diffusion elements. Most of these also show concentration dependence and diffusion system dependence. Table 6-4 shows a collection of diffusivity data, the calculated diffusivities at 1000°C and 300°C, and the mobilities at 300°C, under an electric field across the oxide layer. Figure 6-16 shows the calculated diffusivities in SiO$_2$ based on the values given in Table 6-4. These data were collected by Bartholomew[26] in 1969 and they have not been updated.

TABLE 6-3

DIFFUSIVITIES IN SiO$_2$

ELEMENT	REF.	$\dfrac{D_o}{(cm^2/s)}$	E (eV)	D(900°C) (cm^2/s)	C$_s$ (cm^{-3})	SOURCE & AMBIENT
Boron	23	7.23×10^{-6}	2.38	4.4×10^{-16}	10^{19} to 2×10^{20}	B$_2$O$_3$ vapor, O$_2$ + N$_2$
	23	1.23×10^{-4}	3.39	3.4×10^{-19}	6×10^{18}	B$_2$O$_3$ vapor, Ar
	23	3.16×10^{-4}	3.53	2.2×10^{-19}	Below 3×10^{20}	Borosilicate
Gallium	23	1.04×10^5	4.17	1.3×10^{-13}	—	Ga$_2$O$_3$ vapor, H$_2$ + N$_2$ + H$_2$O
Phosphorus	23	5.73×10^{-5}	2.30	7.7×10^{-15}	8×10^{20} to 10^{21}	P$_2$O$_5$ vapor, N$_2$
	24	1.86×10^{-1}	4.03	9.3×10^{-19}	8×10^{17} to 8×10^{19}	Phosphosilicate, N$_2$
Arsenic	25	67.25	4.7	4.5×10^{-19}	$<5 \times 10^{20}$	Ion Implant, N$_2$
	25	3.7×10^{-2}	3.7	4.8×10^{-18}	$<5 \times 10^{20}$	Ion Implant, O$_2$
Antimony	23	1.31×10^{16}	8.75	3.6×10^{-22}	5×10^{19}	Sb$_2$O$_5$ vapor, O$_2$ + N$_2$

Note: C$_s$ = Surface concentration on silicon after diffusion from the specified source and ambient in the absence of an oxide barrier.

TABLE 6-4

DIFFUSION COEFFICIENTS IN AMORPHOUS SiO$_2$ [26]

MATERIAL	$\dfrac{D}{(cm^2/s)}$	D at 1000°C (cm^2/s)	D at 300°C (cm^2/s)	E (eV)	μ at 300°C (cm^2/V-s)
H+	1	1.01×10^{-3}	2.05×10^{-7}	0.73	4.13×10^{-6}
He	3.0×10^{-4}	5×10^{-7}	3×10^{-8}	0.24	6.1×10^{-7}
H$_2$	5.65×10^{-4}	9×10^{-6}	6.3×10^{-8}	0.446	1.2×10^{-6}
D$_2$	5.01×10^{-4}	—	4×10^{-8}	0.455	8.1×10^{-6}
H$_2$O	1.0×10^{-6}	7×10^{-10}	1.1×10^{-13}	0.79	2.2×10^{-12}
OH	9.5×10^{-4}	2×10^{-7}	9.5×10^{-10}	0.68	1.9×10^{-8}
O$_2$	2.7×10^{-4}	1×10^{-7}	1.6×10^{-14}	1.16	3.3×10^{-13}
P	5.3×10^{-8}	8.1×10^{-14}	1×10^{-25}	1.46	2.02×10^{-24}
B	1.7×10^{-5}	1.6×10^{-18}	4.0×10^{-30}	3.37	8.1×10^{-29}
S	6×10^{-5}	2.5×10^{-15}	3×10^{-24}	2.6	6.1×10^{-23}
Ga	3.8×10^{5}	8.0×10^{-12}	2.5×10^{-31}	4.15	5.03×10^{-30}
Au	8.2×10^{-10}	5.54×10^{-15}	2.5×10^{-19}	0.8	5.0×10^{-18}
Au	8.5×10^{3}	2.17×10^{-12}	8.0×10^{-23}	3.7	1.6×10^{-21}
Au	1.52×10^{-7}	4.65×10^{-16}	2.0×10^{-28}	2.14	4.0×10^{-27}
Na	6.9	5.2×10^{-5}	2.8×10^{-11}	1.30	5.6×10^{-10}
Na*	2.13	—	—	0.92	—
Na†	0.398	—	6.0×10^{-11}	1.12	1.2×10^{-9}
Na‡	3.44×10^{-2}	1.0×10^{-5}	—	1.22	—
Na§	4.3×10^{-8}	4.8×10^{-16}	8.25×10^{-26}	2.0	1.6×10^{-24}
Pt	1.2×10^{-13}	4.5×10^{-17}	2.0×10^{-20}	0.75	4.03×10^{-19}

* 170-250°C

† 250-573°C

‡ 573-1000°C

§ in Si$_3$N$_4$

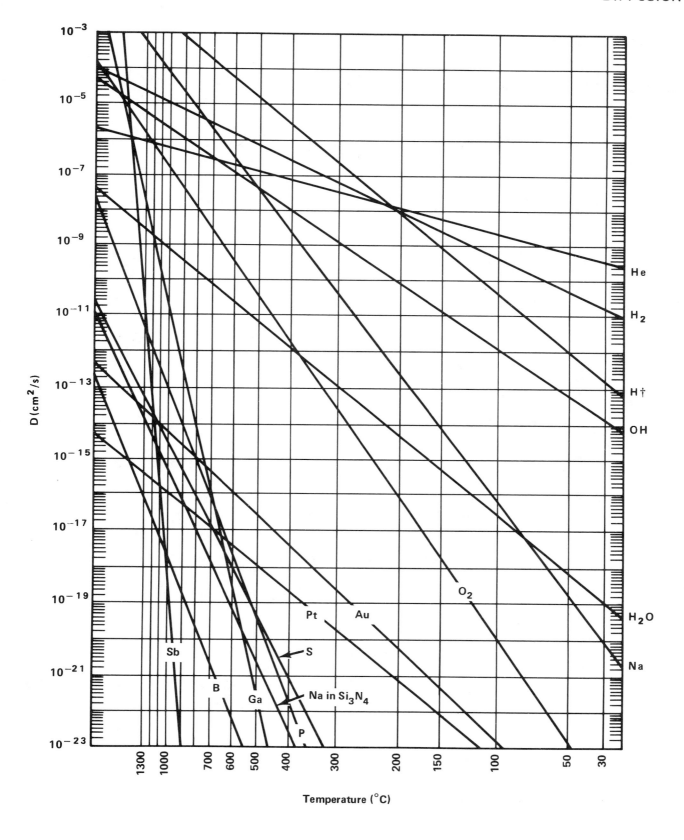

Figure 6-16 — Diffusion Coefficients for Various Substances in SiO$_2$ [26]

6.5 Diffusivities in Polycrystalline Silicon Films

Table 6-5 shows a few examples of diffusivities in polycrystalline silicon films. These films were produced by a chemical vapor deposition technique over a temperature range of 600°C to 1000°C. The film thicknesses are generally less than one micron. In general, the diffusion of the dopant elements shown, proceeds along grain boundaries of the single crystallites of varying sizes (from less than 1000Å to a few tens of microns). Inside each crystallite, the diffusivities are similar to those found in single crystals. The diffusivities shown in Table 6-5 are for the grain boundary diffusions. They showed strong dependence on the texture of the polysilicon films. Arsenic and phosphorus also showed trapping at grain boundaries, which affects the diffusion and the electrical properties of the polysilicon film. Boron atoms did not appear to segregate at the grain boundaries.

TABLE 6-5

EXAMPLES OF DIFFUSIVITIES IN POLYCRYSTALLINE SILICON FILMS

ELE-MENT	REF.	$\dfrac{D_o}{cm^2/s}$	E (eV)	$\dfrac{D}{cm^2/s}$ @	T °C
As	27	8.6×10^4	3.9	2.4×10^{-14}	800
As	28	0.63	3.2	3.2×10^{-14}	950
B	29	1.5×10^{-3} to 6×10^{-3}	2.4 to 2.5	9×10^{-14}	900
B	30			4×10^{-14}	925
P	31			6.9×10^{-13}	1000
P	31			7×10^{-12}	1000

REFERENCES

1. K.C. Nomura, "Analytic Solution of the Double-Diffusion Problem", *J. Appl. Phys.,* Vol. 32 (1961), pp. 1167-1168.

2. C.P. Wu, E.C. Douglas, and C.W. Mueller, "Redistribution of Ion-Implanted Impurities in Silicon During Diffusion in Oxidizing Ambients", *IEEE Trans. on Electron Dev.,* Vol. ED-23 (September 1976), pp. 1095-1097.

3. R.B. Fair, "Profile Estimation of High-Concentration Arsenic Diffusions in Silicon", *J. Appl. Phys.,* Vol. 43, No. 3 (March 1972), pp. 1278-1280.

4. R.B. Fair and J.C.C. Tsai, "Profile Parameters of Implanted-Diffused Arsenic Layers in Silicon", *J. Electrochem. Soc.: Solid-State Science and Technology,* Vol. 123, No. 4 (April 1976), pp. 583-586.

5. R.B. Fair, "Boron Diffusion in Silicon-Concentration and Orientation Dependence, Background Effects, and Profile Estimation", *J. Electrochem. Soc.: Solid-State Science and Technology,* Vol. 122, No.6 (June 1975), pp. 800-805.

6. A.M. Lin, R.W. Dutton, and D. A. Antoniadis, unpublished work.

7. R.B. Fair and J.C.C. Tsai, "Theory and Direct Measurement of Boron Segregation in SiO_2 During Dry, Near Dry, and Wet O_2 Oxidation", *J. Electrochem. Soc.: Solid-State Science and Technology,* Vol. 125, No. 12 (December 1978), pp. 2050-2058.

8. R.B. Fair and J.C.C. Tsai, "A Quantitative Model for the Diffusion of Phosphorus in Silicon and the Emitter Dip Effect", *J. Electrochem. Soc.: Solid-State Science and Technology,* Vol. 124, No. 7 (July 1977), pp. 1107-1108.

9. R.B. Fair, "Analysis of Phosphorus-Diffused Layers in Silicon", *J. Electrochem. Soc.: Solid-State Science and Technology,* Vol. 125, No. 2 (February 1978), pp. 323-327.

10. R.B. Fair, "Quantified Conditions for Emitter-Misfit Dislocation Formation in Silicon", *J. Electrochem. Soc.: Solid-State Science and Technology,* Vol. 125, No. 6 (June 1978) pp. 923-926.

11. E. Guerrero et al, "Generalized Model for the Clustering of As Dopants in Si," *J. Electrochem. Soc.: Solid-State Science and Technology,* Vol. 129, No. 8 (August 1982), p. 1826.

12. H. Ryssel et al, "High Concentration Effects of Ion-Implanted Boron in Silicon," *Appl. Phys.,* Vol. 22, No. 1 (May 1980), p. 35.

13. D.L. Kendall and D.B. DeVries, "Diffusion in Silicon," *Semiconductor Silicon 1969,* ed. R.R. Haberect and E.L. Kern (New York: Electrochem. Soc. Inc., 1969), p. 414.

14. R.B. Fair, "Concentration Profiles of Diffused Dopants in Silicon," *Impurity Doping Process in Silicon,* ed. F.F.Y. Wang (New York: North-Holland, 1981), Chapter 7.

15. T.L. Chiu and H.N. Ghosh, "Diffusion Model for Arsenic in Silicon," *IBM J. Res. Develop.,* Vol. 15 (1971), p. 472.

16. R.B. Fair and J.C.C. Tsai, "The Diffusion of Ion-Implanted Arsenic in Silicon," *J. Electrochem. Soc.,* Vol. 122 (1975), p. 1689.

17. B. J. Masters and J. M. Fairfield, "Arsenic Isoconcentration Diffusion Studies in Silicon," *J. Appl. Phys.*, Vol. 40 (1969), p. 2390.

18. B. L. Sharma, "Diffusion in Semiconductors," *Trans. Tech. Pub.*, Germany, (1970), pp. 87-117.

19. R. F. Bailey and T. G. Mills, "Diffusion Parameters of Platinum in Silicon," *Semiconductor Silicon 1969*, ed. R. R. Habrect and E. L. Kern (New York: Electrochem. Soc. Inc., 1969), p. 481.

20. W. Wurker, K. Roy, and J. Hesse, "Diffusion and Solid Solubility of Chromium in Silicon," *Mater. Res. Bull.* (U.S.A.) Vol. 9 (1974), p. 971.

21. H. Kitagawa and K. Hashimoto, "Diffusion Coefficient of Cobalt in Silicon," *Japan. J. Appl. Phys.*, Vol. 16 (1977), No. 1, p. 173.

22. J. C. Mikkelsen, Jr., "Diffusivity of Oxygen in Silicon During Steam Oxidation," *Appl. Phys. Lett.*, Vol. 40 (1982), p. 336.

23. M. Ghezzo and D. M. Brown, "Diffusivity Summary of B, Ga, P, As, and Sb in SiO_2," *J. Electrochem. Soc.: Solid State Science and Technology,* Vol. 120, No. 1 (January 1973), pp. 146-148.

24. R. N. Ghoshtagore, "Silicon Dioxide Masking of Phosphorus Diffusion in Silicon," *Solid-State Electron.,* Vol. 18 (1975), p. 399.

25. Y. Wada and D. A. Antoniadis, "Anomalous Arsenic Diffusion in Silicon Dioxide," *J. Electrochem. Soc.: Solid State Science and Technology,* Vol. 128, No. 6 (June 1981), pp. 1317-1320.

26. C. Y. Bartholomew, private communication.

27. B. Swaminathan, K. C. Saraswat, R. W. Dutton, and T. I. Kamins, "Diffusion of Arsenic in Polycrystalline Silicon," *Appl. Phys. Lett.,* Vol. 40 (1982), p. 795.

28. K. Tsukamoto, Y. Akasaka, and K. Horie, "Arsenic Implantation into Polycrystalline Silicon and Diffusion to Silicon Substrate," *J. Appl. Phys.,* Vol. 48 (1977), p. 1815.

29. S. Horiuchi and R. Blanchard, "Boron Diffusion in Polycrystalline Silicon Layers," *Solid-State Electron.,* Vol. 18 (1975), p. 529.

30. C. J. Coe, "The Lateral Diffusion of Boron in Polycrystalline Silicon and Its Influence on the Fabrication of Sub-Micron Mosts," *Solid-State Electron.,* Vol. 20 (1977), p. 985.

31. T. I. Kamins, J. Manolin, and R. N. Tucker, "Diffusion of Impurities in Polycrystalline Silicon," *J. Appl. Phys.,* Vol. 43 (1972), p. 83.

7. ION IMPLANTATION

LIST OF SYMBOLS

Symbol	Definition
$C(x)$	Concentration
C_{max}	Maximum ion concentration
C_s	Surface concentration
D	Diffusion coefficient
R_p	Ion range
ΔR_p	Standard deviation of Gaussian distribution
R_s	Sheet resistance
t	Time
x	Distance
Φ	Ion dose
X_j	Junction depth
$<X>$	Average ion range
$<\Delta X>$	Standard deviation
$<\Delta Y>$	Standard deviation of lateral ion spread
θ	Angle
$\sigma = \Delta R_p$	Standard deviation
β	Kurtosis of ion profile
γ	Skewness of ion profile

Introduction

The technique of ion implantation has been accepted for some time for impurity doping control in the fabrication of silicon integrated circuits and discrete devices. The most commonly used impurity ions are antimony, arsenic, boron, and phosphorus. They are used to create buried collectors, bases, emitters, and resistors or to achieve surface charge control. Since device properties depend on intricate details of the process technology under consideration, an exhaustive presentation of design curves is nearly impossible. Hence, this section presents a few basic curves and pertinent data as a guide for rough estimations on the sheet resistance, junction depth, or threshold voltage control. Subsection 7.1 presents equations of ion distribution for implantation into amorphous materials. Often, the experimental results deviate from the simple expressions due to ion channelling effects, impurity segregation, and precipitation of impurity atoms at high concentration levels. Therefore, one has to be aware of such possibilities when using these equations. Section 6 of this manual presents a few curves on diffusion results from ion-implanted samples. These curves are not repeated in this section.

7.1 Basic Expressions

First Order Approximation - Gaussian Distribution[1]

Theoretical calculations based on the LSS theory[2] of ion distributions assume that the target material is amorphous and extends to infinity in both directions. The ion distribution can then be described by a Gaussian function (Equation 7.1). Figure 7-1 shows a linear plot of the Gaussian distribution. The ion distribution tail to the left of the origin is considered to be the portion of the ions that are reflected from the target surface. This is of importance at ion energies <200 keV for boron.[3]

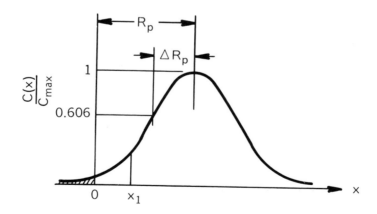

Figure 7-1 — Gaussian Distribution

Table 7-1 shows the position of the concentration at various decades of the peak concentration.[4]

TABLE 7-1

NUMERICAL VALUES FOR A NORMALIZED GAUSSIAN DISTRIBUTION

$C\left(\dfrac{x-R_p}{\Delta R_p}\right)/C_{max}$	0.6065	0.5	10^{-1}	10^{-2}	10^{-3}	10^{-4}
$\dfrac{x-R_p}{\Delta R_p}$	± 1	± 1.2	± 2	± 3	± 3.7	± 4.3
$C\left(\dfrac{x-R_p}{\Delta R_p}\right)/C_{max}$	10^{-5}	10^{-6}	10^{-7}			
$\dfrac{x-R_p}{\Delta R_p}$	± 4.8	± 5.3	± 5.7			

Equations describing the Gaussian distribution are:

$$C(x) = \frac{\phi}{\sqrt{2\pi}\ \Delta R_p} \exp\left[-\frac{(x-R_p)^2}{2\Delta R_p^2}\right],\qquad\qquad 7.1$$

$$C_{max} = C(x = R_p) = \frac{\phi}{\sqrt{2\pi}\ \Delta R_p} \cong 0.4\phi/\Delta R_p,\qquad\qquad 7.2$$

$$C(R_p \pm \Delta R_p) = 0.6065\ C_{max},\qquad\qquad 7.3$$

where

$C(x)$ = ion concentration at x (atoms/cm^3)

ϕ = ion dose (ions/cm^2)

R_p = projected ion range (cm)

ΔR_p = projected standard deviation, or projected straggle distance (cm).

$$\text{Total atoms from } x_1 \text{ to } \infty = \int_{x_1}^{\infty} C(x)\ dx = \frac{\phi}{2}\ \text{erfc}\left(\frac{x_1 - R_p}{\sqrt{2}\Delta R_p}\right).\qquad\qquad 7.4$$

One consequence of the approximation of the implanted-ion dose by a Gaussian distribution is that a portion of the distribution lies outside the surface at $x = 0$ (the shaded area of Figure 7-1) and represents a total dose of $\phi/2\ \text{erfc}\ (R_p/\sqrt{2}\ \Delta R_p)$. This term is less than 5 percent of the total when $\dfrac{R_p}{\sqrt{2}\Delta R_p} \geqslant 1.5$.

Range Statistics of Implantation into Amorphous Si, SiO$_2$, and Si$_3$N$_4$

The range statistics for amorphous material have been calculated using the LSS theory by Gibbons, Johnson, and Mylorie.[5] A portion of their tables for Si, SiO$_2$, and Si$_3$N$_4$ are reproduced in Tables 7-2, 7-3, and 7-4, respectively. Both the assumed Gaussian distribution and the calculated range statistics are for amorphous silicon. However, SIC devices are fabricated on single crystal silicon. One has to keep these conditions in mind when using Tables 7-2 through 7-4. In practice, the silicon wafer surface is tilted 7° to 10° from the ion beam direction to increase the near-surface atom density. This is called implantation into a random direction or a random equivalent direction. For SiO$_2$ and Si$_3$N$_4$ films, the surfaces are considered to be amorphous.

An amorphous surface can be produced on silicon by bombarding the wafer surface with silicon, neon, or argon ions prior to implanting the ions of interest. The minimum doses required to produce an amorphous silicon surface are shown in Table 7-5. Figure 7-2 presents the minimum dose required to produce an amorphous layer in silicon as a function of the wafer temperature during implantation at a low dose rate where the ion beam does not increase the wafer temperature, (i.e., the wafer temperature was controlled separately).[6]

Lateral Spread of Implanted Ions

The ion beam has a distribution in the plane perpendicular to the beam direction. When the ion beam strikes a target, ions are scattered both laterally and longitudinally (i.e., in the direction of the ion beam). The projected range, R$_p$, and the projected standard deviation describe the distribution of ions in the longitudinal direction. The lateral distribution can be obtained by assuming that the incident ion beam is directed at an angle, θ, from the normal of the target surface.[7] Then

$$C(x) = \frac{\phi}{\sqrt{2\pi}} \frac{\cos\theta}{<\Delta x>} \exp\left[-\frac{(x-<x>)^2}{2<\Delta x>^2}\right],$$
7.5

where

$$<\Delta x>^2 = \Delta R_p^2 \cos^2\theta + <\Delta Y>^2 \sin^2\theta,$$
7.6

and

$$<x> = R_p$$

R$_p$ = the projected range in the longitudinal direction

ΔR$_p$ = the standard deviation of longitudinal spread

$<\Delta Y>$ = the standard deviation of the lateral spread.

Figure 7-3 shows the coordinate system for the determination of lateral spread of incident ions.

TABLE 7-2

PROJECTED RANGE STATISTICS FOR ION IMPLANTATION INTO AMORPHOUS Si

ENERGY (keV)	ANTIMONY PROJECTED RANGE (μm)	ANTIMONY PROJECTED STANDARD DEVIATION (μm)	ARSENIC PROJECTED RANGE (μm)	ARSENIC PROJECTED STANDARD DEVIATION (μm)	BORON PROJECTED RANGE (μm)	BORON PROJECTED STANDARD DEVIATION (μm)	PHOSPHORUS PROJECTED RANGE (μm)	PHOSPHORUS PROJECTED STANDARD DEVIATION (μm)
10	0.0088	0.0026	0.0097	0.0036	0.0333	0.0171	0.0139	0.0069
20	0.0141	0.0043	0.0159	0.0059	0.0662	0.0283	0.0253	0.0119
30	0.0187	0.0058	0.0215	0.0080	0.0987	0.0371	0.0368	0.0166
40	0.0230	0.0071	0.0269	0.0099	0.1302	0.0443	0.0486	0.0212
50	0.0271	0.0084	0.0322	0.0118	0.1608	0.0504	0.0607	0.0256
60	0.0310	0.0096	0.0374	0.0136	0.1903	0.0556	0.0730	0.0298
70	0.0347	0.0107	0.0426	0.0154	0.2188	0.0601	0.0855	0.0340
80	0.0385	0.0118	0.0478	0.0172	0.2465	0.0641	0.0981	0.0380
90	0.0421	0.0130	0.0530	0.0189	0.2733	0.0677	0.1109	0.0418
100	0.0457	0.0140	0.0582	0.0207	0.2994	0.0710	0.1238	0.0456
110	0.0493	0.0151	0.0634	0.0224	0.3248	0.0739	0.1367	0.0492
120	0.0529	0.0162	0.0686	0.0241	0.3496	0.0766	0.1497	0.0528
130	0.0564	0.0172	0.0739	0.0258	0.3737	0.0790	0.1627	0.0562
140	0.0599	0.0183	0.0791	0.0275	0.3974	0.0813	0.1757	0.0595
150	0.0634	0.0193	0.0845	0.0292	0.4205	0.0834	0.1888	0.0628
160	0.0669	0.0203	0.0898	0.0308	0.4432	0.0854	0.2019	0.0659
170	0.0704	0.0213	0.0952	0.0325	0.4654	0.0872	0.2149	0.0689
180	0.0739	0.0224	0.1005	0.0341	0.4872	0.0890	0.2279	0.0719
190	0.0773	0.0234	0.1060	0.0358	0.5086	0.0906	0.2409	0.0747
200	0.0808	0.0244	0.1114	0.0374	0.5297	0.0921	0.2539	0.0775
220	0.0878	0.0264	0.1223	0.0407	0.5708	0.0950	0.2798	0.0829
240	0.0947	0.0283	0.1334	0.0439	0.6108	0.0975	0.3054	0.0880
260	0.1017	0.0303	0.1445	0.0470	0.6496	0.0999	0.3309	0.0928
280	0.1086	0.0322	0.1558	0.0502	0.6875	0.1020	0.3562	0.0974
300	0.1156	0.0342	0.1671	0.0533	0.7245	0.1040	0.3812	0.1017

TABLE 7-3

PROJECTED RANGE STATISTICS FOR ION IMPLANTATION INTO SiO_2

ENERGY (keV)	ANTIMONY PROJECTED RANGE (μm)	ANTIMONY PROJECTED STANDARD DEVIATION (μm)	ARSENIC PROJECTED RANGE (μm)	ARSENIC PROJECTED STANDARD DEVIATION (μm)	BORON PROJECTED RANGE (μm)	BORON PROJECTED STANDARD DEVIATION (μm)	PHOSPHORUS PROJECTED RANGE (μm)	PHOSPHORUS PROJECTED STANDARD DEVIATION (μm)
10	0.0071	0.0020	0.0077	0.0026	0.0298	0.0143	0.0108	0.0048
20	0.0115	0.0032	0.0127	0.0043	0.0622	0.0252	0.0199	0.0084
30	0.0153	0.0042	0.0173	0.0057	0.0954	0.0342	0.0292	0.0119
40	0.0188	0.0052	0.0217	0.0072	0.1283	0.0418	0.0388	0.0152
50	0.0222	0.0061	0.0260	0.0085	0.1606	0.0483	0.0486	0.0185
60	0.0254	0.0070	0.0303	0.0099	0.1921	0.0540	0.0586	0.0216
70	0.0286	0.0078	0.0346	0.0112	0.2228	0.0590	0.0688	0.0247
80	0.0316	0.0086	0.0388	0.0125	0.2528	0.0634	0.0792	0.0276
90	0.0347	0.0094	0.0431	0.0138	0.2819	0.0674	0.0896	0.0305
100	0.0377	0.0102	0.0473	0.0151	0.3104	0.0710	0.1002	0.0333
110	0.0406	0.0110	0.0516	0.0164	0.3382	0.0743	0.1108	0.0360
120	0.0436	0.0118	0.0559	0.0176	0.3653	0.0774	0.1215	0.0387
130	0.0465	0.0126	0.0603	0.0189	0.3919	0.0801	0.1322	0.0412
140	0.0494	0.0133	0.0646	0.0201	0.4179	0.0827	0.1429	0.0437
150	0.0523	0.0141	0.0690	0.0214	0.4434	0.0851	0.1537	0.0461
160	0.0552	0.0149	0.0734	0.0226	0.4685	0.0874	0.1644	0.0485
170	0.0581	0.0156	0.0778	0.0239	0.4930	0.0895	0.1752	0.0507
180	0.0610	0.0164	0.0823	0.0251	0.5172	0.0914	0.1859	0.0529
190	0.0639	0.0171	0.0868	0.0263	0.5409	0.0933	0.1966	0.0551
200	0.0668	0.0178	0.0913	0.0275	0.5643	0.0951	0.2073	0.0571
220	0.0726	0.0193	0.1003	0.0299	0.6100	0.0983	0.2286	0.0611
240	0.0784	0.0208	0.1095	0.0323	0.6544	0.1013	0.2498	0.0649
260	0.0842	0.0222	0.1187	0.0347	0.6977	0.1040	0.2709	0.0685
280	0.0900	0.0237	0.1280	0.0370	0.7399	0.1065	0.2918	0.0719
300	0.0958	0.0251	0.1374	0.0394	0.7812	0.1087	0.3125	0.0751

TABLE 7-4

PROJECTED RANGE STATISTICS FOR ION IMPLANTATION INTO Si_3N_4

ENERGY (keV)	ANTIMONY		ARSENIC		BORON		PHOSPHORUS	
	PROJECTED RANGE (μm)	PROJECTED STANDARD DEVIATION (μm)	PROJECTED RANGE (μm)	PROJECTED STANDARD DEVIATION (μm)	PROJECTED RANGE (μm)	PROJECTED STANDARD DEVIATION (μm)	PROJECTED RANGE (μm)	PROJECTED STANDARD DEVIATION (μm)
10	0.0056	0.0015	0.0060	0.0020	0.0230	0.0111	0.0084	0.0037
20	0.0090	0.0024	0.0099	0.0033	0.0480	0.0196	0.0154	0.0065
30	0.0119	0.0033	0.0135	0.0045	0.0736	0.0267	0.0226	0.0092
40	0.0147	0.0040	0.0169	0.0056	0.0990	0.0326	0.0300	0.0118
50	0.0173	0.0047	0.0202	0.0066	0.1239	0.0377	0.0376	0.0143
60	0.0197	0.0054	0.0235	0.0077	0.1482	0.0422	0.0453	0.0168
70	0.0222	0.0061	0.0268	0.0087	0.1719	0.0461	0.0532	0.0192
80	0.0246	0.0067	0.0301	0.0097	0.1950	0.0496	0.0612	0.0215
90	0.0269	0.0074	0.0334	0.0108	0.2176	0.0527	0.0693	0.0237
100	0.0292	0.0080	0.0367	0.0118	0.2396	0.0555	0.0774	0.0259
110	0.0315	0.0086	0.0400	0.0127	0.2610	0.0581	0.0856	0.0280
120	0.0338	0.0092	0.0433	0.0137	0.2820	0.0605	0.0939	0.0301
130	0.0360	0.0098	0.0467	0.0147	0.3025	0.0627	0.1022	0.0321
140	0.0383	0.0104	0.0500	0.0157	0.3226	0.0647	0.1105	0.0340
150	0.0405	0.0110	0.0534	0.0167	0.3424	0.0666	0.1188	0.0358
160	0.0428	0.0116	0.0568	0.0176	0.3617	0.0684	0.1271	0.0377
170	0.0450	0.0122	0.0603	0.0186	0.3807	0.0700	0.1354	0.0394
180	0.0472	0.0128	0.0637	0.0195	0.3994	0.0716	0.1437	0.0411
190	0.0495	0.0134	0.0672	0.0205	0.4178	0.0731	0.1520	0.0428
200	0.0517	0.0139	0.0706	0.0214	0.4358	0.0744	0.1602	0.0444
220	0.0562	0.0151	0.0776	0.0233	0.4712	0.0770	0.1767	0.0475
240	0.0606	0.0162	0.0847	0.0252	0.5056	0.0793	0.1931	0.0505
260	0.0651	0.0174	0.0918	0.0270	0.5390	0.0815	0.2094	0.0533
280	0.0696	0.0185	0.0990	0.0289	0.5717	0.0834	0.2255	0.0559
300	0.0741	0.0196	0.1063	0.0307	0.6037	0.0852	0.2415	0.0584

TABLE 7-5

ION DOSE REQUIRED FOR AMORPHORIZATION
OF Si SINGLE CRYSTAL

ELEMENT	THEORETICAL VALUES		EXPERIMENTAL VALUES (IONS/cm^2)
	LOW TEMP (IONS/cm^2)	300 K (IONS/cm^2)	
Al	—	—	$>10^{14}$
Ar	—	—	$(3\text{-}5) \times 10^{14}$ @ 30KeV
As	10^{14}	2×10^{14}	2×10^{14}
B	9×10^{14}	4×10^{17}	$>2 \times 10^{16}$
Bi	4×10^{13}	6×10^{13}	5×10^{13}
Ne	—	—	10^{15}
P	2×10^{14}	8×10^{14}	5×10^{14}
Sb	6×10^{13}	10^{14}	10^{14}
Si	—	—	10^{16}

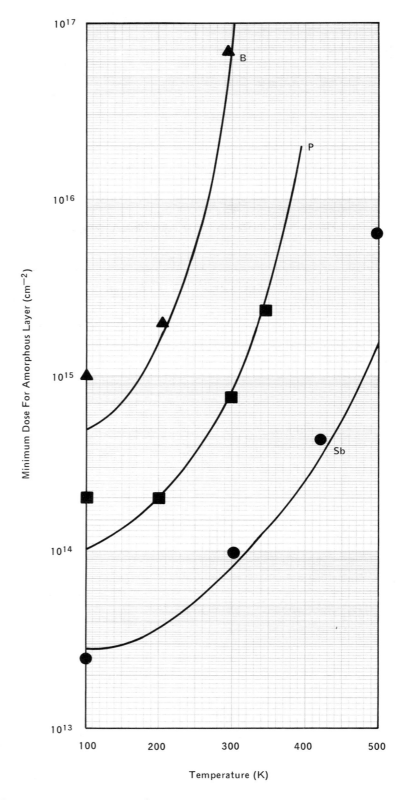

Figure 7-2 — Minimum Dose Required to Produce an Amorphous Layer in Si at a Low Dose Rate (0.2μA cm^{-2})[6] (Reprinted with permission from Gordon and Breach Science Publishers, Ltd.)

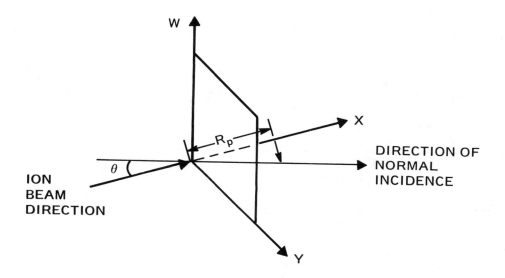

Figure 7-3 — Ion Beam Incident at an Angle θ From the Target Surface Normal

Furukawa et al[8] presented a theoretical study on the lateral spread of implanted ions. Their values of R_p, $\triangle R_p$, and $\triangle Y$ are given in Table 7-6.

TABLE 7-6

CALCULATED RESULTS OF R_p, $\triangle R_p$, AND $\triangle Y$ IN (Å) FOR VARIOUS IONS IMPLANTED INTO Si

ENERGY (keV)		20	40	60	80	100	120	140	160	180	200
Al	R_p	288	563	846	1137	1430	1724	2023	2323	2617	2907
	$\triangle R_p$	129	234	329	422	504	578	650	720	783	840
	$\triangle Y$	112	208	300	385	470	552	630	702	773	843
As	R_p	150	262	368	473	577	682	788	894	1001	1109
	$\triangle R_p$	56	96	133	169	204	237	269	302	334	368
	$\triangle Y$	41	69	96	121	146	171	196	220	243	266
B*	R_p	658	1277	1847	2380	2887	3362	3812	4242	4654	5050
	$\triangle R_p$	270	423	526	605	669	721	764	801	833	862
	$\triangle Y$	290	483	638	761	857	942	1018	1083	1140	1191
Ga	R_p	154	271	384	494	606	718	831	944	1059	1175
	$\triangle R_p$	59	102	142	180	217	252	287	323	359	395
	$\triangle Y$	43	74	103	130	158	185	212	238	263	288
In	R_p	131	222	303	381	456	530	604	677	750	822
	$\triangle R_p$	41	70	95	119	142	164	186	207	229	250
	$\triangle Y$	31	50	68	84	100	115	130	145	159	173
N	R_p	535	1089	1653	2199	2728	3246	3759	4259	4740	5207
	$\triangle R_p$	246	430	579	702	802	891	972	1046	1112	1171
	$\triangle Y$	251	457	637	799	948	1079	1192	1293	1389	1480
P	R_p	253	488	729	974	1226	1479	1732	1988	2247	2506
	$\triangle R_p$	114	201	288	367	445	516	581	642	702	762
	$\triangle Y$	94	175	249	323	393	462	531	596	659	718
Sb	R_p	130	220	299	375	448	520	592	663	733	803
	$\triangle R_p$	39	68	92	115	137	158	179	200	220	240
	$\triangle Y$	30	49	66	82	97	111	125	139	153	166

*The electronic stopping constant, $S_e = 2.06 \times 10^{13} E^{1/2}$, as measured by Eisen was used in this calculation. (Reprinted with permission from the Japanese Journal of Applied Physics)

The expression for the lateral and normal distributions of ions is given in Equation 7.7.[9] The coordinates are shown in Figure 7-4.

$$C(x,y,z) = \frac{\phi}{(2\pi)^{3/2}\,\Delta R_p \Delta Y \Delta Z}\,\exp\left[-\frac{(x-R_p)^2}{2\Delta R_p^2}\right]$$

$$\times \int_{-d}^{d}\int_{-\infty}^{\infty}\exp\left[-\frac{(Y-z)^2}{2\Delta Y^2}-\frac{(Z-y)^2}{2\Delta Z^2}\right]dz\,dy$$

$$= \frac{\phi}{2(2\pi)^{1/2}\,\Delta R_p}\,\exp\left[-\frac{(x-R_p)^2}{2\Delta R_p^2}\right]\mathrm{erfc}\left(\frac{Y-d}{\sqrt{2}\Delta Y}\right),\qquad\qquad 7.7$$

where

$$Y \geqslant 0$$

$$-\infty > z \geqslant \infty$$

Sample surface plane is at $x = 0$ and the window width $= 2d$ in the y-direction.

ϕ = ion dose $(cm)^{-2}$

R_p = projected range in the x-direction (the direction of the ion beam) (cm)

ΔR_p = projected standard deviation in the x-direction (cm)

Y = projected range in the y-direction beyond the window edge (cm)

ΔY = projected standard deviation in the y-direction beyond the window edge. (cm)

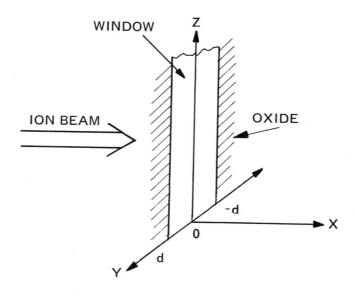

Figure 7-4 — Coordinate System for Implantation Through a Strip Window

Experimental measured values are given in Table 7-7.[7] [9]

TABLE 7-7

MEASURED VALUES OF ΔY, ΔR_p, AND R_p FOR P AND B

ELEMENTS	ENERGY (keV)	ΔY (Å)	ΔR_p (Å)	R_p (Å)
^{31}P	145	570	675	1600
^{31}P	260	825	825	2950
^{11}B	150	1100	775	4500
^{11}B [9]	150	1613	1227	4010

For self-aligned MOS devices, the lateral spread of source and drain implantation will affect the threshold voltage for short channel devices. Tsuchiya[10] utilized the threshold voltage measurement to determine the lateral spread of the boron ion under a mask edge. He gave an expression for the two-dimensional ion distribution that included an ion beam tilt angle and an angle between the mask edge and the wafer surface. This is shown in Figure 7-5. The coordinate system is similar to that shown in Figure 7-4. Figure 7-6 shows the cross-sectional view of a self-aligned structure and the calculated two-dimensional ion distribution. The calculation assumes a 40-keV boron ion beam implanted at tilt angles of 0° and 8° through an SiO_2 layer of 300Å thickness and a gate mask of 0.5μm thickness (either polysilicon or SiO_2). It is further assumed that the masking material has the same ion range statistics as silicon.

The equations for the profile calculations are

$$C(x, y) = C_{max} \frac{1}{(2\pi)^{1/2}\, \Delta Y} \int_{-\infty}^{\infty} \exp\left\{ -\frac{(y-\xi)^2}{2\Delta Y^2} - \frac{[x + d_m(\xi) - R_p]^2}{2\Delta R_p^2} \right\} d\xi, \quad 7.8$$

where

$$d_m(y) = y \tan\theta_t + d \sec\theta_t \qquad \text{for } y < \xi_1$$

$$d_m(y) = -y \tan(\theta_e - \theta_t) \qquad \text{for } \xi_1 < y < 0$$

$$d_m(y) = y \tan\theta_t \qquad \text{for } y > 0,$$

and where

C_{max} = the maximum concentration of the distribution (atoms/cm^3)

R_p = the mean projected range (cm)

ΔR_p = the standard deviation in the direction perpendicular to the wafer surface (cm)

ΔY = the standard deviation in the direction parallel to the wafer surface (cm)

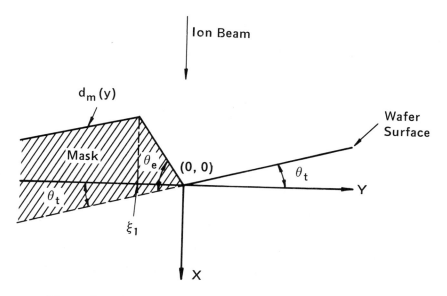

Figure 7-5 — The Coordinate System[10]
(Reprinted with permission from the American Institute of Physics)

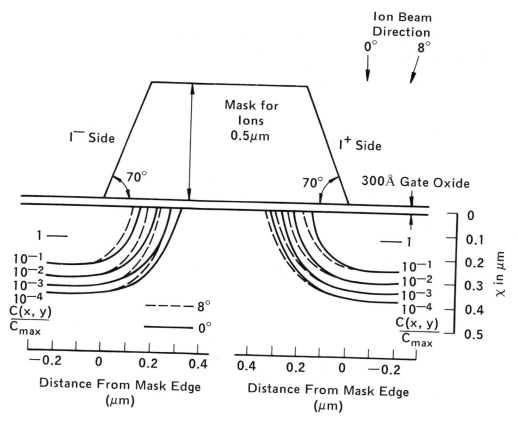

Figure 7-6 — Comparison of Calculated Lateral Spread of Boron Ions Underneath a Mask From Implantation Through a 300Å Gate Oxide at 40 keV. The Tilt Angles of the Wafer With Respect to Ion Beam are 0° and 8°. Assuming R_p = 1566Å, $\triangle R_p$ = 580Å, and $\triangle X$ = 653Å.[10] (Reprinted with permission from the American Institute of Physics.)

X = the coordinate parallel to the ion beam direction

Y = the coordinate perpendicular to the x-coordinate; when $\theta_t = 0°$, y is parallel to the wafer surface

ξ_1 = the distance between the top and the bottom edge of the mask along the wafer surface (cm)

d_m = the mask thickness (cm)

θ_t = the tilt angle—the angle between the incident ion beam and the normal of the wafer surface

θ_e = the angle between the wafer surface and the mask edge.

The range statistics for boron used in the calculations of Figure 7-6 are different from those shown in Table 7-6. This is due to the fact that the electronic stopping constant as measured by Eisen was used in Table 7-6.

Figure 7-7 shows the calculated lateral spread at the wafer surface in the I^+ side. Figure 7-8 shows the calculated distance between the mask edge and the position where the surface concentration under the mask is one-tenth of the peak concentration.

Asymmetrical Distribution

The symmetrical Gaussian distribution of implanted ions in an amorphous target is a first-order approximation. In practice, when the measured samples have been annealed at low temperatures, they often show asymmetrical distributions. Samples implanted with various ions often show asymmetrical profiles without low-temperature annealing. (For most profile measurements done either by an electrical method or by neutron-activation analysis that involves layer removal by anodic oxidation of the sample and etching, a low-temperature annealing is required.) Gibbons and Mylroie[11] presented simple expressions for non-symmetrical distributions by including a third-moment term. They used this technique to match an experimental profile from a silicon sample that was implanted with antimony and determined a third moment to describe this profile.

Hofker[3] determined the values of the first four moments from experimental profiles of boron that was implanted into amorphous and polycrystalline silicon. His results are summarized in Table 7-8. The definitions of these four moments are given as follows:

$$\mu \text{ (average range } \langle x \rangle) = R_p = \int_{-\infty}^{\infty} x\, f(x)\, dx. \tag{7.9}$$

$$\sigma \text{ (standard deviation)} = \Delta R_p = \int_{-\infty}^{\infty} (x - \langle x \rangle)^2\, f(x)\, dx. \tag{7.10}$$

$$\gamma \text{ (skewness, dimensionless)} = \frac{1}{\sigma^3} \left[\int_{-\infty}^{\infty} (x - \langle x \rangle)^3\, f(x)\, dx \right]. \tag{7.11}$$

$$\beta \text{ (kurtosis, dimensionless)} = \frac{1}{\sigma^4} \left[\int_{-\infty}^{\infty} (x - \langle x \rangle)^4\, f(x)\, dx \right]. \tag{7.12}$$

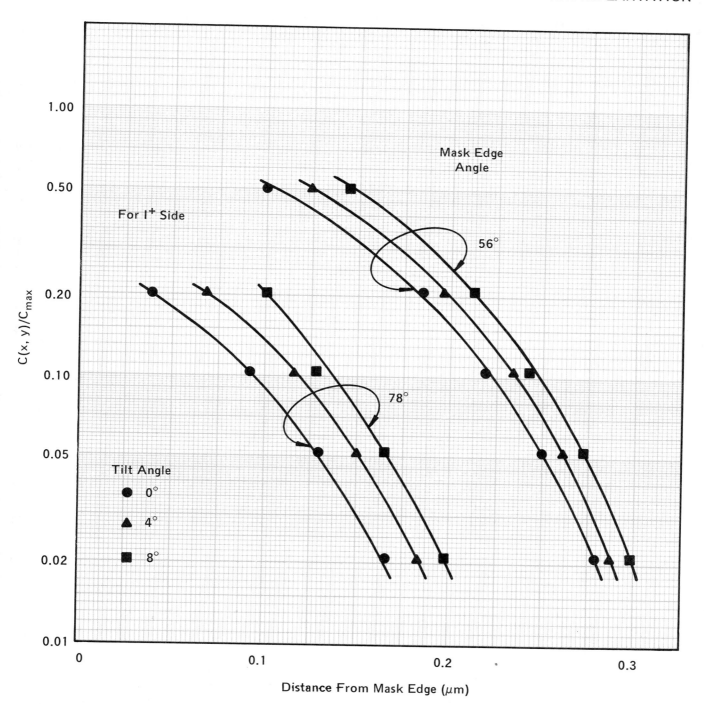

Figure 7-7 —Calculated Lateral Spread Distributions at Wafer Surface in the I⁺ Side For Mask Edge Angles of 56° and 78°, Where Tilt Angles Are 0°, 4°, and 8° [10]
(Reprinted with permission from the American Institute of Physics.)

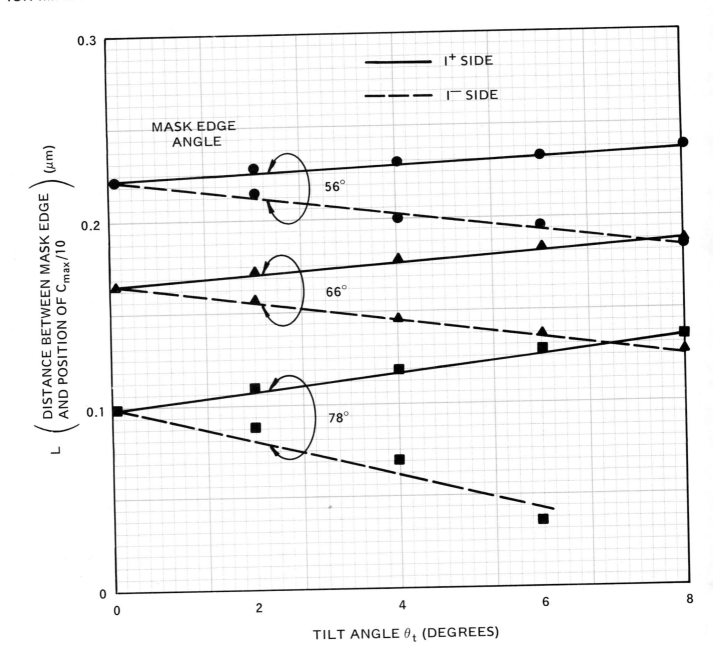

Figure 7-8 —Calculated Tilt Angle Dependences of L', For the I$^+$ and I$^-$ Sides. L' is the Distance Between the Mask Edge and the Position of C$_{max}$/10 at Substrate Surface, ●, 56° Mask Edge Angle; ▲, 66°; ■, 78°.[10] (Reprinted with permission from the American Institute of Physics.)

TABLE 7-8

THE AVERAGE RANGE $<x>$, STANDARD DEVIATION σ, SKEWNESS γ, AND KURTOSIS β OF DISTRIBUTIONS OF BORON IMPLANTED AT DIFFERENT ENERGIES IN AMORPHOUS AND POLYCRYSTALLINE SILICON

TARGET MATERIAL	ENERGY (keV)	$<x>$ (Å)	σ (Å)	γ	β
amorphous Si	30	1015	425	—0.55	3.6
amorphous Si	50	1645	570	—1.0	5.5
amorphous Si	70	2330	680	—1.0	5.3
amorphous Si	100	3150	790	—1.3	7.0
amorphous Si	150	4420	960	—2.05	15
amorphous Si	200	5340	1060	—2.4	22
polycrystalline Si	70	2260	700	—1.0	5.5
polycrystalline Si	100	3100	845	—1.3	7.5
polycrystalline Si	200	5560	1080	—2.15	17.5
polycrystalline Si	300	7355	1160	—2.25	19
polycrystalline Si	400	9020	1270	—2.75	32
polycrystalline Si	600	12375	1295	—2.6	28
polycrystalline Si	800	15150	1465	—3.3	60

Furukawa and Ishiwara[12] applied the Edgeworth distributions, with some success, to determine these moments from the experimental profiles. However, the Edgeworth distribution often gave negative values in the profile calculation. Hofker applied a Pearson distribution to calculate these constants successfully, as given in Table 7-8.

Recently, Ryssel et al[13] used the same Pearson distribution to calculate the range parameters of boron. Their experimental data were obtained from the $^{10}B(n, a)^7Li$ nuclear reaction. They used <111>-and <100>-oriented, float-zone, single-crystal silicon samples that were tilted by 7° and rotated so that the <100> flat coincided with the tilting axis to minimize ion channeling during implantation. Their values of R_p and $\triangle R_p$, γ, and β are given in Figures 7-9, 7-10, and 7-11, respectively.

For ion implant energy at 30 keV, the boron profile can be approximated with a Gaussian function, using the calculated values of R_p and $\triangle R_p$ by Gibbons, Johnson, and Mylroie.[5] Detailed discussions of the application of the Pearson distribution for boron-implanted profiles are given in references 3 and 13.

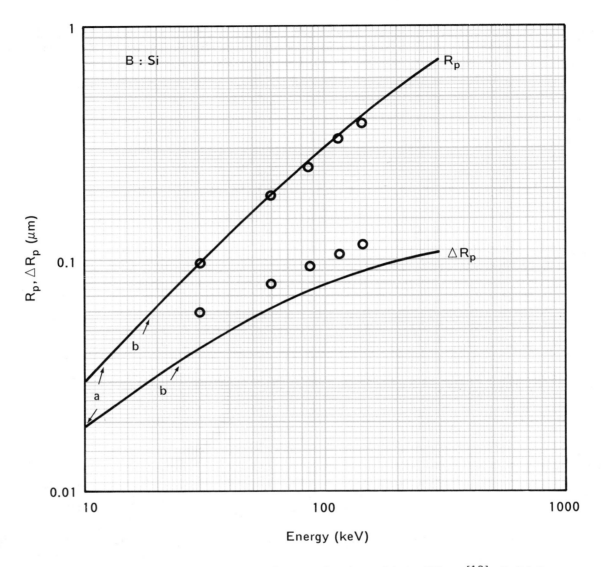

Figure 7-9 — Range and Range Straggling of Boron Implanted Into Silicon[13]   Solid Curves
Calculated Values a:^{10}B ; b:^{11}B.   (Reprinted with permission from Springer-Verlag Inc.)

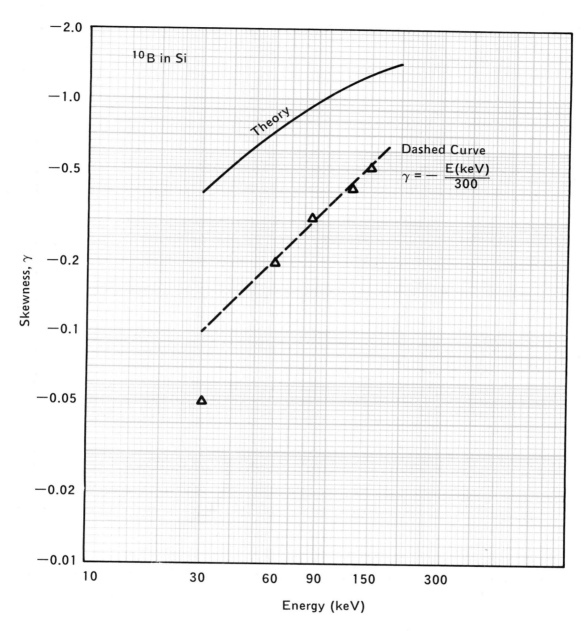

Figure 7-10 — Skewness of Boron Implanted Into Silicon[13] (Reprinted with permission from Springer-Verlag Inc.)

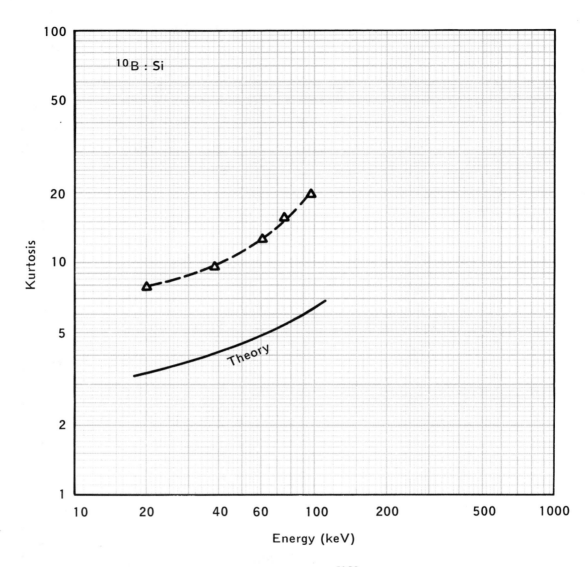

Figure 7-11 —Kurtosis of Boron Implanted Into Silicon[13]   (Reprinted with permission from Springer-Verlag Inc.)

Diffusion from an Ion-Implanted Layer

If the ion-implanted profile can be described by a Gaussian function, as in Equation 7.1, and if the diffusion is in a nonoxidizing ambient and there is no loss of impurity atoms during diffusion, the diffusion profile can be approximated by the following equation:

$$C(x, t) = \frac{\phi}{\sqrt{\pi} \ (2\Delta R_p^2 + 4 \ Dt)^{1/2}} \ \exp \left[- \frac{(x - R_p)^2}{2\Delta R_p^2 + 4 \ Dt} \right], \qquad 7.13$$

where

$$\phi = \text{ion dose (cm}^{-2})$$

$$R_p = \text{projected ion range (cm)}$$

$$\Delta R_p = \text{the standard deviation (cm)}$$

$$D = \text{the diffusion constant (cm}^2/s)$$

$$t = \text{the diffusion time (s)}$$

7.2 Table of Impurity Energy Levels in Silicon

The impurity energy level for ion-implanted annealed silicon shows slight difference from the energy level resulting from doping in silicon melt. The results of Zorin et al[14] are given in Table 7-9.

TABLE 7-9

ENERGY LEVELS OF IMPURITIES INTRODUCED INTO Si BY ION IMPLANTATION[14]

IMPURITY	r_i/r_{Si}	T_{an1} °C	ΔE_1, eV (IMPLANTATION)	ΔE_2, eV (DOPING FROM MELT)
Substitutional impurities				
B in p-Si	0.75	500	0.08	0.045
B in n-Si	—	500	0.055	—
B in n-Si	—	450	0.065	—
In	1.28	900	0.16	0.16
P	0.94	500	0.05	0.044
As	1.01	500	0.05	0.049
Sb	1.16	700-800	0.039	0.039
Interstitial impurities				
Li	0.53	410	0.029	0.033
Na	0.85	670	0.032	0.028-0.032
Na	—	720	0.032	—
K	1.14	620	0.073	—
K	—	670	0.10	—
K	—	720	0.04	—

Note: Here, r_i and r_{Si} are the covalent radii of the impurity and silicon atoms, respectively. In the case of Li, Na, and K, the ratio is r_i/r_{int}, where r_i is the ionic radius of an impurity and r_{int} is the radius of an interstice.

7.3   Critical Angles for Random Implantation into Single-Crystal Silicon

Silicon integrated circuits (SIC) and discrete devices are fabricated on single-crystal wafers with <100> or <111> orientation. When an incident ion beam is perpendicular to the wafer surface, the doping profile deviates from the Gaussian distribution. This is partly due to the fact that a portion of the ions penetrate deeply due to a channeling mechanism. The effect of higher moments in the ion range statistics also contributes to this deviation. In order to obtain a reproducible and a Gaussian-like function distribution, the silicon wafer is tilted with respect to the ion beam direction and is also rotated around an axis perpendicular to the wafer surface. This is often referred to as implantation into random directions. In the application of ion implantation for ion pre-deposition and diffusion, where the junction depths are deep, a simple tilting of the wafer surface 7 to 10 degrees between the ion direction and the normal of the wafer surface is sufficient.   For devices which require critical low dose implantation and depth control, the channeled ions will cause significant variations in the electrical properties of these devices. One method to reduce channeling is to rotate the wafer about the normal axis.[52]   The magnitude of rotation depends on the wafer primary flat orientation, the ion scanning characteristics, and the wafer sizes. Another method is to implant through a thin insulation film.

The calculated critical angles for channeling are given in Table 7-10. When the tilting of the wafer is more than 6°, the ion beam will not align with a major channel direction.

TABLE 7-10

CRITICAL ANGLES FOR CHANNELING OF SELECTED IONS IN SILICON[15]

ION	ENERGY (KeV)	CHANNEL DIRECTION		
		<110>	<111>	<100>
Arsenic	30	5.9°	5.0°	4.5°
	50	5.2°	4.4°	4.0°
Boron	30	4.2°	3.5°	3.3°
	50	3.7°	3.2°	2.9°
Nitrogen	30	4.5°	3.8°	3.5°
	50	4.0°	3.4°	3.0°
Phosphorus	30	5.2°	4.3°	4.0°
	50	4.5°	3.8°	3.5°

(© 1968 IEEE)

7.4   Calculated Sheet Resistance Versus Ion Dose Curves for Implanted Silicon Wafers Without Annealing

Figures 7-12 through 7-15 show the calculated sheet resistance as a function of ion dose for antimony, arsenic, boron, and phosphorus in amorphous silicon. The calculations are based on the assumptions that the implant profile is a Gaussian function with the range statistics as calculated by the LSS theory and that the implant damage has no effect on the electrical activity of the implanted ions, i.e., 100 percent electrical activity.[16]   These curves can be used as a starting point for estimating the ion dose for a required sheet resistance. In general, after high temperature annealing and/or diffusion, the sheet resistances are lower than the calculated values shown in Figures 7-12 through 7-15. However, when the annealing temperature is not high enough, only a portion of the implanted ions is electrically active, and the measured sheet resistance could be higher than the calculated values. Since antimony has very shallow ranges in silicon, the electrical activity is difficult to determine for annealing at temperatures below 900°C and no experimental datum is shown in Figure 7-12. Examples of diffusion results are given in Section 6.4.

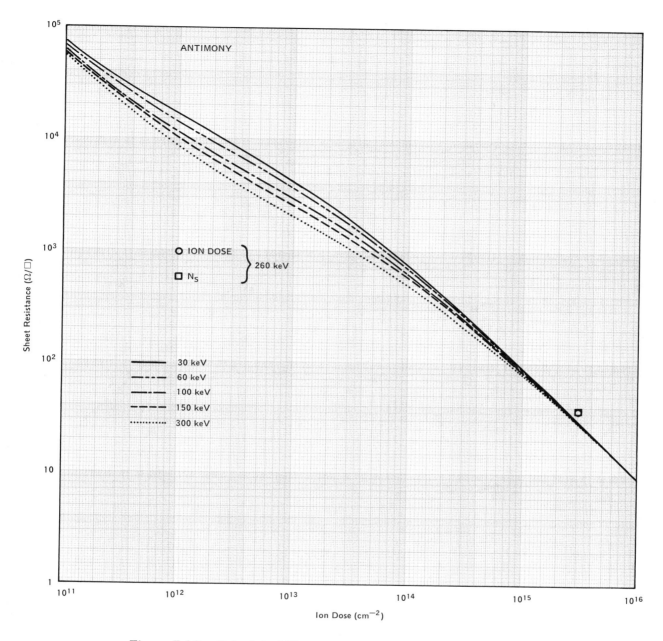

Figure 7-12—Calculated Sheet Resistance Versus Ion Dose for Sb

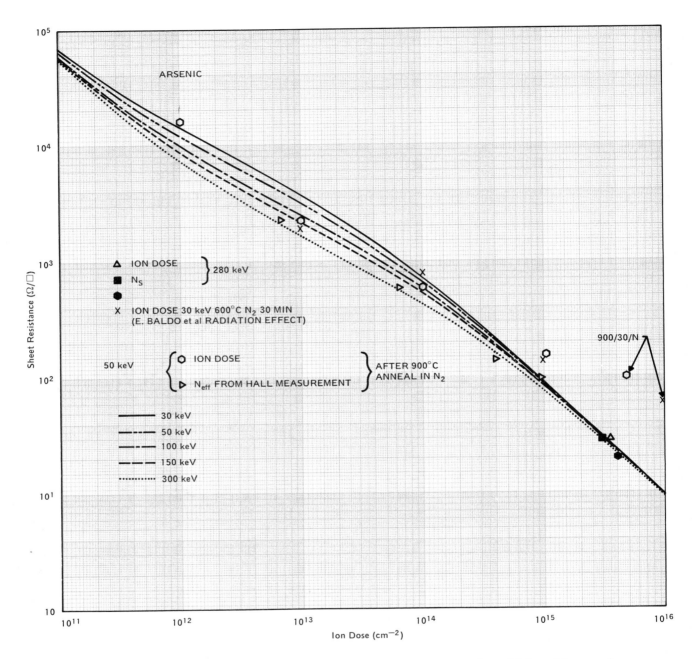

Figure 7-13 — Calculated Sheet Resistance Versus Ion Dose for As — Values Measured After 900°C Annealing[17] [18]

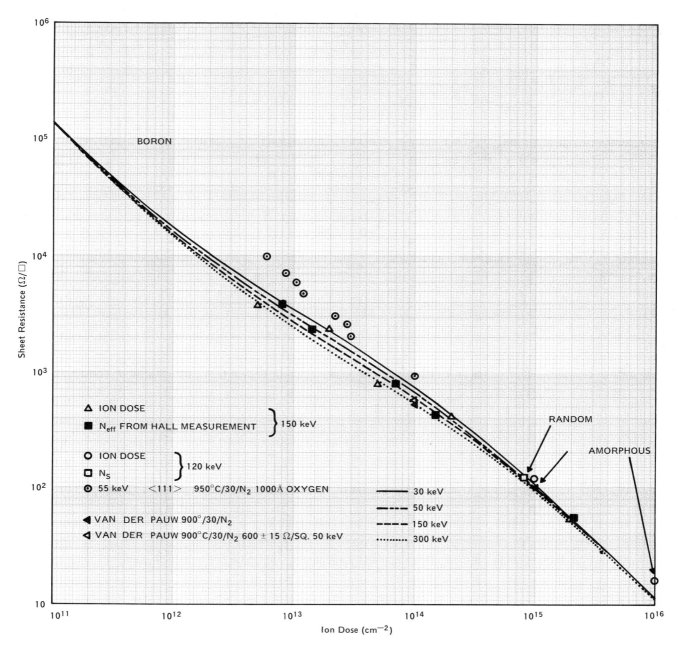

Figure 7-14 — Calculated Sheet Resistance Versus Ion Dose for B — Values Measured
After 900°C Annealing[19]

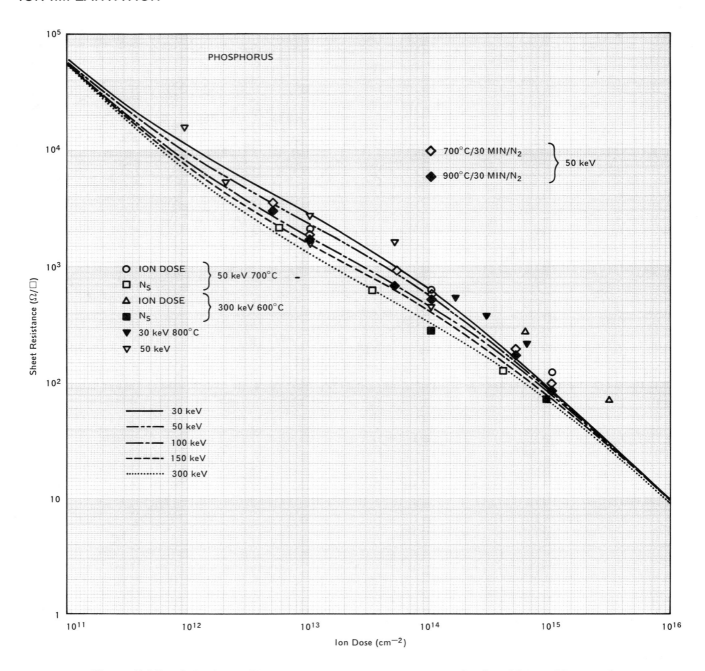

Figure 7-15 —Calculated Sheet Resistance Versus Ion Dose for P — Values Measured
After Annealing[16]

7.5   Measured Sheet Resistance Versus Ion Dose Curves

The application of ion implantation to silicon integrated circuits often requires precise sheet resistance control after implantation and diffusion. This has been referred to as "ion predeposition" and controls the total dopant atoms introduced into the silicon crystal. The control of sheet resistance is improved from the chemical diffusion source to a sigma in the range of 1 to 2 percent (i.e., 3 sigma ±6 to 7 percent) for a large number of implantation runs over a long period of time for ion doses 100 times greater than the ion current measuring noise. However, the implanted resistor values depend not only on the total dopant atoms but also on the oxide window width control, thermal treatment (diffusion and oxidation), and possible compensating contaminants that are inadvertently introduced into the wafer during heat treatments. This subsection presents results for sheet resistance after implantation and diffusion under commonly encountered conditions. Additional data and curves are contained in Section 6 of this manual and will not be repeated here. The electrical results of implanted-diffused wafers are sensitive to the specific device technology and facility location. Hence, one has to be aware of these factors in applying the curves or tables given in this section.

Some of the curves are labeled as "annealing" results. This term is used rather loosely. When the annealing temperature is greater than $600°C$ (in general), diffusion of the implanted atoms will occur. The extent of the diffusion phenomenon depends on the temperature, time, and ambient. Hence, diffusion and annealing of the implanted atoms often occur simultaneously.

Since laser- and electron-beam annealing are still being actively investigated at many laboratories, and no application to SIC fabrication has been established, we have not included results from laser- or electron-beam annealing in this issue of QRM. When sufficient data for device applications are available, they will be added in future revisions. Rapid thermal annealing by halogen lamps or arc lamps produced results comparable to those shown in Figures 7-13 through 7-15 for As, B, and P, respectively.

Annealing and Diffusion Results for Antimony in Silicon[20]

Antimony is mainly used as a buried layer dopant in thin film epitaxy (1 to 3 μm) devices where auto doping from the buried layer dopant is of concern. Auto doping is a lesser problem for antimony than arsenic or phosphorus. Figure 7-16 shows the sheet resistance, junction depth, and surface concentration versus ion dose of antimony implanted/diffused into (111)-oriented silicon wafers.

The surface concentrations were measured by the differential sheet resistance (DSR) method, the infrared plasma resonance method (IR) or the fast IR method (PRET IR). The DSR method gives more accurate values of surface concentration but it is destructive and time consuming. The IR plasma minimum method measures the average concentration of antimony near the surface and it is nondestructive.[21]

Figures 7-17 and 7-18 show similar data for diffusion in oxygen at $1250°C$ for 3.5 hours and 7 hours, respectively. Figure 7-19 shows C_s, R_s, and X_j for diffusion in oxygen at $1200°C$ for 4 hours. Figure 7-20 shows a curve of the surface concentration of antimony versus $(R_s X_j)^{-1}$. Figures 7-21 and 7-22 show curves of the sheet resistance versus ion dose for different amounts of silicon removal by HCl etching prior to epitaxial film growth. Figure 7-23 shows the amount of masking oxide film required to prevent antimony atoms from diffusing through the masking oxide at $1250°C$ for various diffusion times. Figure 7-24 shows examples of antimony and arsenic implant/diffusion profiles after diffusion in oxygen at $1250°C$.

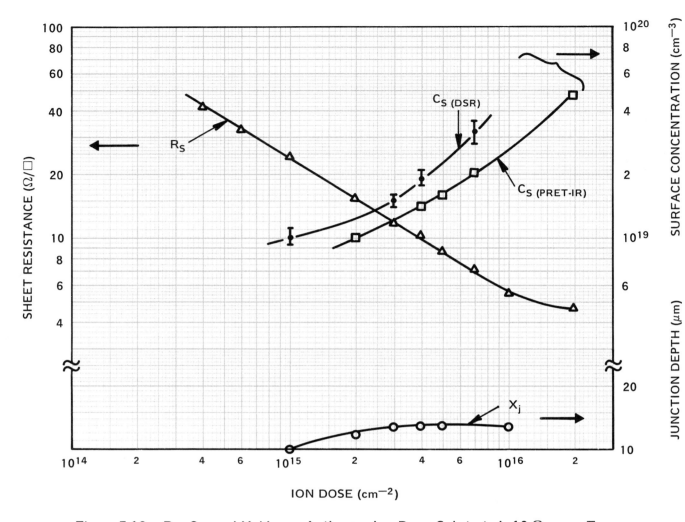

Figure 7-16 — R_S, C_S, and X_j Versus Antimony Ion Dose; Substrate is 10 $\Omega \bullet$ cm p-Type (111) Silicon, Implanted at 150 keV, and Diffused at 1250°C for 2 Hours in O_2 and 8 Hours in N_2

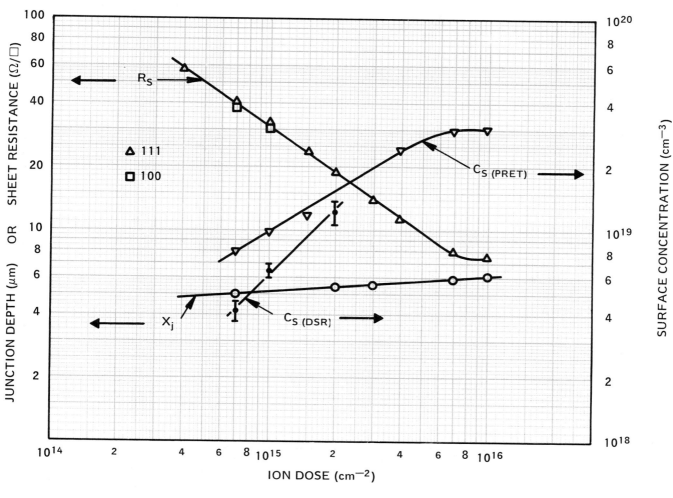

Figure 7-17 — R_S, C_S, and X_j Versus Antimony Ion Dose; Substrate is 10 $\Omega \bullet$cm p-Type
(111) Silicon, Implanted at 150 keV, and Diffused at 1250°C for 3.5 Hours
in O_2

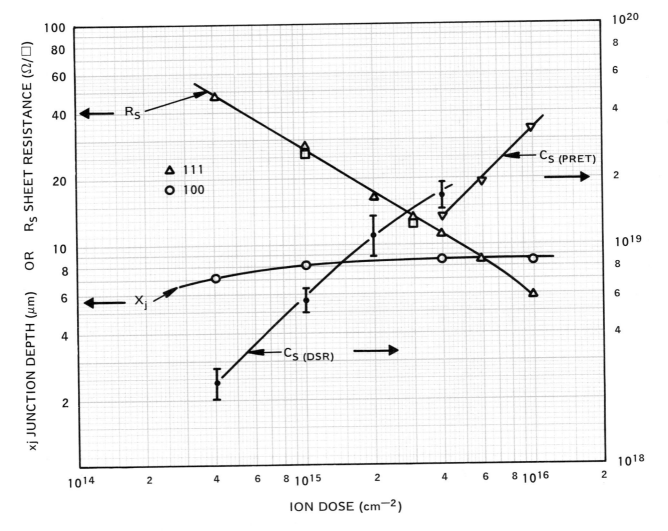

Figure 7-18 — R_S, C_S, and X_j Versus Antimony Ion Dose; Substrate is 10 $\Omega \bullet$cm p-Type (111) Silicon, Implanted at 150 keV, and Diffused at 1250°C for 7 Hours in O_2

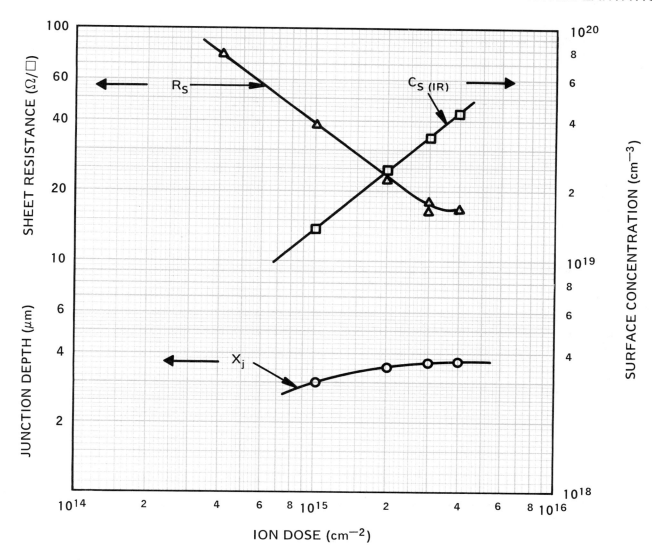

Figure 7-19 — R_S, C_S, and X_j Versus Antimony Ion Dose; Substrate is 10 $\Omega \cdot$cm p-Type (111) Silicon, Implanted at 150 keV, and Diffused at 1250°C for 4 Hours in O_2

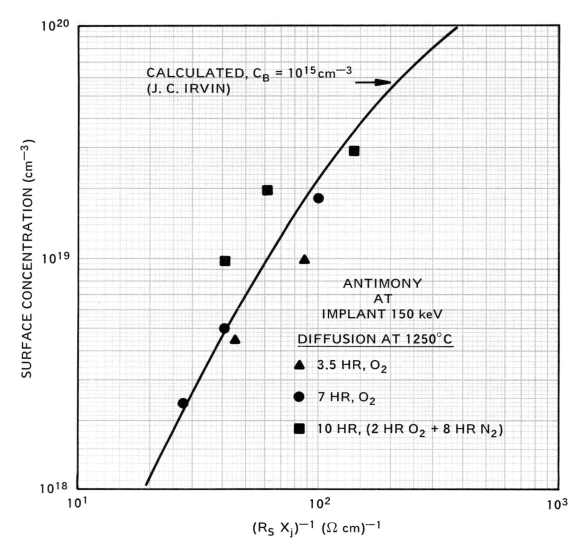

Figure 7-20 — Antimony Surface Concentration Versus $(R_S X_j)^{-1}$

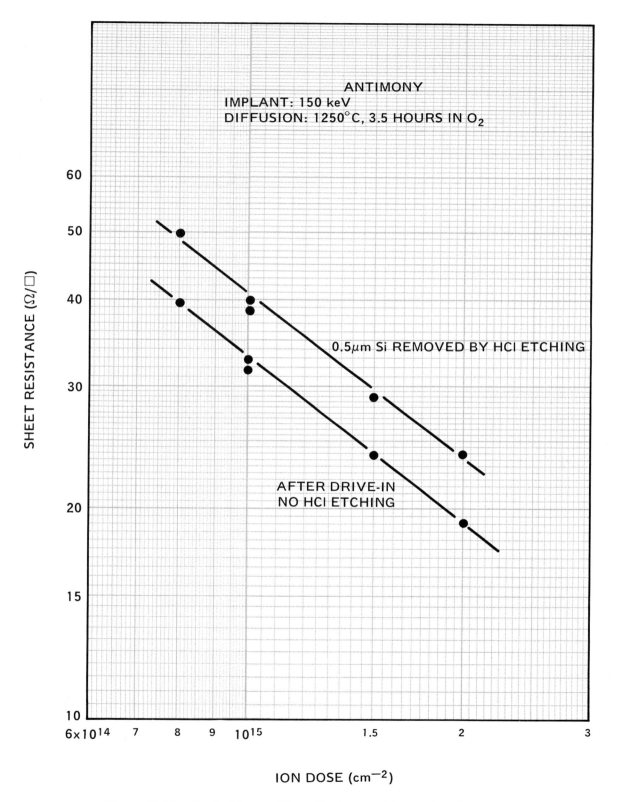

Figure 7-21 —Buried Layer Sheet Resistance Versus Ion Dose

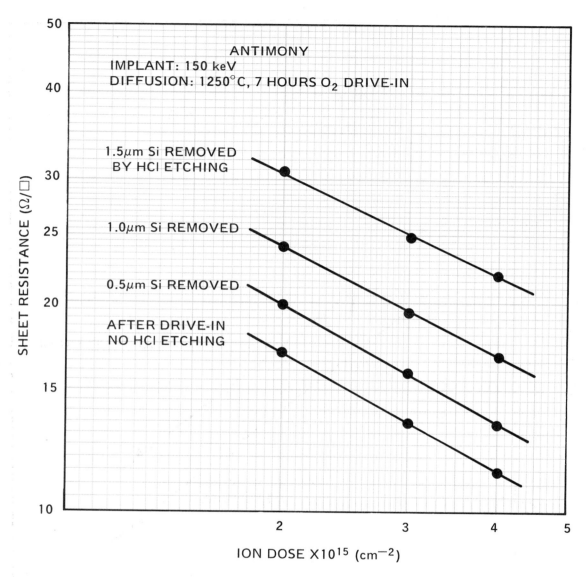

Figure 7-22—Buried Layer Sheet Resistance Versus Ion Dose as a Function of HCl Etching

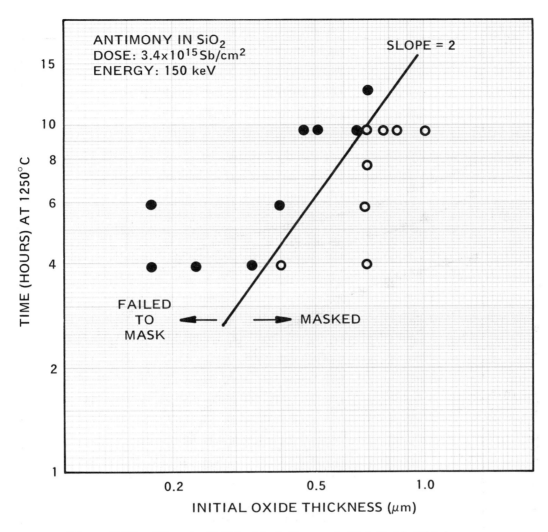

Figure 7-23 —Time to Fail to Mask Versus Initial Oxide Thickness

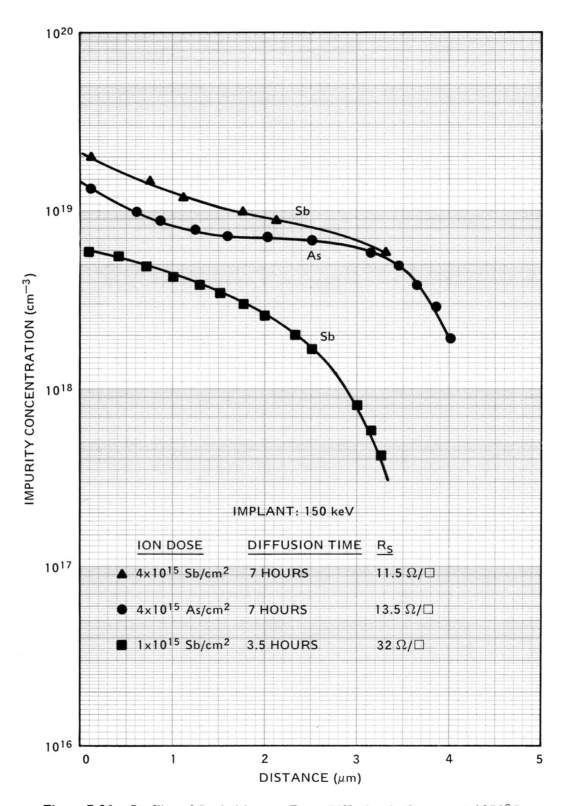

Figure 7-24 —Profiles of Buried Layers From Diffusion in Oxygen at 1250°C

Annealing and Diffusion Results for Arsenic in Silicon

Many experimental studies on annealing or diffusion of arsenic in silicon have been published. Only a few examples of results are given as a guide. Results from implantation through an oxide or nitride layer are also given in this subsection.

Figure 7-25 shows the sheet resistance versus arsenic ion dose as a function of diffusion time and ambient at 1270°C.[22] Arsenic ions were implanted into (100) silicon in a random direction at 30 keV.

Figure 7-26 shows the sheet resistance versus annealing temperature for different arsenic implant doses.[23] The implant was into (111)-oriented silicon wafers in a random direction at 150 keV and the annealing was in nitrogen for 30 minutes.

Figure 7-27 shows the sheet resistance versus diffusion time for a random implant of arsenic into (111)-oriented silicon at 150 keV.[24] The diffusion was at 1200°C in oxygen.

Figure 7-28 shows the sheet resistance versus diffusion temperature for arsenic implant into bare silicon at 200 keV, through 300Å of Si_3N_4 at 300 keV, and through 500Å of SiO_2 at 300 keV for different ion doses.[25] The implant was into a 1 to 2 $\Omega \bullet cm$ (111)-oriented wafer at 8° off the <111> direction, and the diffusion was in nitrogen for 20 minutes.

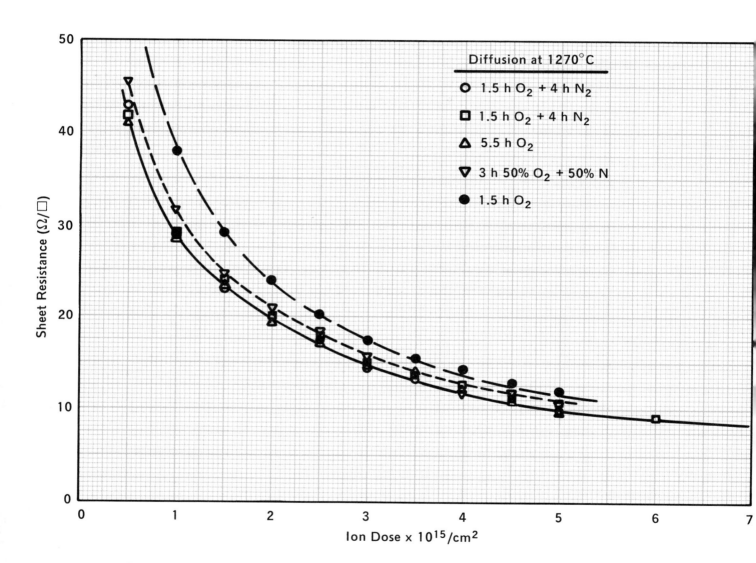

Figure 7-25 —Sheet Resistance Versus As Dose Implanted Into (100) Si at 30 keV[22]

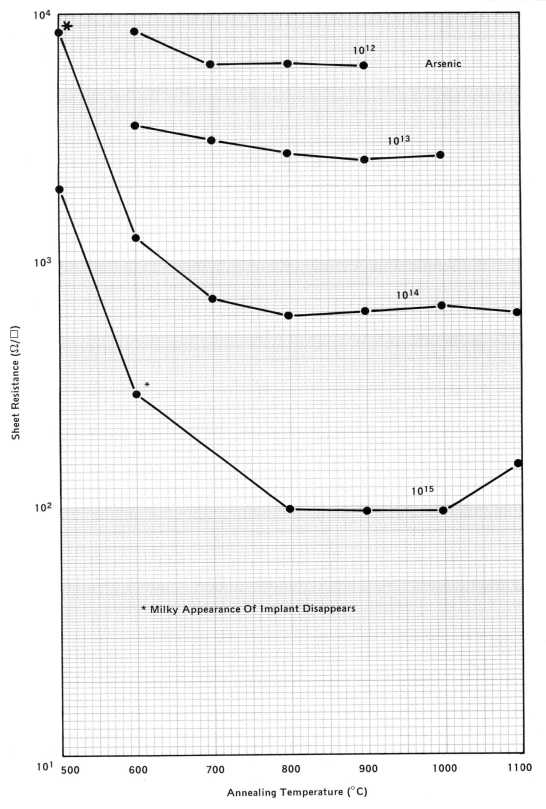

Figure 7-26 — Measured Sheet Resistance Versus Annealing Temperature for As Random
Implant Into (111) Si at 150 keV, Annealing in N_2 for 30 Minutes[23]

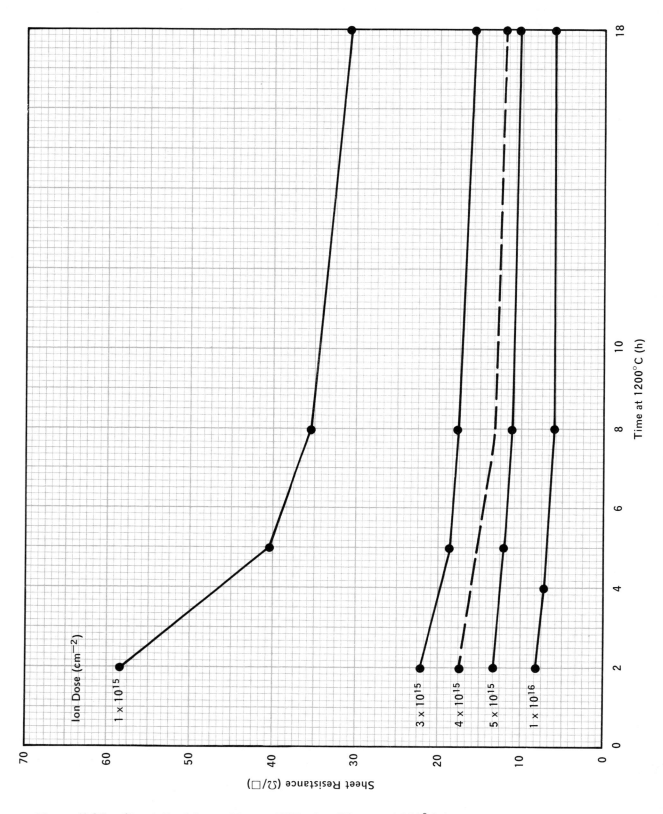

Figure 7-27 —Sheet Resistance Versus Diffusion Time at 1200°C for As Random Implant Into (111) Si at 150 keV[24]

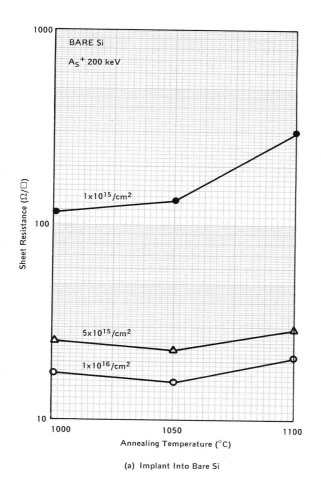

(a) Implant Into Bare Si

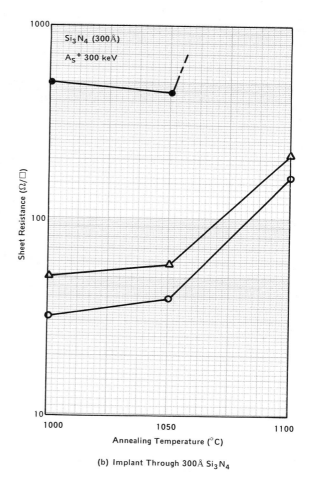

(b) Implant Through 300Å Si₃N₄

Figure 7-28 — Sheet Resistance Versus Diffusion Temperature for Arsenic Annealing in Nitrogen 20 Mins.[25] (Reprinted with permission from Plenum Publishing Corp.)

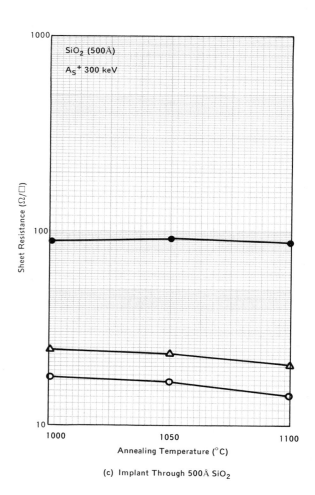

(c) Implant Through 500Å SiO₂

Figure 7-28 — Sheet Resistance Versus Diffusion Temperature for Arsenic Annealing in Nitrogen 20 Mins.[25] (Reprinted with permission from Plenum Publishing Corp.)

Annealing and Diffusion Results for Boron in Silicon

A few examples of sheet resistance versus ion dose for boron-implanted silicon wafers are given in this subsection. Additional information for $^{11}B^+$ implantation can be found in Section 6 and in subsection 7-6 of this section.

Figure 7-29 shows sheet resistance versus annealing or diffusion temperature for $^{11}B^+$ and $^{49}BF_2^+$ implanted silicon wafers.[26] The implant energy was 50 keV. The silicon wafers were (100)-oriented, n-type and 1 to 2 Ω•cm. The annealing was in nitrogen for 30 minutes. When BF_3 gas is used as the boron ion source, the ion current of BF_2 approaches twice that of the ^{11}B ion current. Hence, to reduce the implant time for high-dose implantation runs, BF_2^+ ions are often implanted into SIC wafers. A secondary ion mass spectrometer (SIMS) profile analysis shows that fluorine remains in silicon when the diffusion temperature is below 1150°C. BF_2^+ implant has been used as the collector tub dopant for SIC devices. When the diffusion subsequent to the implant is above 1150°C and no adverse electrical effect has been observed. However, little is known of the effect of fluorine on the electrical property of a p-n junction which has been implanted with BF_2^+ ions and annealed at lower temperatures.

Figure 7-30 shows sheet resistance versus annealing temperature for a $^{11}B^+$ implant into a (111)-oriented wafer. The implant was at 150 keV and in a random direction. The annealing was in nitrogen for 30 minutes.[27]

Figure 7-31 shows the sheet resistance versus ion dose for two terminal resistors.[28] The annealing of the implant is achieved during a silicon nitride deposition process. Since the process details are considerably different, this curve is only useful as a guide. Figure 7-32 shows similar curves for (111)-oriented wafers.[29]

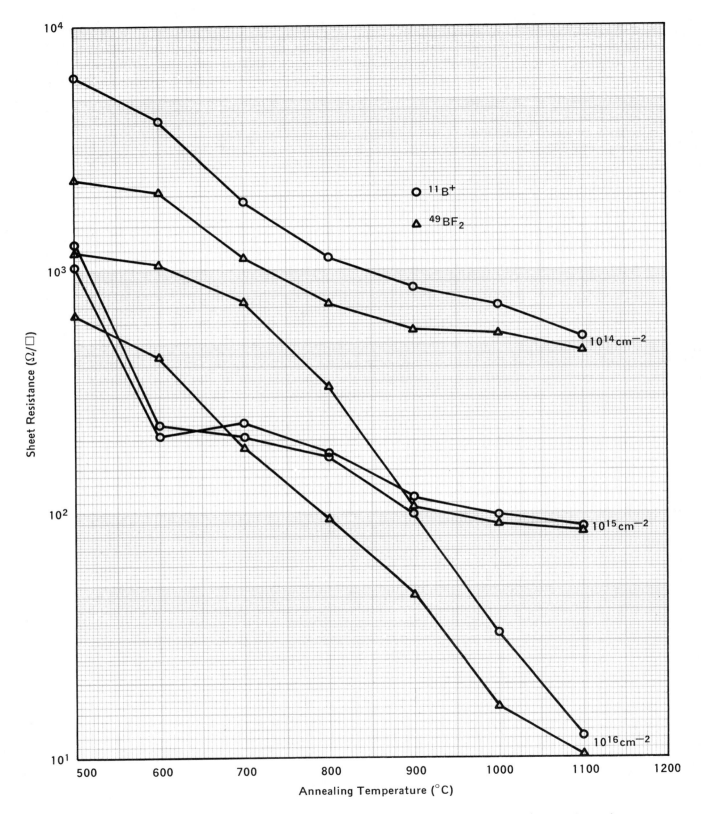

Figure 7-29 —Sheet Resistance Versus Annealing Temperature for $^{11}B^+$ and $^{49}BF_2^+$
Implanted Into 1 to 2 $\Omega \bullet cm$ n-Type (100) Silicon at 50 keV[26]

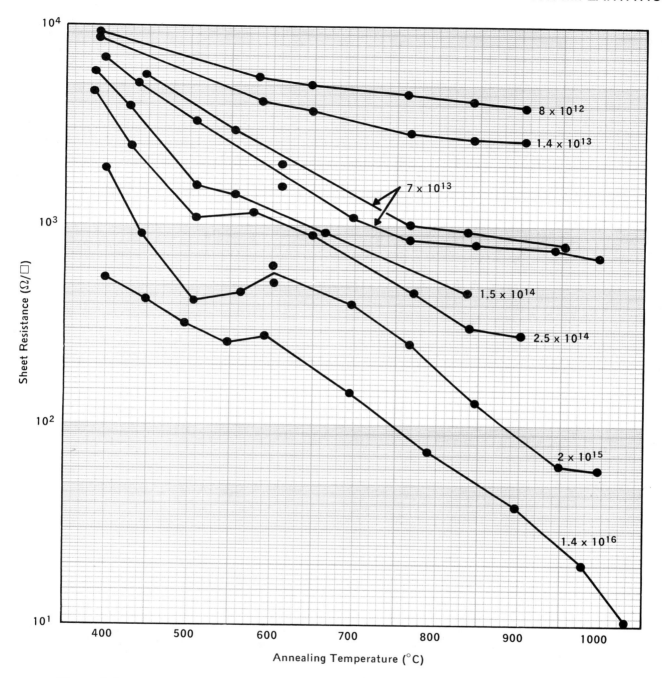

Figure 7-30 —Measured Sheet Resistance Versus Annealing Temperature for B Random
Implant Into (111) Si at 150 keV, Annealing in N_2 for 30 Minutes[27]

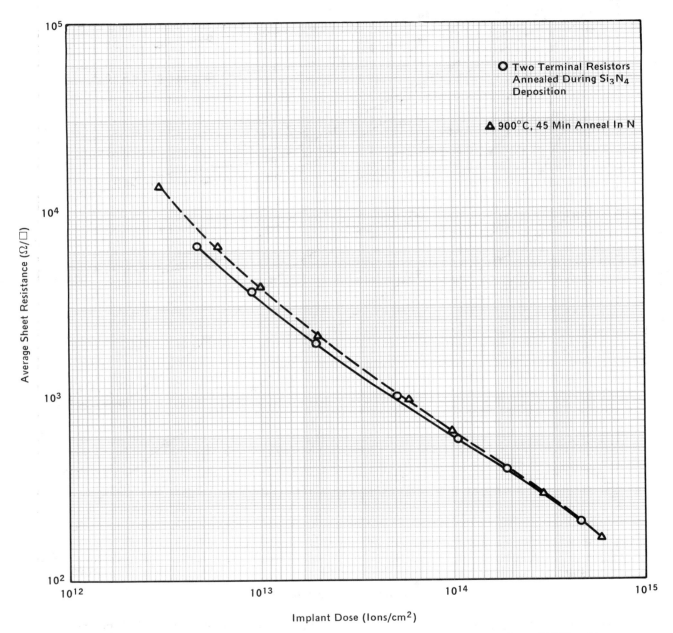

Figure 7-31 —Sheet Resistance Versus B Dose for Random Implant Into (100) n-Type Si at 50 keV[28]

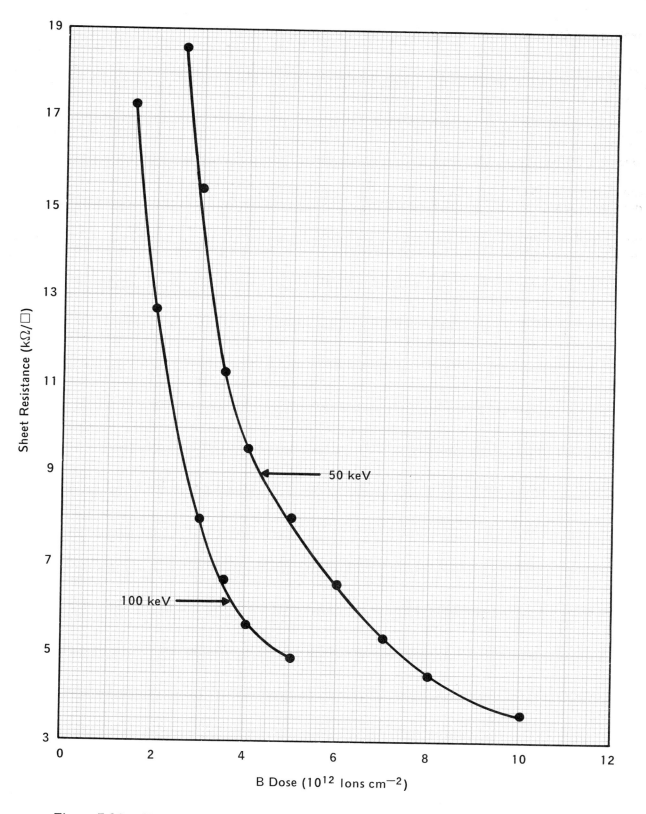

Figure 7-32 —Sheet Resistance Versus B Dose for Random Implant Into 2 to 4 Ω•cm
n-Type (111) Si, Annealed During Si_3N_4 Deposition at 900°C[29]

Annealing and Diffusion Results for Phosphorus in Silicon

Curves of sheet resistance for phosphorus implanted into (111)-oriented silicon as a function of annealing temperature and ion dose are shown in Figure 7-33. The implant was in a random direction and the annealing was in nitrogen for 30 minutes. Data from two sources are shown in this figure.[30]

Figure 7-34 shows curves of sheet resistance versus annealing temperature for phosphorus implanted into (111)-oriented silicon in a channelled direction.[31] The implant was at 300 KeV and the annealing was in nitrogen for 30 minutes.

Figure 7-35 shows sheet resistance versus ion dose for phosphorus implanted into (100)-oriented silicon in a random direction at 50 KeV and diffused at 1050°C in wet nitrogen and at 1200°C in oxygen.[32]

Figure 7-36 shows junction depth versus diffusion time for (100) silicon implanted at 50 KeV and diffused at 1200°C in oxygen.[32]

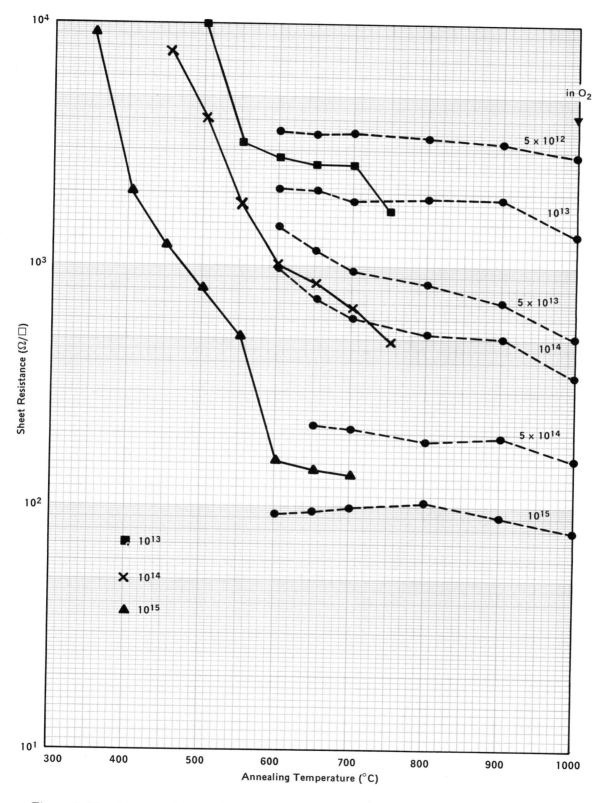

Figure 7-33 —Measured Sheet Resistance Versus Annealing Temperature for Phosphorus
Random Implant Into (111) Si at 50 keV, Annealing in N_2 for 30 Minutes

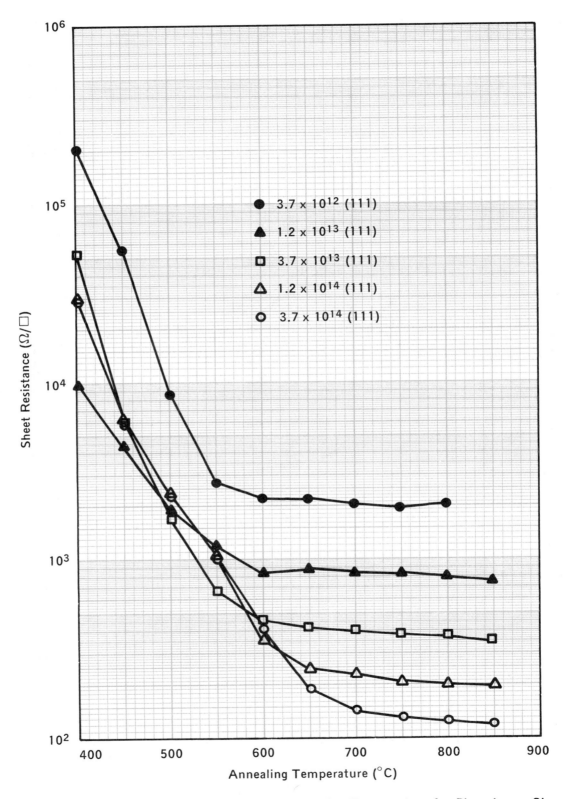

Figure 7-34 —Measured Sheet Resistance Versus Annealing Temperature for Phosphorus-Channeled Implant Into (111) Si at 300 keV, Annealing in N_2 for 30 Minutes[31]

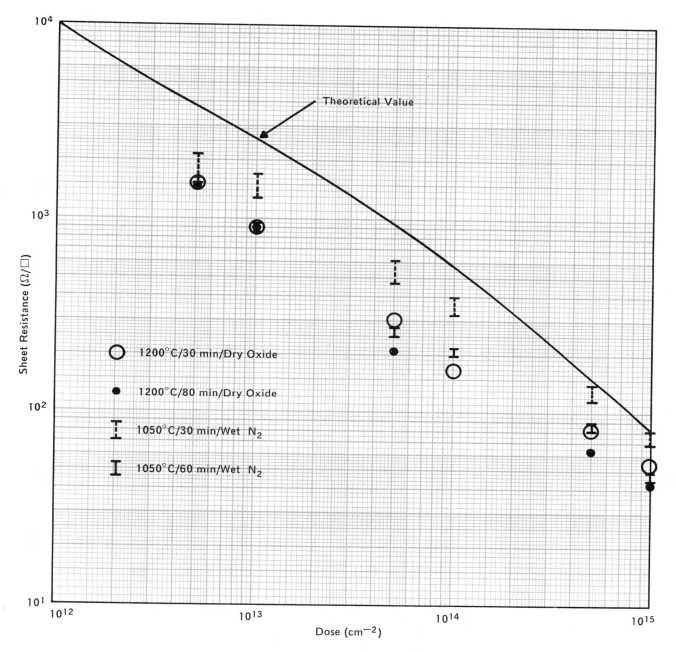

Figure 7-35 —Sheet Resistance Versus Ion Dose for Phosphorus Random Implant Into
(100) Si at 50 keV[32]

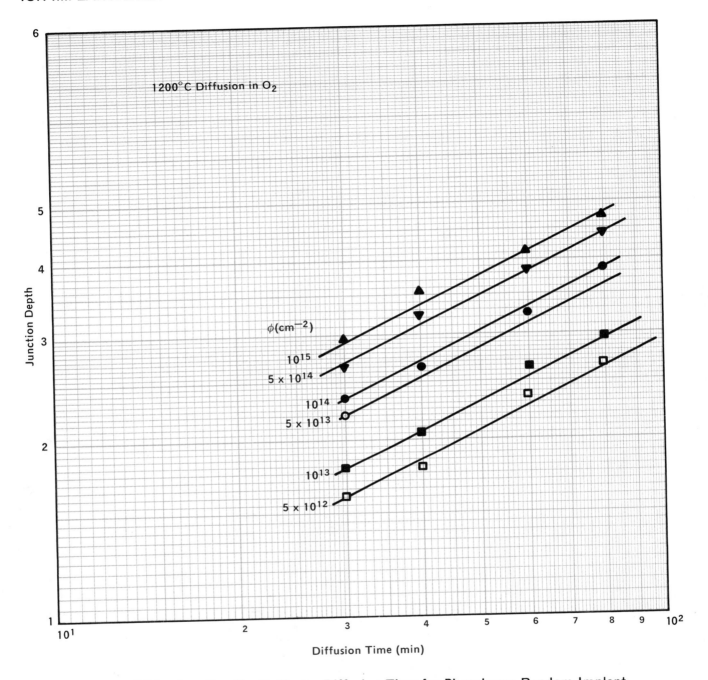

Figure 7-36 — Junction Depth Versus Diffusion Time for Phosphorus Random Implant Into (100) Si at 50 keV[32]

Implant Through an Insulating Film

In MOS device fabrication, the threshold voltage of an MOS transistor is often adjusted by implantation of dopants through the gate oxide at appropriate energies and doses. Another application of ion implantation through an insulating film is for the protection of the silicon surface during diffusion after boron implantation. Diffusion of ion-implanted boron in a nonoxidizing ambient is necessary. This is due to the fact that annealing and diffusion of boron-implanted silicon in an ambient which contains more than 0.1 percent of oxygen will generate dislocations to such a high density as to affect the electrical yield of n-p-n transistors.[33] However, it is very difficult to maintain the diffusion ambient with oxygen concentration in the 0.1 percent range in the existing 3-inch-diameter or 4-inch-diameter diffusion furnaces. To get around this problem, some devices are made with boron implantation through a thin oxide layer (100 to 200Å). Diffusion is performed in nitrogen or argon at temperatures above 1100°C and diffusion times longer than 30 minutes. Diffusion in nitrogen or argon without a small percent of oxygen or without a thin oxide layer may cause the silicon surface to form a thin layer of nitride. The surface sometimes shows stain colors that could be a suboxide of silicon or an uneven surface presumed to be due to thermal etching of silicon. The stains are difficult to remove and can affect electrical contact formation. Some curves for implanted arsenic, boron, or phosphorus ions through an oxide layer are given in this subsection.

Figure 7-37 shows the fraction of boron dose reaching silicon through a masking oxide versus ion energy for different thicknesses of the oxide film.[34] Figure 7-38 shows oxide thickness versus ion energy when 90 percent of the boron ions or when 0.001 percent of boron ions reach the silicon under the oxide. The latter condition is the point at which boron ceases to reach the silicon surface. The curves shown in Figure 7-38 are obtained from information given in reference 34.

Figure 7-39 shows examples of threshold voltage shift from boron and phosphorus implantation through an oxide layer as a function of boron, phosphorus, or silicon ion dose.[51] Ions of $^{11}B^+$ were implanted through a 750Å-thick gate oxide at 30 KeV; ions of $^{28}Si^+$ and $^{31}P^+$ were implanted through a 500Å-thick gate oxide at 50 KeV. The results are normalized for a final 1000Å-thick gate oxide.

Figure 7-40 shows sheet resistance versus annealing temperature for arsenic implantation through various oxide thicknesses. The implantation was at 300 keV for a dose of $5 \times 10^{15} cm^{-2}$ in a random direction into (111)-oriented silicon. The annealing was in nitrogen for 20 minutes. The masking oxide layer was removed prior to diffusion for annealing at 1150°C.[25] G. Nakamura et al also presented transmission electron microscope (TEM) data on the defect annealing and recoil implantation of oxygen from high energy and high dose arsenic implantation. One should take these factors into consideration when using the curves shown in Figure 7-40.

Figure 7-41 shows sheet resistance versus oxide thickness for boron implant at several energy and dose combinations.[34] The annealing was performed at 925°C for 15 minutes in nitrogen.

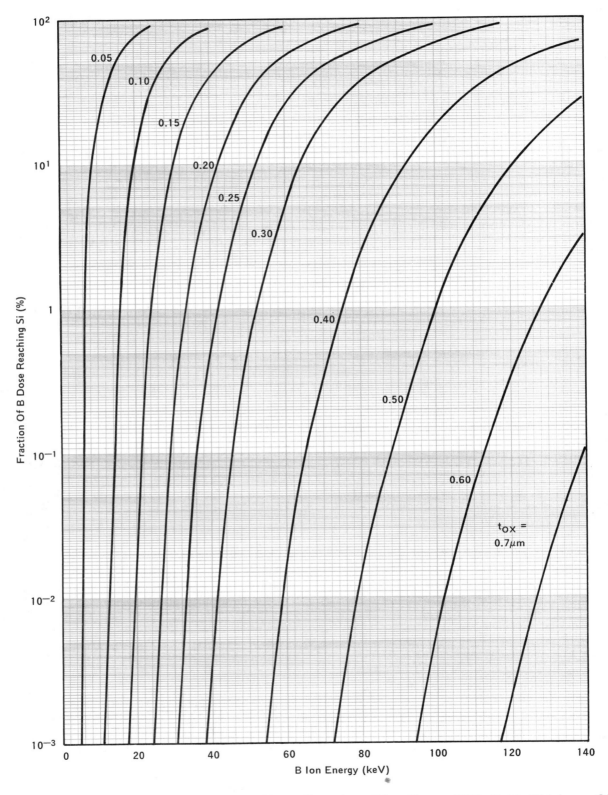

Figure 7-37 — Fraction of B Dose Reaching Si as a Function of Ion Energy With Oxide Thickness Given Parametrically[34] (Reprinted with permission from Pergamon Press, Ltd.)

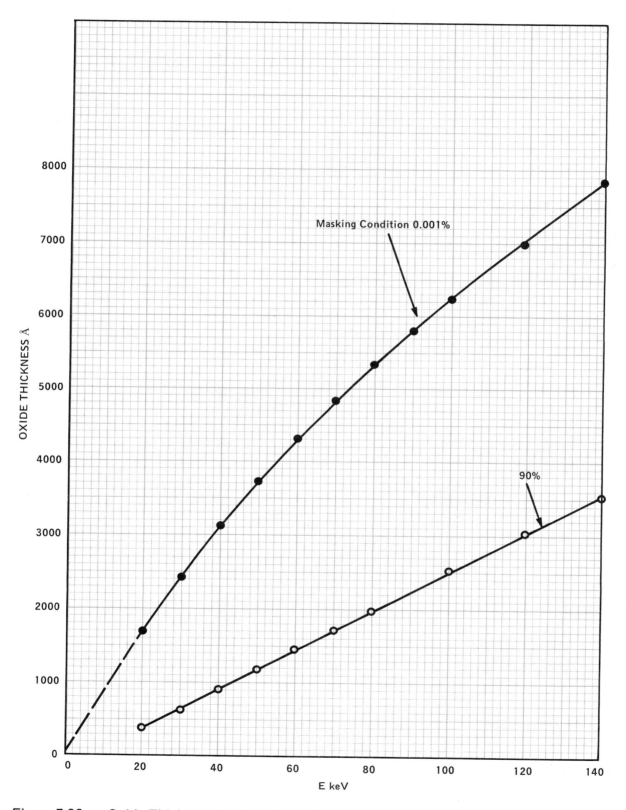

Figure 7-38 — Oxide Thickness For Fraction of Boron Dose Reaching Si Versus Ion Energy[34]
(Reprinted with permission from Pergamon Press, Ltd.)

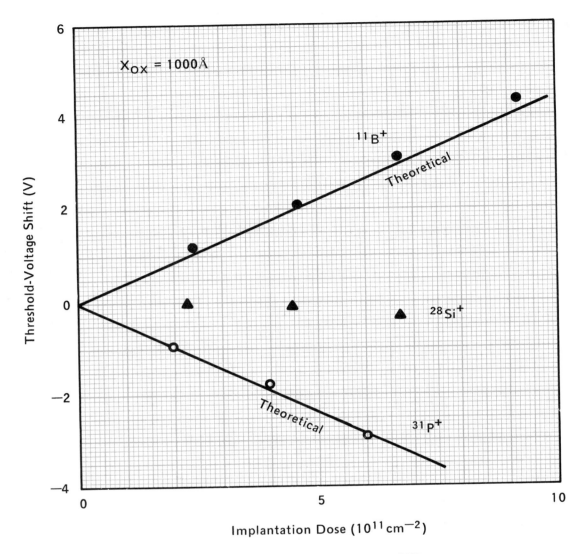

Figure 7-39 — Threshold-Voltage Shift in Ion-Implanted MOSFET[52] (Reprinted with permission from the Japan Society of Applied Physics)

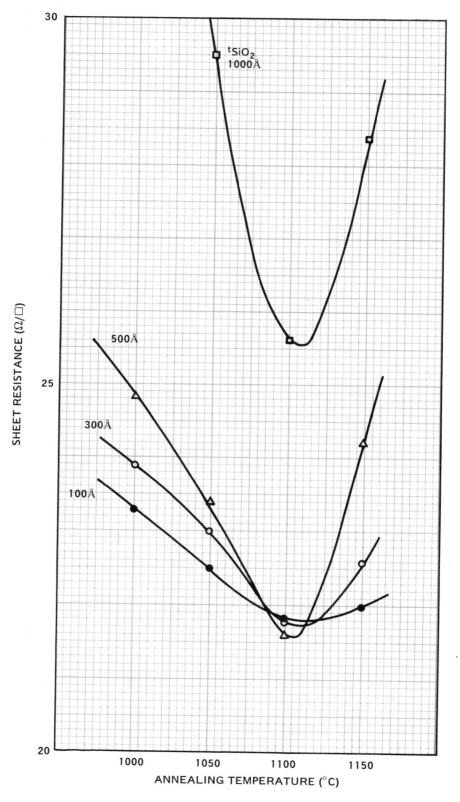

Figure 7-40 — Isochronal Anneal of Sheet Resistance of Si Implanted By As[+] With Various Oxide Thicknesses.[25] (Reprinted with permission from Plenum Publishing Corp.)

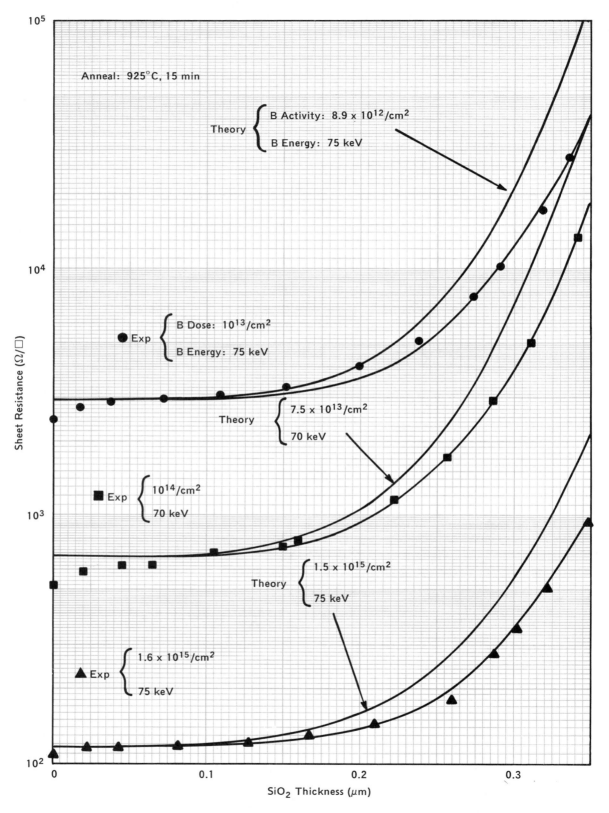

Figure 7-41 — Sheet Resistance of B Implant Versus SiO$_2$ Thickness, Annealing at 925°C, 15 min in N$_2$ [34] (Reprinted with permission from Pergamon Press, Ltd.)

7.6 Annealing Results on Implantation into Polycrystalline Silicon

Sheet Resistance Versus Annealing Temperature[35]

Figure 7-42 shows the results for arsenic implanted into poly Si and annealed in nitrogen at different temperatures. The implantation was at 150 keV. The poly Si was deposited in a Nitrox reactor at atmospheric pressure. North, Adam, and Richards[35] found that poly Si films deposited in a Nitrox reactor and a low-pressure chemical vapor deposition (LPCVD) reactor gave the same sheet resistance characteristics as a function of annealing temperature. Figure 7-43 shows similar curves for boron and phosphorus implanted at 150 and 300 keV, respectively. Figure 7-44 shows annealing results for boron implanted at 30, 70, and 150 keV to a dose level of $1 \times 10^{15} cm^{-2}$. Data for implantation into both single crystal and polycrystalline silicon are given.

Sheet Resistance Versus Ion Dose[35]

Figure 7-45 shows sheet resistance versus arsenic ion dose for implantation into both single crystal and polycrystalline silicon wafers. The implant energy, the annealing temperature, and time are given in the figure. Similar curves for boron and phosphorus are shown in Figures 7-46 and 7-47, respectively.

Effect of Preannealing

North[35] [36] found that the electrical resistance of boron- or phosphorus-implanted polysilicon depends on the annealing temperature of the poly Si film prior to the ion implantation and diffusion. The poly Si film was deposited at temperatures below 900°C and annealed at 1000°, 1100°, and 1200°C for 30 minutes in nitrogen. Boron or phosphorus was implanted into these annealed poly Si films and heat treated at different temperatures. The results for boron and phosphorus are shown in Figures 7-48 and 7-49, respectively.

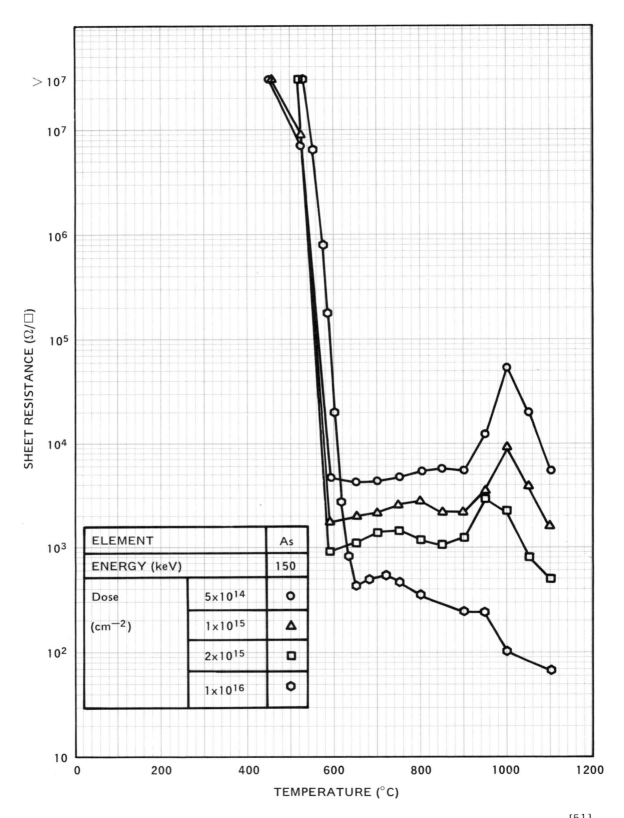

Figure 7-42 —Sheet Resistance Versus Temperature for Arsenic Implant Into Poly Si[51]
(This Figure was originally presented at the Fall 1978 Meeting of the Electrochemical Society, Inc.,
held in Pittsburgh, Pennsylvania)

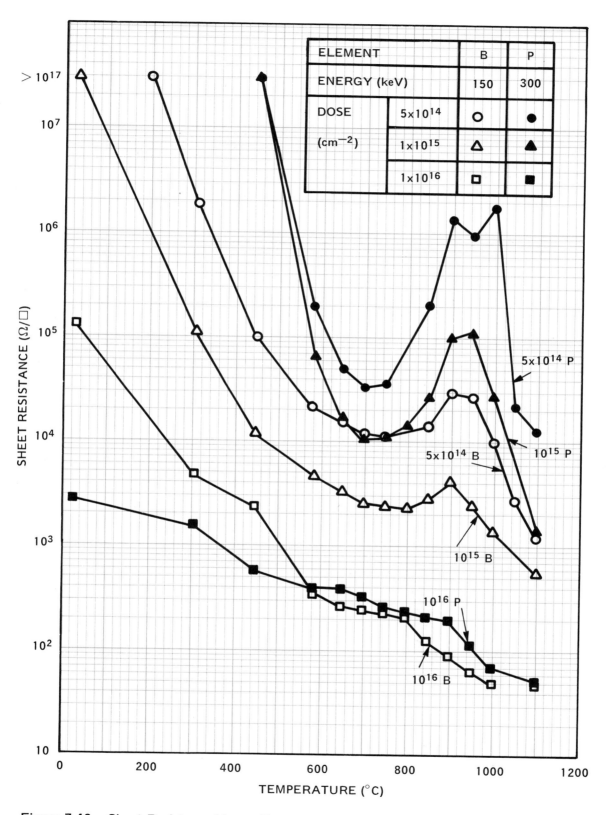

Figure 7-43 —Sheet Resistance Versus Temperature for Boron or Phosphorus Implant Into Poly Si[51] (This Figure was originally presented at the Fall 1978 Meeting of the Electrochemical Society, Inc., held in Pittsburgh, Pennsylvania)

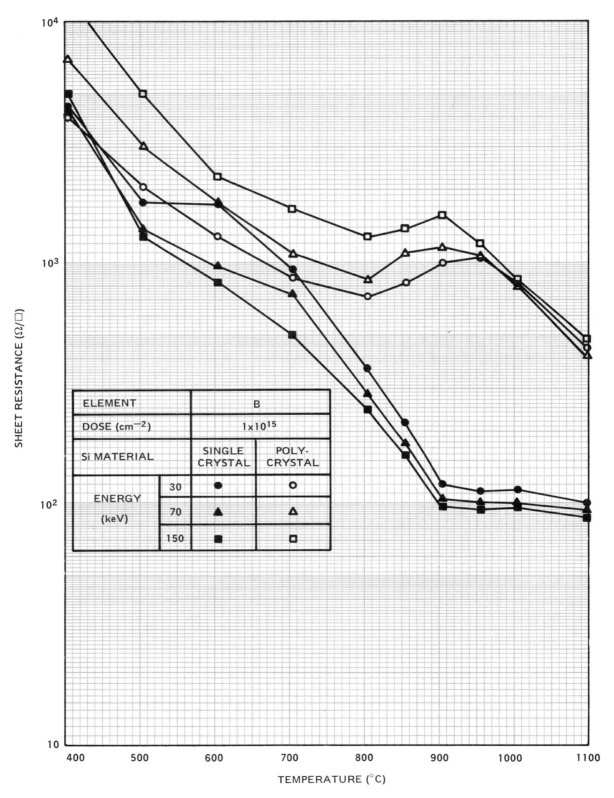

Figure 7-44 —Sheet Resistance Versus Temperature for Boron Implant Into Poly Si[51]
(This Figure was originally presented at the Fall 1978 Meeting of the Electrochemical
Society, Inc., held in Pittsburgh, Pennsylvania)

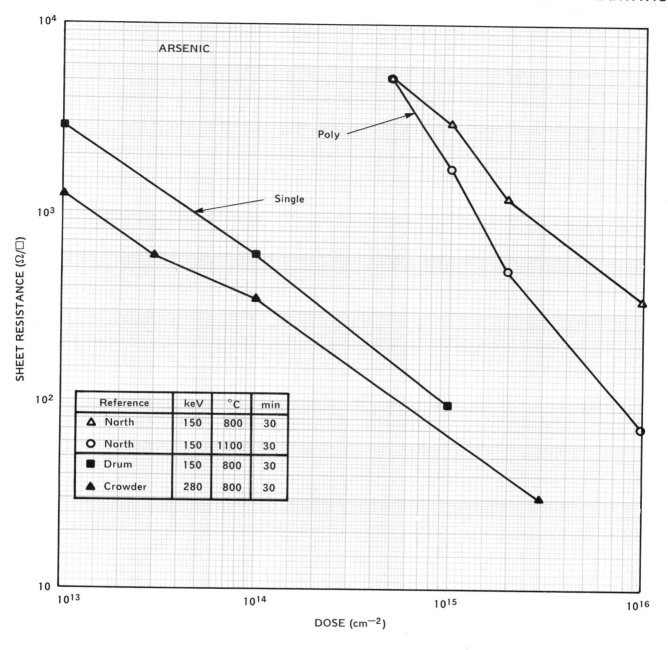

Figure 7-45 — Sheet Resistance Versus Arsenic Dose[35]

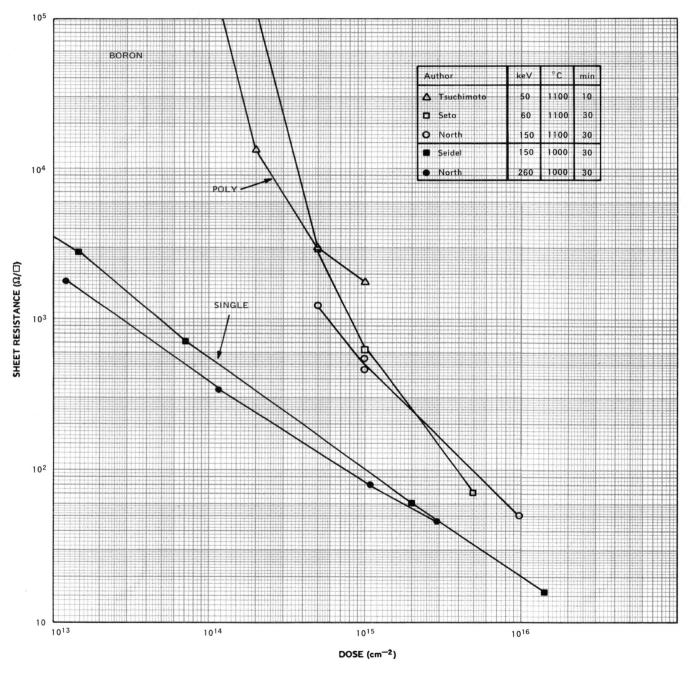

Figure 7-46 —Sheet Resistance Versus Boron Dose[35]

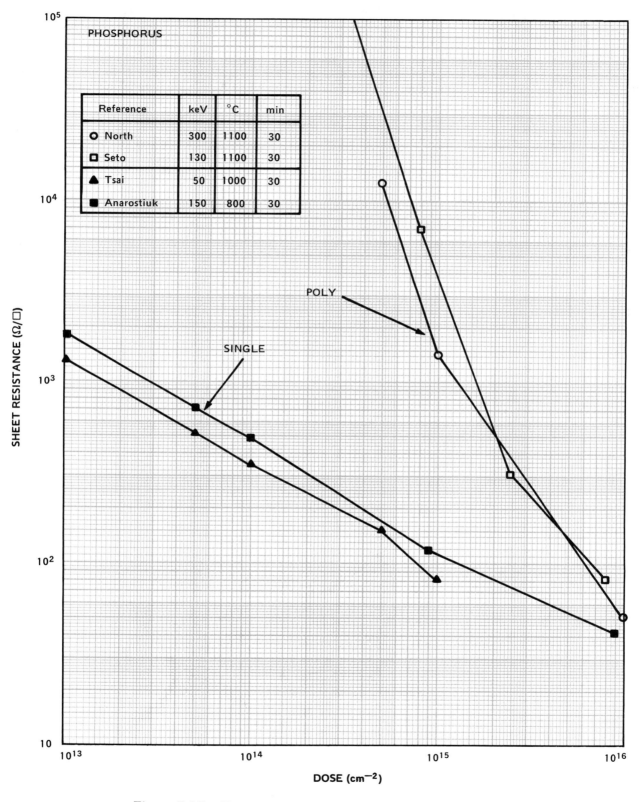

Figure 7-47 —Sheet Resistance Versus Phosphorus Dose[35]

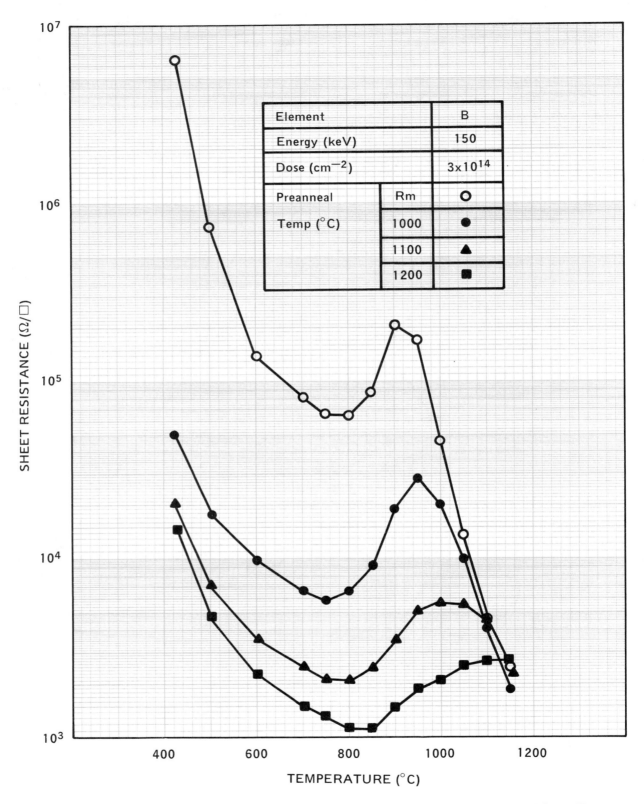

Figure 7-48 —Sheet Resistance Versus Temperature for Poly Si Film Annealed Prior To Boron Implantation and Diffusion[36]

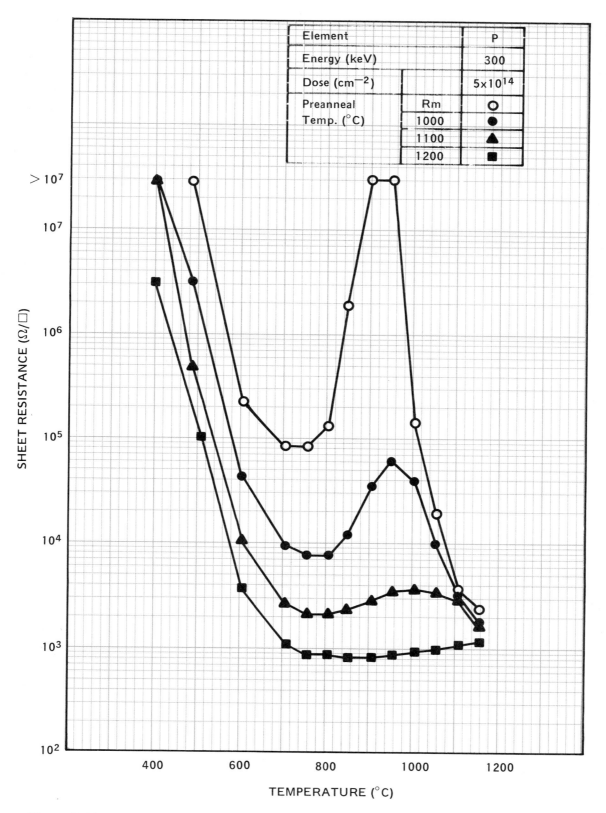

Figure 7-49 —Sheet Resistance Versus Temperature for Poly Si Film Annealed Prior To
Phosphorus Implantation and Diffusion[36]

Ion Implantation Into Doped Polycrystalline Silicon for MOS Devices

In the process of developing self-aligned poly Si gate MOS integrated circuits, the poly Si is doped with phosphorus, arsenic, or antimony to reduce the resistivity of the poly Si film. The dopant can be introduced by ion implantation and annealing. Andrews[37] studied the sheet resistance variation as a function of ion dose for poly Si implanted with Sb, As, and P. All implantations were at 30 keV. The poly Si film was deposited at 700°C in a Nitrox reactor to a thickness of 5000Å on 5000Å-thick thermal oxide. After implantation, the samples were diffused at 1100°C for 20 minutes in nitrogen, followed by oxidation in oxygen for 30 minutes and annealing in argon for another 30 minutes. Figure 7-50 shows the measured sheet resistance and implant dose.

Polycrystalline Silicon Resistor[38]

An example of using a doped poly Si film as a high-value resistor is given in this subsection. Boron was implanted into 5000Å-thick LPCVD poly Si film at 30 keV. Figure 7-51 shows sheet resistance versus boron ion dose after an annealing at 900°C for 30 minutes in nitrogen. Figure 7-52 shows the sheet resistance versus annealing temperature for an ion dose of $2.5 \times 10^{14} cm^{-2}$. The annealing was in nitrogen for 30 minutes.

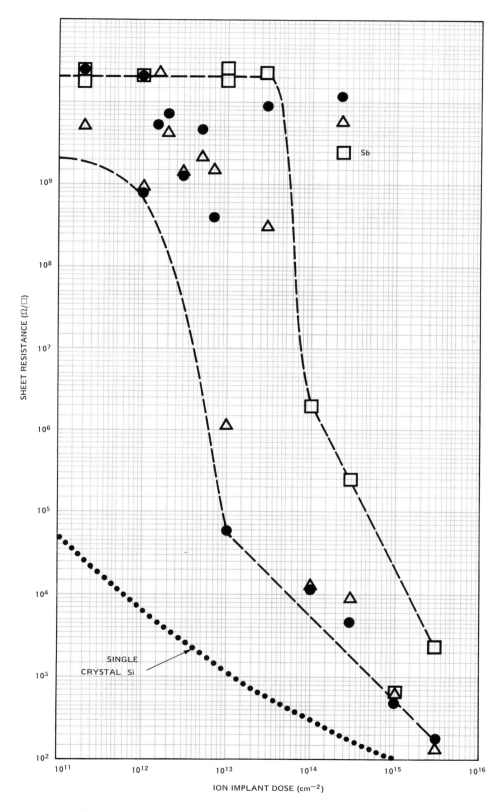

Figure 7-50 —Sheet Resistance Versus Ion Dose; Implanted Into 5000Å Poly Si Film at 30 keV,
Diffused at 1100°C, 20 Minutes N$_2$ + 30 Minutes O$_2$ + 30 Minutes Argon[37]
(Reprinted with permission from Plenum Publishing Corp.)

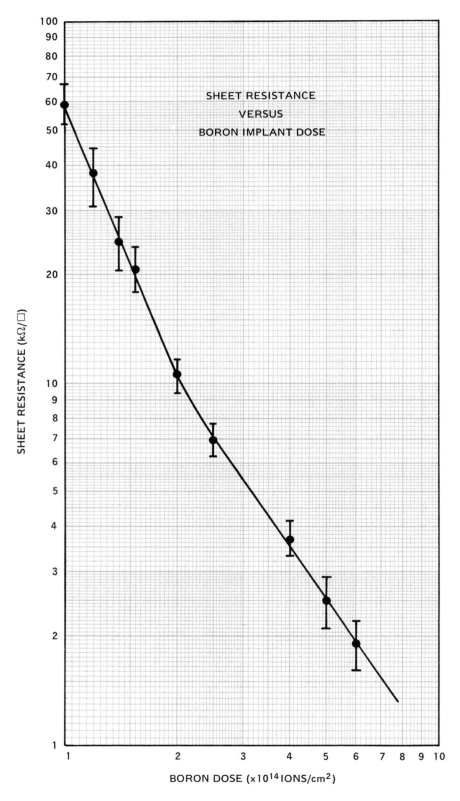

Figure 7-51 —Sheet Resistance Versus Boron Implant Dose Implanted at 30 keV Into 5000Å
LPCVD Poly Si Film, Annealed at 900°C For 30 Minutes in N$_2$ [38]

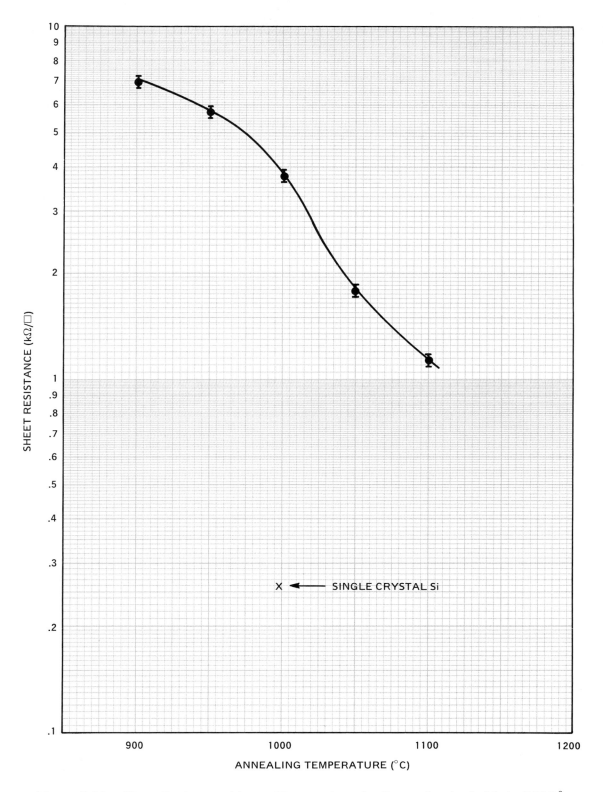

Figure 7-52 —Sheet Resistance Versus Temperature for Boron Implanted Into 5000Å
Poly Si Film at 30 keV and 2.5×10^{14} Ions/cm^2 and Annealed at 30 Minutes
in N$_2$ [38]

7.7   Sheet Resistance for Three-Inch-Diameter Wafers with Furnace Ramping

Most of the experimental data on ion-implanted/diffused samples were obtained from small size samples or from silicon wafers less than 3 inches in diameter. The heat treatments were usually done in diffusion furnaces without furnace ramping. When an SIC is fabricated on 3-inch-diameter wafers, heat treatments are done in furnaces that ramp slowly from 700-850°C up to the diffusion or oxidation temperature and then down to the 800°C range. The use of furnace ramping increases the length of time a wafer is at the diffusion or oxidation temperature. Thus, the ion dose for a given sheet resistance has to be adjusted accordingly. Some established values are given in Tables 7-11 and 7-12 for (100)- and (111)- oriented wafers, respectively. Since the results depend on individual furnace conditions, these data should be used as a guide.

TABLE 7-11

SHEET RESISTANCE DATA FOR 3-INCH-DIAMETER (100) SILICON WAFERS WITH FURNACE RAMPING[39]

SUBSTRATE		IMPLANTATION			DIFFUSION				SHEET RESISTANCE (Ω/sq)	JUNCTION DEPTH (μm)
TYPE (n OR p)	RESISTIVITY (Ω•cm)	ELEMENT	DOSE (cm^{-2})	ENERGY (keV)	TEMP. (°C)	TIME (min)	RAMP RATE (°C/min)	OXYGEN* (%)		
p-bulk	0.5 — 1	^{75}As	3×10^{15}	30	1270	330	9.3	50	16**	7.0
n-bulk	0.5 — 1	^{11}B	1.5×10^{15}	30	1200	40	10	0.1	55**	3.5
n-epi	2 — 4	^{11}B	5.9×10^{14}	30	1150	55	8.6	0.1	200	2.5
n-epi	0.0135	^{11}B	8×10^{15}	30	1100	60	7.1	0.1	12	1.5
n-epi	0.23	^{11}B	2.6×10^{13}	30	1200	120	10	100	3000	4.0
n-base	—	^{11}B	5×10^{15}	30	1100	80	17	0.1	40	2.4
	—	^{11}B	8×10^{15}	30	1100	54	17	0.1	23±1.5	2.4
	—	^{11}B	1×10^{16}	30	1100	40	17	0.1	20	2.4
p-bulk	4 — 15	^{31}P	3×10^{13}	30	1270	330	9.3	50	220**	11.8
n-bulk	0.5 — 1	^{31}P	3×10^{14}	30	1150	70 — 90	8.6	0.1	92 — 100	2.7
	0.5 — 1	^{31}P	2×10^{13}	30	1150	120	8.6	0.1	105	2.9
	0.5 — 1	^{31}P	2.6×10^{13}	30	1150	110	8.6	0.1	105	2.7
	0.5 — 1	^{31}P	3.4×10^{13}	30	1150	100	8.6	0.1	105	2.4

* Balance with nitrogen

** Measured on control wafer

TABLE 7-12

SHEET RESISTANCE DATA FOR 3-INCH-DIAMETER (111) SILICON WAFERS WITH FURNACE RAMPING[40]

SUBSTRATE		IMPLANTATION			DIFFUSION			SHEET RESISTANCE (Ω/sq)	JUNCTION DEPTH (μm)
TYPE (n OR p)	RESISTIVITY ($\Omega \bullet$cm)	ELEMENT	DOSE (cm^{-2})	ENERGY (keV)	TEMP. ($^\circ$C)	TIME (min)**	OXYGEN* (%)		
All p-type bulk Si	8-20	122Sb	2.5x10^{15}	150	1250	120	30	21	5.0
	8-20	122Sb	2.5x10^{15}	30-150	1250	420	10	17	~7
	8-20	122Sb	7x10^{15}	30-150	1250	420	10	7.8	~8
	8-20	75As	4x10^{15}	30	1100	20	0.1	24	—
	8-20	31P	1x10^{13}	150	1250	180	10	650	7.0
	8-20	31P	1x10^{13}	150	1250	120	10	650	5.4
	8-20	31P	1x10^{13}	150	1250	60	10	700	4.0
	8-20	31P	1x10^{13}	150	1250	20	10	720	2.9

* Balance with nitrogen

** 12°C/Min. ramp up and 2.5°C/Min. Ramp Down

7.8   Implant Profiles in Insulating Films

In SIC processing, one is sometimes interested in knowing the implant profile of dopants in SiO_2 or Si_3N_4 layers.   However, the measurement of impurity profiles is rather difficult in these insulating films.   Very few publications present data from direct measurements.   All the evidence indicates that these profiles are asymmetrical from the peak concentration.

Schimko et al[41] measured the arsenic and boron profiles in SiO_2.   From curve fitting of these profiles with the Edgeworth approximation, they calculated the moments of these profiles.   Their results are given in Table 7-13.   Measured peak ion ranges agree reasonably well with the theoretical calculations by Gibbons, Johnson, and Mylroie.[5]

Figures 7-53 and 7-54 show examples of boron implanted through an oxide layer where the peak concentration is close to the SiO_2-Si interface or slightly inside Si.[42] [43]   Figure 7-55 shows a phosphorus implant profile and an implant diffusion profile in SiO_2.   The implantation was at 30 keV with an ion dose of $3\times10^{15} cm^{-2}$.   The diffusion was at $1100°C$ for 60 minutes.   The implant profile is slightly skewed toward the right-hand side.   The curve fitting of the diffusion profile gives a diffusion constant close to $3\times10^{-16} cm^2/s$.[44]   Figure 7-56 shows a phosphorus profile in Si_3N_4 film.   The implantation was at 30 keV with a dose of $3\times10^{15} cm^{-2}$.[44]   This profile is skewed slightly toward the right-hand side of a Gaussian distribution.   Heat treatment at $1100°C$ for 60 minutes in nitrogen produced undiscernible diffusion of the phosphorus atoms.   The measured profile can be fitted to a Gaussian profile near the peak concentration region.   Hirao et al[45] measured profiles of boron, phosphorus, and arsenic implanted through and into Si_3N_4 films.   They also determined the recoiled nitrogen profiles.   Refer to their paper for details.

TABLE 7-13

MEASURED MOMENTS FOR ARSENIC AND BORON[41]
(Reprinted with permission from Physica Status Solidi)

ION	E (keV)	ϵ	$<x>$ (Å)	σ (Å)	CM_3 (Å)	γ_1	γ_2
$^{11}B^+$	40	6	1361	478	360	−0.39	2.68
$^{11}B^+$	50	7.5	1744	550	428	−0.49	2.96
$^{11}B^+$	70	10.5	2448	674	580	−0.638	3.49
$^{11}B^+$	100	15	3448	803	783	−0.922	4.44
$^{11}B^+$	150	22.5	5180	1030	1095	−1.20	5.04
$^{11}B^+$	300	45	8544	1221	1711	−1.96	9.1
$^{75}As^+$	40	0.21	352	125	85	0.31	3.3
$^{75}As^+$	80	0.42	517	176	108	0.23	3.1
$^{75}As^+$	150	0.78	830	290	209	0.35	3.1

Key:

$<x>$ = the mean range

σ = the range scattering = $<(x-<x>)^2>^{1/2}$

CM_3 = the skewness = $<(x-<x>)^3>^{1/3}$

γ_1 = the reduced skewness = $\dfrac{<(x-<x>)^3>}{<(x-<x>)^2>^{3/2}}$

γ_2 = the reduced excess = $\dfrac{<(x-<x>)^4>}{<(x-<x>)^2>^2}$

ϵ = the reduced energy

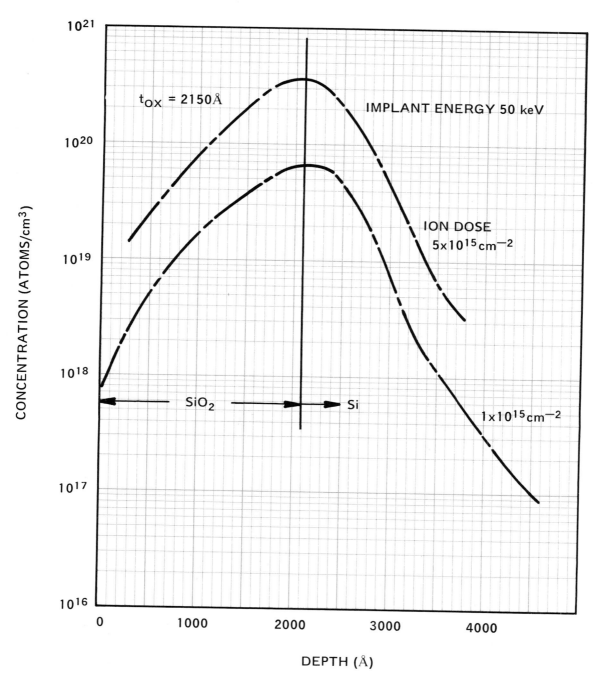

Figure 7-53 —Boron-Implanted Profile in SiO_2 and Si[42]

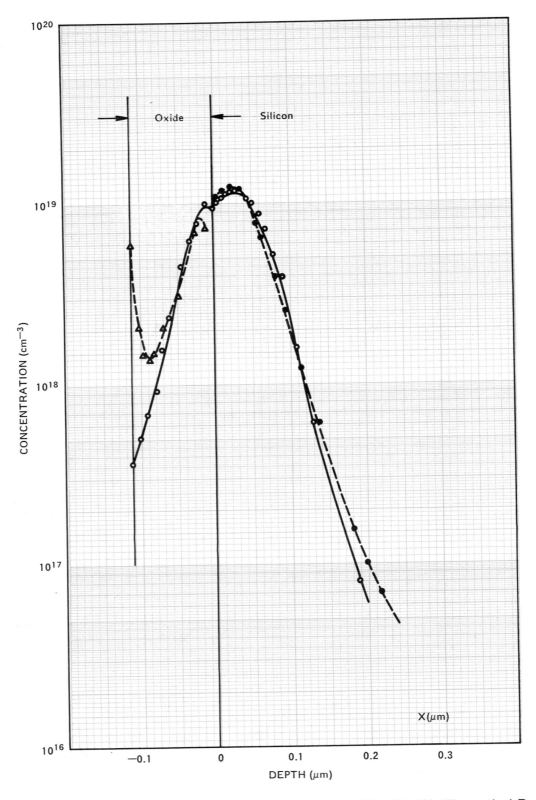

Figure 7-54 —Boron Profile in SiO$_2$-Si Sample. Experimental Profile ($\triangle$); Theoretical Results From Furukawa's Theory (O). Implantation at 40 keV and Boron Dose of 9 × 10^{13}cm^{-2}.[43] (Reprinted with permission from Pergamon Press, Ltd.)

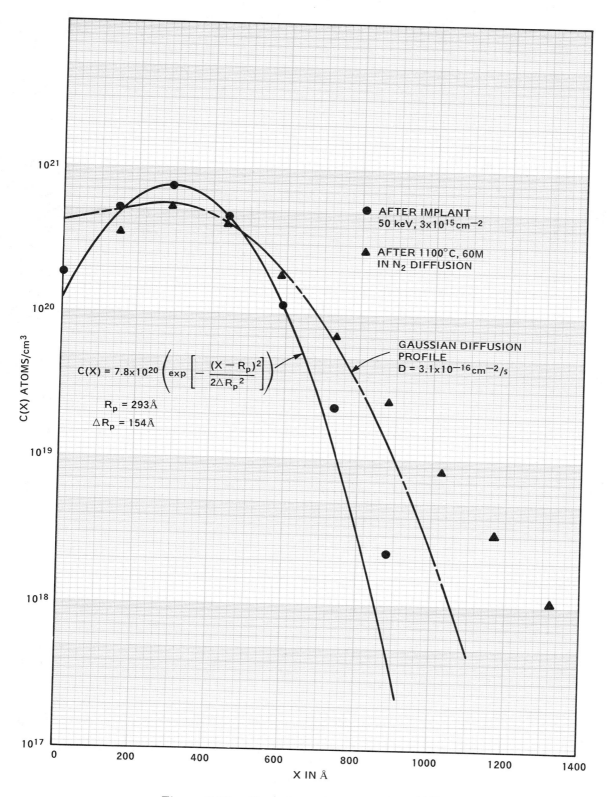

Figure 7-55 —Phosphorus Profile in SiO$_2$ [44]

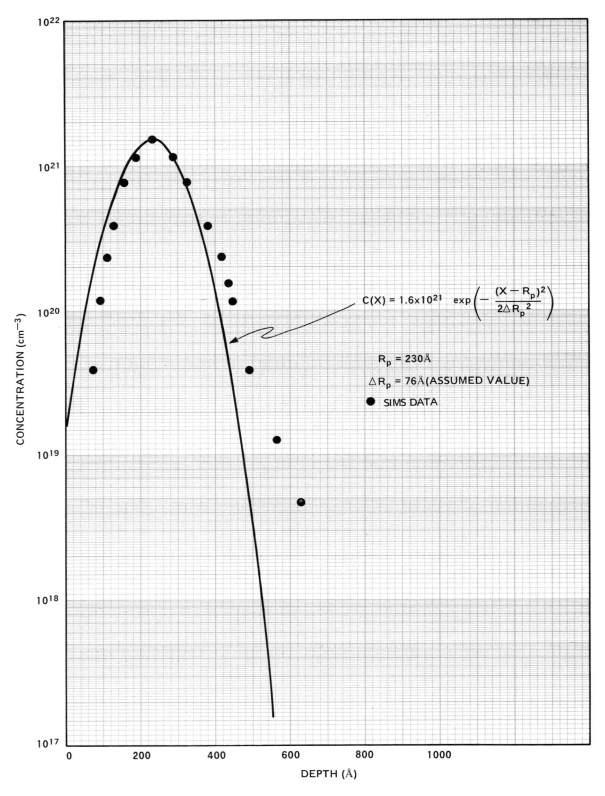

Figure 7-56 — Phosphorus Profile in Si_3N_4 [44]

7.9 Processing Results

This subsection presents some measurements from device processing that are not readily available in the current literature. Figure 7-57 shows an example of experimental estimation of boron surface concentration for a boron-implanted/diffused base as a function of steam oxidation time.[42] They were determined from SIMS profile measurements and extrapolated to the oxide silicon interface. Figure 7-58 shows the arsenic profiles for low-dose implants after various processing steps.[44] These profiles were determined by a C-V technique on Schottky diodes. Curve A is measured from a sample processed through the following steps:

(a) Steam oxidation at 1050°C for 2 hours.

(b) Boron diffusion with BN source at 1140°C for 60 minutes.

(c) Drive-in at 1200°C for 60 minutes in 1% O_2 and 99% N_2.

Curve B represents a sample which, in addition to the processing steps of Curve A, includes the following steps:

(d) Steam oxidation at 1050°C for 90 minutes.

(e) Phosphorous diffusion in $POCl_3$ at 1100°C for 60 minutes.

(f) Drive-in at 1200°C for 4 hours in 1% O_2 and 99% N_2.

Curve C represents a sample which, in addition to the following steps of Curves A and B, includes the following steps:

(g) Steam oxidation at 1050°C for 30 minutes.

(h) Diffusion in O_2 at 1150°C for 30 minutes.

(i) Steam oxidation at 1050°C for 2 hours.

(j) Phosphorous diffusion in $POCl_3$ at 1040°C for 45 minutes.

(k) Steam oxidation at 900°C for 45 minutes.

Figure 7-59 shows the pinch-off voltage for a p-channel JFET in CBIC technology as a function of boron dose.[46]

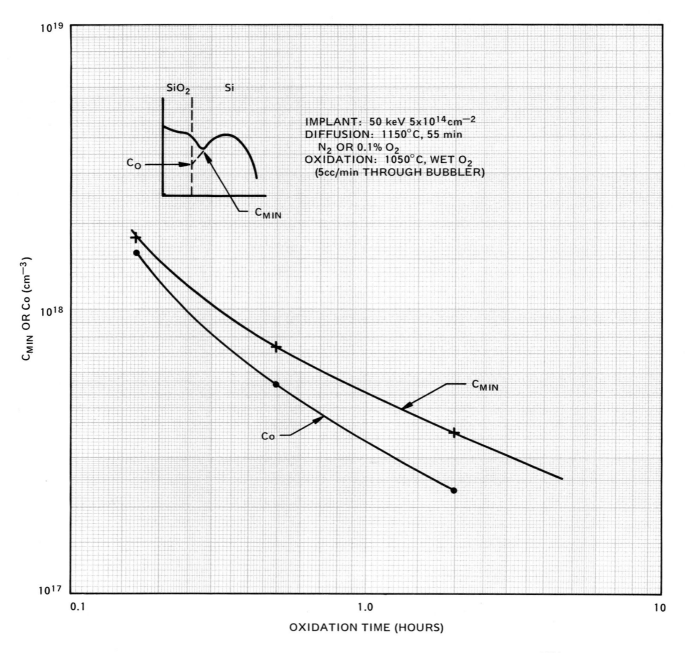

Figure 7-57 —Boron Surface Concentration Versus Oxidation Time[42]

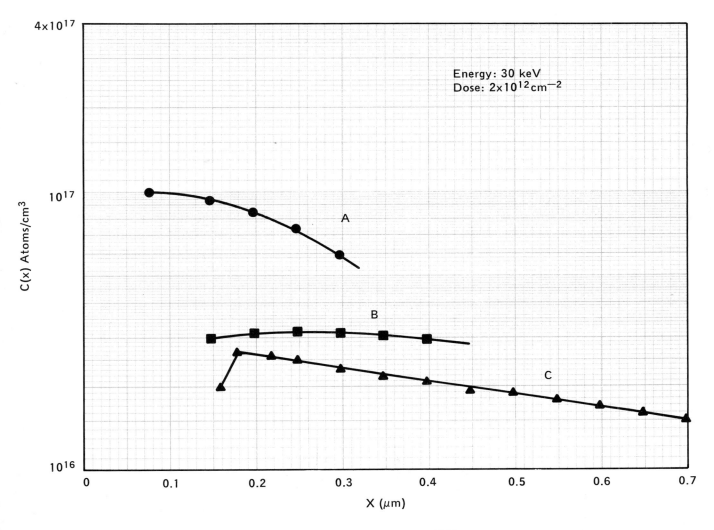

Figure 7-58 —Arsenic Profiles[44]

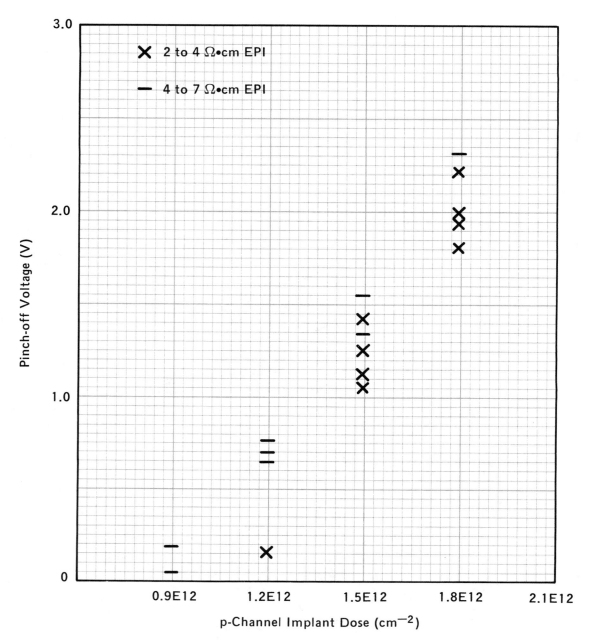

Figure 7-59 —Pinch-off Voltage of a p-Channel JFET[46]

7.10 Miscellaneous Information

Silicon Wafer Heating During Implantation

Figure 7-60 shows the calculated temperature rise due to heating of the ion beam during ion implantation.[47] Wada and Nishimatsu[48] (Figure 7-61) showed experimental data on the temperature rise on the back side of a silicon wafer and the back side of a wafer holder at different ion implant powers. For high-beam current implant at moderate ion energies, the thermal effect of the ion beam on the silicon wafer should be considered.

Photoresist Film as an Ion Implantation Mask

There is an increased interest in using photoresist as a mask for ion implantation in microwave devices where registration at a close tolerance on small-dimensioned windows is important. Using SiO_2 as an implant mask would require the application of photoresist, etching, and re-alignment of the structures. In some fabrication processes for self-aligned MOS devices, photoresist film has also been used as a mask for implantation. However, ion implantation into a photoresist film at a moderately high energy or high dose releases gaseous species during implantation. This can cause poor vacuum and lead to electrical discharge and interruption of the implantation run. The implanted ion also polymerizes the photoresist film so that it becomes difficult to remove. Little information has been published on photoresist as an ion implant mask. Baccarani and Pickar[49] studied the range statistics of boron in KTFR photoresist film. Figure 7-62 shows the KTFR thickness versus ion energy, with transmission coefficient as a parameter. Range statistics for various ions in AZ111 are tabulated by Gibbons, Johnson, and Mylroie.[5] Baicu and Steen[50] presented their results on using Shipley AZ1470 to mask implantation of arsenic at 180 keV for a dose of $6 \times 10^{15}\,cm^{-2}$. The photoresist was one micron thick. Adverse effects from photoresist masking were also discussed.

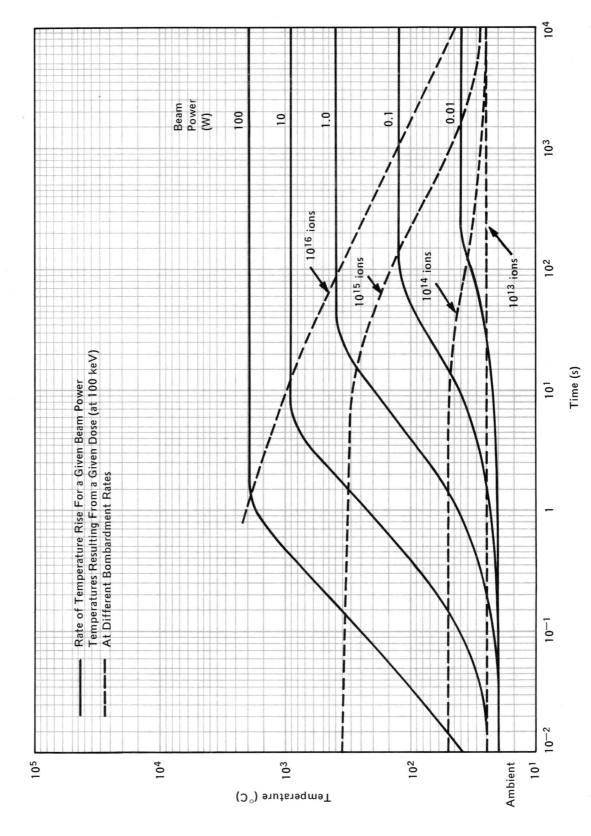

Figure 7-60 — Si Substrate Temperature Rise Due To Ion-Beam Heating During Ion Implantation[47]
(Reprinted with permission from North-Holland Publishing Company.)

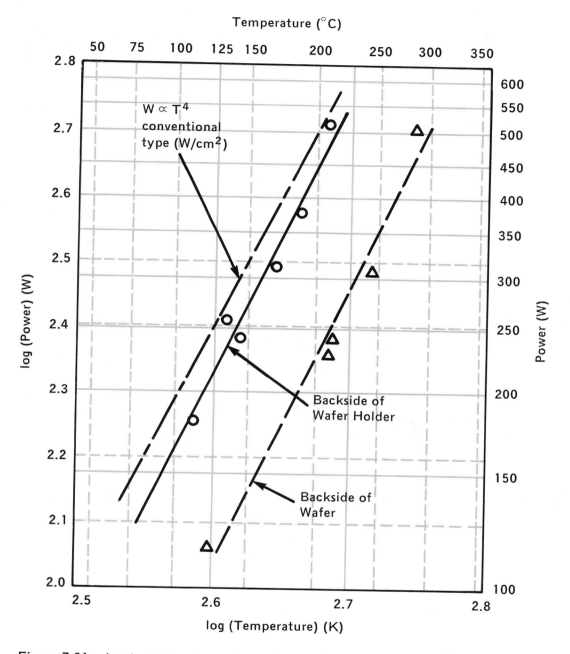

Figure 7-61 — Implantation Power Dependence of Temperature Rise[48]   (Reprinted
with permission from the Japanese Journal of Applied Physics)

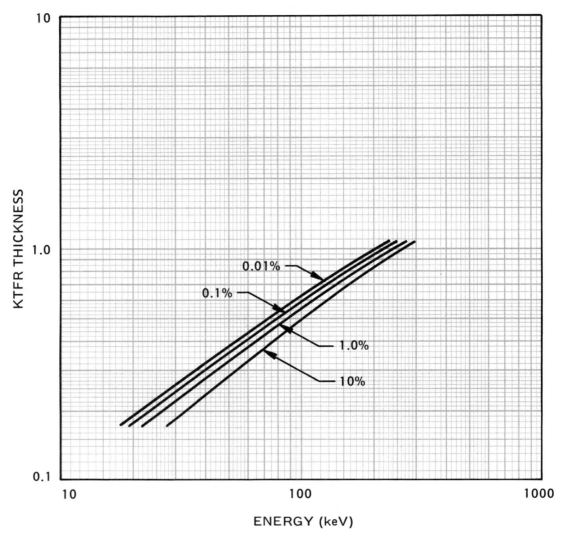

Figure 7-62 —KTFR Thickness Versus $^{11}B^+$ Energy With Transmission Coefficient as a Parameter[49]

REFERENCES

1. G. Carter and W. A. Grant, "Ion Ranges," *Ion Implantation of Semiconductors.,* New York: John Wiley & Sons, 1976, pp. 42-67.

2. J. Lindhard, M. Scharff, and H. E. Schiott, "Range Concepts and Heavy Ion Ranges," *Maf. Fys. Medd. Dan. Vid. Selsk.,* Vol. 33 (1963), pp. 1-42.

3. W. K. Hofker, "Implantation of Boron in Silicon," *Philips Res. Repts.,* Suppl. No. 8 (1975), p. 42.

4. J. F. Gibbons, "Ion Implantation in Semiconductors — Part 1 Range Distribution Theory and Experiments," *Proceedings of IEEE,* Vol. 56, No. 3 (March 1968), pp. 295-319.

5. J. F. Gibbons, W. S. Johnson, and S. W. Mylroie, "Projected Range Statistics," *Semiconductors, and Related Materials,* 2nd Ed. Dowden, Hutchinson and Ross, Inc., 1975.

6. F. F. Morehead, Jr. and B. L. Crowder, "A Model for the Formation of Amorphous Si by Ion Bombardment," *Radiation Effects,* Vol. 6 (1970), p. 30.

7. H. Okabayashi and D. Shinoda, "Lateral Spread of ^{31}P and ^{11}B Ions Implanted in Silicon," *J. Appl. Phys.,* Vol. 44, No. 9 (September 1973), pp. 4220-4221.

8. S. Furukawa, H. Matsumura, and H. Ishiwara, "Theoretical Considerations on Lateral Spread of Implanted Ions," *Japan. J. Appl. Phys.,* Vol. 11, No. 2 (February 1972), p. 138.

9. Y. Akasaka, K. Horie, and S. Kawazu, "Lateral Spread of Boron Ions Implanted in Silicon," *Appl. Phys. Letters,* Vol. 21, No. 4 (August 1972), pp. 128-129.

10. T. Tsuchiya, "Tilt Angle and Mask Edge Angle Dependences of Lateral Spread of Implanted Boron Ions Under Mask Edge," *J. Appl. Phys.,* Vol. 51, No. 11 (November 1980), pp. 5773-5780.

11. J. F. Gibbons and S. Mylroie, "Estimation of Impurity Profiles in Ion-Implanted Amorphous Targets Using Joined Half-Gaussian Distribution." *Appl. Phys. Letters,* Vol. 22 No. 11 (June 1973), pp. 568-569.

12. S. Furukawa and H. Ishiwara, "Range Distribution Theory Based on Energy Distribution of Implanted Ions," *J. Appl. Phys.,* Vol. 43, No. 3 (March 1972), pp. 1268-1273.

13. H. Ryssel et al, "Range Parameters of Boron Implanted Into Silicon," *Appl. Phys.,* Vol. 24 (1981), pp. 39-43.

14. E. I. Zorin et al, "Position of the Energy Levels of Ion-Implanted Impurities in Silicon," *Soviet Phys. Semiconductors,* Vol. 7, No. 10 (April 1974), p. 1340.

15. J. F. Gibbons, op cit, p. 307.

16. J. C. C. Tsai, private communication.

17. B. L. Crowder, "The Role of Damage in the Annealing Characteristics of Ion Implanted Si," *J. Electrochem. Soc.: Solid-State Science and Technology,* Vol. 177, No. 5 (May 1970), pp. 671-674.

18. R. J. Scavuzzo, private communication.

19. T. E. Seidel, private communication.

20. C. M. Drum, "Diffusion of Implanted Antimony and Arsenic in Silicon." Electrochem. Soc. Spring Meeting, San Francisco, CA., *Extended Abstr.* (1974), p. 210.

21. P. A. Schumann, Jr., "Plasma Resonance Calibration Curves for Silicon, Germanium, and Gallium Arsenide," *Solid-State Tech.,* Vol. 13, No. 1 (January 1970), pp. 50-52.

22. R. D. Plummer, private communication.

23. C. M. Drum, private communication.

24. C. M. Drum and P. Miller, private communication.

25. G. Nakamura et al, "Anomalous Annealing Behavior of Secondary Defects in Si Implanted with As Ions Through Dielectric Layer," *Ion Implantation in Semiconductors 1976,* ed. by F. Chernow, J. A. Borders, and D. K. Brice, New York: Plenum Press, 1977, pp. 493-498.

26. R. E. Ahrens, private communication.

27. T. E. Seidel and A. U. MacRae, private communication.

28. N. F. Krohn, private communication.

29. C. M. Drum and R. A. Porter, private communication.

30. A. Anderson and G. Swanson, "Electrical Measurements on Silicon Layers Implanted with 50 KeV Phosphorus Ions," *Radiation Effects,* Vol. 15, 1972, pp. 231-241, and J. C. C. Tsai, unpublished data.

31. R. A. Moline and G. W. Reutlinger, private communication.

32. J. C. C. Tsai, private communication.

33. T. E. Seidel et al, "Transistors with Boron Bases Predeposited by Ion Implantation and Annealed in Various Oxygen Ambients," *IEEE Trans. Electron Dev.,* Vol. ED-24, No. 6 (June 1977), pp. 717-723.

34. L. O. Bauer et al, "Properties of Silicon Implanted with Boron Ions Through Thermal Silicon Dioxide," *Solid-State Electron.,* Vol. 16 (1973), pp. 289-300.

35. J. C. North, A. C. Adam, and G. F. Richards, private communication.

36. J. C. North, private communication.

37. J. M. Andrews, "Electrical Conduction in Implanted Polycrystalline Silicon," *J. Electron. Mat.,* Vol. 8, No. 3 (1979), p. 232.

38. A. M. Gottlieb, private communication.

39. B. E. Eiche, R. F. Hornberger, and H. D. Seidel, private collection.

40. C. M. Drum, private communication.

41. R. Schimko et al, "Implanted Arsenic and Boron Concentration Profiles in SiO_2 Layers," *Phys. Stat. Sol.,* Vol. 28 (1975), p. 89.

42. R. A. Clapper and J. C. C. Tsai, private communication.

43. A. Essaid et al, "Boron Profile in Implanted SiO$_2$ on Silicon Substrate: Comparison Between Theory and Experiment," *Solid-State Electronics,* Vol. 24, No. 8 (1981), p. 787.

44. J. C. C. Tsai, private communication.

45. T. Hirao et al, "The Concentration Profiles of Projectiles and Recoiled Nitrogen in Si After Ion Implantation Through Si$_3$N$_4$ Films," *J. Appl. Phys.,* Vol. 50, No. 1 (January 1979), pp. 193-201.

46. D. E. Bien and R. L. Minear, private communication.

47. G. Dearnaley et al, "Production and Manipulation of Ion Beams," *Ion-Implantation,* Amsterdam: North-Holland Publishing Co., 1973, p. 426.

48. Y. Wada and S. Nishimatsu, "Application of High Current Arsenic Ion Implantation on Dynamic MOS Memory LSI's," *Japan. J. Appl. Phys.,* Vol. 18 (1978) Supplement 18-1, p. 248.

49. G. Baccarani and K. A. Pickar, private communication.

50. B. Baicu and L. Steen, Preliminary Application Report, Applied Implant Technology, A Subsidiary of Applied Materials, Inc., June 15, 1981.

51. J. C. North, A. C. Adam, and G. F. Richards, "Ion Implantation Doping of Polycrystalline Silicon," Electrochem. Soc. Fall Meeting, Pittsburgh, PA., *Extended Abstr.* 78-2 (1978), p. 542.

52. T. Tokuyama, "Invited: Ion Implantation into MOS Structures," J. Japan Soc. Appl. Phys., Vol. 43 (1974), Supplement 12-5, p. 501.

8. PROCESS DATA

8.1 General

This section provides the processing curves required to achieve the passivation and diffusion of silicon semiconductor devices. Subsection 8.2 is a brief description of the kinetics of oxidation. Subsection 8.3 contains process data on phosphorus and boron diffusions. Subsection 8.4 provides a table of pertinent data on low-pressure chemical vapor deposition of various films. Subsection 8.5 is a chart of furnace operation.

8.2 Oxidation

This subsection consists of a brief description of the kinetics of oxidation followed by the process curves for the following operations. A more detailed description may be found in reference 1.

(a) Steam oxidation (pyrogenic) (Figure 8-4).

(b) Dry oxidation (Figure 8-5).

(c) Oxidation of heavily doped phosphorus-diffused silicon (Figure 8-6).

(d) Oxidation of heavily doped boron-diffused silicon (Figure 8-7).

Kinetics of Oxidation

In order for the oxidizing species to reach the silicon surface it must go through three consecutive steps:

(a) It must be transported from the bulk of the gas to the gas-oxide interface.

(b) It must diffuse across the oxide layer already present.

(c) It must react at the silicon surface.

This process is represented schematically in Figure 8-1[2] with the corresponding fluxes.

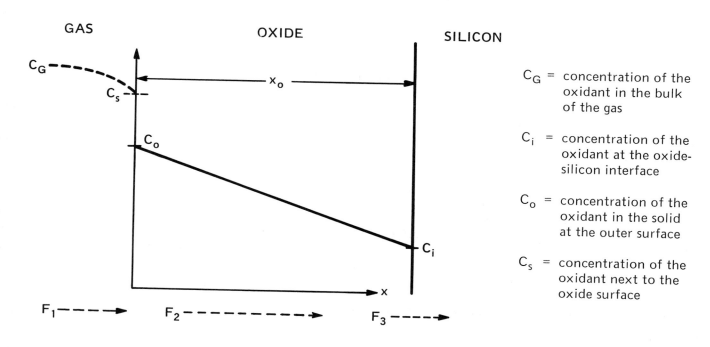

C_G = concentration of the oxidant in the bulk of the gas

C_i = concentration of the oxidant at the oxide-silicon interface

C_o = concentration of the oxidant in the solid at the outer surface

C_s = concentration of the oxidant next to the oxide surface

Figure 8-1 — Model for the Thermal Oxidation of Silicon[2]

The gas phase flux (F_1) is given by the following relationship:

$$F_1 = h_G (C_G - C_s),$$

8.1

where

$$h_G = \text{mass-transfer coefficient.}$$

Assume Henry's Law holds true; then

$$C_o = HP_s,$$

where

H = Henry's Law constant
P_s = partial pressure of oxidant
  next to the oxide surface.

Define an equilibrium concentration $C*$ (using Henry's Law again),

$$C* = HP_G,$$

where

P_G = partial pressure in the bulk of the gas.

Assume the oxidant gas to be ideal, then

$$C_G = P_G/KT,$$

$$C_s = P_s/KT.$$

Equation 8-1 then becomes

$$F_1 = h(C* - C_o),$$

where

h = gas-phase mass-transfer
  coefficient in terms of
  concentration in the solid,
  given by $h = h_G/HKT$.

The flux across the oxide (F_2) is given as a diffusive flux, defined by

$$F_2 = D \frac{C_o - C_i}{X_o},$$

where

D = diffusivity of the oxidizing
  species in the oxide layer

C_o = concentration of the oxidant at
  the gas-oxide interface

C_i = concentration of the oxidant at
  the oxide-silicon interface.

The rate of reaction taking place at the oxide-silicon interface is given by

$$F_3 = K_s C_i,$$

where

$$K_s = \text{chemical surface-reaction rate constant for oxidation.}$$

Under steady state conditions

$$F_1 = F_2 = F_3,$$

$$F_3 = F_2, \text{ so } K_s C_i = D/X_o (C_o - C_i),$$

$$F_3 = F_1, \text{ so } K_s C_i = h (C^* - C_o).$$

Eliminate C_o and solve for C_i so that

$$C_i = \frac{C^*}{1 + \dfrac{K_s}{h} + \dfrac{K_s X_o}{D}} . \qquad 8.2$$

Solve for C_o so that

$$C_o = \frac{(1 + \dfrac{K_s X_o}{D}) \; C^*}{1 + \dfrac{K_s}{h} + \dfrac{K_s X_o}{D}} . \qquad 8.3$$

Now consider the two limiting forms of Equations 8.2 and 8.3. When the diffusivity is very small, $C_i \rightarrow o$ and $C_o \rightarrow C^*$ (diffusion-controlled). When the diffusivity is very large, C_i and C_o will be equal and given by

$$C^* / (1 + K_s/h) \text{ (reaction-controlled)}.$$

To calculate the rate of oxide growth we must define N_1, where N_1 is the number of oxidant molecules (depending on the type of reaction) incorporated into a unit volume of oxide.

    (a)   In a dry oxide reaction, $Si + O_2 \rightleftarrows SiO_2$.

    (b)   In a steam oxide reaction, $Si + 2H_2O \rightleftarrows SiO_2 + 2H_2$.

The flux of oxidant reaching the oxide-silicon interface is

$$N_1 \frac{dX_o}{dt} = K_s C_i = \frac{K_s C^*}{1 + \dfrac{K_s}{h} + \dfrac{K_s X_o}{D}} .$$

Solve the above differential equation subject to the initial condition

$$X_o (o) = X_i,$$

where

$$X_i = \text{the thickness of an oxide layer grown in an earlier oxidation step.}$$

Then

$$X_o^2 - X_i^2 + A(X_o - X_i) = Bt,$$

$$A = 2D \left(\frac{1}{K_s} + \frac{1}{h} \right),$$

$$B = \frac{2DC^*}{N_1}.$$

In another form

$$X_o^2 + AX_o = B (t + \tau), \qquad\qquad 8.4$$

where

$$\tau = \frac{X_i^2 + A X_i}{B}.$$

Solving Equation 8.4 for oxide thickness as a function of time results in

$$\frac{X_o}{A/2} = \sqrt{1 + \frac{t + \tau}{A^2/4B}} - 1.$$

For the two limiting cases of this general relationship we get, for large times, i.e., $t \gg A^2/4B$,

$$X_o^2 = Bt,$$

or the so called parabolic relationship where B = parabolic rate constant; for small times, i.e.,

$(t + \tau) \ll A^2/4B$,

$$X_o = (B/A) (t + \tau),$$

or the so called linear region where B/A = the linear rate constant given by

$$B/A = \frac{K_s h}{K_s + h} \frac{C^*}{N_1}.$$

The two reactions are shown schematically in Figure 8-2 [3].

The general relationship for the oxidation of silicon and its two limiting forms is shown in Figure 8-3 [4].

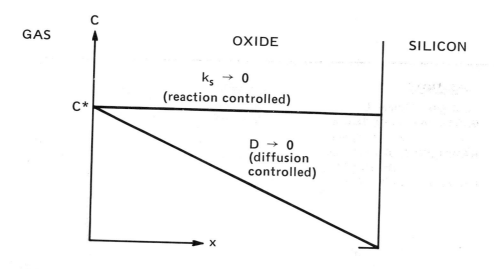

Figure 8-2 — Distribution of the Oxidizing Species in the Oxide Layer for the Two Limiting Cases of Oxidation [3]

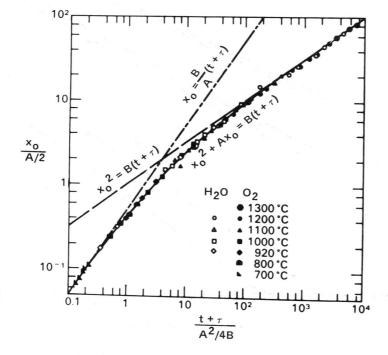

Figure 8-3 — The General Relationship for Silicon Oxidation and its Two Limiting Forms [4]

The factors influencing oxidation rates are

(a) Oxidant species.

(b) Temperature.

(c) Oxidant gas pressure.

(d) Crystallographic orientation of Si substrate.

(e) Substrate doping.

(f) Impurities or additives.

    (1) In gas ambient.

    (2) On Si surface.

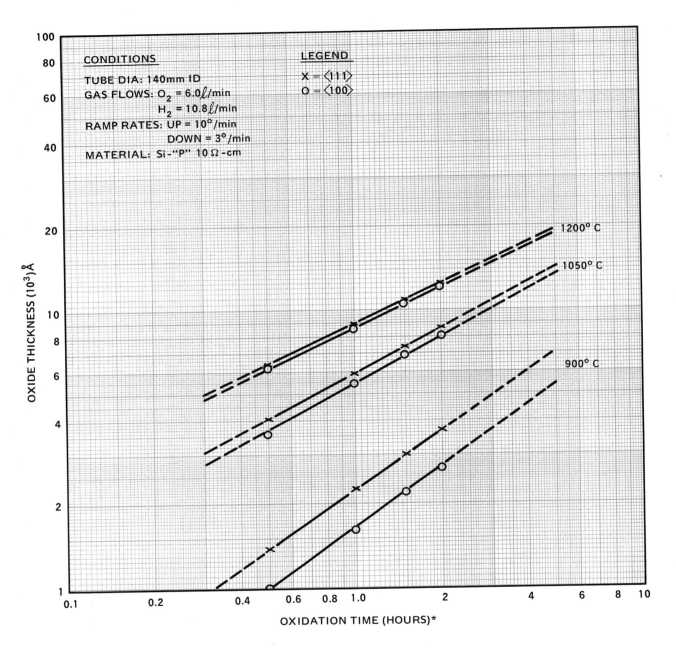

Figure 8-4 — Pyrogenic Oxide Thickness Versus Time

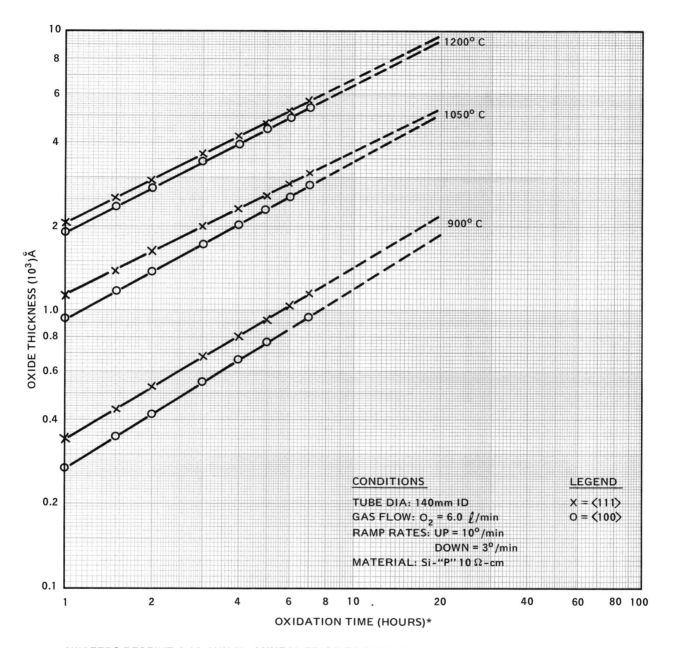

Figure 8-5 — Dry Oxide Thickness Versus Time

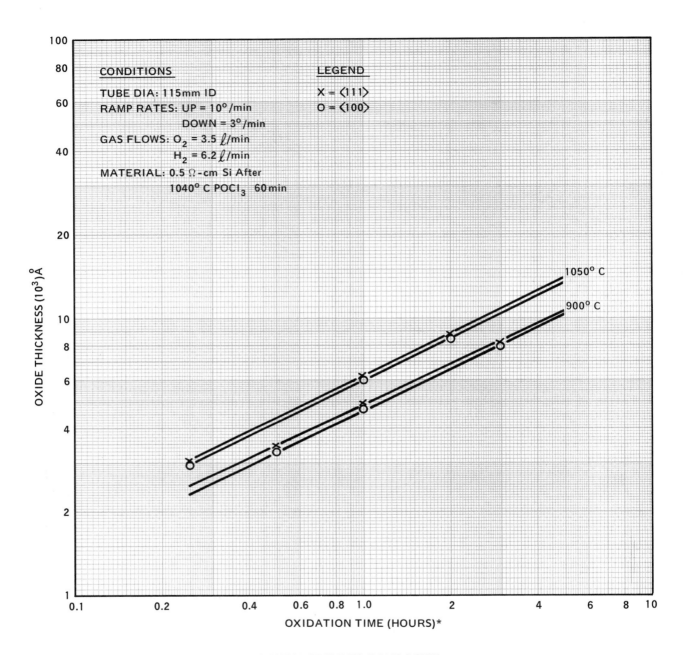

Figure 8-6 — Pyrogenic Oxidation of Phosphorus-Diffused Silicon

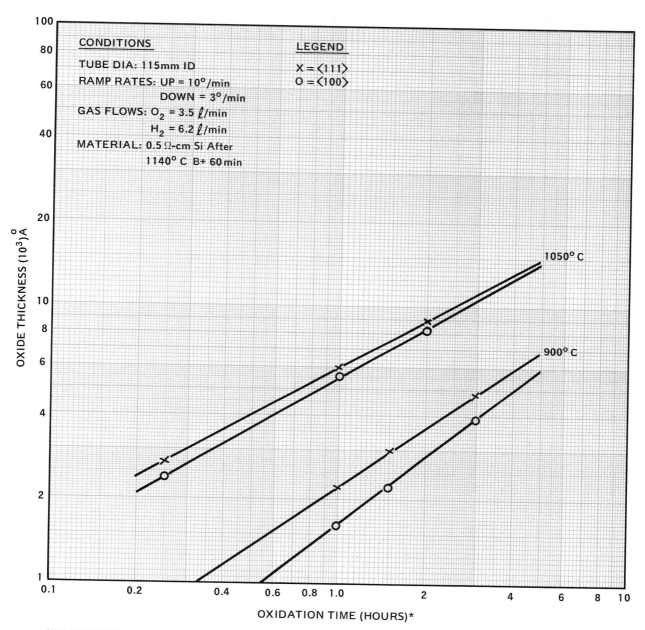

Figure 8-7 — Pyrogenic Oxidation of Boron-Diffused Silicon

8.3 Diffusion

Information on the kinetics of diffusion can be obtained from Section 6 of this manual. Included in this subsection are process data on phosphorus and boron diffusions along with glass removal curves for various phosphorus and boron diffusion times.

(a) $POCl_3$ sheet resistance (Figure 8-8).

(b) $POCl_3$ junction depth (Figure 8-9).

(c) B^+ sheet resistance (Figure 8-10).

(d) B^+ junction depth (Figure 8-11).

(e) Phosphorus-glass removal rate in 15:1 DI:HF (Figure 8-12).

(f) Boron-glass removal rate in 15:1 DI:HF (Figure 8-13).

8.4 Low-Pressure Chemical Vapor Deposition (LPCVD)

Table A provides pertinent data for the LPCVD of the following films.

(a) Polycrystalline silicon.

(b) Semi-insulating polycrystalline silicon (SIPOS).

(c) Si_3N_4.

(d) SiO_2.

TABLE A

CONDITIONS FOR LOW-PRESSURE CHEMICAL VAPOR DEPOSITION OF SELECTED FILMS

	POLY	SIPOS*	Si_3N_4	SiO_2
SiH_4 (cc/m)	190	190	—	—
SiH_2Cl_2 (cc/m)	—	—	30	30
N_2O (cc/m)	—	42	—	300
NH_3 (cc/m)	—	—	300	—
N_2 (cc/m)	400	400	—	—
T ($^\circ$C)	644	644	892	892
P (torr)	0.35	0.35	0.2	1.0
R (Å/m)	147	64	110	62

* For ~ 22 Atomic % O

8.5 Furnace Operation

Figure 8-14 schematically represents the ramp-up, dwell, and ramp-down cycles for typical drive-in, oxidation, and $POCl_3$ diffusions.

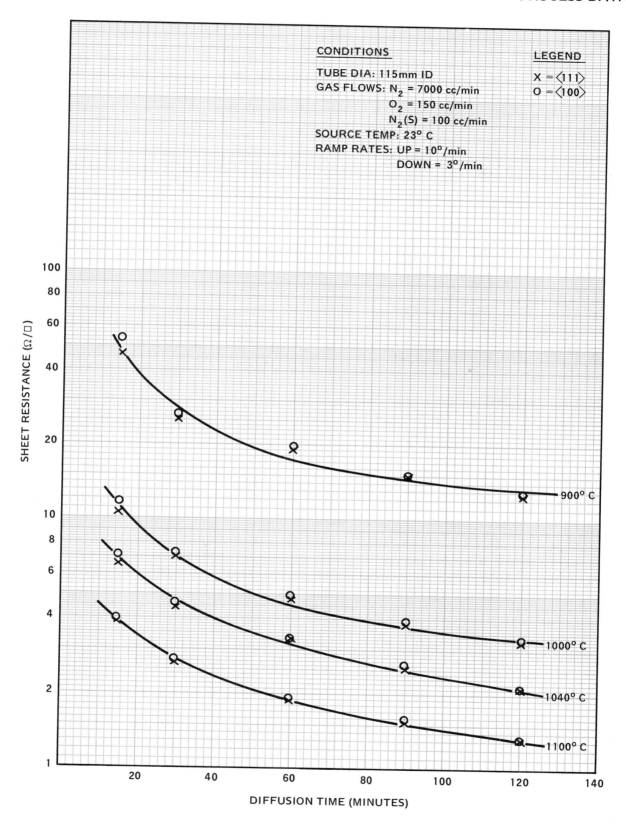

Figure 8-8 — POCl$_3$ Sheet Resistance Versus Time

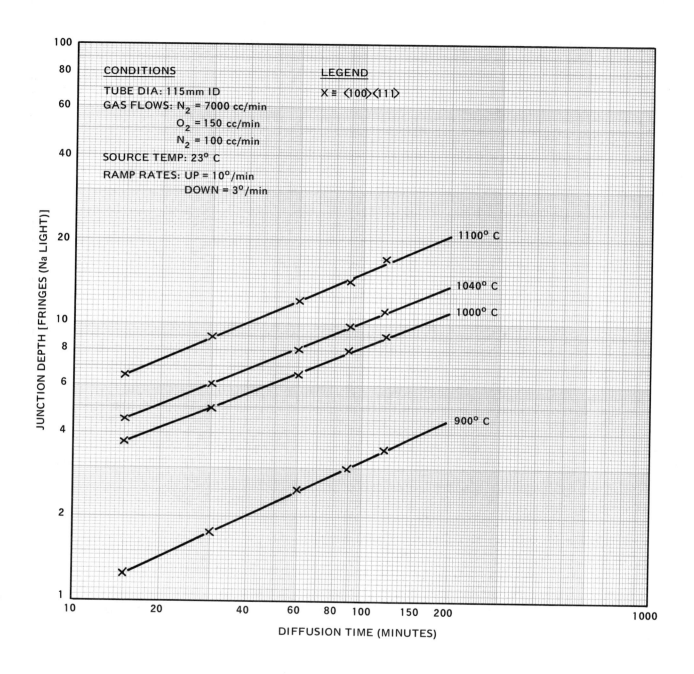

Figure 8-9 — POCl$_3$ Junction Depth Versus Time

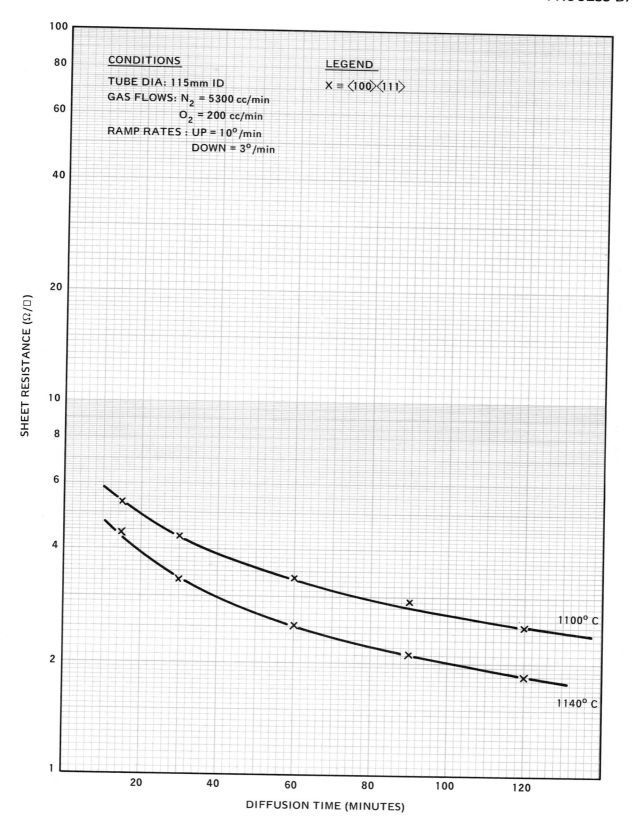

Figure 8-10 — B+ Sheet Resistance Versus Time

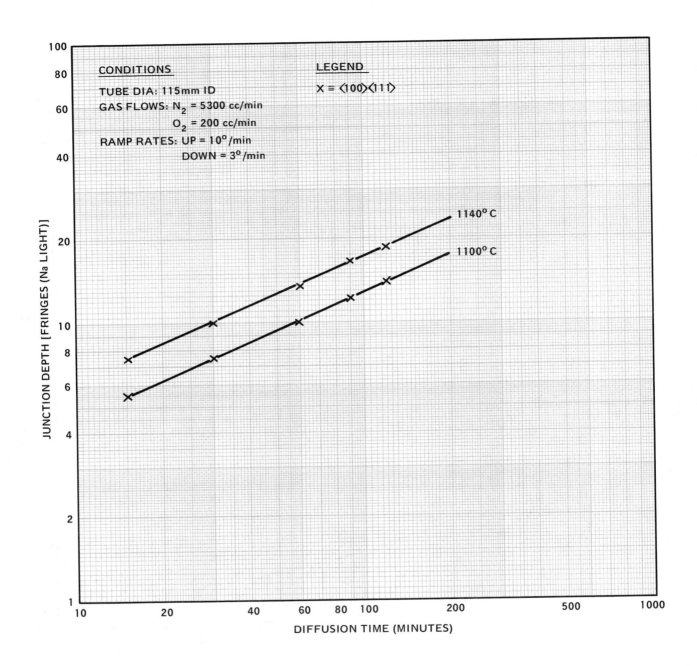

Figure 8-11 — B+ Junction Depth Versus Time

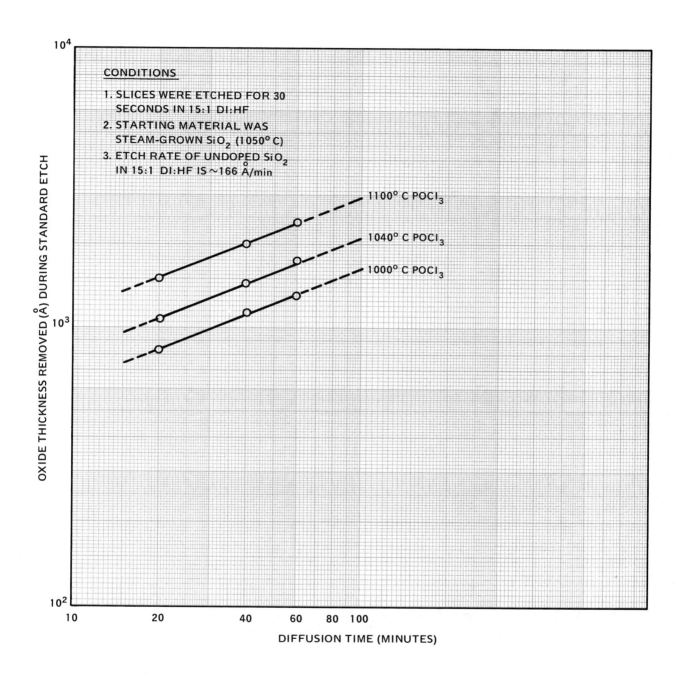

Figure 8-12 — Oxide Thickness Removed During Standard Etch Versus POCl₃ Diffusion Time

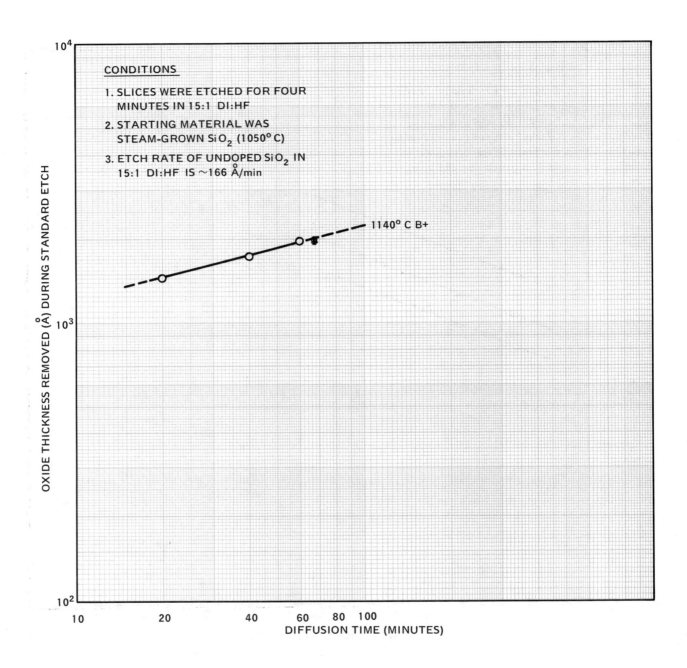

Figure 8-13 — Oxide Thickness Removed During Standard Etch Versus B+ Diffusion Time

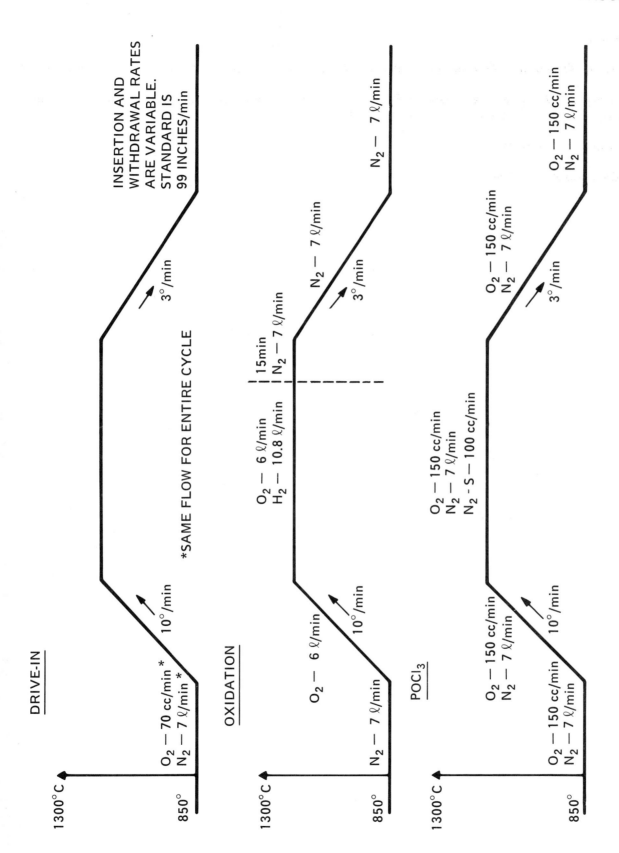

Figure 8-14 — Operating and Ramping Conditions

REFERENCES

1. A. S. Grove, *Physics and Technology of Semiconductor Devices.* New York: John Wiley & Sons, 1967.

2. B. E. Deal and A. S. Grove, "General Relationship for the Thermal Oxidation of Silicon," *J. Appl. Phys.,* Vol. 36, No. 12 (December 1965), p. 3771.

3. A. S. Grove, op cit, p. 26.

4. B. E. Deal and A. S. Grove, op cit, p. 3777.

9. IC STRUCTURES AND WAFER PROPERTIES

9.1   GENERAL

This section provides reference material on two aspects of silicon integrated circuit (SIC) design. Subsection 9.2 provides background and curves for calculating IC yield with respect to defect density D_0. Koehler, Josenhans, and Shichman[1], and Koehler and Chi[2] have described models to depict IC yield versus defect density. Subsection 9.3 is an overview of some representative SIC physical structures. Abbreviated representative process sequences which give some background for current IC technology are also included in this subsection. Subsection 9.4 is a glossary of SIC terminology and acronyms.

9.2   CALCULATION OF IC YIELD VERSUS DEFECT DENSITY, D_0

IC Chip Yield as a Function of Chip Size

As the size of an IC chip increases, the probability of a defect occurring on a chip increases, thus reducing the overall yield. The resulting dependency of yield versus chip area depends on the state of the art, the design rules, the technology, and the process; it is also subject to many other variations, especially variations with time (such as learning, etc.).

Various expressions have been used to describe the relationship of yield Y as a function of chip area A in such a way that the expression is of general validity and that only the numerical values are determined by the specific conditions stated above. The parameters of interest in these expressions are:

(a) D_0 = mean defect density, expressed in number of defects per area (commonly per cm^2)

(b)   D may not be the same for all chips under consideration (within a wafer, lot, or code). If it is distributed, a specific shape of the distribution must be assumed, which leads to a specific equation of Y = f(A); if the distribution is considered variable, the deviation σ or variance $\lambda = \sigma^2$ is described within the equation.

(c)   If, within a fraction of the population, no yield is obtained ("wipeout"), the yield-producing fraction of the population, over which randomly distributed defects are observed, is called Y_0.

Yield Models

Figure 9-1 presents a table of some yield models*, the underlying shapes of the distributions of defects, and the values of the variances of these distributions. The last model, which is based on a gamma distribution of defects[3][4] (which in contrast to the gaussian distribution does not allow negative values of D), leads to a particularly simple expression:

$$\frac{Y}{Y_0} = \frac{1}{(1+\lambda D_0 A)^{1/\lambda}} \, .$$

In this expression, λ is the variance of the distribution of D and is a parameter; λ is normalized with respect to D_0^2. Three special cases are noteworthy:

(a)   For $\lambda = 0$, the expression becomes $\frac{Y}{Y_0} = \exp{(-D_0 A)}$

(b)   For $\lambda \approx 0.2$, the expression is almost identical to the second model in Figure 9-1, since the Y = f(A) relationship is not very sensitive to the specific shape of the distribution. This triangular distribution model has frequently been called the "Murphy Model." Note, however, that Murphy[5] proposed this model, together with the rectangular distribution, as an example for including the spread of the defect density, D_0.

*One model, used occasionally, which uses the expression $Y/Y_0 = 1/(1+D_0 A)^n$, is not included in the table since it is erroneously based on certain statistical considerations that do not apply here.

DISTRIBUTION OF D	$\dfrac{Y}{Y_0} =$	$\lambda =$
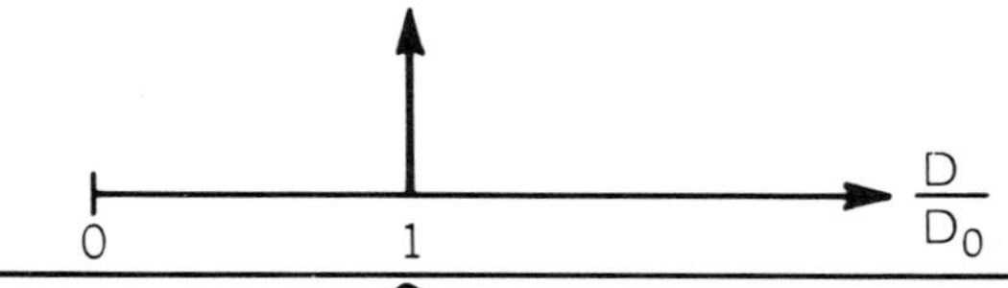	$e^{-D_0 A}$	0
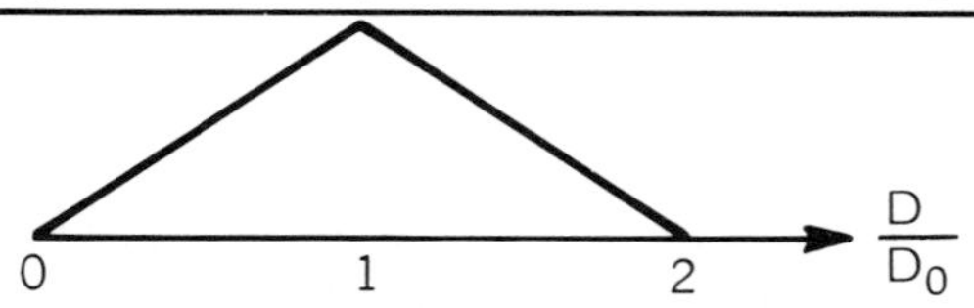	$\left(\dfrac{1-e^{-D_0 A}}{D_0 A}\right)^2$	0.22 (±0.02)
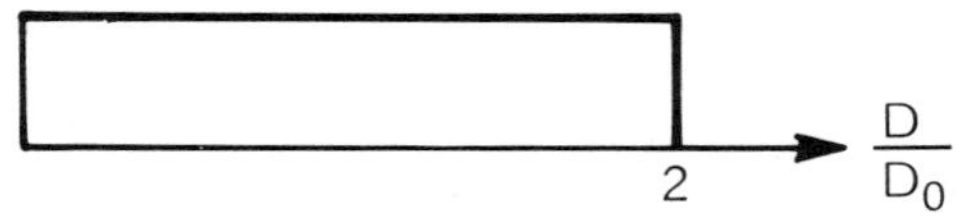	$\dfrac{1-e^{-2D_0 A}}{2 D_0 A}$	0.5 (±0.1)
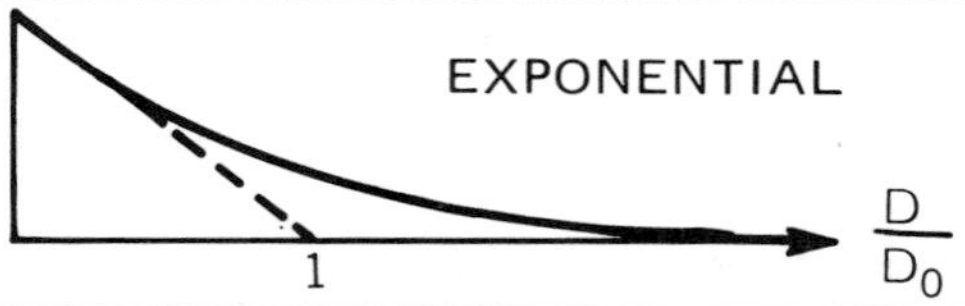 EXPONENTIAL	$\dfrac{1}{1+D_0 A}$	1
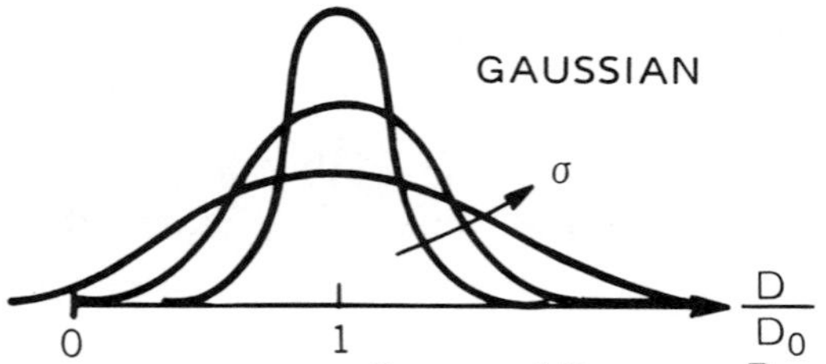 GAUSSIAN $\dfrac{1}{\sqrt{2\pi}\sigma}\ \exp\left[-\dfrac{1}{2}\left(\dfrac{\frac{D}{D_0}-1}{\sigma}\right)^2\right]$	$\dfrac{1}{2}\exp\left(-AD_0 + \dfrac{\sigma^2 A D_0}{2}\right) \quad \}$ $1+\mathrm{erf}\left[\dfrac{1}{\sqrt{2}}\left(\dfrac{1}{\sigma}-\sigma D_0 A\right)\right] \quad \}$	$\approx \sigma^2$ (for small σ)
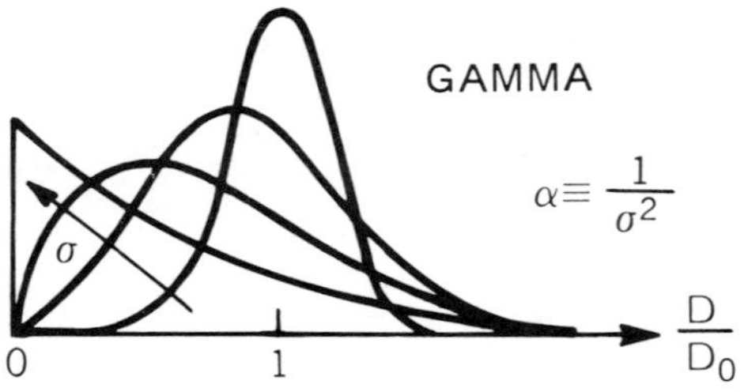 GAMMA $\alpha \equiv \dfrac{1}{\sigma^2}$ $\dfrac{\alpha}{\Gamma(\alpha)}\left(\alpha\dfrac{D}{D_0}\right)^{\alpha-1}\exp\left(-\alpha\dfrac{D}{D_0}\right)$	$\dfrac{1}{(1+\sigma^2 D_0 A)^{1/\sigma^2}} \quad\equiv$ $\boxed{\dfrac{1}{(1+\lambda D_0 A)^{1/\lambda}}}$	σ^2

σ = standard deviation normalized with respect to D_0

Figure 9-1 — Yield Models

(c) For $\lambda = 1$, the expression becomes identical to the exponential distribution proposed by Seeds.[6]

The 3-parameter yield model,

$$\frac{Y}{Y_0} = \frac{1}{(1+\lambda D_0 A)^{1/\lambda}} \, ,$$

is represented in semilogarithmic form in Figure 9-2 as $\frac{Y}{Y_0} = f(D_0 A)$. As the spread λ of D within the population considered (wafer, lot, etc) increases, the yield for devices of larger area and/or more defects becomes better since the areas with low D contribute primarily to the yield.

While this representation of $\frac{Y}{Y_0} = f(D_0 A)$ is useful in showing the effects of the variations of the various parameters, a plot of $Y/Y_0 = f(A)$ is more practical for many uses. A set of graphs for various values of λ is given in Figures 9-3 through 9-10. Note again that the graph of $\lambda = 0.2$ (Figure 9-5) is practically identical with the one for the triangular distribution that has been in wide use (the "Murphy Model"). To determine yield from these curves, the values for λ, Y_0, and D_0 must be chosen in accordance with the process used and the related past experience.

Depending on the values of the parameters chosen, the equation can describe a technology, a code, a lot, a wafer, or a part of a wafer, such as a ring within a wafer.

A curve giving yield versus defect density for beam lead chips specialized for specific beam sizes is shown in Figure 9-11. This curve assumes $\lambda \sim 0.2$.

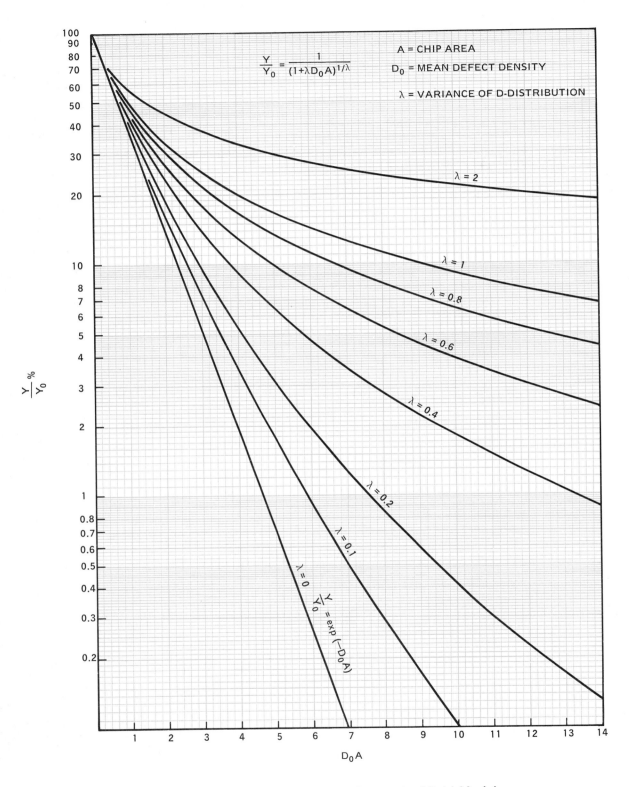

Figure 9-2 — Three-Parameter Yield Model

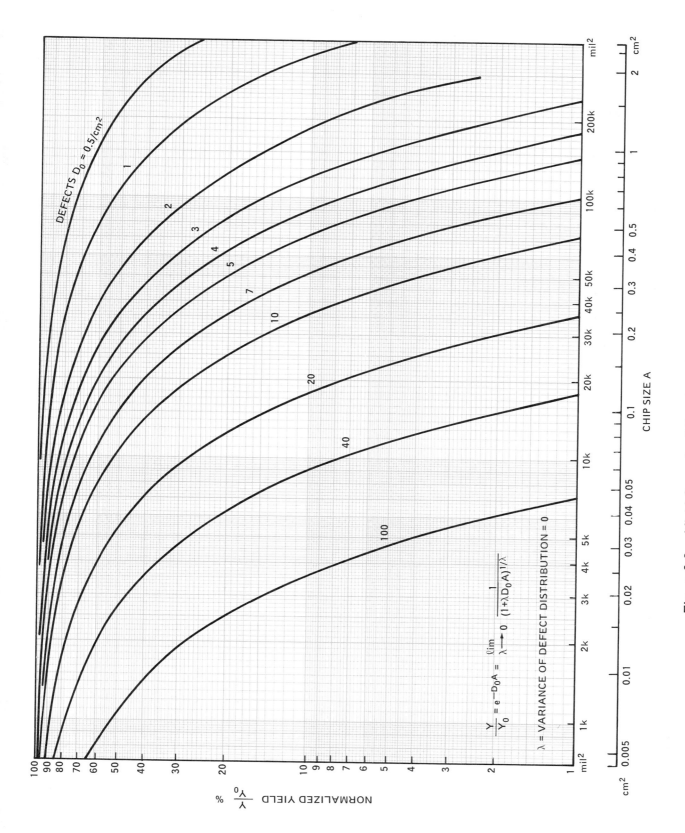

$$\frac{Y}{Y_0} = e^{-D_0A} = \lim_{\lambda \to 0} \frac{1}{(1+\lambda D_0 A)^{1/\lambda}}$$

λ = VARIANCE OF DEFECT DISTRIBUTION = 0

Figure 9-3 — Yield Curve for 3-Parameter Yield Model, $\lambda = 0$

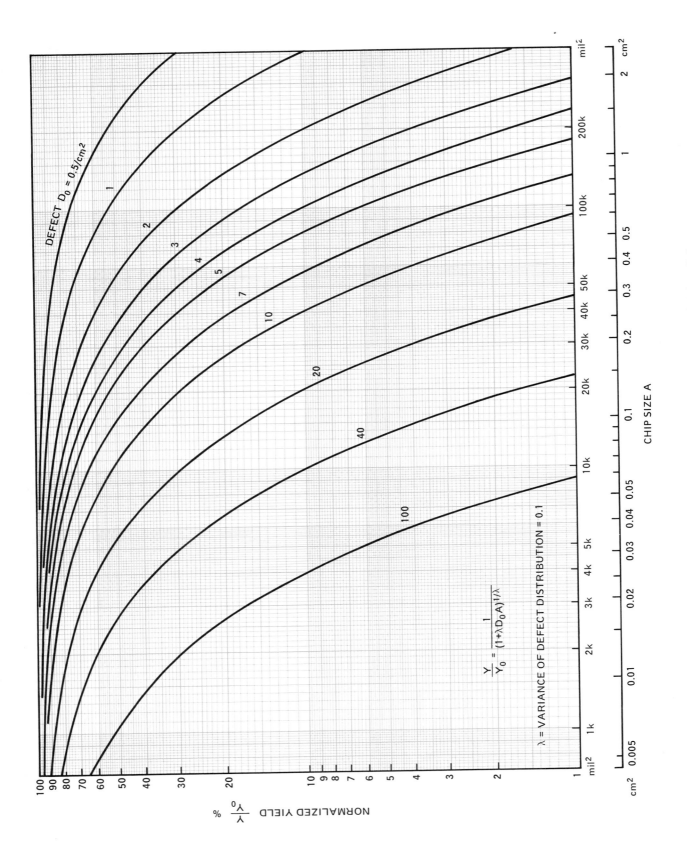

$$\frac{Y}{Y_0} = \frac{1}{(1+D_0A)^{1/\lambda}}$$

λ = VARIANCE OF DEFECT DISTRIBUTION = 0.1

Figure 9-4 — Yield Curve for 3-Parameter Yield Model, $\lambda = 0.1$

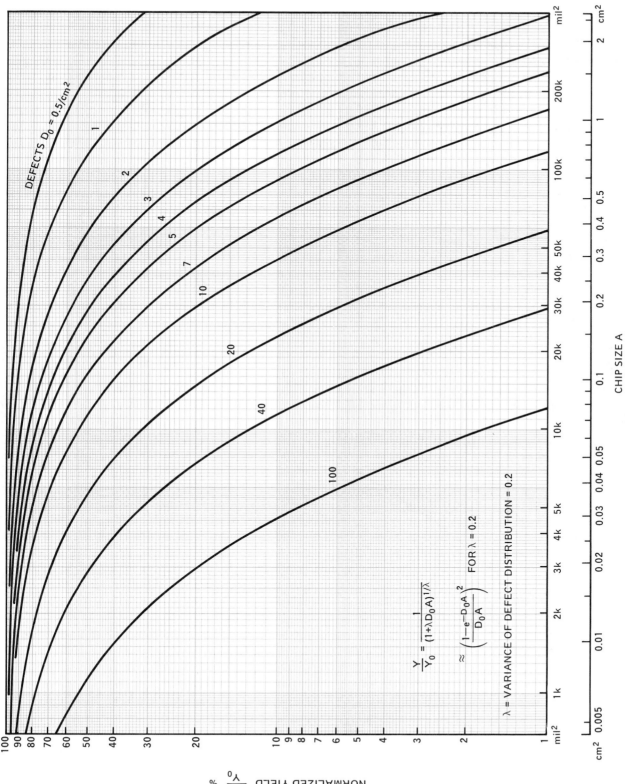

$$\frac{Y}{Y_0} = \frac{1}{(1+\lambda D_0 A)^{1/\lambda}}$$

$$\approx \left(\frac{1-e^{-D_0 A}}{D_0 A}\right)^2 \quad \text{FOR } \lambda = 0.2$$

λ = VARIANCE OF DEFECT DISTRIBUTION = 0.2

Figure 9-5 — Yield Curve for 3-Parameter Yield Model, $\lambda = 0.2$

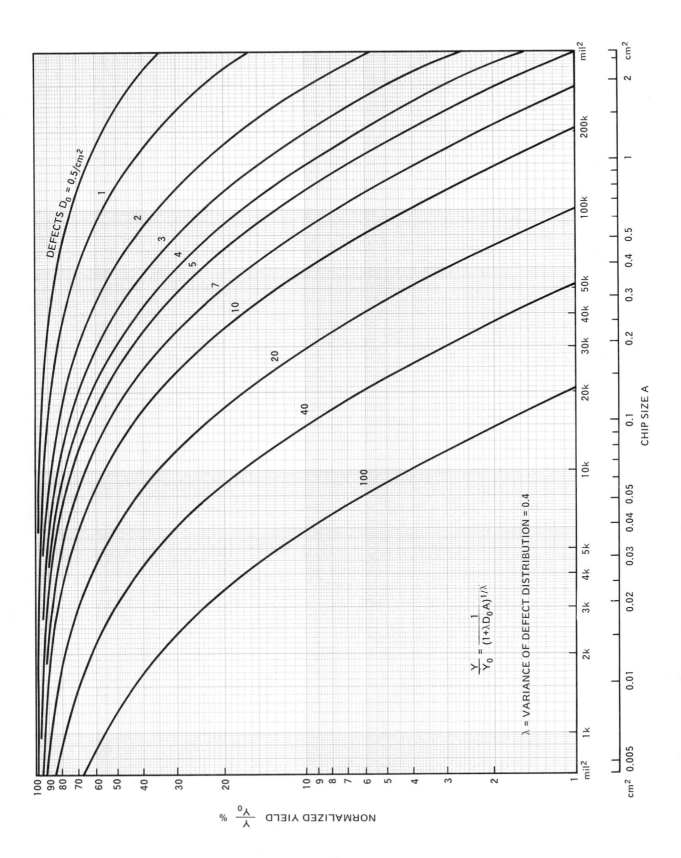

$$\frac{Y}{Y_0} = \frac{1}{(1+\lambda D_0 A)^{1/\lambda}}$$

λ = VARIANCE OF DEFECT DISTRIBUTION = 0.4

Figure 9-6 — Yield Curve for 3-Parameter Yield Model, λ = 0.4

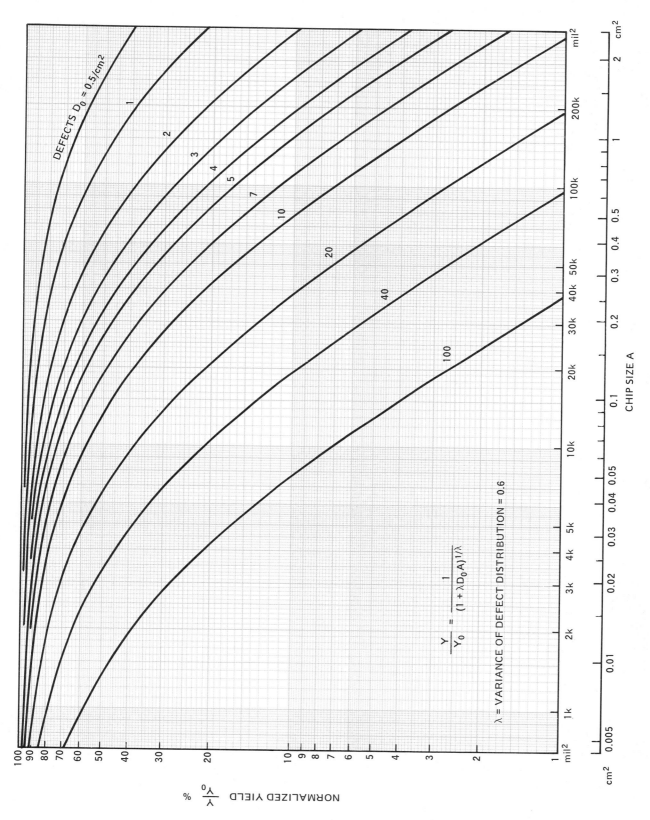

$$\frac{Y}{Y_0} = \frac{1}{(1 + \lambda D_0 A)^{1/\lambda}}$$

λ = VARIANCE OF DEFECT DISTRIBUTION = 0.6

Figure 9-7 — Yield Curve for 3-Parameter Yield Model, λ = 0.6

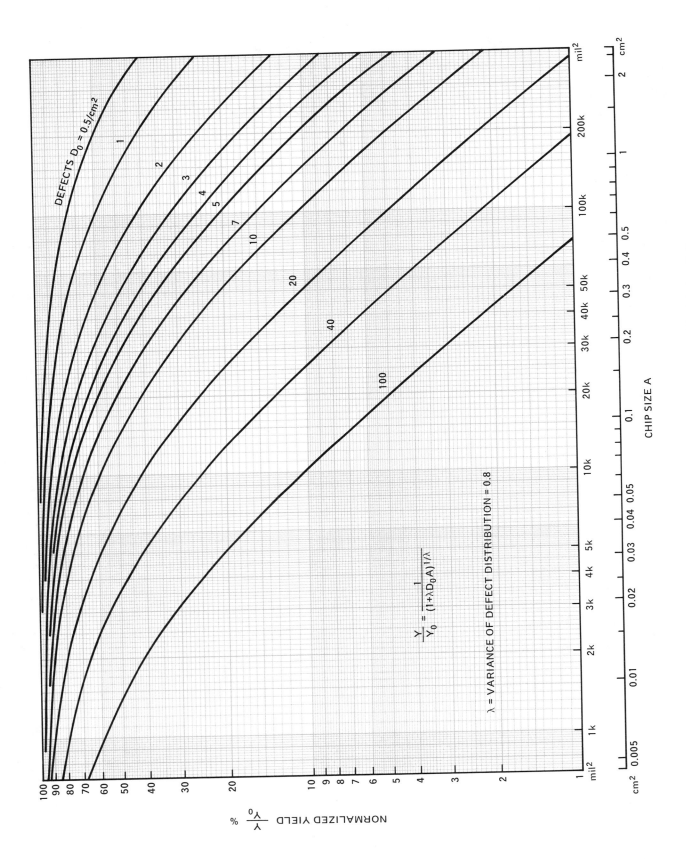

$$\frac{Y}{Y_0} = \frac{1}{(1+\lambda D_0 A)^{1/\lambda}}$$

λ = VARIANCE OF DEFECT DISTRIBUTION = 0.8

Figure 9-8 — Yield Curve for 3-Parameter Yield Model, $\lambda = 0.8$

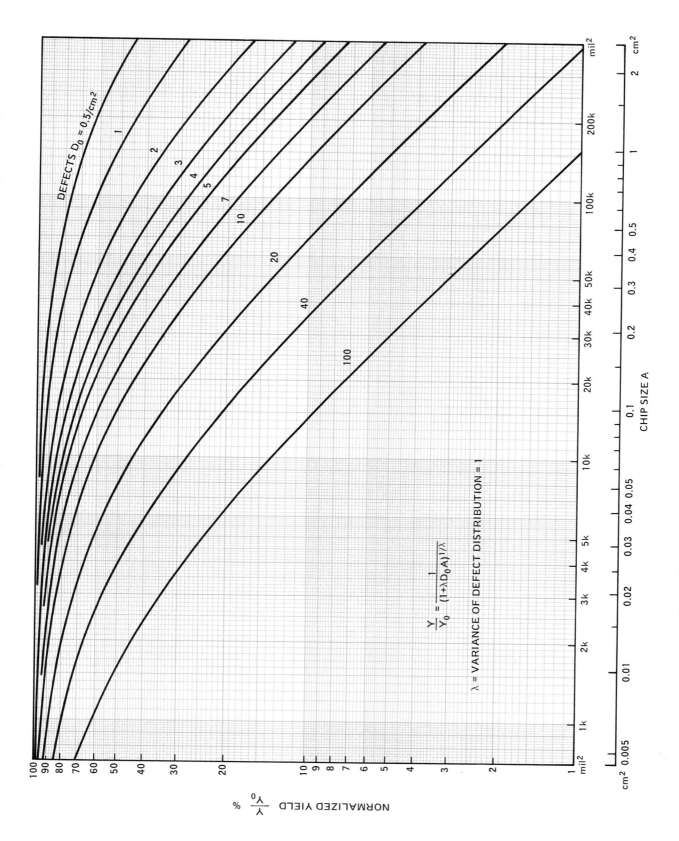

$$\frac{Y}{Y_0} = \frac{1}{(1+\lambda D_0 A)^{1/\lambda}}$$

λ = VARIANCE OF DEFECT DISTRIBUTION = 1

Figure 9-9 — Yield Curve for 3-Parameter Yield Model, $\lambda = 1$

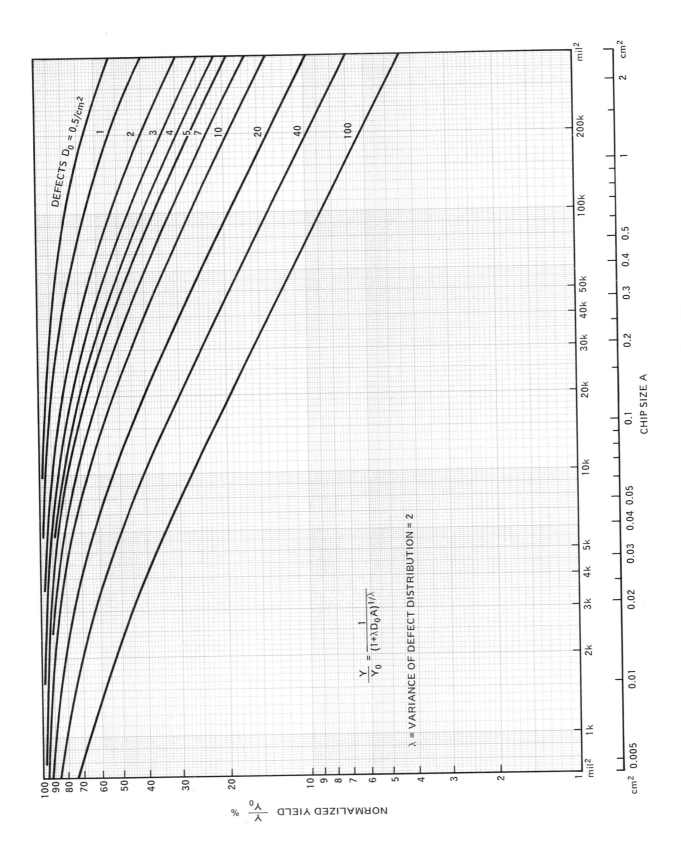

$$\frac{Y}{Y_0} = \frac{1}{(1+\lambda D_0 A)^{1/\lambda}}$$

λ = VARIANCE OF DEFECT DISTRIBUTION = 2

Figure 9-10 — Yield Curve for 3-Parameter Yield Model, $\lambda = 2$

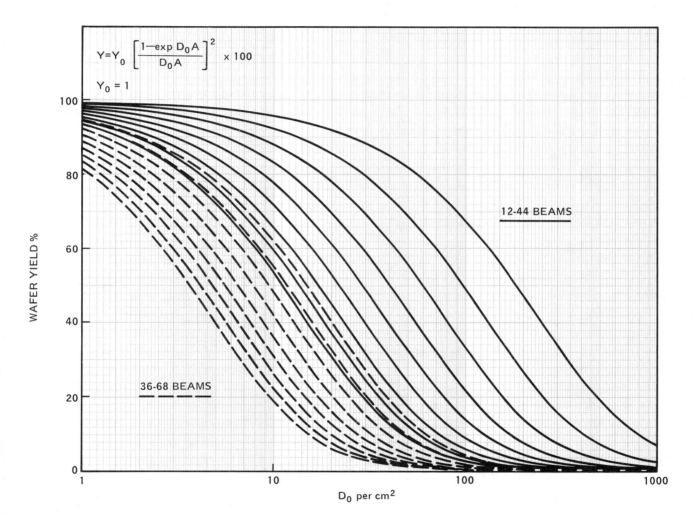

$$Y = Y_0 \left[\frac{1 - \exp D_0 A}{D_0 A} \right]^2 \times 100$$

$$Y_0 = 1$$

Figure 9-11 — Yield Versus Defect Density for Beam Lead Chips

9.3   SILICON INTEGRATED CIRCUIT (SIC) STRUCTURES

Scale of Integration

Figure 9-12   presents a chronological overview showing the increase in the number of components per IC chip from 1960 to 1980 and a projection of the further increase in components through the year 2000. Also shown in this figure is a graph showing the decrease in the minimum feature length from 1960 to 1980 and a projection of the further decrease in minimum size through the year 2000.

Basic IC Components

Silicon integrated circuit technology can be classified into two main categories depending on the basic transistor type used: bipolar or MOS unipolar transistors. A cross section of an oxide-isolated npn bipolar transistor is shown in Figure 9-13.   Figure 9-14 depicts the change in base thickness and emitter stripe width for high-speed bipolar transistors up to 1980 and the projected change in base thickness and emitter stripe width through the year 1990.

A cross section of a n-channel MOSFET is shown in Figure 9-15. A generalized guide for MOSFET scaling is shown in Figure 9-16.

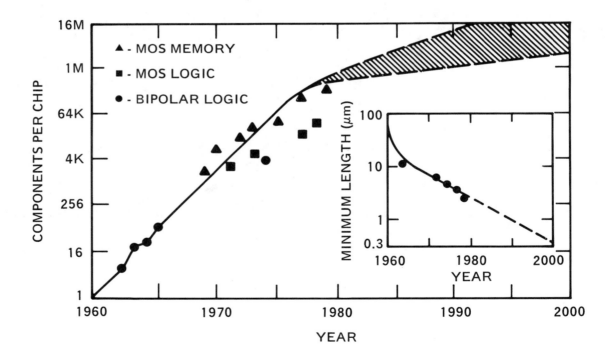

Figure 9-12 — Chronology of Change in IC Component Count and Minimum Length[7]

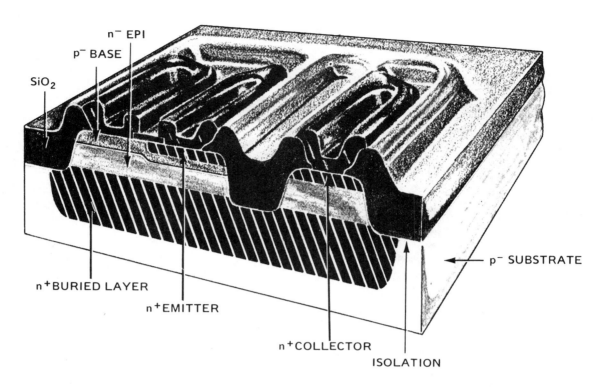

Figure 9-13 — Oxide-Isolated npn Bipolar Transistor—Cross Section

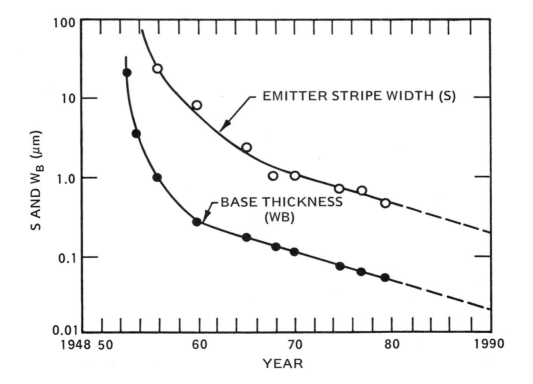

Figure 9-14 — Chronology of Change in Base Thickness and Emitter Stripe Width[7]

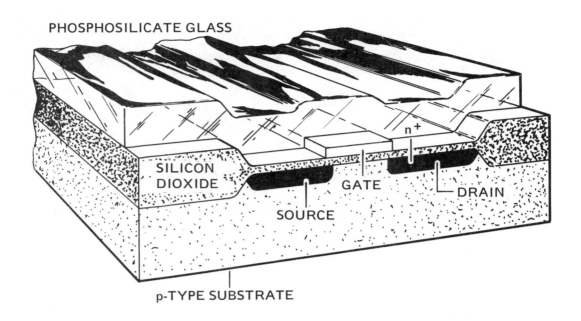

Figure 9-15 — n-Channel MOSFET—Cross Section

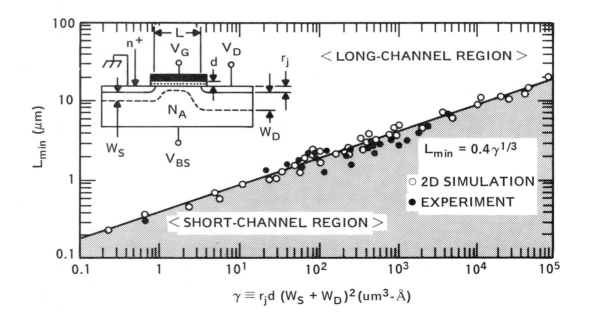

Figure 9-16 — Generalized Guide for MOSFET Scaling [7]

Technology Classifications

Verhofstadt[8] presents a discussion of IC technologies in general. The presentation in this manual is limited to some typical technologies. A graphical breakdown of monolithic silicon integrated circuit technologies is given in Figure 9-17. The two main categories, bipolar and MOS unipolar, are based on the transistor type used: bipolar or MOSFET. The next lower level of technology classification is determined by the method of transistor isolation: lateral or vertical. The common integrated circuit technologies included in these broad categories are the next level of organization. Finally, the significant circuit specializations associated with each of these IC technologies makes up the lowest level of organization. The descriptions of the basic technologies that follow will adhere to this organizational pattern. Definitions of acronyms will be found in the glossary in subsection 9.4.

A. Bipolar SIC

The bipolar transistor must be electrically isolated within an IC. The two basic isolation techniques are pn junction isolation and dielectric (oxide) isolation. Since these two techniques are used both separately and in combination, it is convenient to break the method of isolation classification into two levels: lateral and vertical. As these names imply, lateral isolation separates the devices one from another across the surface of the chip while vertical isolation separates the device from the substrate.

Bipolar, Junction-Isolated SIC

Consider the junction-isolated branch of Figure 9-17. Here, vertical isolation is accomplished either by epitaxial techniques or by a surface-diffused junction. Consider first the epitaxial technology.

SBC

The basic junction-isolated epitaxial bipolar IC device, the standard buried collector (SBC) npn transistor, is illustrated in Figure 9-18. The heavily doped n^+ buried layer and the associated n^+ collector region provide a low-resistance collector contact accessible from the surface of the transistor. The transistor collector region is an n-type tub isolated by a surrounding pn junction. The pn junction forming the tub walls is fabricated by a p^+ diffusion from the surface of the device which penetrates to the p-type substrate. It provides the lateral junction isolation for the transistor. The substrate is the p-type bottom of the tub. The epi to p-substrate (or n^+ buried layer to p-substrate) diode provides vertical isolation.

A Schottky clamped SBC process sequence is illustrated by Figure 9-19.

As indicated in Figure 9-17, the SBC structure is the basic structure for a wide variety of logic and linear circuit configurations. Integrated injector logic, I^2L,[9] is an innovative extension of the SBC technology. The surface topology and cross section of a merged pnp-npn logic cell are shown in Figure 9-20.

CDI

Collector diffused isolation (CDI) (Figure 9-21) is a variation of SBC epitaxial bipolar technology. This technology, while retaining the epi and buried layer, eliminates the p-isolation and p-base diffusions of SBC.

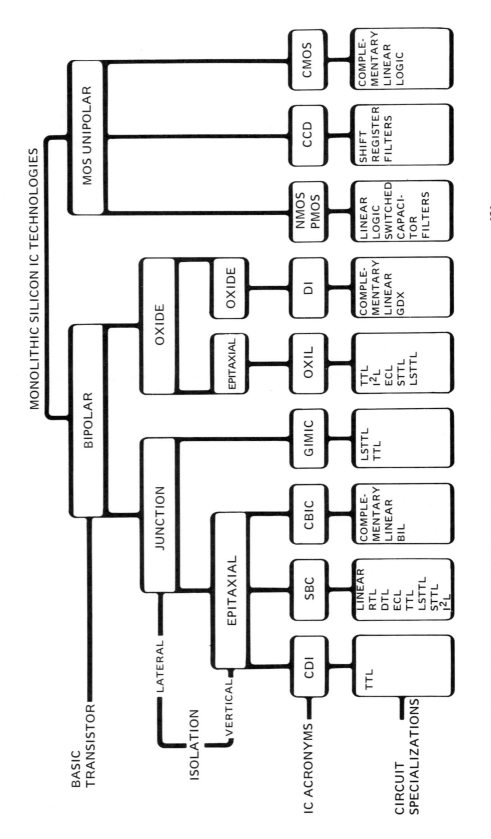

Figure 9-17 — Monolithic Silicon Integrated Circuit Technologies (After Verhofstadt[8])(© 1976 IEEE)

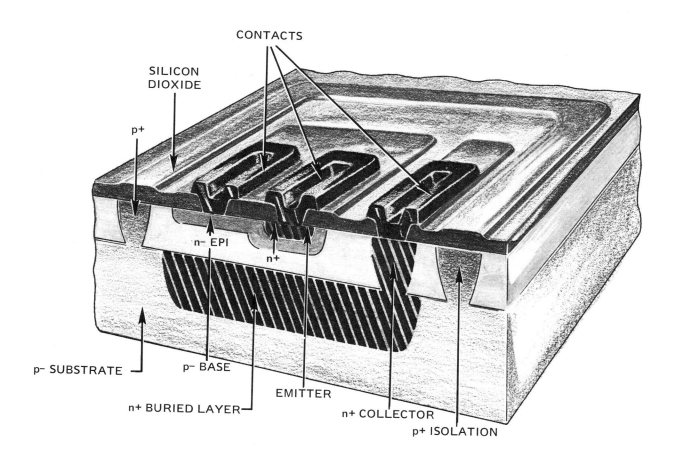

Figure 9-18 — SBC (Standard Buried Collector) npn Transistor Cross Section

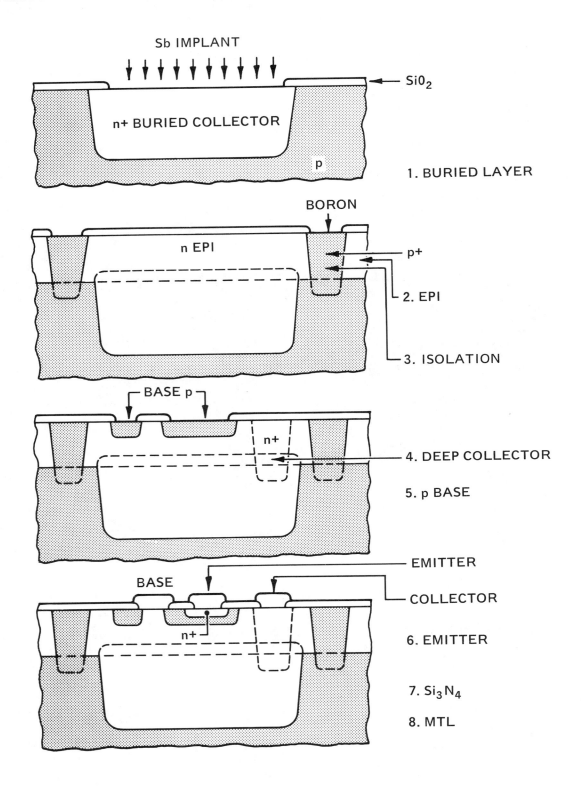

Figure 9-19 — SBC Schottky Clamped Process Sequence

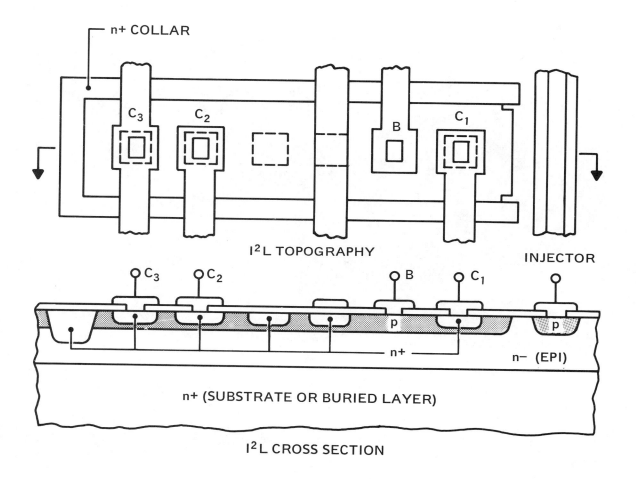

Figure 9-20 — I²L Cross Section and Topography

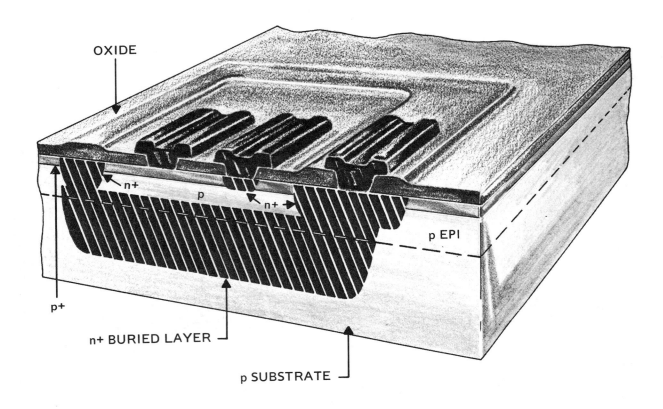

Figure 9-21 — CDI Cross Section

GIMIC-0

An outline of the guard-ring isolated monolithic integrated circuit (GIMIC-0) process[10] is shown in Figure 9-22. This process sequence is provided since it is a modern extension of the original diffused collector isolation scheme. In GIMIC-0, the diffused n— collector region, or tub, provides both lateral and vertical isolation from the π (lightly doped p-type) substrate, thus avoiding the buried layer and epitaxial complications of SBC. It is used extensively in low-power Schottky clamped TTL.

CBIC/BIL

The most sophisticated linear bipolar technology is CBIC, or complementary bipolar integrated circuit. It provides truly complementary high-speed npn and pnp transistors. (See Figure 9-23.) It also includes a compact I^2L type of logic cell compatible with 30-volt linear functions. An isometric view of this buried injector logic (BIL)[11] is shown in Figure 9-24.

Bipolar, Oxide-Isolated SIC

OXIL

Oxide isolated logic (OXIL)[12] (Figure 9-25) is an example of a bipolar oxide-isolated epitaxial technology. The lateral isolation is provided by oxide walls. The vertical isolation is provided by the n^+ buried collector to p-substrate junction. OXIL combines high-frequency conventional down-transistors with inverted up-transistors which are fabricated in a common process on the same chip site. The up-transistor is especially designed to optimize I^2L circuits for high packing density, speed, and performance. High-pressure-steam oxide isolation and an up-diffused active base are combined to fabricate the up-transistor with $f_t > 500\,MHz$ and $\beta = 100$, which allows I^2L delays down to 3 ns at FO = 1 and 7 ns at FO = 6. The down-transistor is an oxide-isolated implanted-base transistor with an As emitter. It exhibits gains of 100 to 150 at $f_t = 2\,GHz$ and supports subnanosecond current mode logic (CML), high-current buffer circuitry, and linear interfacing.

GDS

The final bipolar transistor category is dielectrically isolated SIC. An example is the gated diode switch (GDS) of Figure 9-26.[13] This technology provides bipolar devices in tubs totally isolated by oxide, with a polysilicon substrate. Since the oxide-isolated technology provides nearly ideal electrical isolation with a great deal of device flexibility, it has been used for some time for specialized high-performance complementary linear circuits. In addition, it extends the voltage capability of bipolar SIC.

B. MOSFET SIC

The MOSFET is a self-isolated device. The source, drain, and channel regions are normally biased such that no current flows to external points.

NMOS

Figure 9-27 shows a typical single-level polysilicon NMOS process sequence.[14] Figure 9-28 shows two variations of the basic MOSFET structure. The first of these is VMOS, in which an external groove is used to define the channel regions. The second variation, DMOS, uses a double diffusion process to realize the MOSFET channel. These technologies are finding extensive use in high-speed, high-power switching. Figure 9-29 is a process sequence for a two-level n-channel polysilicon gate MOS memory structure.[15]

CMOS

Where both n- and p-channel MOSFETs are used, complementary MOS (CMOS) tubs are required to house the devices. Figure 9-30 shows a twin-tub CMOS cross section where the appropriate MOSFET tub is formed in a ν (lightly doped n-type) epi layer by either n- and p- type surface diffusions.[16]

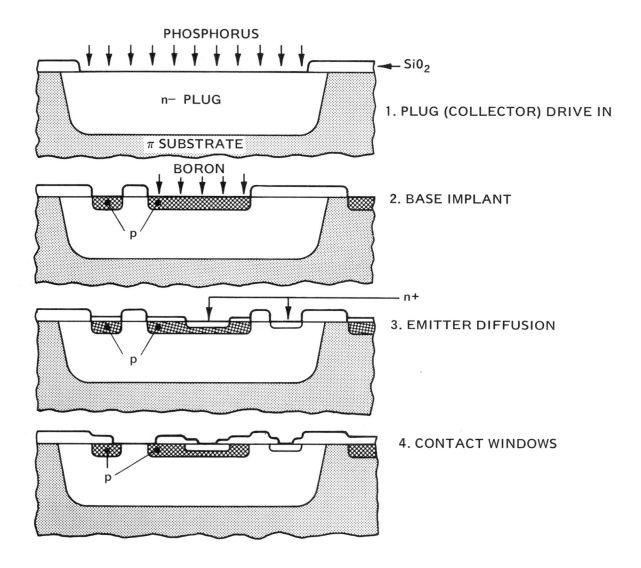

Figure 9-22 — GIMIC-0 Process Sequence[10] (© 1974 IEEE)

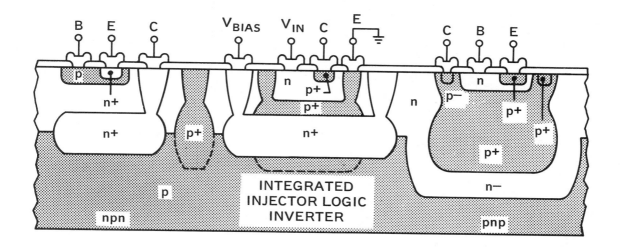

Figure 9-23 — CBIC Buried Injection Logic Structure

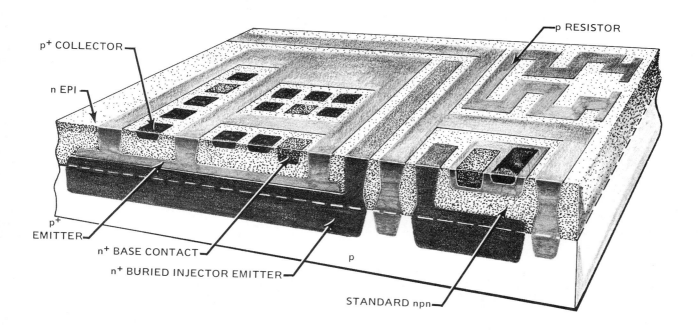

Figure 9-24 — Buried Injector Logic[11] (© 1978 IEEE)

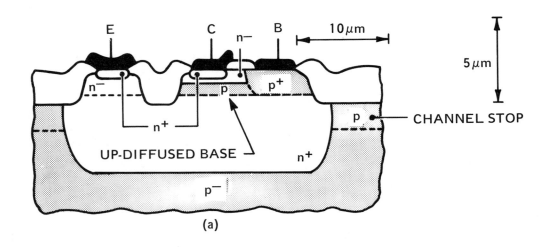

(a)

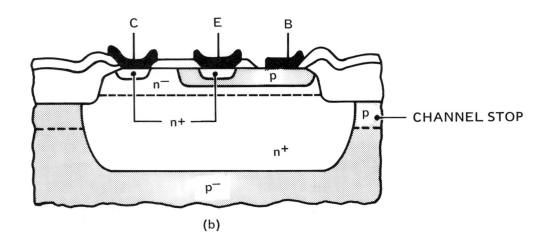

(b)

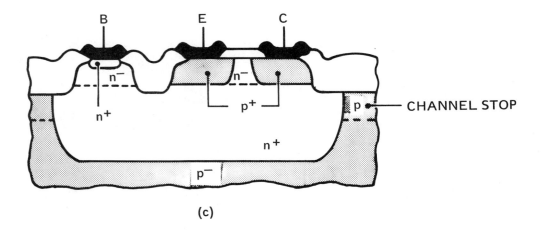

(c)

Figure 9-25 — Cross Section of (a) Up-Transistor, (b) Down-Transistor, and
(c) Lateral pnp Transistor in the OXIL Technology[12] (© 1980 IEEE)

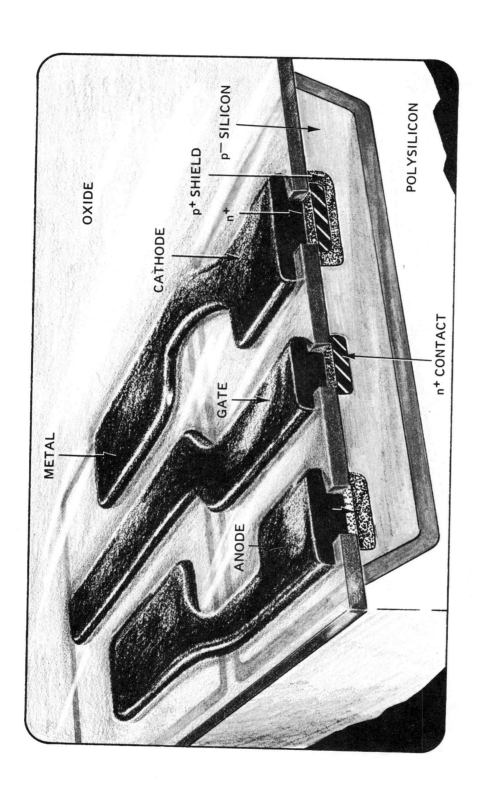

Figure 9-26 — Cross Section of the Gated Diode Switch[13] (© 1980 IEEE)

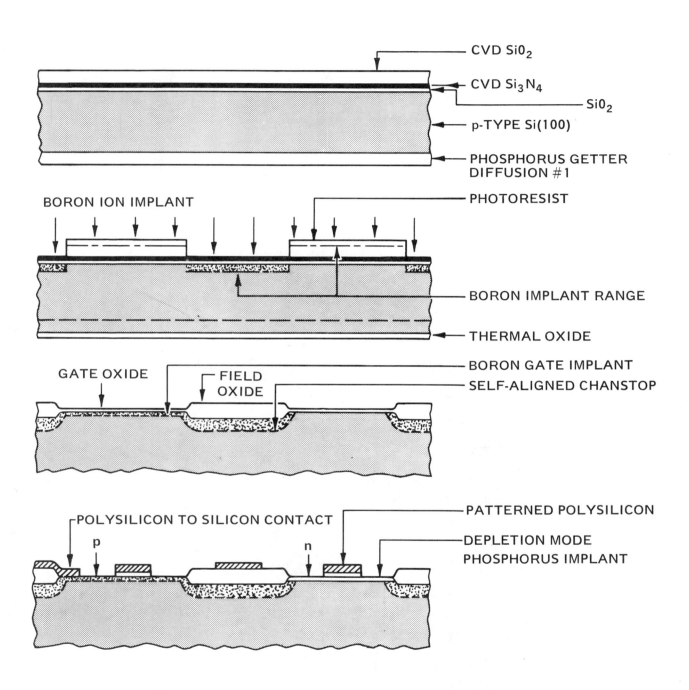

Figure 9-27 — Typical MOS Process Sequence (Sheet 1 of 2)[14] (© 1975 IEEE)

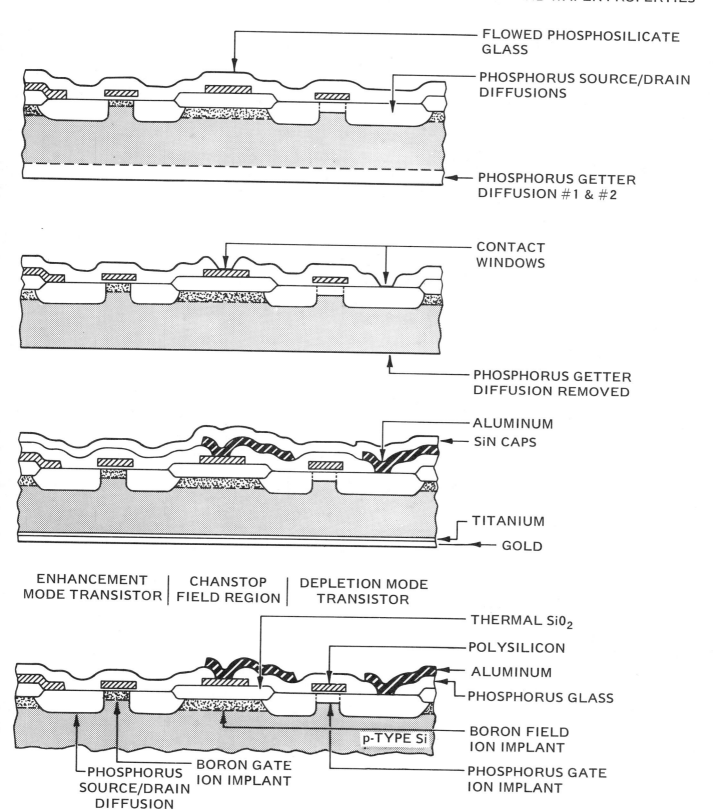

Figure 9-27 — Typical MOS Process Sequence (Sheet 2 of 2)[14] (© 1975 IEEE)

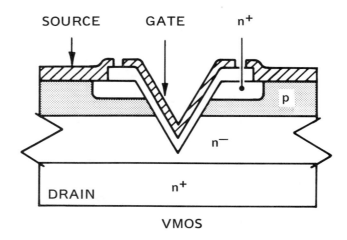

VMOS

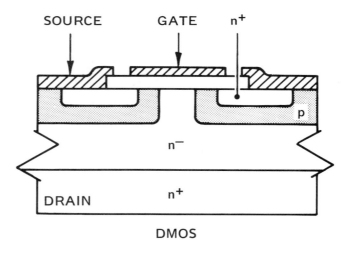

DMOS

Figure 9-28 — VMOS & DMOS Mosfet Cross Sections

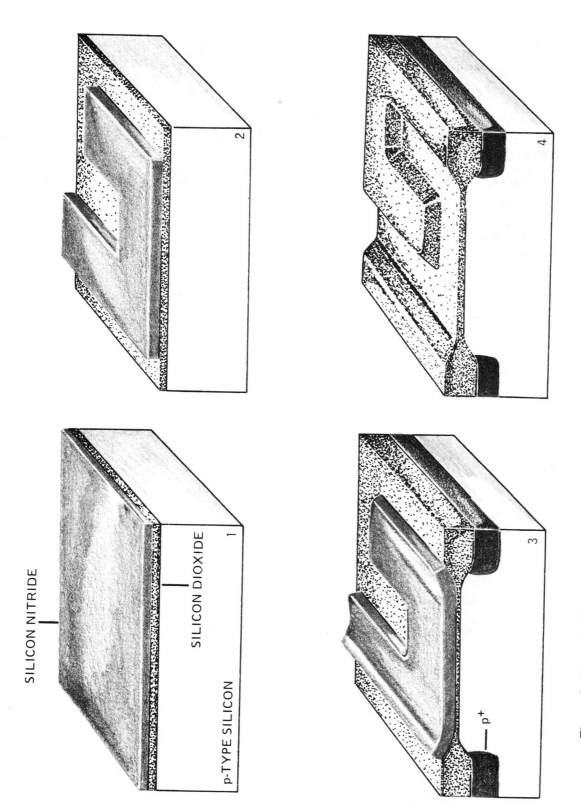

SILICON NITRIDE

SILICON DIOXIDE

p-TYPE SILICON

p+

Figure 9-29 — Fabrication Sequence for a 2-Level n-Channel Polysilicon-Gate Metal-Oxide Semiconductor (MOS) Circuit Element[15] (© 1977 by Scientific American Inc. All rights reserved.) (Sheet 1 of 2)

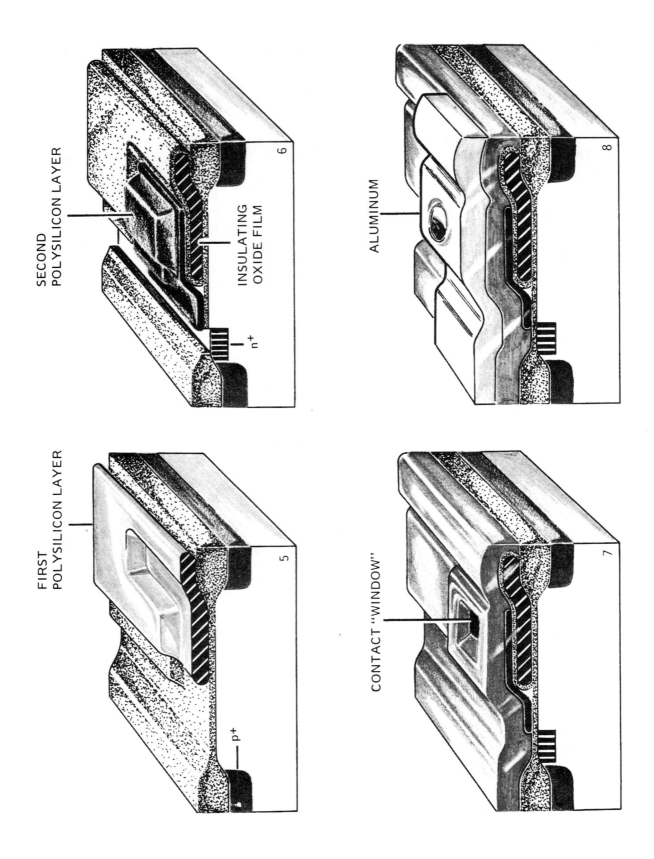

SECOND
POLYSILICON LAYER

INSULATING
OXIDE FILM

n+

6

FIRST
POLYSILICON LAYER

p+

5

ALUMINUM

8

CONTACT "WINDOW"

7

Figure 9-29 — Fabrication Sequence for a 2-Level n-Channel Polysilicon-Gate Metal-Oxide Semiconductor (MOS) Circuit Element[15] (© 1977 by Scientific American Inc. All rights reserved.) (Sheet 2 of 2)

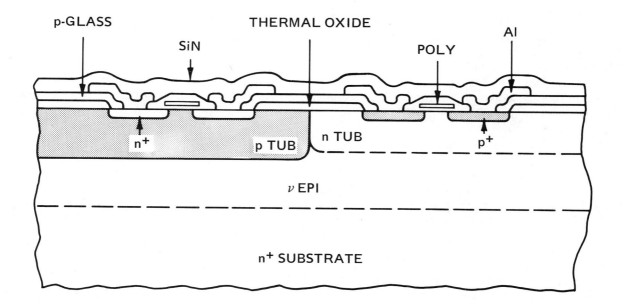

Figure 9-30 — Twin Tub CMOS[16]

9.4 GLOSSARY OF SILICON DEVICE TERMS

avalanche diode — A diode that conducts current only above a certain "breakdown" voltage, by virtue of high-field impact ionization (avalanche).

avalanche multiplication — A high-field in semiconductors which leads to an increase in current. High energy charge carriers create additional carriers by impact ionization of valence electrons.

BIGFET, bipolar insulated gate field-effect transistor — A simple integrated circuit combining a field-effect transistor and a bipolar transistor.

BIL — buried injector logic

CBIC — complementary bipolar integrated circuit

CCD, charge-coupled device — A semiconductor device whose action depends on the storage of an electric charge within a semiconductor by an insulated electrode on its surface, with the possibility of selectively moving the charge to another electrode by proper manipulation of voltages on the electrode.

CDI, collector diffusion isolation — A process for fabricating integrated circuits whereby the collector diffusion also electrically isolates the transistors from one another.

CML — current mode logic

CMOS — complementary MOS

DDC — dual dielectric charge storage cells

DIP, dual in-line package — A circuit package somewhat longer than it is wide, with the leads coming out of the two long sides.

DMOS — diffused MOS IC

DSAMOS — diffusion self-aligned MOS

ECL — emitter-coupled logic

EFL — emitter-follower logic

FAMOS — floating gate avalanche injection MOS

FET, field-effect transistor — A transistor consisting of a source, gate, and drain, whose action depends on the flow of majority carriers past the gate from source to drain. The flow is controlled by the transverse electric field under the gate. See **IGFET**.

FIC — film integrated circuit

GDS — gated diode switch

GIMIC — guard ring isolated monolithic integrated circuit

HIC, hybrid integrated circuit — An assembly of one or more semiconductor devices and a thin-film integrated circuit on a single substrate, usually of ceramic.

IGFET — A field-effect transistor whose gate is insulated from the semiconductor by a thin intervening layer of insulation, usually thermal oxide.

IMPATT diode, impact avalanche and transit time diode — An avalanche diode used as a high-frequency oscillator or amplifier. Its negative resistance depends upon the transit time of charge carriers through the depletion layer.

ISL — integrated Schottky logic

I^2L — integrated-injector logic

JFET, junction field-effect transistor (see FET) — A transistor in which the gate electrode is formed by a pn junction.

LSTTL — low-power Schottky transistor-transistor logic

MOS — A metal-oxide semiconductor structure.

MOSFET — A metal-oxide field-effect transistor containing a metal gate over thermal oxide over silicon. The MOSFET structure is one way to make an IGFET.

NMOS — n-channel MOS IC

OXIL — oxide isolated logic

OXIM, oxide-isolated monolith — A method of making an integrated circuit in which an oxide layer is introduced to insulate semiconductor regions from each other.

PIN diode — A diode constructed of an intrinsic layer between p and n layers and used in high-speed or high-power microwave switching.

PMOS — p-channel MOS IC

PNIP transistor — A semiconductor crystal structure consisting of layers that are p-type, n-type, intrinsic, and p-type. Used for high-voltage, high-frequency bipolar transistors.

pnpn — A semiconductor crystal structure in which the layers are successively p-type, n-type, p-type and n-type. When ohmic contacts are made to the various layers, a silicon-controlled rectifier (SCR) or thyristor, results.

SBC, standard buried collector — A method of making integrated circuits in which diffused collector areas are "buried" by overlying layers.

Schottky barrier — A potential barrier formed between a metal and a semi-conductor.

SCR, silicon-controlled rectifier — A pnpn device useful for switching and power applications because it can have both high breakdown voltage and high current-carrying capability. See **pnpn**.

SIC, silicon integrated circuit — An integrated circuit in which all the elements such as transistors, diodes, resistors, and capacitors are successively fabricated in or on the silicon and interconnected.

SIPOS — semi-insulating polycrystalline oxygen-doped silicon

STL — Schottky transistor logic

TRAPATT, trapped plasma avalanche transit time diode — A diode used as a microwave oscillator in a manner analogous to the IMPATT diode.

TTL — transistor-transistor logic

VMOS — vertical MOS IC

REFERENCES

1. D. Koehler, J. G. Josenhans, and H. Schichman, private communication.

2. M. C. Chi and D. Koehler, private communication.

3. A. G. F. Dingwall, "High-Yield Processed Bipolar LSI Arrays," paper presented at *International Electron Devices Meeting,* Washington, D.C., 1968.

4. G. B. Herzog et al, "Large Scale Integrated Circuit Arrays," *Interim Technical Report No. 11,* Air Force Avionics and RCA Laboratories, (AF33 (615) 3491), (November 1968).

5. B. T. Murphy, "Cost-size Optima of Monolithic Integrated Circuits," *Proceedings of the IEEE,* Vol. 52, No. 12 (December 1964), pp. 1537-1545.

6. R. B. Seeds, "Yield, Economic, and Logistic Models for Complex Digital Arrays," *IEEE International Convention Record,* New York (1967), part 6, pp. 60-61.

7. S. M. Sze, private communication.

8. P. W. J. Verhofstadt, "Evaluation of Technology Options for LSI Processing Elements," *Proceedings of the IEEE,* Vol. 64, No. 6 (June 1976), pp. 842-851.

9. *Integrated Injection Logic,* ed. J. E. Smith, A Volume in the IEEE PRESS Selected Reprint Series (New York: John Wiley & Sons, 1980).

10. P. T. Panousis and R. L. Pritchett, "GIMIC-0 — A Low Cost Non-Epitaxial Bipolar LSI Technology Suitable for Application to TTL Circuits," *International Electron Devices Meeting,* Washington, D.C., December, 9-11, 1974, (New York: IEEE 1974), p. 518.

11. A. A. Yiannoulos, "Buried Injector Logic: Second Generation I^2L Performance," *1978 IEEE International Solid-State Circuits Conference,* San Francisco, CA, February 15-17, 1978, (New York: IEEE 1978), p. 12.

12. J. Agraz-Güereña, P. T. Panousis, and B. L. Morris, "OXIL, a Versatile Bipolar VLSI Technology," *IEEE Trans. on Electron Dev.,* Vol. ED-27, No. 8 (August 1980), pp. 1397-1401.

13. P. W. Schackle et al, "A 500 V Monolithic Bidirectional 2x2 Crosspoint Array," *1980 IEEE International Solid-State Circuits Conference,* San Francisco, CA, February 13-15, 1980, (New York: IEEE 1980), p. 170.

14. J. T. Clemens, R. H. Doklan, and J. J. Nolen, "An N-Channel Si-Gate Integrated Circuit Technology," *1975 International Electron Devices Meeting,* Washington, D.C., December 1-3, 1975, (New York: IEEE 1975), p. 301.

15. W. G. Oldham, "The Fabrication of Microelectronic Circuits," *Scientific American,* Vol. 237, No. 3, September 1977, p. 122.

16. R. S. Payne et al, private communication.

10. CONDUCTIVITY OF DIFFUSED LAYERS

LIST OF SYMBOLS

Symbol	Definition
C, C_j	Concentration
C_{AO}, C_S	Electrically Active Surface Concentration
C_B	Substrate Concentration
C_0	Surface Concentration
C_{TO}	Total Arsenic Surface Concentration
D	Diffusivity
D_i	Intrinsic Diffusivity
I	Current
N_i, n_i	Intrinsic Electron Concentration
Q	Total Impurity Under Gaussian Profile
Q_T	Implant Dose
q	Electron Charge
R_s, R_i	Sheet Resistance
t	Time
V	Voltage
X, dX, ΔX	Distance
X_j	Junction Depth
μ, μ_i	Mobility
μ_{eff}	Effective Mobility
ρ_s	Sheet Resistance
$\overline{\sigma}$	Effective Conductivity

10.1 General

The common Gaussian and erfc doping profiles can be characterized by two parameters: typically, the surface concentration, C_O, and a diffusion length, $\sqrt{4Dt}$. (See Section 6 for expressions.) The routine experimental technique for the evaluation of doped layers is to diffuse the layer into a substrate of the opposite doping type (p layer into n substrate) to form a pn junction. This junction serves two purposes. The first is to provide a single point evaluation of the diffusion profile. The junction depth, X_j, of the diffused layer corresponds to that distance from the surface at which the diffused impurity concentration is equal to that of the substrate, C_B. Hence, measurement of X_j locates the point at which $C(X_j) = C_B$. This supplies one of the two measurements necessary to characterize the layer. See Figure 10-1(a). The junction depth is usually determined by angle-lapping the sample and delineating the junction with a stain etch. (HF with a small percentage of HNO_3 under strong illumination is a common stain for Si junctions.) The depth of the junction is then measured using optical interference techniques.[1] The second purpose of the pn junction is to provide electrical isolation of the diffused layer from the substrate. This electrical isolation allows the sheet resistance of the diffused layer to be measured easily using standard 4-point probe techniques. This reading is an integral evaluation of the doping in the layer. Consider the doping profile of Figure 10-1(b). Recalling that the sheet conductance $(1/R_s)$ of a layer of constant doping concentration C_i and width ΔX is

$$\frac{1}{R_i} = q\mu_i C_i \Delta X \ldots,$$

10.1.1

the total sheet conductance of the diffused layer can be expressed by summing all these layers, i.e.,

$$\frac{1}{R_s} = q(\mu_1 C_1 + \mu_2 C_2 + \ldots \mu_n C_n)\Delta X \ldots$$

10.1.2

$$= q\int_0^{X_j} \mu(X)C(X)\,dX \ldots$$

10.1.3

Since the mobility is a function of concentration, the mobility-concentration product is the significant element as indicated.

It is obvious that the sheet resistance by itself can say very little about how the impurity is distributed in the layer. However, by utilizing the junction depth, an effective conductivity $\bar{\sigma}$ can be defined, which is significant. The effective conductivity of the diffused layer is

$$\bar{\sigma} = \frac{1}{R_s X_j} \ldots$$

10.1.4

$$= \frac{1}{X_j} q\int_0^{X_j} \mu(X)C(X)\,dX \ldots$$

10.1.5

Equation 10.1.5 has been numerically evaluated for both Si and Ge by J. Irvin[2] and D. Cuttriss[3] for various diffusion profiles and background impurity concentrations. They have also calculated the effective conductivity of erfc and Gaussian distributions of subsurface layers between $X(X \leqslant X_j)$ and X_j, which can be useful for device calculations. These curves for Si are presented in Figures 10-2 through 10-53. A listing of these curves is shown on the following page. Equation 10.1.5 can also be expressed in terms of an effective mobility, μ_{eff}, which is defined as

$$\mu_{eff} = \frac{\int_0^{X_j} \mu(X)C(X)\,dX}{\int_0^{X_j} C(X)\,dX} \ldots$$

10.1.6

Thus, 10.1.5 becomes

$$\bar{\sigma} = q/X_j \, \mu_{eff}\int_0^{X_j} C(X)\,dX.$$

10.1.7

CONDUCTIVITY OF DIFFUSED LAYERS

(a) Surface Concentration, C_O, Versus Effective Conductivity, $\overline{\sigma}$, for Silicon

IMPURITY DISTRIBUTION	RANGE (cm^{-3})	FIGURE NUMBER n-TYPE	p-TYPE
erfc	$10^{17} \geqslant C_O \geqslant 10^{14}$	10-2	10-12
	$10^{20} \geqslant C_O \geqslant 10^{17}$	10-3	10-13
	$10^{22} \geqslant C_O \geqslant 10^{19}$	10-4	10-14
Gaussian	$10^{17} \geqslant C_O \geqslant 10^{14}$	10-22	10-32
	$10^{20} \geqslant C_O \geqslant 10^{17}$	10-23	10-33
	$10^{22} \geqslant C_O \geqslant 10^{19}$	10-24	10-34
Exponential	$10^{17} \geqslant C_O \geqslant 10^{14}$	10-42	10-45
	$10^{20} \geqslant C_O \geqslant 10^{17}$	10-43	10-46
	$10^{22} \geqslant C_O \geqslant 10^{14}$	10-44	10-47
Linear	$10^{17} \geqslant C_O \geqslant 10^{14}$	10-48	10-51
	$10^{20} \geqslant C_O \geqslant 10^{17}$	10-49	10-52
	$10^{22} \geqslant C_O \geqslant 10^{19}$	10-50	10-53

(b) Effective Conductivity of Subsurface Layers Between X and X_j for Silicon

IMPURITY DISTRIBUTION	BACKGROUND DOPING	FIGURE NUMBER n-TYPE LAYER	p-TYPE LAYER
erfc	10^{14}	10-5	10-15
	10^{15}	10-6	10-16
	10^{16}	10-7	10-17
	10^{17}	10-8	10-18
	10^{18}	10-9	10-19
	10^{19}	10-10	10-20
	10^{20}	10-11	10-21
Gaussian	10^{14}	10-25	10-35
	10^{15}	10-26	10-36
	10^{16}	10-27	10-37
	10^{17}	10-28	10-38
	10^{18}	10-29	10-39
	10^{19}	10-30	10-40
	10^{20}	10-31	10-41

10.2 Arsenic-Diffused Layers

R.B. Fair has provided curves for both electrically active and total As surface concentration versus diffused layer effective conductivity in n- and p-type silicon. He has studied both implanted and chemical sources. The curves and their corresponding figure numbers are tabulated below.

As SOURCE	SURFACE CONCENTRATION	DIFFUSED LAYER PROPERTY	FIGURE NO.
Implanted/Diffused	Electrically active	Effective sheet resistance	10-54
Chemical	Electrically active	Effective sheet resistance	10-55
Chemical	Total	Average conductivity/ n substrate	10-56
Chemical	Total	Average conductivity/ p substrate	10-57
Implanted/Diffused	Total	Dose/X_J	10-58
Implanted/Diffused	Total	Normalized dose	10-59

Figure 10-60 shows the grade constant versus total As surface concentration divided by X_J for two concentrations.

Additional experimental data for arsenic and phosphorus in silicon is provided in Section 6.

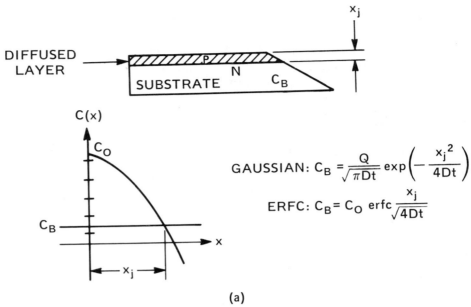

$$\text{GAUSSIAN: } C_B = \frac{Q}{\sqrt{\pi Dt}} \exp\left(-\frac{x_j^2}{4Dt}\right)$$

$$\text{ERFC: } C_B = C_O \text{ erfc} \frac{x_j}{\sqrt{4Dt}}$$

(a)

Metallurgical Junction Depth Measurement

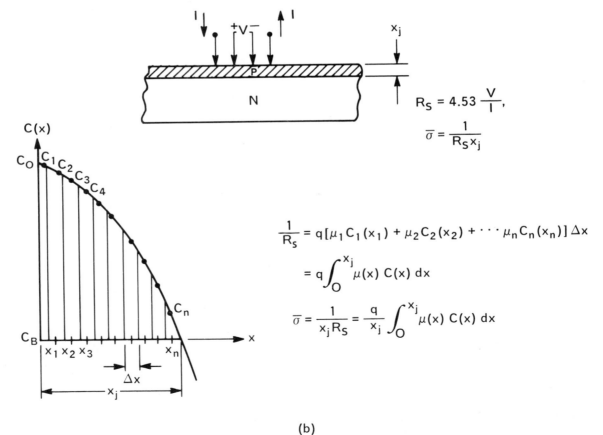

$$R_S = 4.53 \frac{V}{I},$$

$$\bar{\sigma} = \frac{1}{R_S x_j}$$

$$\frac{1}{R_S} = q[\mu_1 C_1(x_1) + \mu_2 C_2(x_2) + \cdots \mu_n C_n(x_n)]\Delta x$$

$$= q \int_O^{x_j} \mu(x) C(x)\, dx$$

$$\bar{\sigma} = \frac{1}{x_j R_S} = \frac{q}{x_j} \int_O^{x_j} \mu(x) C(x)\, dx$$

(b)

Effective Conductivity of the Diffused Layer

Figure 10-1 — Schematic Representation of Effective Conductivity Measurement

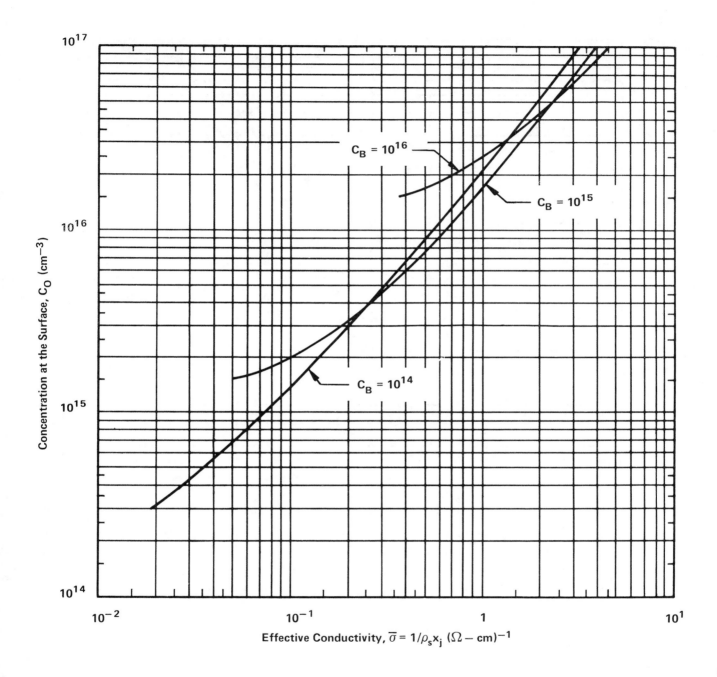

Figure 10-2 — Effective Conductivity of Diffused Layers — Si Erfc, n-Type
Diffusion $10^{17} \geqslant C_O \geqslant 10^{14}$

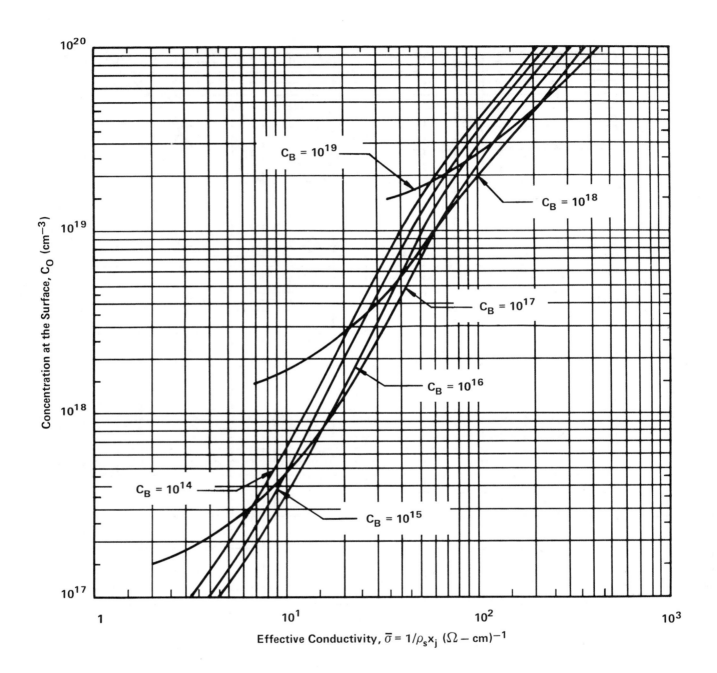

Figure 10-3 — Effective Conductivity of Diffused Layers — Si Erfc, n-Type
Diffusion $10^{20} \geqslant C_O \geqslant 10^{17}$

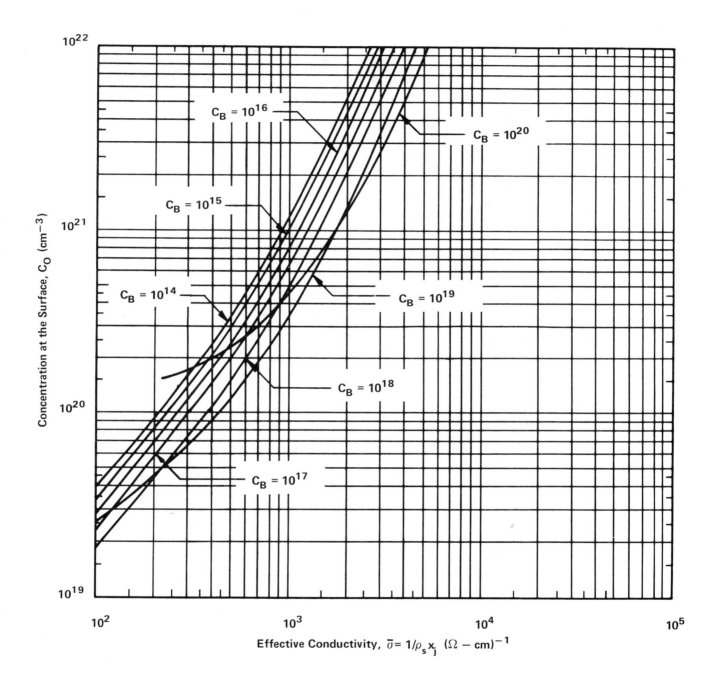

Figure 10-4 — Effective Conductivity of Diffused Layers — Si Erfc, n-Type
Diffusion $10^{22} \geqslant C_O \geqslant 10^{19}$

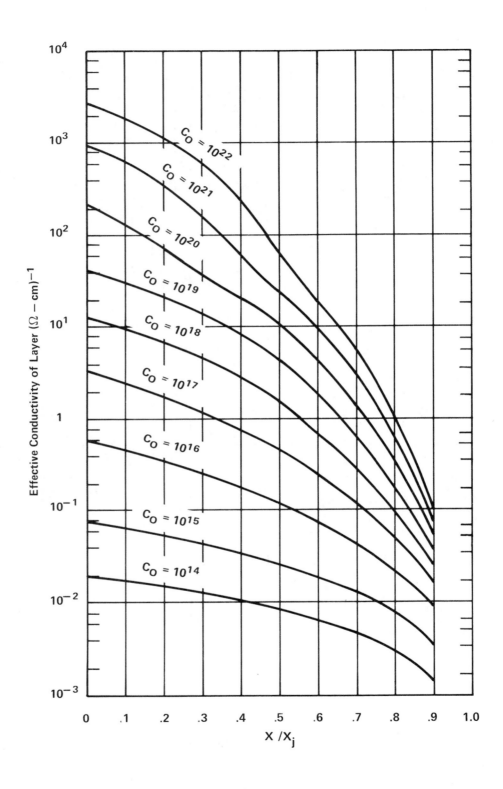

Figure 10-5 — Effective Conductivity of Subsurface Layers — Si Erfc, n-Type
Diffusion $C_B = 10^{14}$

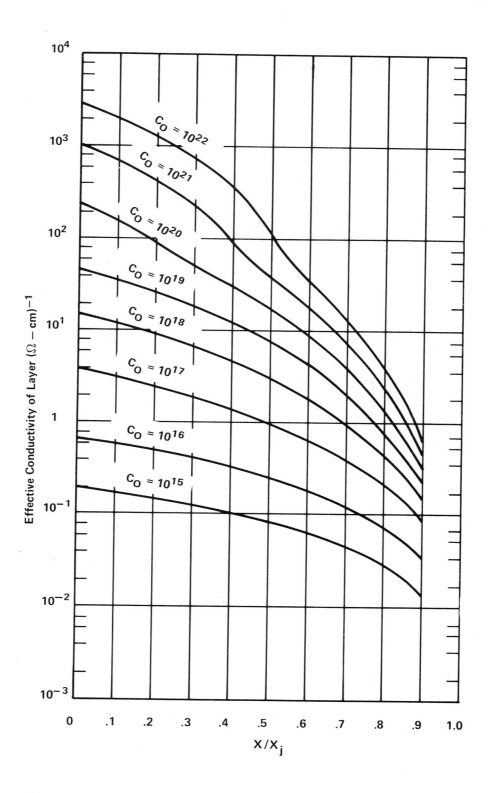

Figure 10-6 — Effective Conductivity of Subsurface Layers — Si Erfc, n-Type Diffusion $C_B = 10^{15}$

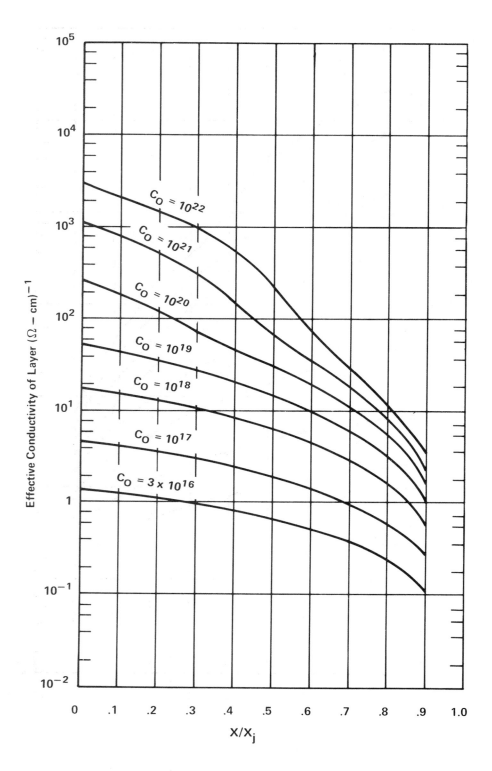

Figure 10-7 — Effective Conductivity of Subsurface Layers — Si Erfc, n-Type Diffusion $C_B = 10^{16}$

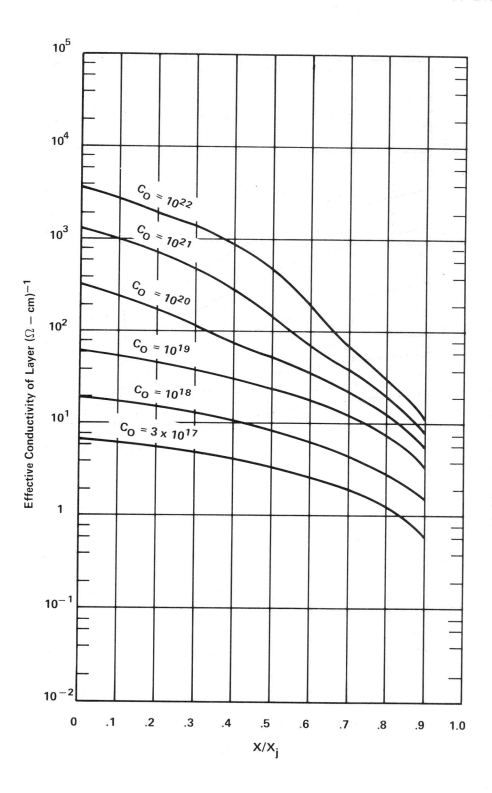

Figure 10-8 — Effective Conductivity of Subsurface Layers — Si Erfc, n-Type
Diffusion $C_B = 10^{17}$

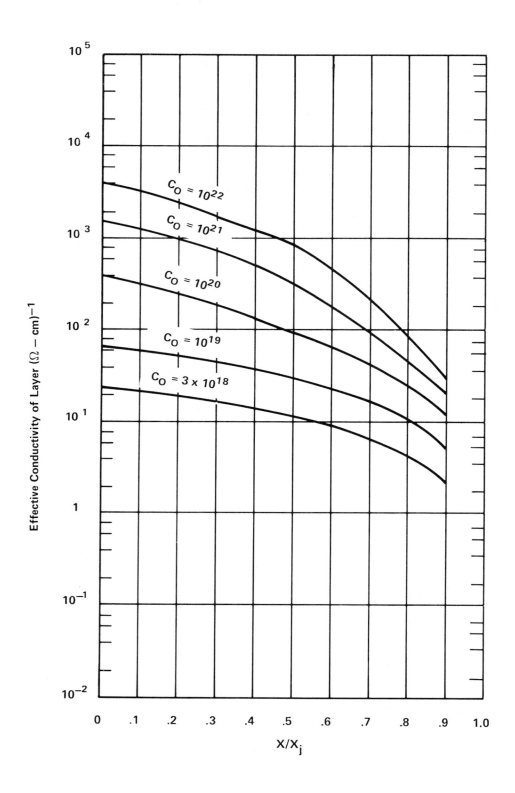

Figure 10-9 — Effective Conductivity of Subsurface Layers — Si Erfc, n-Type
Diffusion $C_B = 10^{18}$

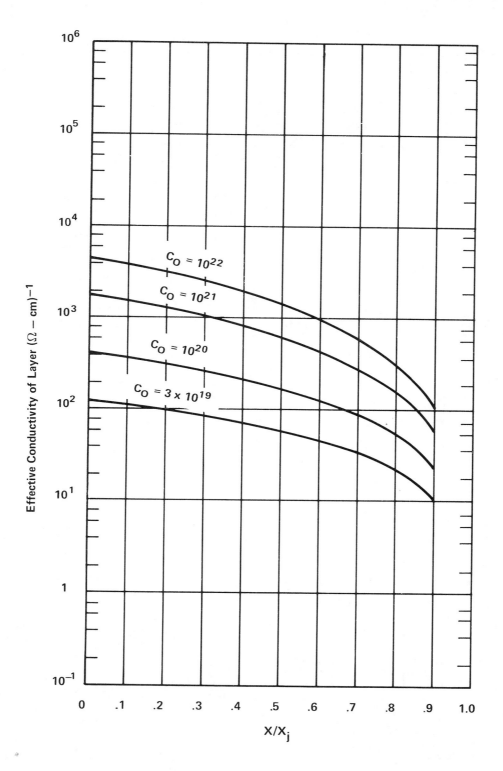

Figure 10-10 — Effective Conductivity of Subsurface Layers — Si Erfc, n-Type
Diffusion $C_B = 10^{19}$

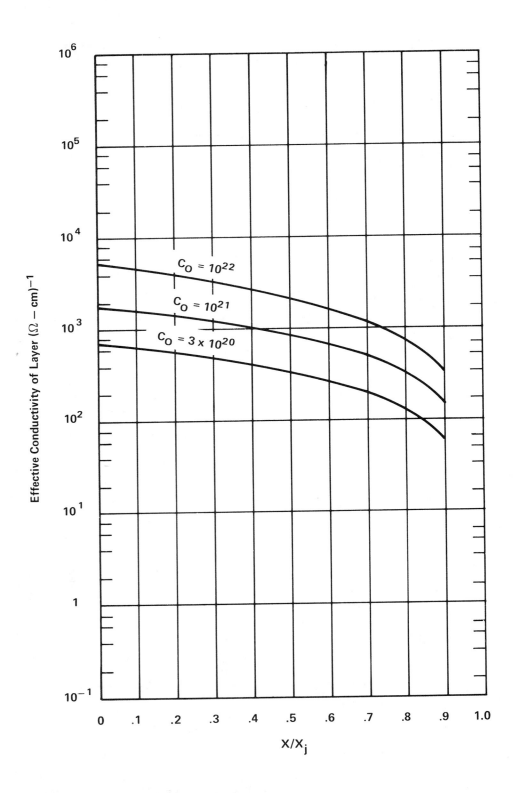

Figure 10-11 — Effective Conductivity of Subsurface Layers — Si Erfc, n-Type
Diffusion $C_B = 10^{20}$

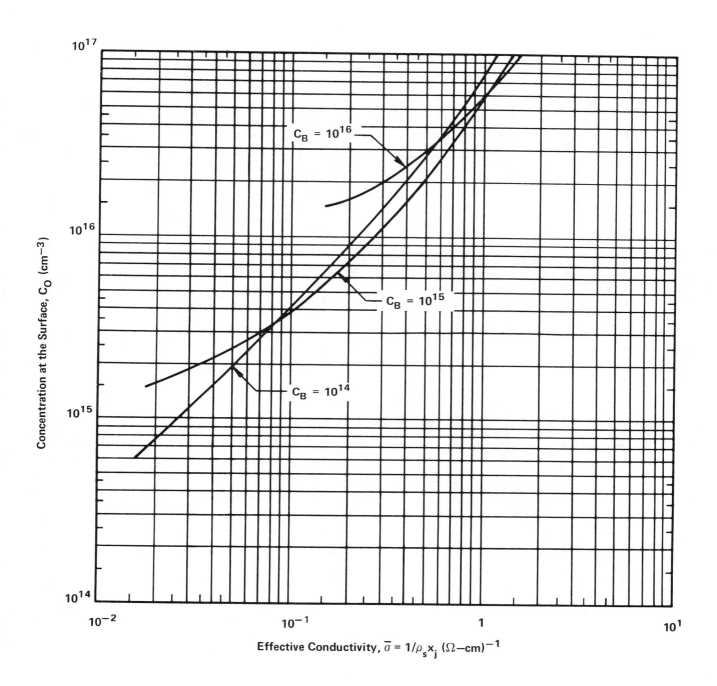

Figure 10-12 — Effective Conductivity of Diffused Layers — Si Erfc, p-Type
Diffusion $10^{17} \geqslant C_O \geqslant 10^{14}$

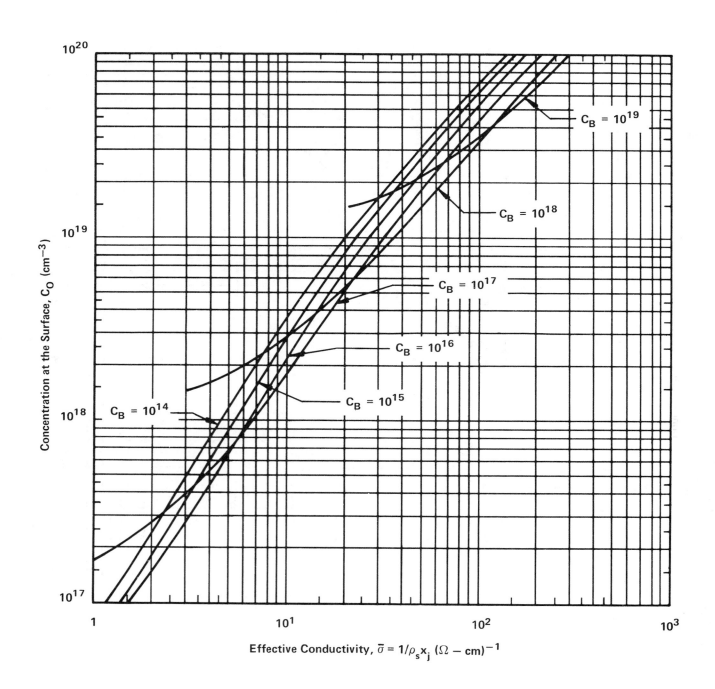

Figure 10-13 — Effective Conductivity of Diffused Layers — Si Erfc, p-Type
Diffusion $10^{20} \geqslant C_O \geqslant 10^{17}$

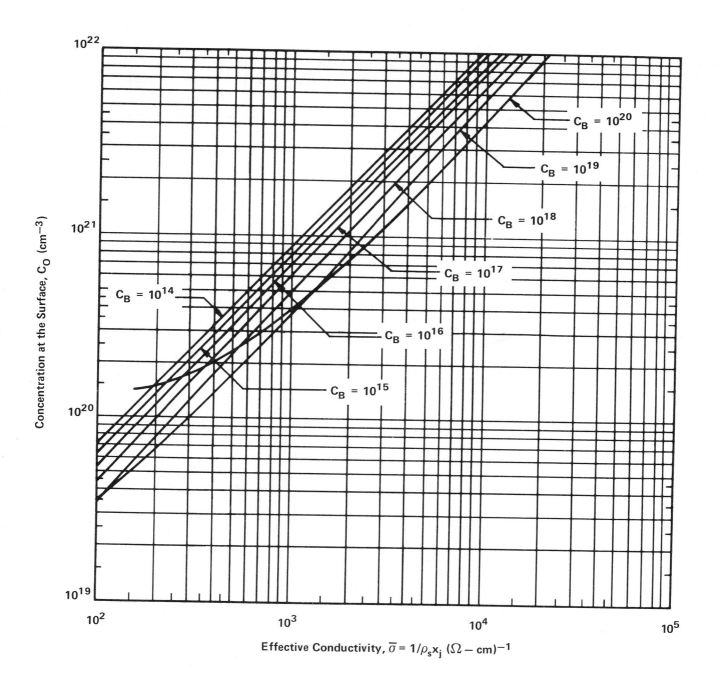

Figure 10-14 — Effective Conductivity of Diffused Layers — Si Erfc, p-Type
Diffusion $10^{22} \geqslant C_O \geqslant 10^{19}$

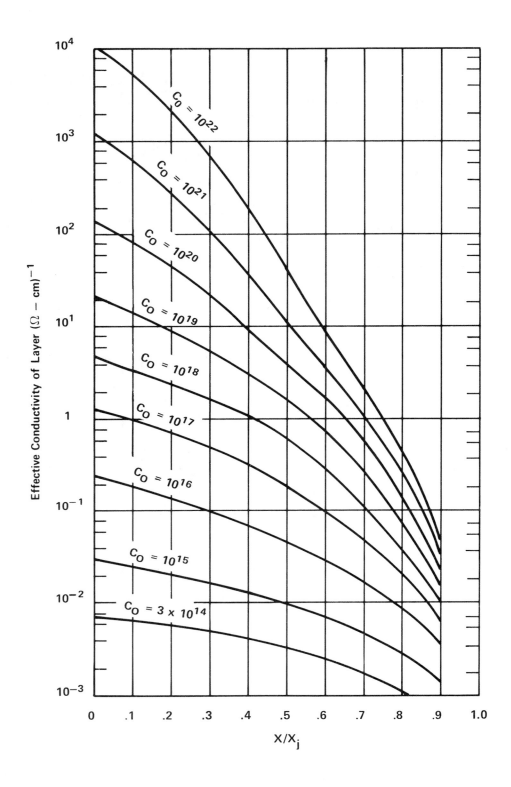

Figure 10-15 — Effective Conductivity of Subsurface Layers — Si Erfc, p-Type Diffusion $C_B = 10^{14}$

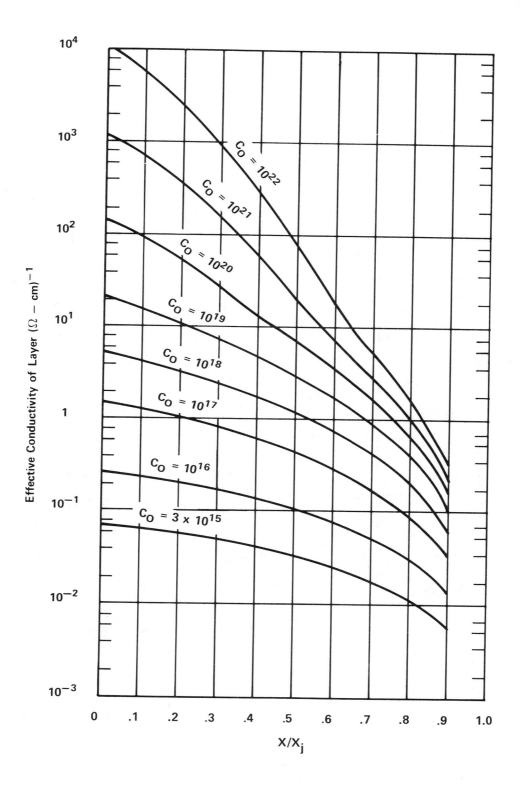

Figure 10-16 — Effective Conductivity of Subsurface Layers — Si Erfc, p-Type Diffusion $C_B = 10^{15}$

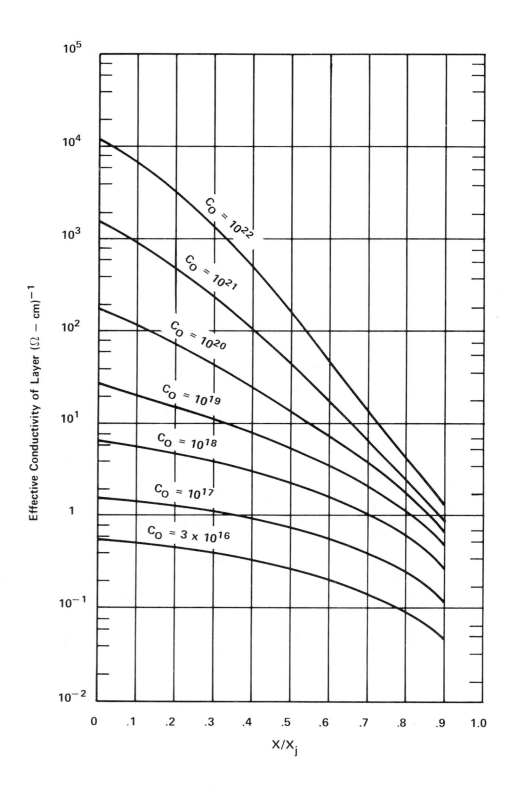

Figure 10-17 — Effective Conductivity of Subsurface Layers — Si Erfc, p-Type
Diffusion $C_B = 10^{16}$

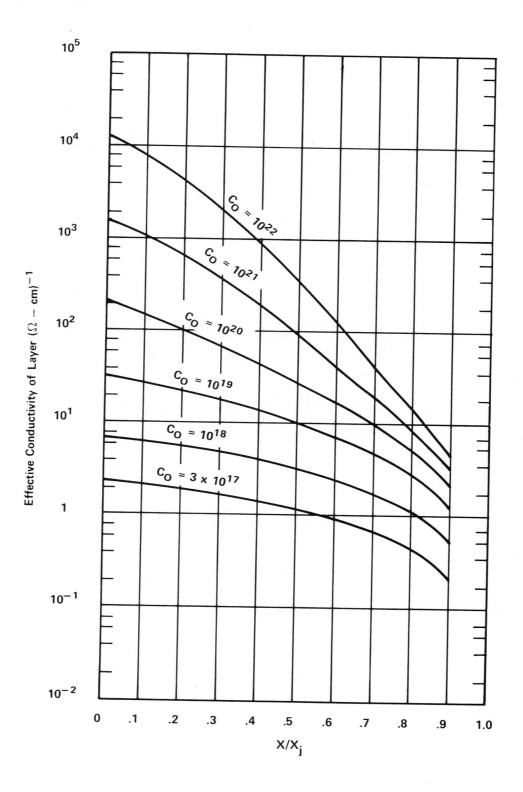

Figure 10-18 — Effective Conductivity of Subsurface Layers — Si Erfc, p-Type
Diffusion $C_B = 10^{17}$

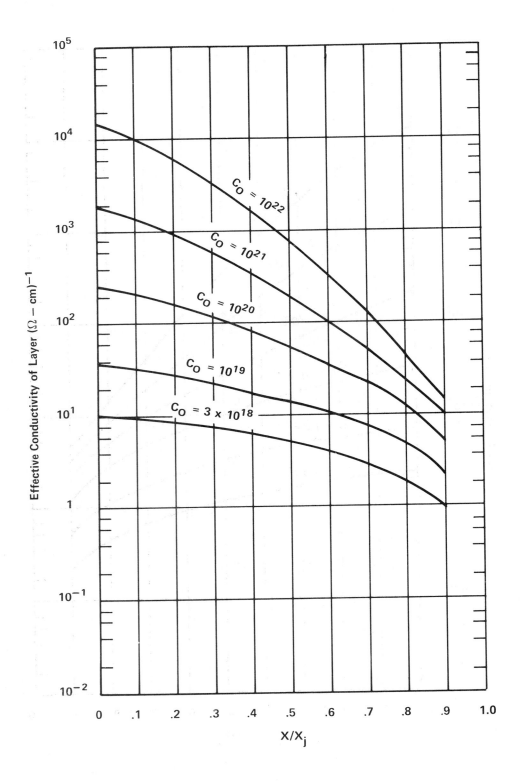

Figure 10-19 — Effective Conductivity of Subsurface Layers — Si Erfc, p-Type Diffusion $C_B = 10^{18}$

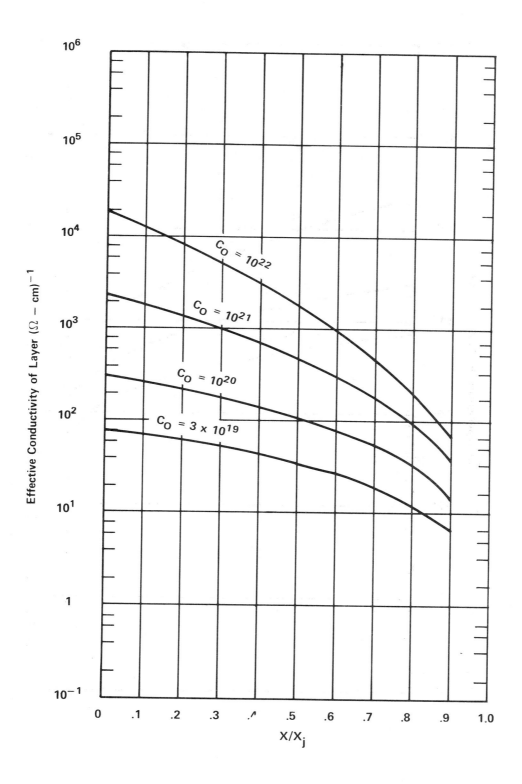

Figure 10-20 — Effective Conductivity of Subsurface Layers — Si Erfc, p-Type Diffusion $C_B = 10^{19}$

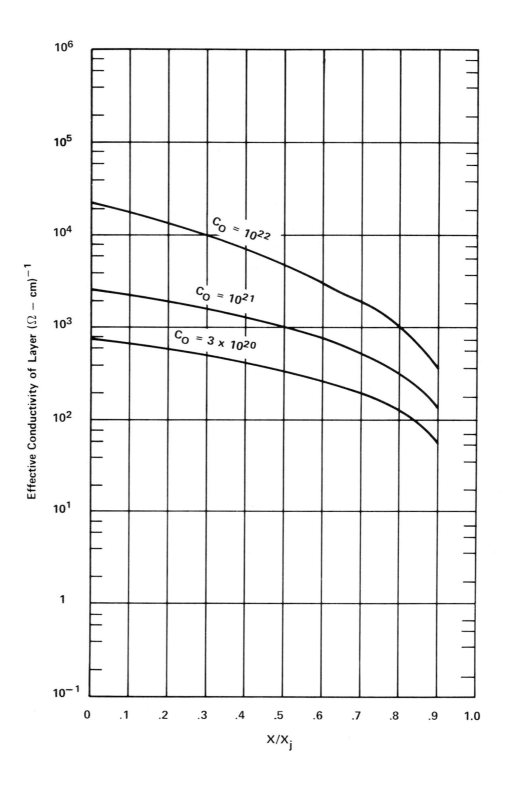

Figure 10-21 — Effective Conductivity of Subsurface Layers — Si Erfc, p-Type Diffusion $C_B = 10^{20}$

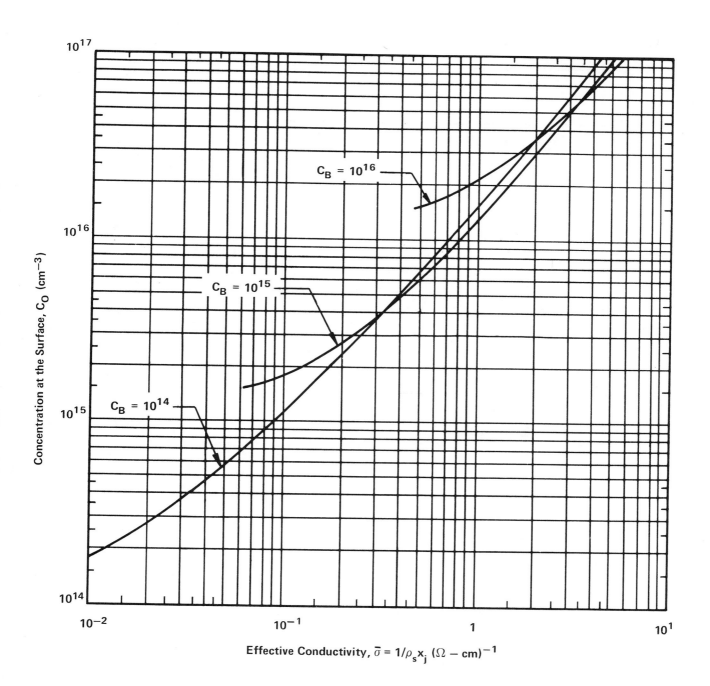

Figure 10-22 — Effective Conductivity of Diffused Layers — Si Gaussian, n-Type
Diffusion $10^{17} \geqslant C_O \geqslant 10^{14}$

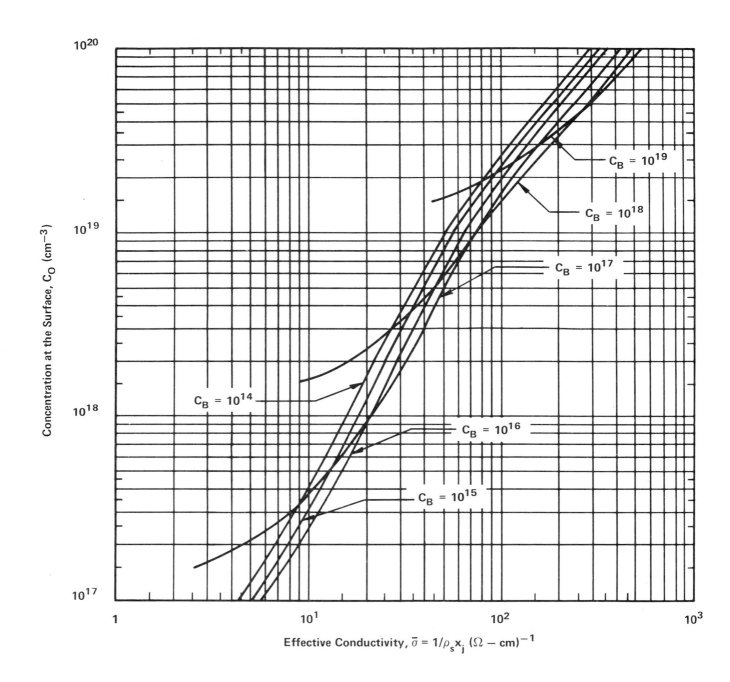

Figure 10-23 — Effective Conductivity of Diffused Layers — Si Gaussian, n-Type
Diffusion $10^{20} \geqslant C_O \geqslant 10^{17}$

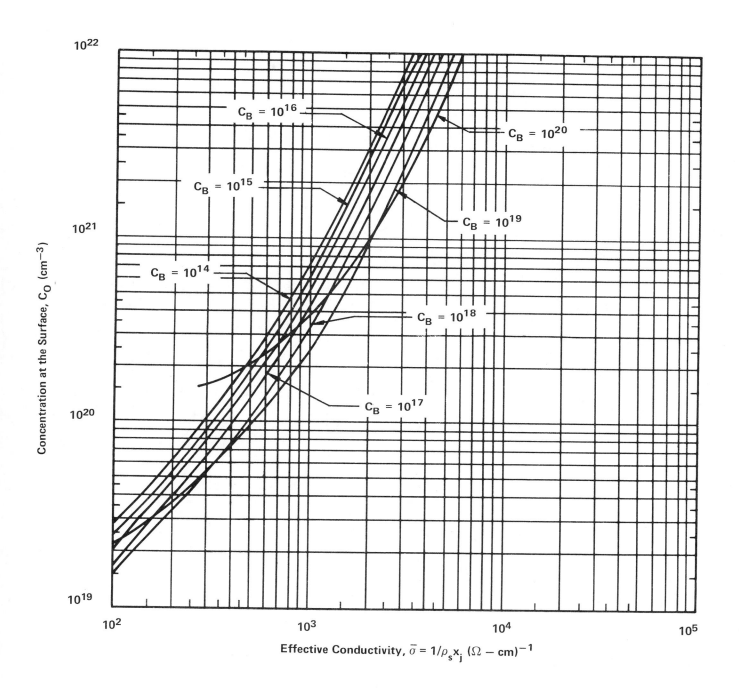

Figure 10-24 — Effective Conductivity of Diffused Layers — Si Gaussian, n-Type Diffusion $10^{22} \geqslant C_O \geqslant 10^{19}$

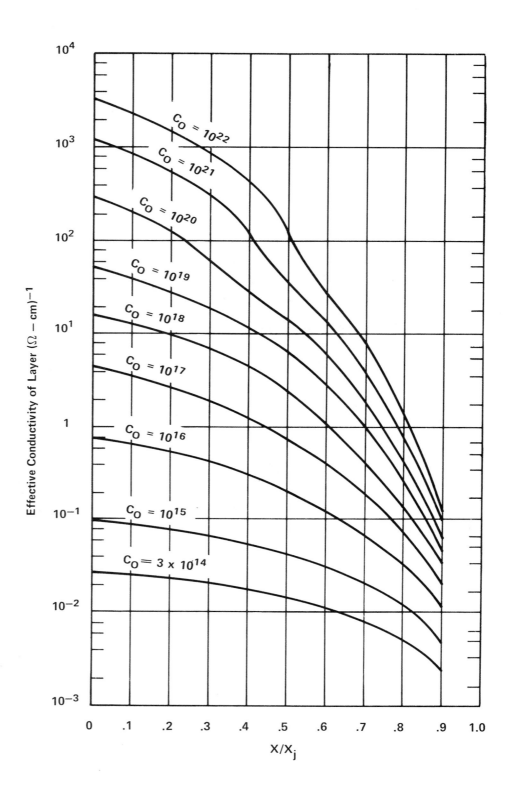

Figure 10-25 — Effective Conductivity of Subsurface Layers — Si Gaussian, n-Type Diffusion $C_B = 10^{14}$

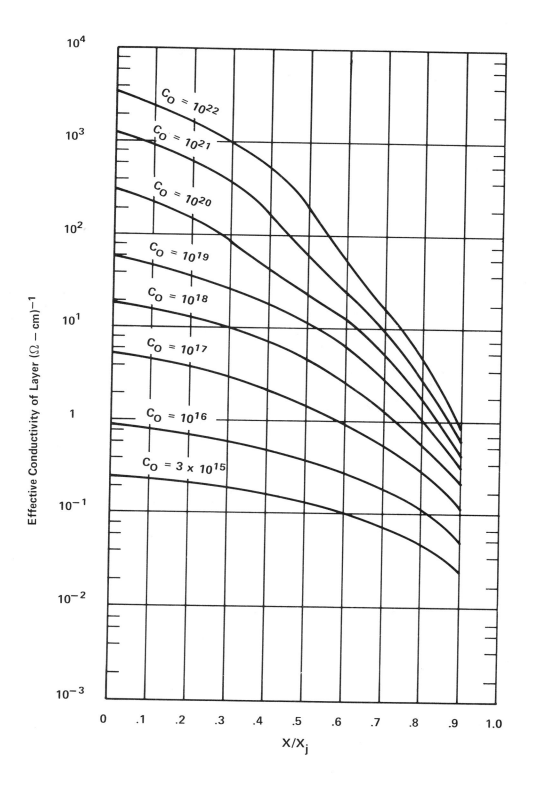

Figure 10-26 — Effective Conductivity of Subsurface Layers — Si Gaussian, n-Type Diffusion $C_B = 10^{15}$

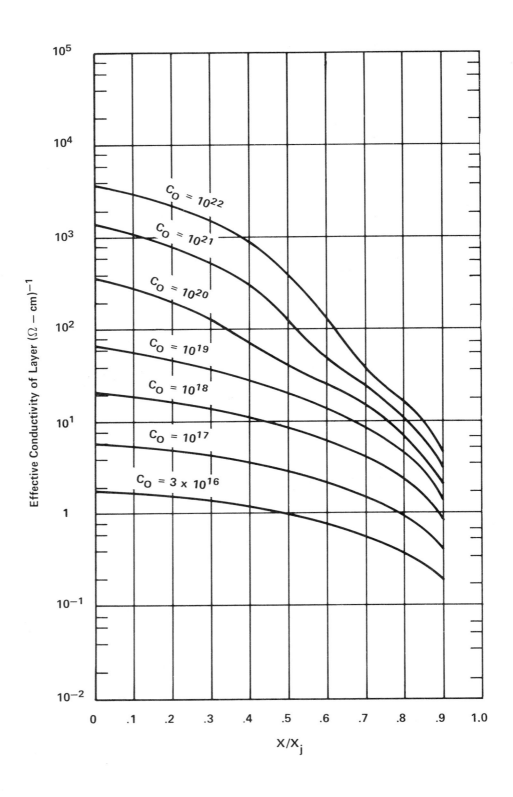

Figure 10-27 — Effective Conductivity of Subsurface Layers — Si Gaussian, n-Type Diffusion $C_B = 10^{16}$

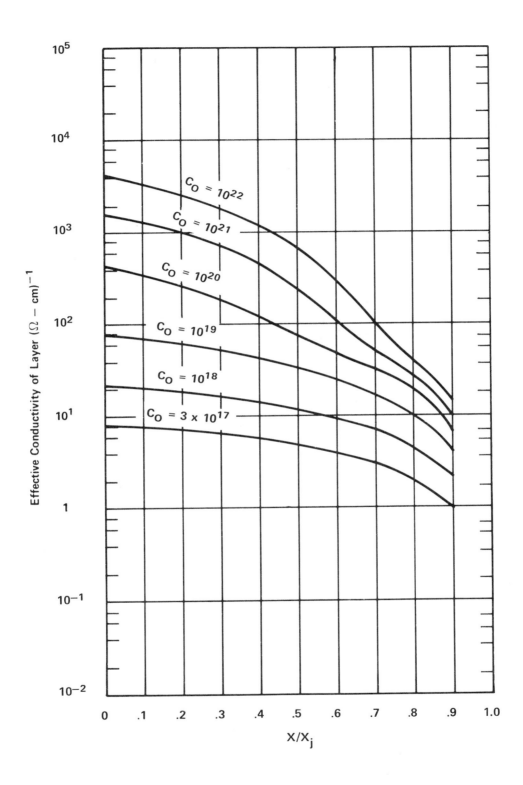

Figure 10-28 — Effective Conductivity of Subsurface Layers — Si Gaussian, n-Type Diffusion $C_B = 10^{17}$

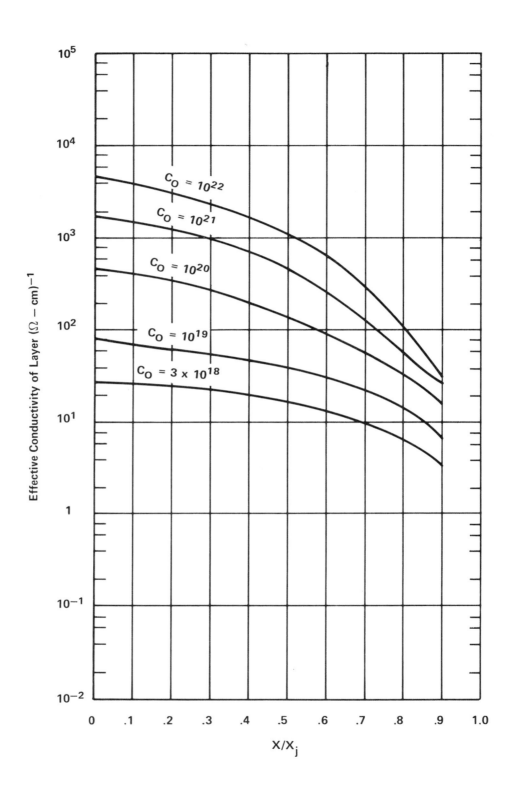

Figure 10-29 — Effective Conductivity of Subsurface Layers — Si Gaussian, n-Type Diffusion $C_B = 10^{18}$

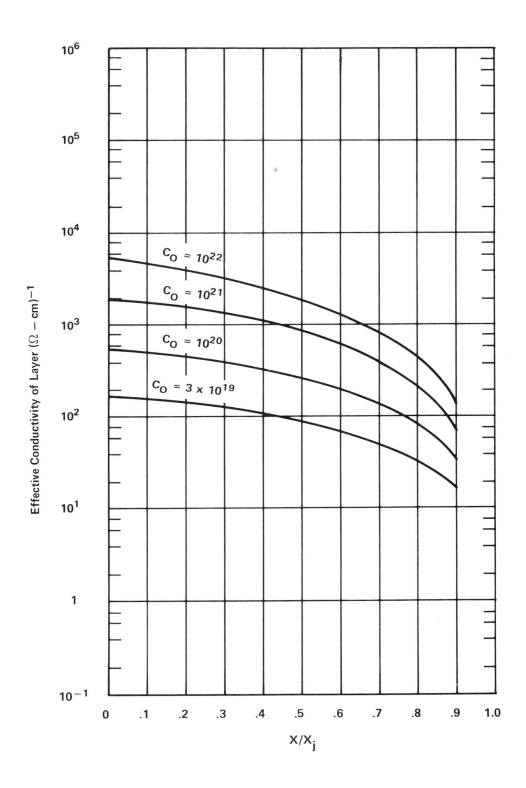

Figure 10-30 — Effective Conductivity of Subsurface Layers — Si Gaussian, n-Type
Diffusion $C_B = 10^{19}$

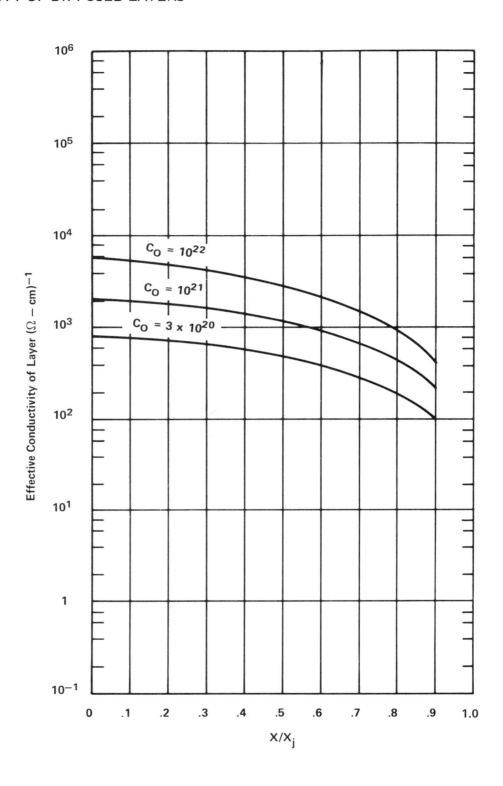

Figure 10-31 — Effective Conductivity of Subsurface Layers — Si Gaussian, n-Type Diffusion $C_B = 10^{20}$

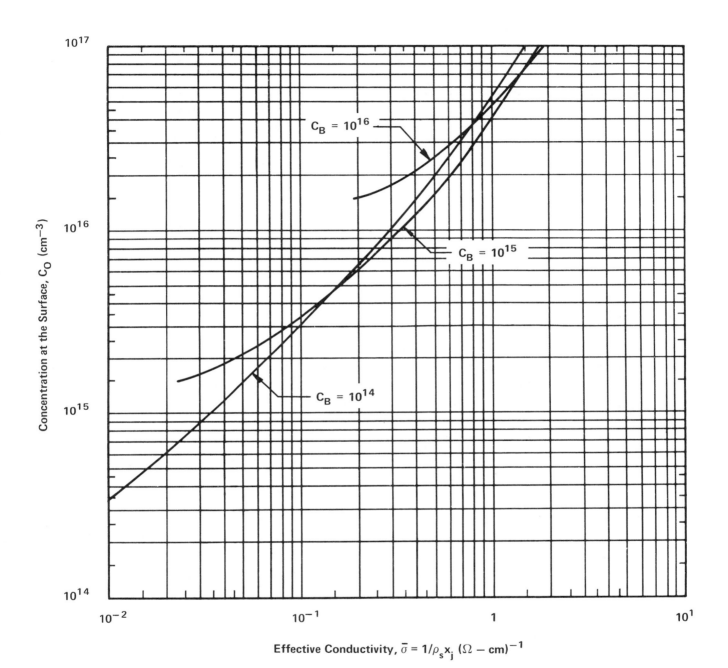

Figure 10-32 — Effective Conductivity of Diffused Layers — Si Gaussian, p-Type Diffusion $10^{17} \geqslant C_O \geqslant 10^{14}$

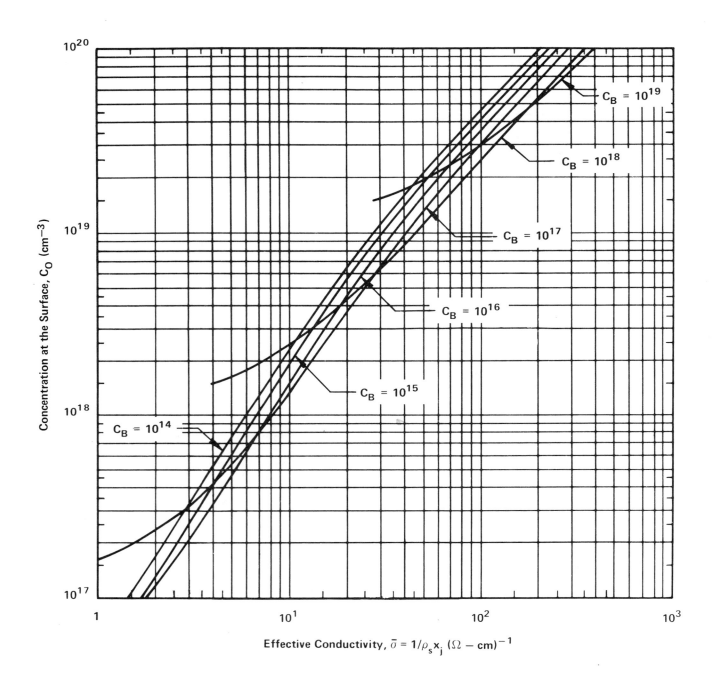

Figure 10-33 — Effective Conductivity of Diffused Layers — Si Gaussian, p-Type
Diffusion $10^{20} \geqslant C_O \geqslant 10^{17}$

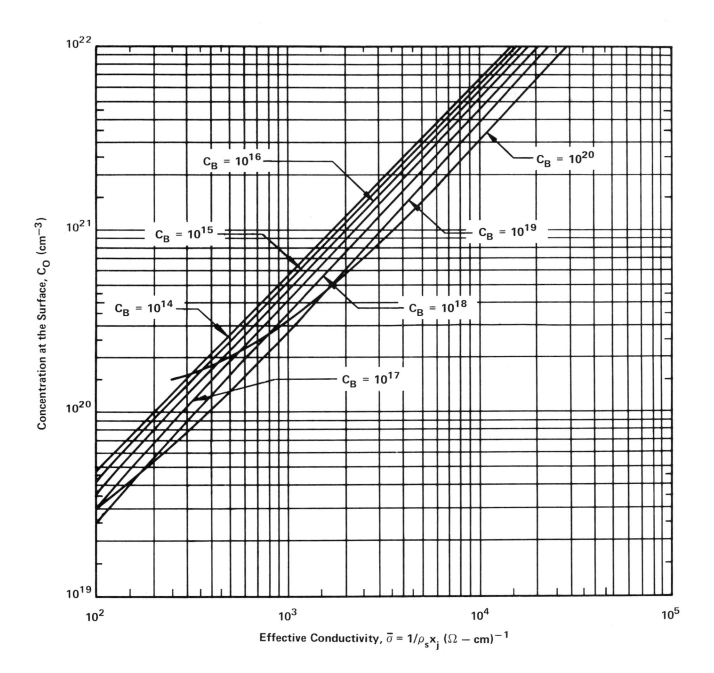

Figure 10-34 — Effective Conductivity of Diffused Layers — Si Gaussian, p-Type
Diffusion $10^{22} \geqslant C_O \geqslant 10^{19}$

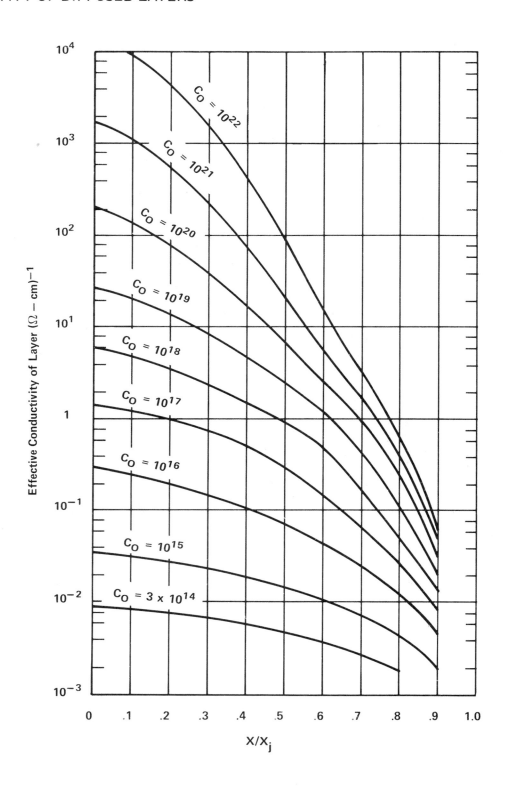

Figure 10-35 — Effective Conductivity of Subsurface Layers — Si Gaussian, p-Type Diffusion $C_B = 10^{14}$

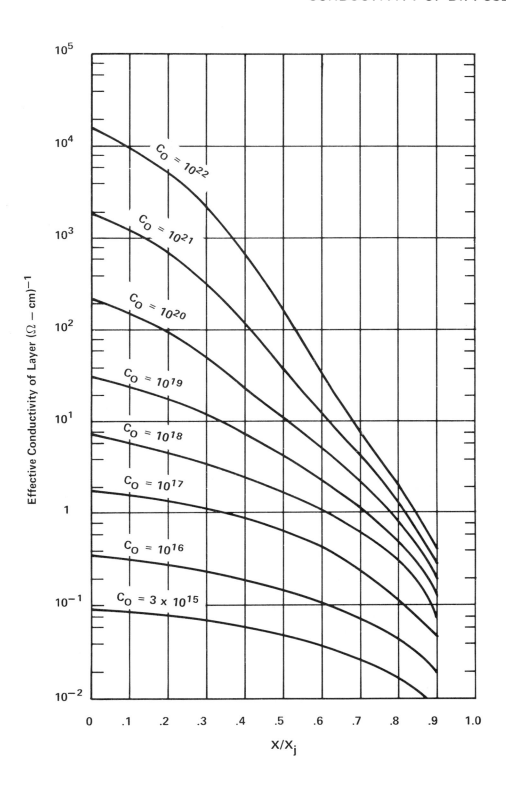

Figure 10-36 — Effective Conductivity of Subsurface Layers — Si Gaussian, p-Type Diffusion $C_B = 10^{15}$

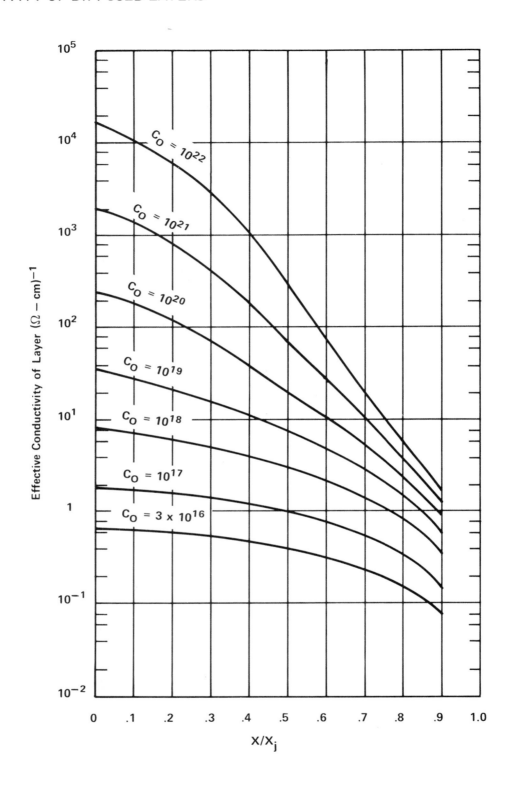

Figure 10-37 — Effective Conductivity of Subsurface Layers — Si Gaussian, p-Type Diffusion $C_B = 10^{16}$

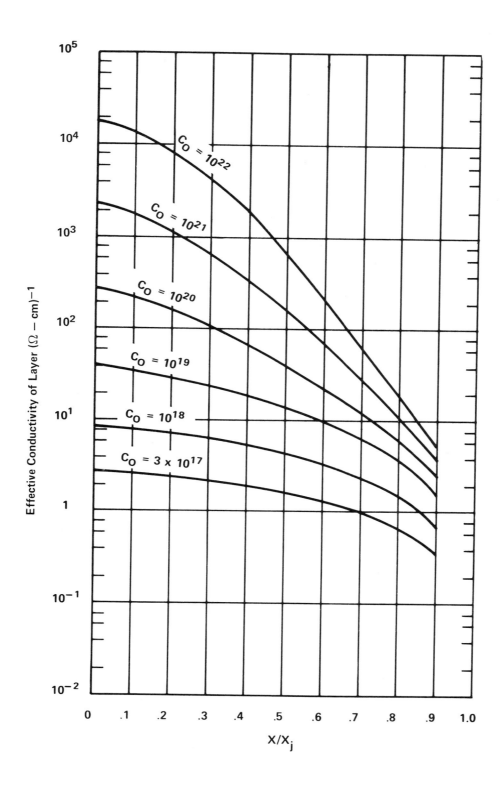

Figure 10-38 — Effective Conductivity of Subsurface Layers — Si Gaussian, p-Type
Diffusion $C_B = 10^{17}$

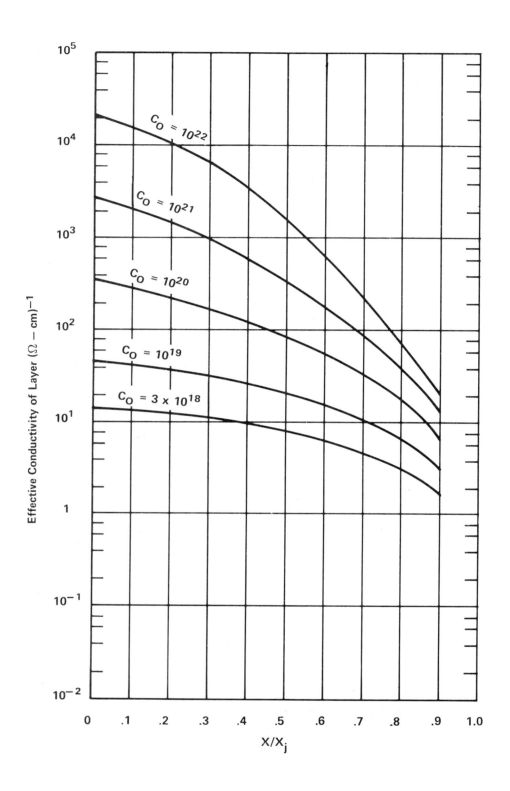

Figure 10-39 — Effective Conductivity of Subsurface Layers — Si Gaussian, p-Type Diffusion $C_B = 10^{18}$

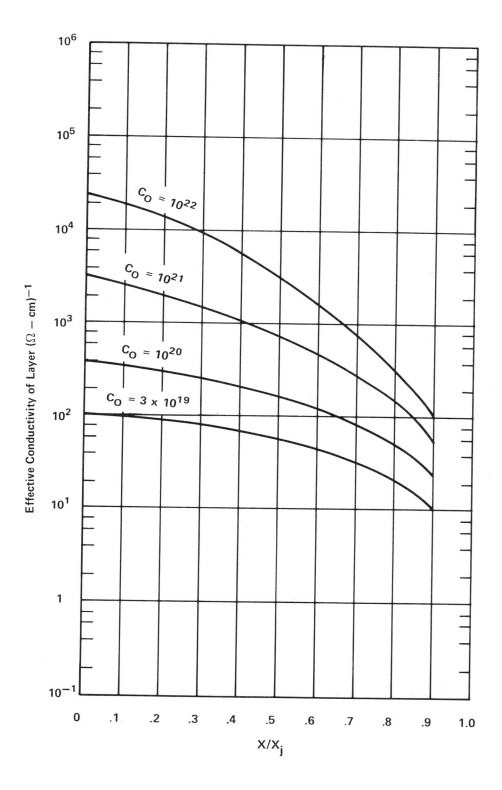

Figure 10-40 — Effective Conductivity of Subsurface Layers — Si Gaussian, p-Type
Diffusion $C_B = 10^{19}$

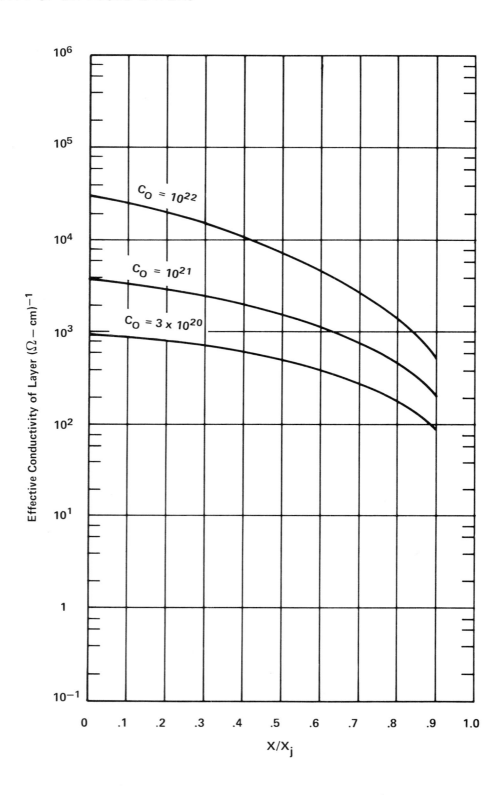

Figure 10-41 — Effective Conductivity of Subsurface Layers — Si Gaussian, p-Type
Diffusion $C_B = 10^{20}$

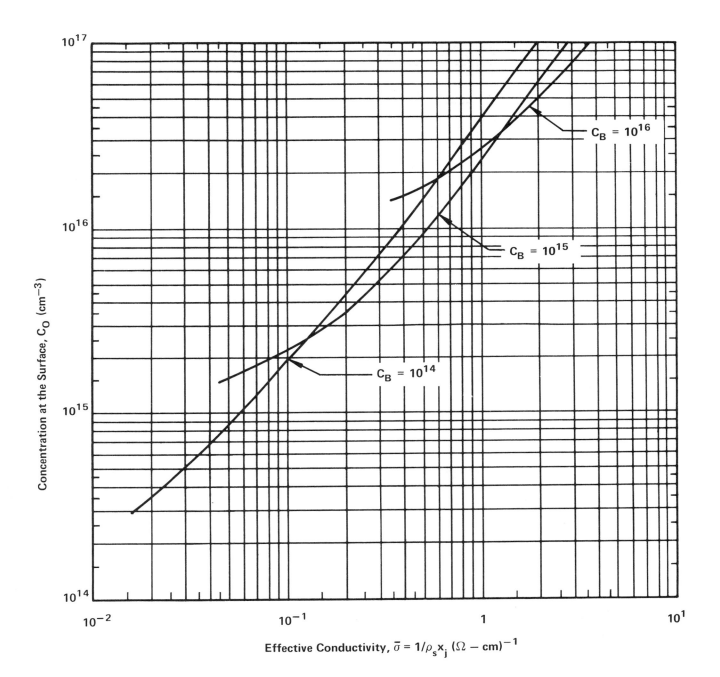

Figure 10-42 — Effective Conductivity of Diffused Layers — Si Exponential, n-Type Diffusion $10^{17} \geqslant C_O \geqslant 10^{14}$

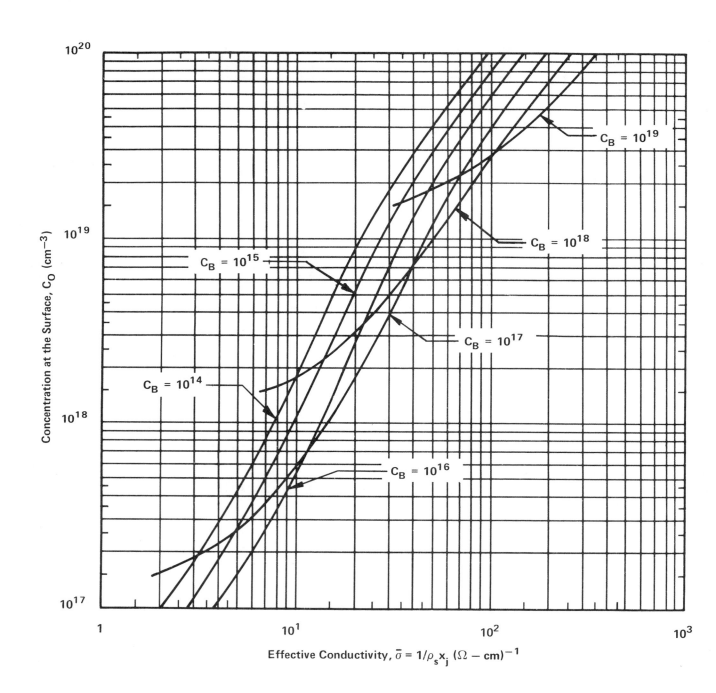

Figure 10-43 — Effective Conductivity of Diffused Layers — Si Exponential, n-Type
Diffusion $10^{20} \geqslant C_O \geqslant 10^{17}$

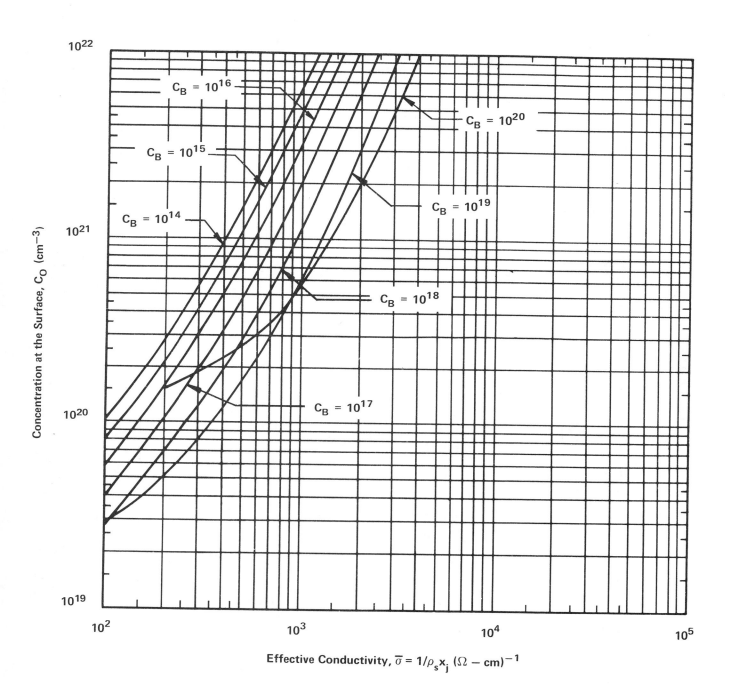

Figure 10-44 — Effective Conductivity of Diffused Layers — Si Exponential, n-Type Diffusion $10^{22} \geqslant C_O \geqslant 10^{19}$

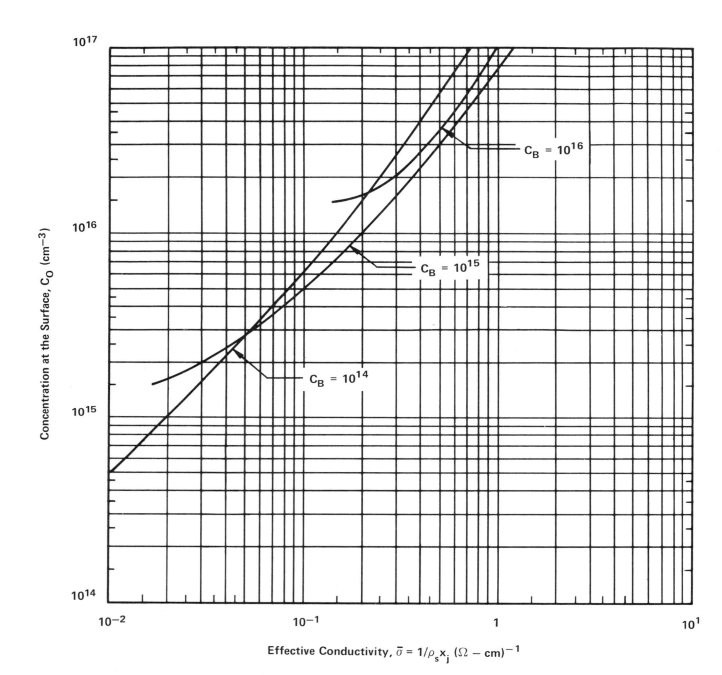

Figure 10-45 — Effective Conductivity of Diffused Layers — Si Exponential, p-Type Diffusion $10^{17} \geqslant C_O \geqslant {}^{14}$

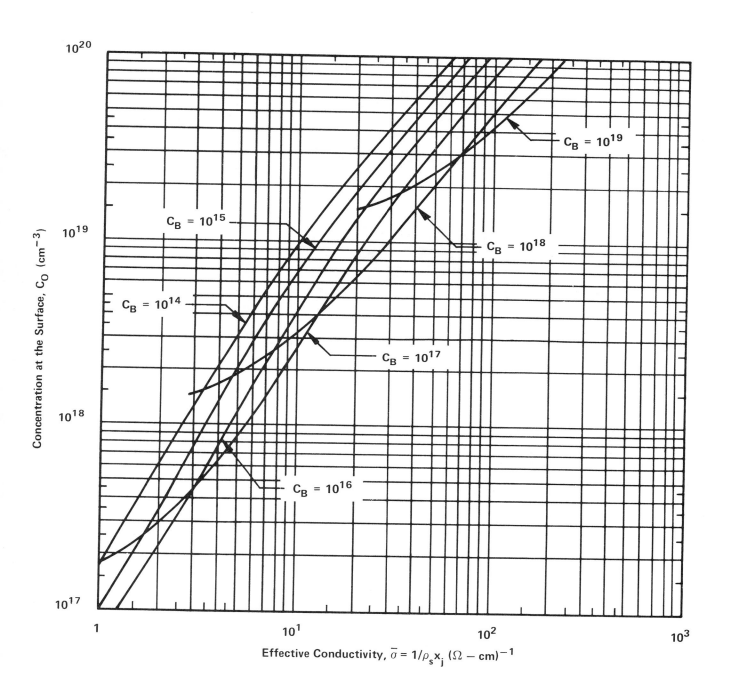

Figure 10-46 — Effective Conductivity of Diffused Layers — Si Exponential, p-Type Diffusion $10^{20} \geqslant C_O \geqslant 10^{17}$

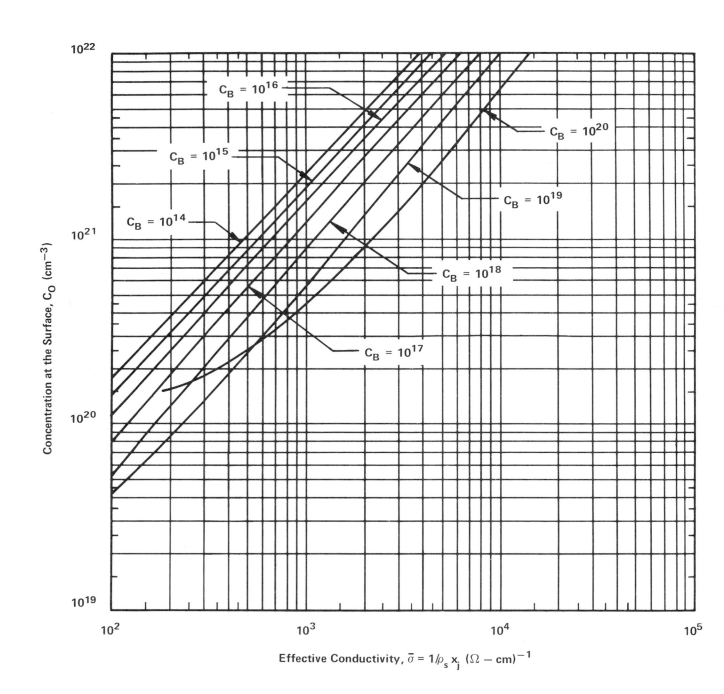

Figure 10-47 — Effective Conductivity of Diffused Layers — Si Exponential, p-Type
Diffusion $10^{22} \geqslant C_O \geqslant 10^{19}$

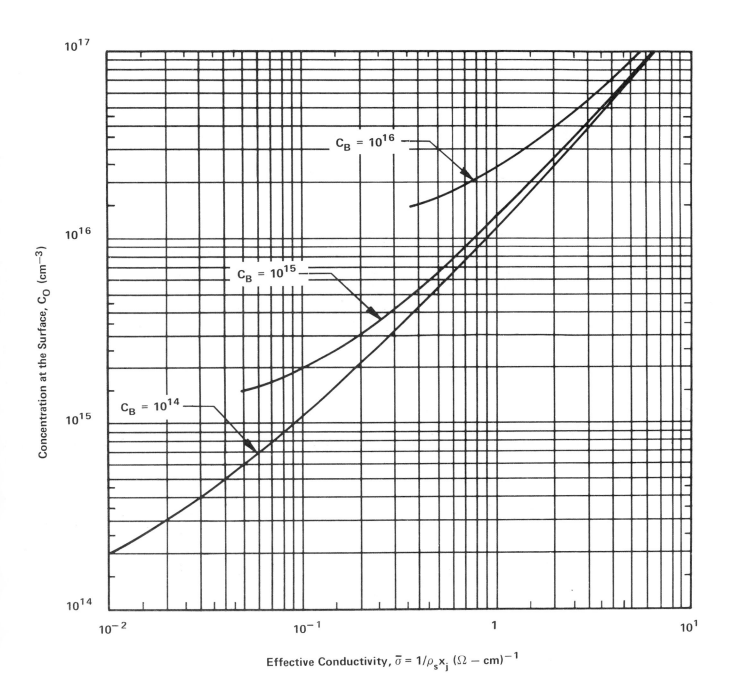

Figure 10-48 — Effective Conductivity of Diffused Layers -- Si Linear, n-Type
Diffusion $10^{17} \geqslant C_O \geqslant 10^{14}$

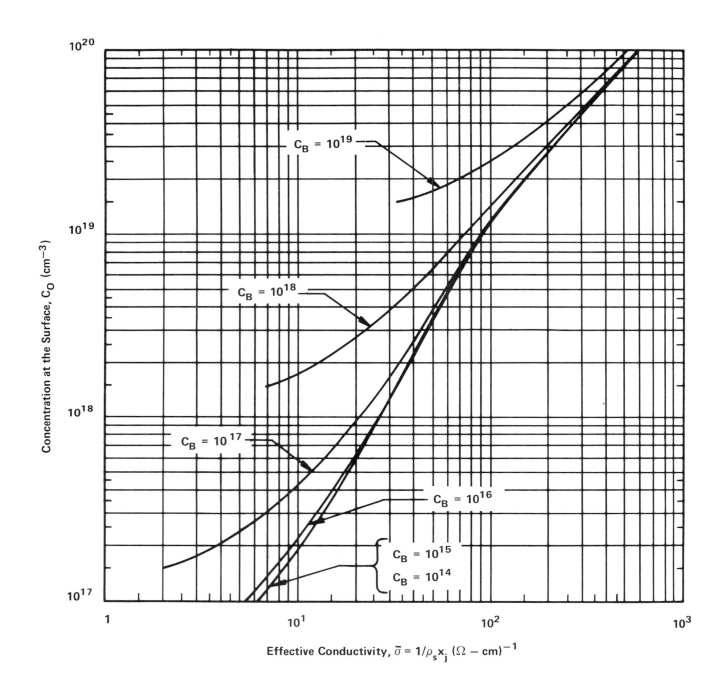

Figure 10-49 — Effective Conductivity of Diffused Layers — Si Linear, n-Type
Diffusion $10^{20} \geqslant C_O \geqslant 10^{17}$

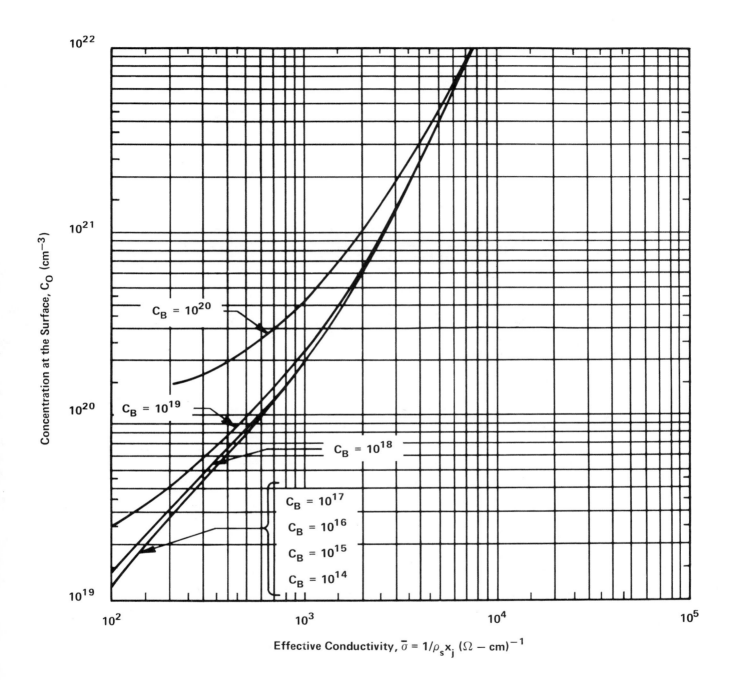

Figure 10-50 — Effective Conductivity of Diffused Layers — Si Linear, n-Type
Diffusion $10^{22} \geqslant C_O \geqslant 10^{19}$

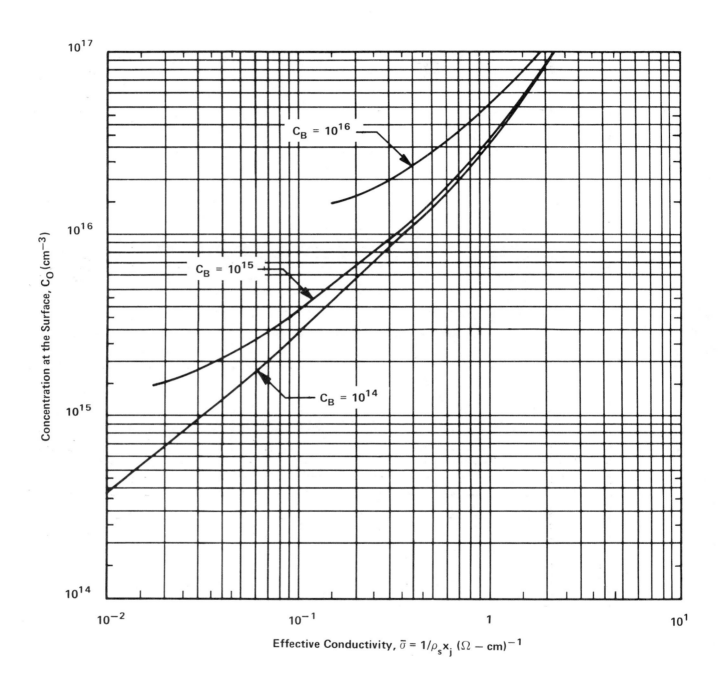

Figure 10-51 — Effective Conductivity of Diffused Layers — Si Linear, p-Type
Diffusion $10^{17} \geqslant C_O \geqslant 10^{14}$

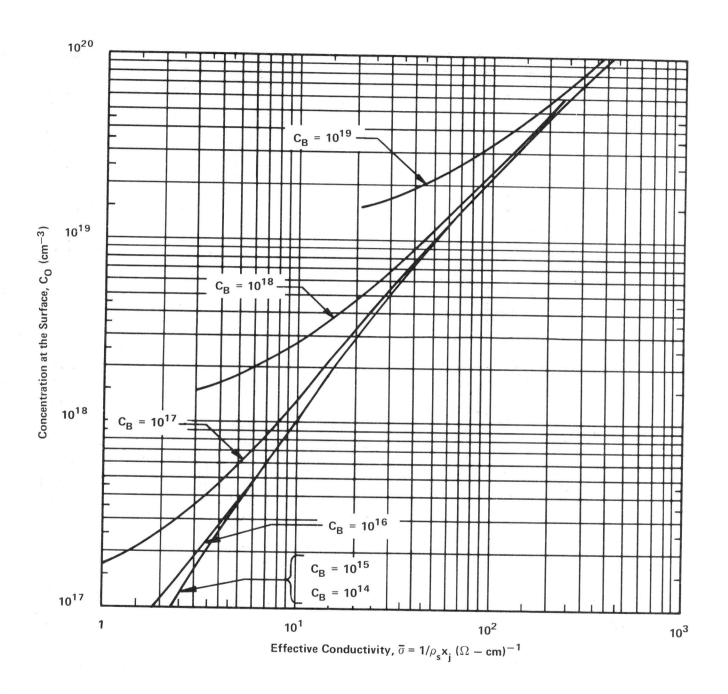

Figure 10-52 — Effective Conductivity of Diffused Layers — Si Linear, p-Type Diffusion $10^{20} \geqslant C_O \geqslant 10^{17}$

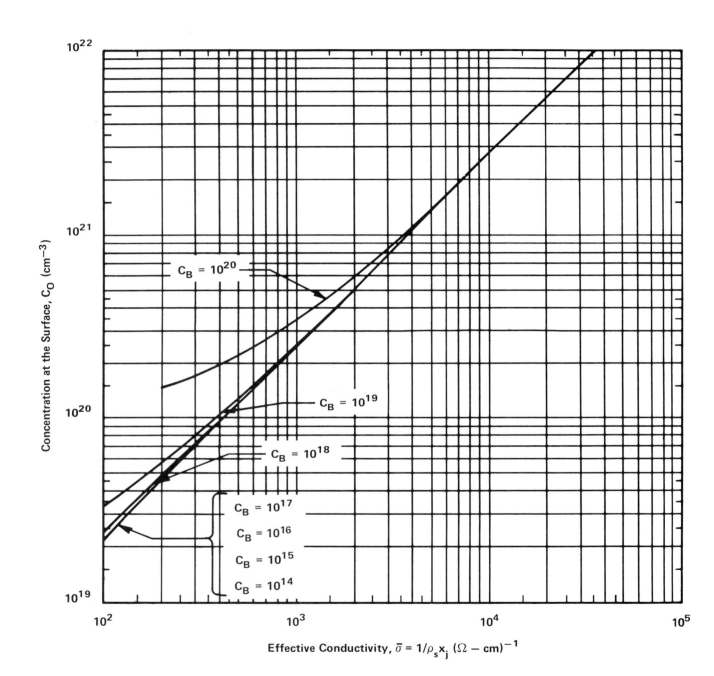

Figure 10-53 — Effective Conductivity of Diffused Layers — Si Linear, p-Type
Diffusion $10^{22} \geqslant C_O \geqslant 10^{19}$

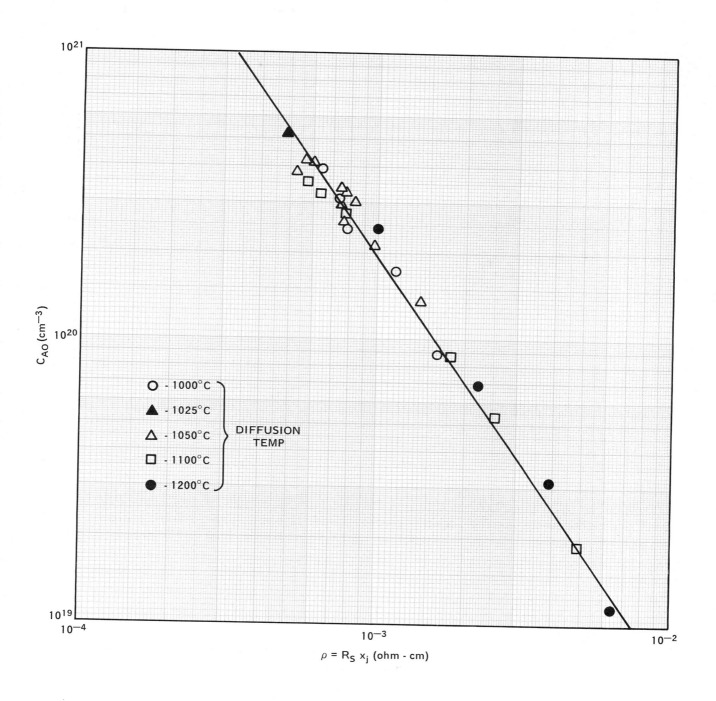

Figure 10-54 — Electrically Active As Surface Concentration (Implanted-Diffused) Versus Effective Resistivity for Si[4] (Reprinted by permission of the publisher, The Electrochemical Society, Inc.)

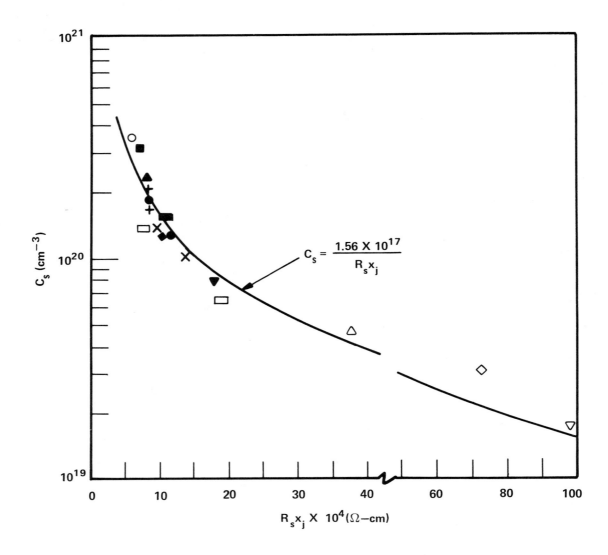

$$C_s = \frac{1.56 \times 10^{17}}{R_s x_j}$$

Figure 10-55 — Electrically Active As (Chemical Sources) Surface Concentration Versus Effective Resistivity for Si [5]

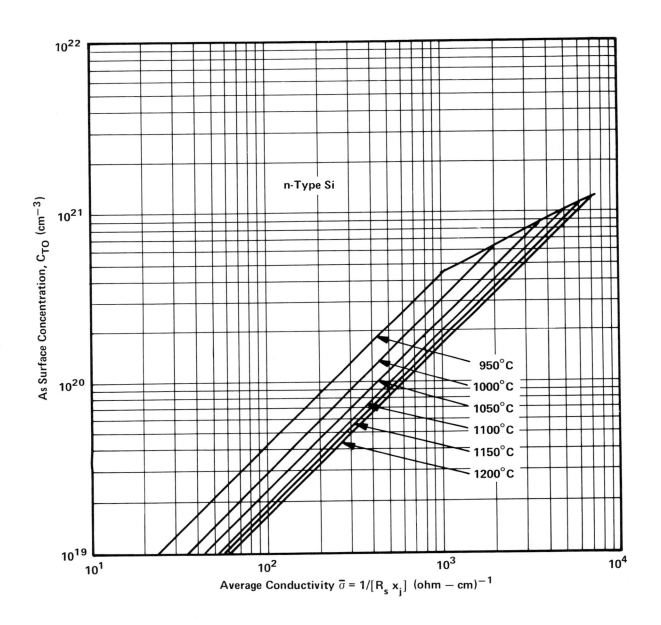

Figure 10-56 -- Total As Surface Concentration (Chemical Sources) Versus Average Conductivity for Diffusions Into n-Type Si ($C_B = 10^{16}/cm^3$) [6]

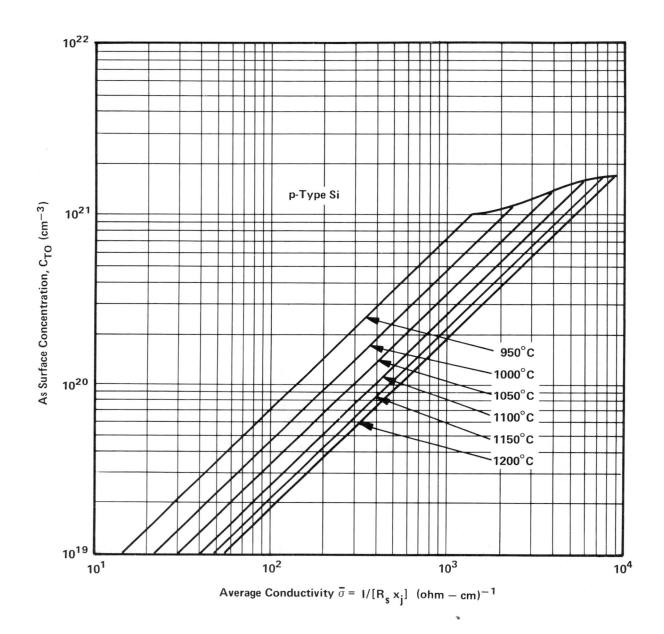

Figure 10-57 — Total As Surface Concentration (Chemical Sources) Versus Average
Conductivity for Diffusions into p-Type Si ($C_B = 10^{16}/cm^3$)[6]

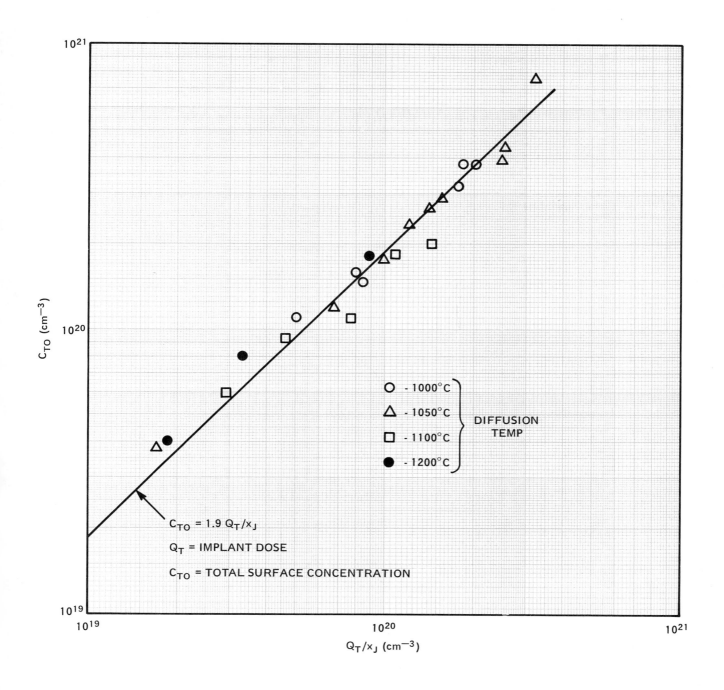

Figure 10-58—Total As Surface Concentration (Implanted-Diffused) Versus Average Doping for Si [4]
(Reprinted by permission of the publisher, The Electrochemical Society, Inc.)

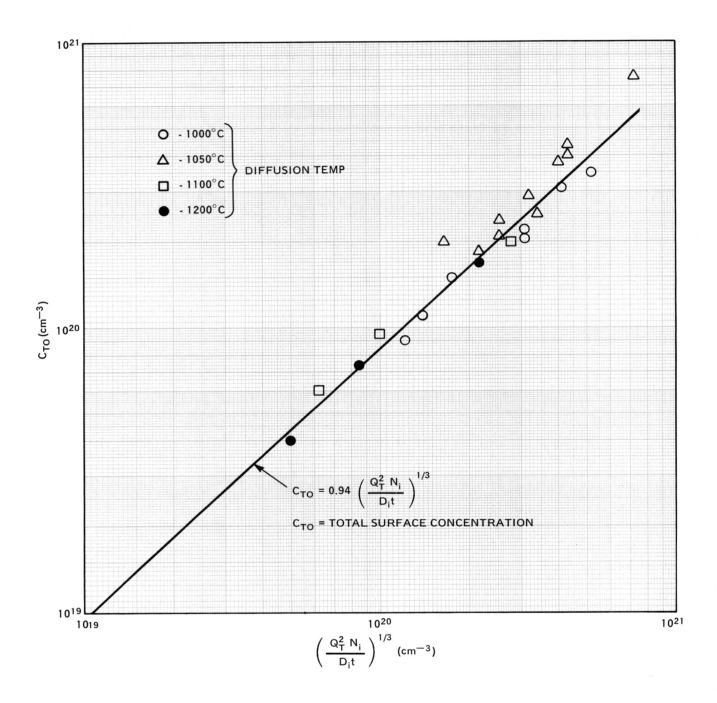

Figure 10-59—Time, Temperature, and Dose Dependence of As Surface Concentration in Si[4]
(Reprinted by permission of the publisher, The Electrochemical Society, Inc.)

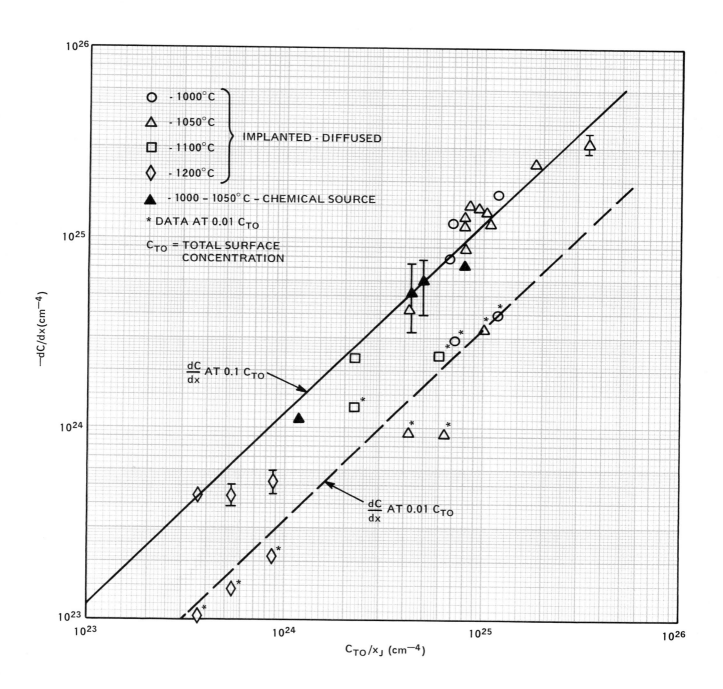

Figure 10-60 — As Profile Gradient Versus C_{TO}/x_j at Two Concentrations[4]
(Reprinted by permission of the publisher, The Electrochemical Society, Inc.)

REFERENCES

1. *Fundamentals of Silicon Integrated Device Technology.* ed. R. M. Burger and R. P. Donovan, Englewood Cliffs, New Jersey: Prentice-Hall, 1967, p. 309.

2. J. C. Irvin, private communication.

3. D. B. Cuttriss, private communication.

4. R. B. Fair and J. C. C. Tsai, "Profile Parameters of Implanted-Diffused Arsenic Layers in Silicon," *J. Electrochem. Soc: Solid-State Science and Technology,* Vol. 123, No. 4 (April 1976), pp. 583-86.

5. R. B. Fair, "Profile Estimation of High-Concentration Arsenic Diffusions in Silicon," *J. Appl. Phys.,* Vol. 43, No. 3 (March 1972), p. 1278.

6. R. B. Fair and G. R. Weber, "Relationship Between Resistivity and Total Arsenic Concentration in Heavily Doped n- and p-Type Silicon," *J. Appl. Phys.,* Vol. 44, No. 1 (January 1973), p. 280.

11. PROPERTIES OF p-n JUNCTIONS

PROPERTIES OF p-n JUNCTIONS

LIST OF SYMBOLS

Symbol	Definition
A	Area
a, α	Concentration Gradient
a_1, a_2, W_p, W_n	Depletion Layer Width
C	Capacitance or Concentration
C_B, N_B	Substrate Concentration
C_S	Surface Concentration
E_g	Band Gap Voltage, Energy Gap
E_{max}, E_m	Maximum Electric Field
N_A	Acceptor Concentration
N_D	Donor Concentration
n, n_p	Electron Concentration
n_i	Intrinsic Carrier Concentration
p, p_n	Hole Concentration
q	Electron Charge
r_j	Junction Curvature
V	Voltage
V_a	Applied Voltage
V_B	Breakdown Voltage
W	Equals the Sum of W_n and W_p
W_0	Space Charge Width at Zero Bias
X	Distance
X_j	Junction Depth
ε	Permitivity
ϕ	Built-in Voltage
ϕ_p, ϕ_n	Fermi Potential

11.1 Depletion Layer Relationships of p-n Junctions

The p-n junction consists of a space charge region of ionized doping atoms depleted of mobile carriers, referred to as a depletion or space charge region. Since this region plays a significant role in the behavior of the p-n junction, a number of curves giving the dependence of depletion layer properties on applied bias, doping, and geometry are presented in this section. All models invoke the abrupt depletion region approximation. This assumes that mobile charge density on either side of the depletion region decreases abruptly to zero in the region.

11.2 Step-Junction (One-Dimensional)

The simplest approximation to the charge density in the depletion region is the step-junction approximation. (See Figure 11-1.) This model assumes that the doping of either side of the metallurgical junction ($N_A = N_D$) is constant, giving rise to constant charge densities on either side of the junction. This charge distribution is shown in Figure 11-1 along with equations describing some depletion layer properties.

If the doping density on one side of the metallurgical junction is much greater than on the other side, the junction properties can be defined entirely in terms of the lightly doped side. This is the one-sided step-junction approximation. This is a practical model for shallow junctions formed by a heavily doped diffusion into a lightly doped region of the opposite doping type, i.e., n^+p or p^+n, and a limiting case for deeper diffused junctions operated at high reverse bias.

Because of the usefulness of the one-sided step-junction, a number of somewhat redundant curves and nomographs are included in this section. The curves are:

FIGURE	JUNCTION PROPERTIES	PARAMETERS	INDEPENDENT VARIABLE
11-2	Built-in voltage	—	Doping concentration
11-3	Space-charge width and capacitance	—	Voltage/doping concentration
11-4	Space-charge width, capacitance, and avalanche breakdown	Doping	Applied voltage

The nomographs included are:

FIGURE	JUNCTION PROPERTIES	INDEPENDENT VARIABLE
11-5	Equivalent electron, proton, and alpha particle energy, depletion width/capacitance, doping concentration/resistivity	Bias voltage
11-6	Depletion width/capacitance, doping/resistivity, breakdown voltage	Bias voltage

11.3 Linear-Graded Junction (One-Dimensional)

A second useful space charge approximation is the linearly graded junction. Here the ionized doping charge density varies linearly across the depleted region, passing through zero at the metallurgical junction as shown in Figure 11-7. The common relationships for the linear-graded junction are also given in Figure 11-7.

A curve of built-in voltage versus doping gradient is provided by Figure 11-8. Figure 11-9 provides depletion layer width, capacitance, and breakdown voltage versus applied voltage with doping gradient as the parameter.

The linear-graded junction is a practical approximation to diffused junctions operated at low applied bias. The main difficulty in using these curves is determining the grade constant. Foxhall[1] has addressed this problem by providing Figures 11-10 and 11-11 which give the grade constant for Gaussian and erfc distributions, respectively, as a function of diffusion paramenters.

11.4 Diffused-Junction (One-Dimensional)

Lawrence and Warner[2] have numerically evaluated one-dimensional diffused junctions to provide depletion region parameters which accurately reflect the details of the doping profile in germanium and silicon. These parameters have been charted in Figures 11-13 through 11-45. The definition of the doping profile and corresponding electric field distribution parameters is provided by Figure 11-12.

The following table lists the curves provided.

PROPERTY	PROFILE	c_B/c_O*	FIGURE
Junction capacitance and total depletion layer thickness	erfc	10^{-1}	11-13
		10^{-2}	11-14
		10^{-3}	11-15
		10^{-4}	11-16
		10^{-5}	11-17
		10^{-6}	11-18
		10^{-7}	11-19
		10^{-8}	11-20
a_1/a_{total}**	erfc	10^{-1}	11-21
		10^{-2}	11-22
		10^{-3}	11-23
		10^{-4}	11-24
		10^{-5}	11-25
		10^{-6}	11-26
		10^{-7}	11-27
		10^{-8}	11-28
Junction capacitance and total depletion layer thickness	Gaussian	10^{-1}	11-29
		10^{-2}	11-30
		10^{-3}	11-31
		10^{-4}	11-32
		10^{-5}	11-33
		10^{-6}	11-34
		10^{-7}	11-35
		10^{-8}	11-36
a_1/a_{total}	Gaussian	10^{-1}	11-37
		10^{-2}	11-38
		10^{-3}	11-39
		10^{-4}	11-40
		10^{-5}	11-41
		10^{-6}	11-42
		10^{-7}	11-43
		10^{-8}	11-44
Peak electric field	erfc & Gaussian	10^{-5}	11-45

---

  * Ratio of background doping to surface concentration.
 ** Ratio of depletion layer width on the diffused side of the metallurgical junction to the total layer width.

11.5 Step- and Linear-Graded Junctions With Curvature

Junction depletion layer properties included thus far have neglected the effects of geometry. Dimensional effects can be very significant for small planar junctions. Planar junctions are formed by implant or diffusion through a window in a masking layer which establishes the lateral geometry of junction. Figure 11-46 illustrates a planar junction. It also shows cylindrical and hemispherical one-dimensional models used to quantify effects of junction curvature. Curves by Lee and Sze[3] and Sze and Gibbons[4] showing the effect of junction curvature on step- and linear-graded junction properties are provided in this section.

The curves included are tabulated below.

JUNCTION PROPERTY	MATERIAL	JUNCTION DOPING PROFILE	GEOMETRY	INDEPENDENT VARIABLE	FIGURE
Capacitance	General	Step	Cylindrical & spherical	Normalized voltage	11-47
	Si	Step	Cylindrical	Voltage	11-48
	Si	Step	Spherical	Voltage	11-49
	General	Linear	Cylindrical & spherical	Normalized voltage	11-50
Depletion layer width and maximum electrical field at breakdown	Ge, Si, GaAs, and GaP	Step	Plane	Voltage	11-51
	Ge, Si, GaAs, and GaP	Linear	Plane	Voltage	11-52
Breakdown voltage	Ge	Step	Plane, cylindrical, spherical	Doping	11-53
Breakdown voltage	Si	Step	Plane, cylindrical, spherical	Doping	11-54
Breakdown voltage	Ge	Linear	Cylindrical	Gradient-constant	11-55
Breakdown voltage	Si	Linear	Cylindrical	Gradient-constant	11-56
Breakdown voltage	Si	Step	Plane, cylindrical	Junction-radius	11-57

Also included is Figure 11-58 which provides some data on plane geometry step-junction breakdown voltage for silicon.

PROPERTY

FORMULA

Depletion-layer width

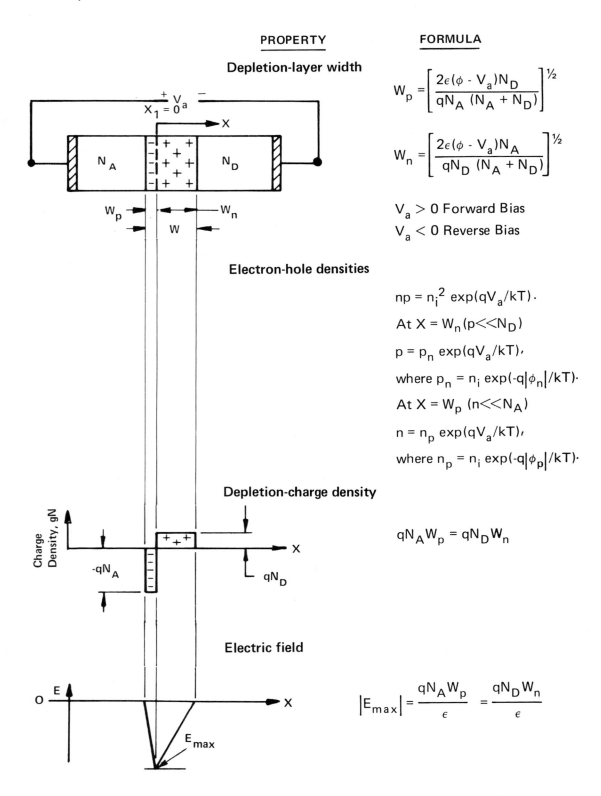

$$W_p = \left[\frac{2\epsilon(\phi - V_a)N_D}{qN_A(N_A + N_D)} \right]^{1/2}$$

$$W_n = \left[\frac{2\epsilon(\phi - V_a)N_A}{qN_D(N_A + N_D)} \right]^{1/2}$$

$V_a > 0$ Forward Bias

$V_a < 0$ Reverse Bias

Electron-hole densities

$np = n_i^2 \exp(qV_a/kT)$.

At $X = W_n \, (p \ll N_D)$

$p = p_n \exp(qV_a/kT)$,

where $p_n = n_i \exp(-q|\phi_n|/kT)$.

At $X = W_p \, (n \ll N_A)$

$n = n_p \exp(qV_a/kT)$,

where $n_p = n_i \exp(-q|\phi_p|/kT)$.

Depletion-charge density

$$qN_A W_p = qN_D W_n$$

Electric field

$$|E_{max}| = \frac{qN_A W_p}{\epsilon} = \frac{qN_D W_n}{\epsilon}$$

Figure 11-1 — Relationships for Step-Junction Diode Depletion Layer
(Sheet 1 of 2)

PROPERTY	FORMULA

Potential distribution

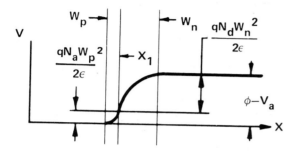

Built-in voltage:

$$\phi = \frac{kT}{q} \ln \frac{N_A N_D}{n_i^2}$$

$$= |\phi_p| + |\phi_n|$$

Capacitance

$$C = \frac{\epsilon A}{W_n + W_p} = \frac{A \left[\dfrac{\epsilon q N_A N_D}{2(N_A + N_D)} \right]^{\frac{1}{2}}}{[\phi - V_a]^{\frac{1}{2}}}$$

Breakdown voltage (one-sided)

$$V_B = E_m \ W/2 = \frac{\epsilon E_m^2}{2q} \ \frac{1}{N_B}$$

$$= V_{OA} \ [N_B(cm^{-3})/10^{16}]^{-0.75}$$

$$= 60(E_g/1.1)^{3/2} (N_B/10^{16})^{-0.75}$$

where:

$$V_{OA} = 25V \text{ for Ge}$$

$$= 60V \text{ for Si}$$

$$= 135V \text{ for GaP}$$

$$= 320V \text{ for GaAs}$$

Figure 11-1— Relationships for Step-Junction Diode Depletion Layer
(Sheet 2 of 2)

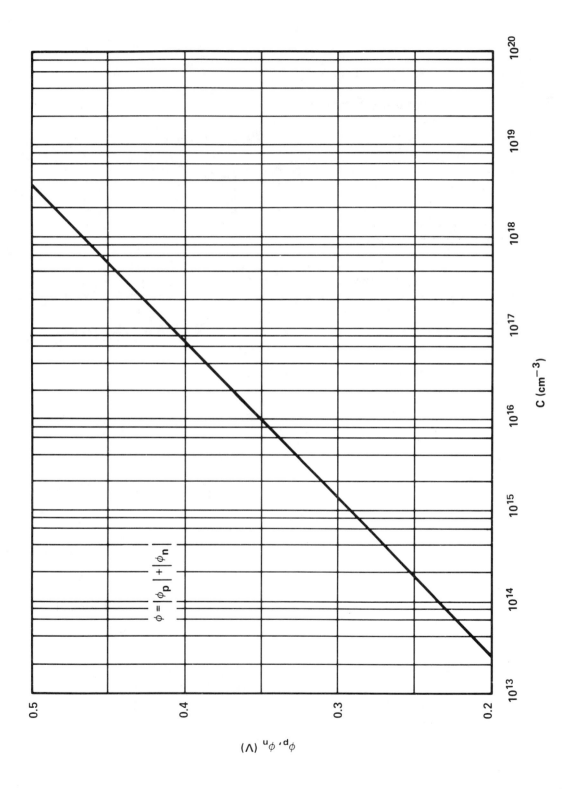

Figure 11-2 — Built-In Voltage Versus Doping for One Side of a Si Step-Junction

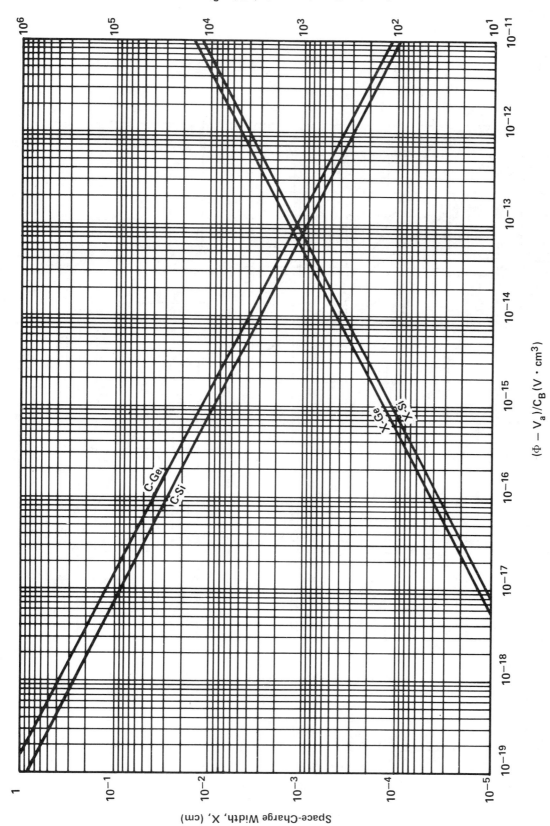

Figure 11-3 — Step-Junction Space-Charge Width and Capacitance Versus Total Voltage to Substrate Doping Ratio

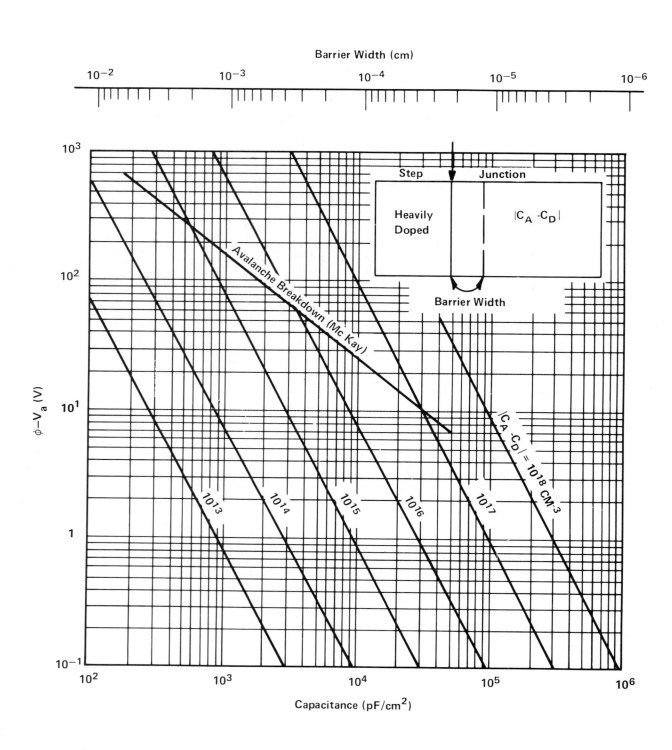

Figure 11-4 — Si Step-Junction Capacitance and Barrier Width as a Function
of Barrier Potential and Doping

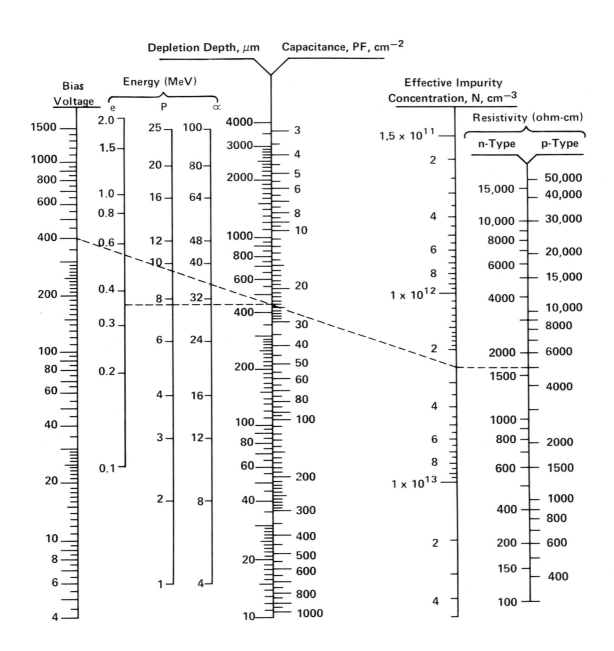

Figure 11-5 — Nomograph for Si Step-Junction

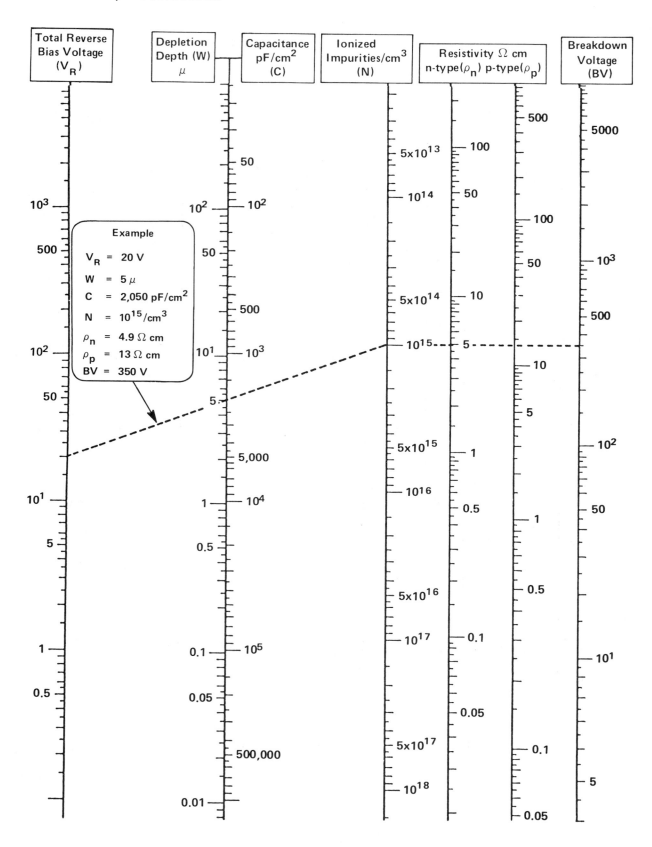

Figure 11-6—Nomograph for Si Step-Junction at 300K [5]

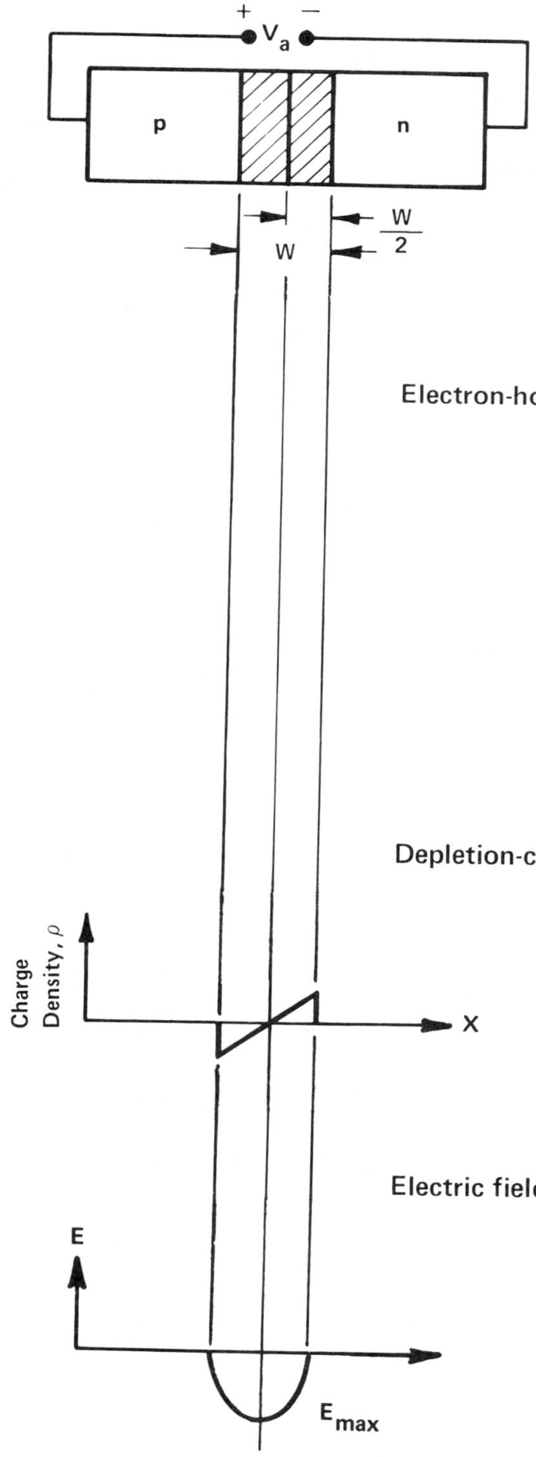

PROPERTY

FORMULA

Depletion-layer width

$$W = \left[\frac{12\epsilon (\phi - V_a)}{qa} \right]^{1/3}$$

$V_a > 0$ Forward Bias

$V_a < 0$ Reverse Bias

$$a \equiv \left| \frac{dN}{dx} \right|_{x = x_j}$$

Electron-hole densities

$np = n_i^2 \exp(qV_a/kT)$.

At $X = \dfrac{W}{2}$ $(p \ll N_D)$

$p = p_n \exp(qV_a/kT)$,

where $p_n = n_i \exp(-q\phi_n /kT)$.

At $X = -\dfrac{W}{2}$ $(n \ll N_A)$

$n = n_p \exp(qV_a/kT)$,

where $n_p = n_i \exp(-q\phi_p /kT)$.

Depletion-charge density

$\rho = ax$

Electric field

$$\left| E_{max} \right| = \frac{aq}{8\epsilon} \; W^2$$

$$= \frac{1.5 (\phi - V_a)}{W}$$

Figure 11-7 — Relationships for Linear-Graded Junction (Sheet 1 of 2)

PROPERTY	FORMULA

Potential distribution

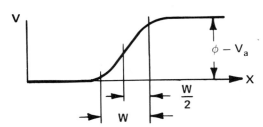

Built-in voltage:

$$\phi = \frac{2kT}{q} \ln \frac{aW_o}{2n_i}$$

See Figure 11-8 for ϕ vs; a for Si

Capacitance

$$C = \frac{\epsilon A}{W} = 0.436 \, (aq)^{1/3} \epsilon^{2/3} A \, (\phi - V_a)^{-1/3}$$

Breakdown voltage

$$V_B = V_{OG} \, [a(cm^{-4})/10^{21}]^{-0.4}$$

where:

$$V_{OG} = 18V \text{ for Ge}$$
$$= 40V \text{ for Si}$$
$$= 80V \text{ for GaP}$$
$$= 145V \text{ for GaAs}$$

$$V_B = 60(E_g/1.1)^{6/5}(a/3 \times 10^{20})^{-2/5},$$

where E_g = bandgap in eV.

Figure 11-7 — Relationships for Linear-Graded Junction (Sheet 2 of 2)

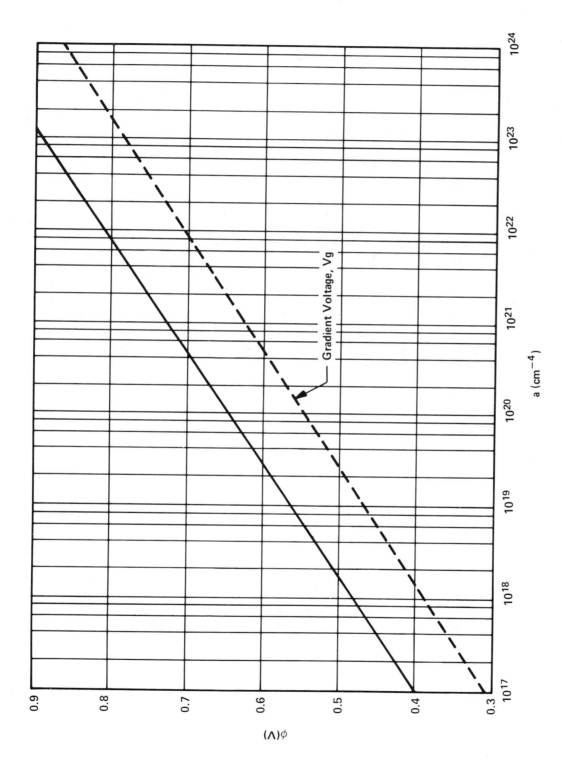

Figure 11-8 — Built-In Voltage Versus Concentration Gradient for a Si Linear-Graded Junction [6]

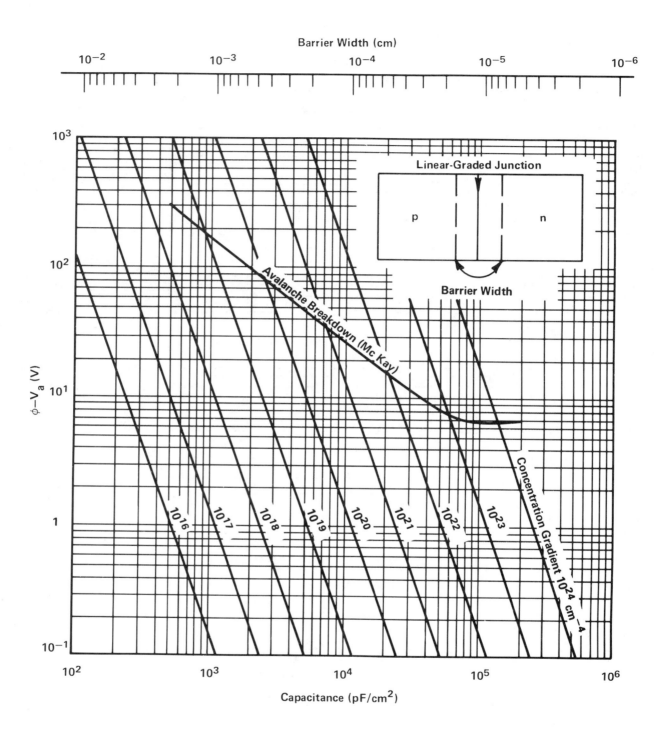

Figure 11-9 — Si Linear-Graded Junction Capacitance and Barrier Width as a Function of Barrier Potential and Concentration Gradient

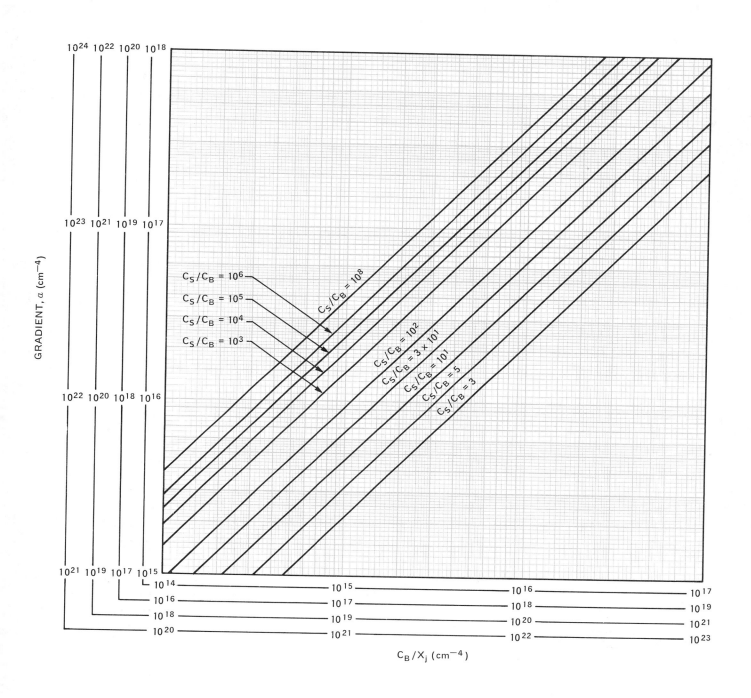

Figure 11-10 — Impurity Gradient From Diffusion Parameters — Gaussian Distribution

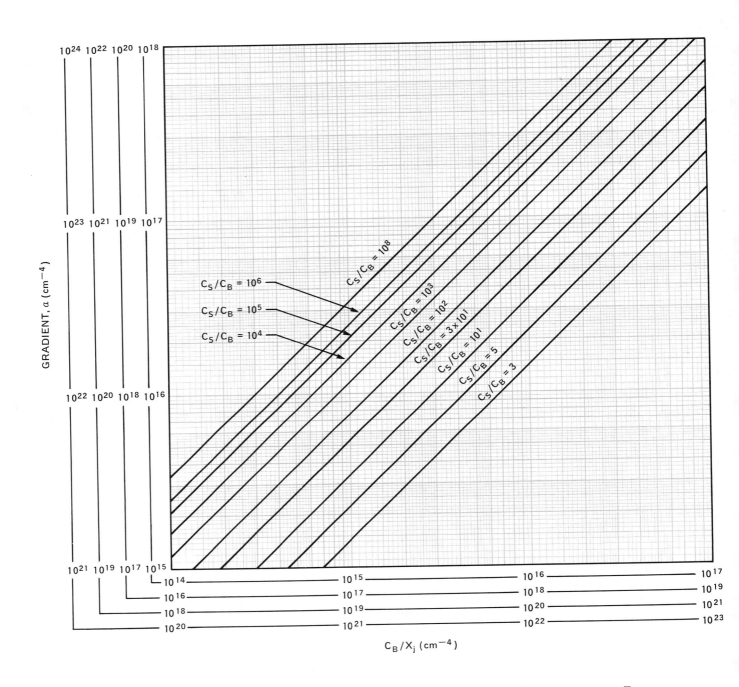

Figure 11-11 — Impurity Gradient From Diffusion Parameters — Complementary Error Function Distribution

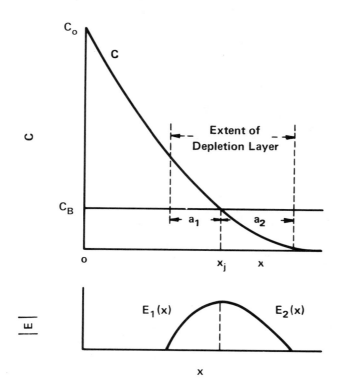

Figure 11-12 — Definition of Diffusion Profile and Corresponding Electric Field
Distribution Parameters in Figures 11-13 Through 11-45 [2]

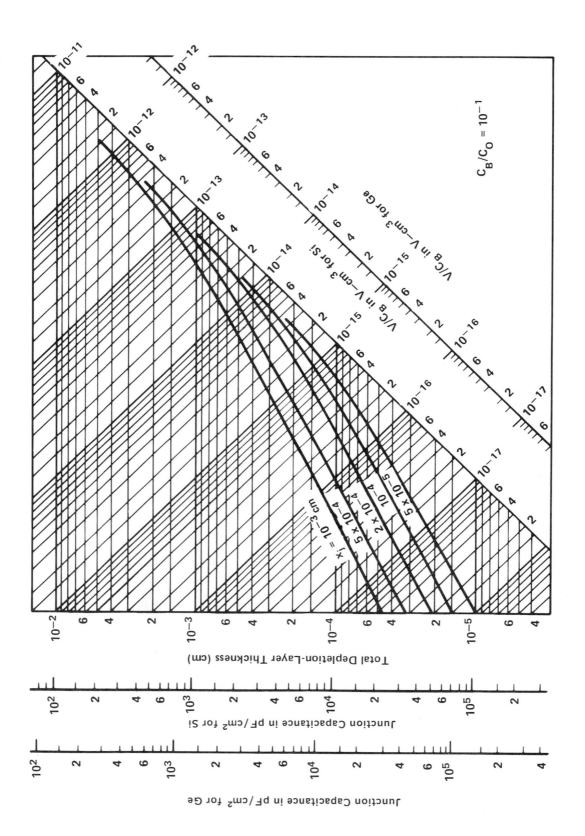

Figure 11-13 — Chart for Use in Range 3×10^{-2} to 3×10^{-1}, Erfc Distribution

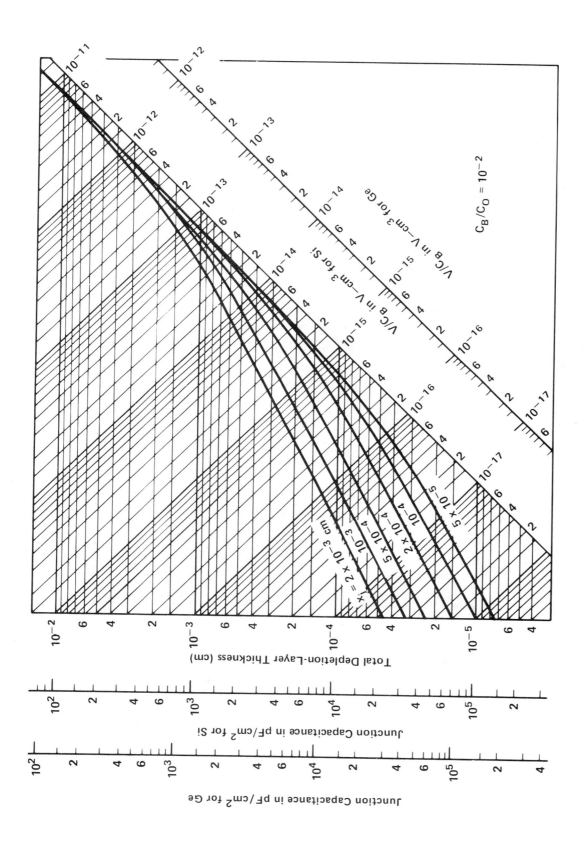

Figure 11-14 — Chart for Use in Range 3 X 10^{-3} to 3 X 10^{-2}, Erfc Distribution

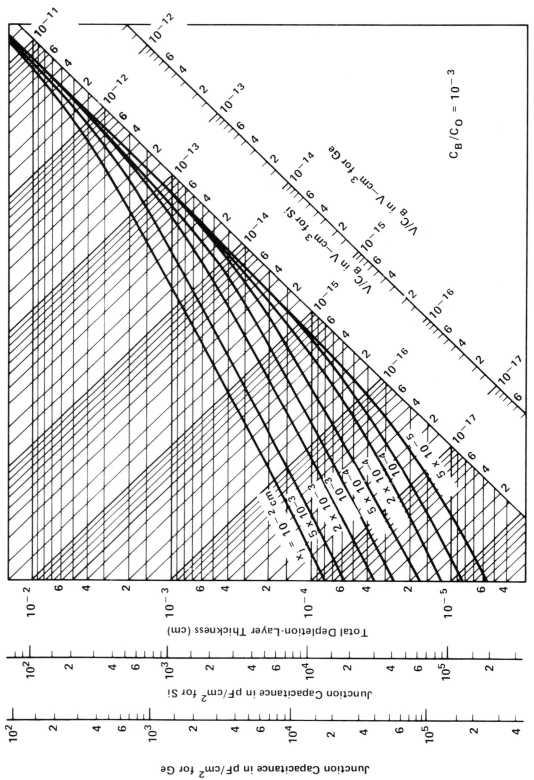

Figure 11-15 — Chart for Use in Range 3×10^{-4} to 3×10^{-3}, Erfc Distribution

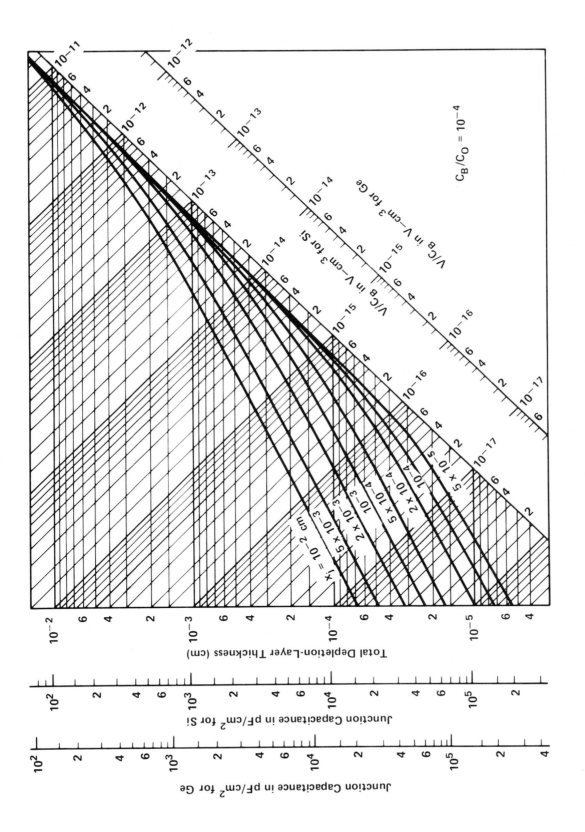

Figure 11-16 — Chart for Use in Range 3 X 10^{-5} to 3 X 10^{-4}, Erfc Distribution

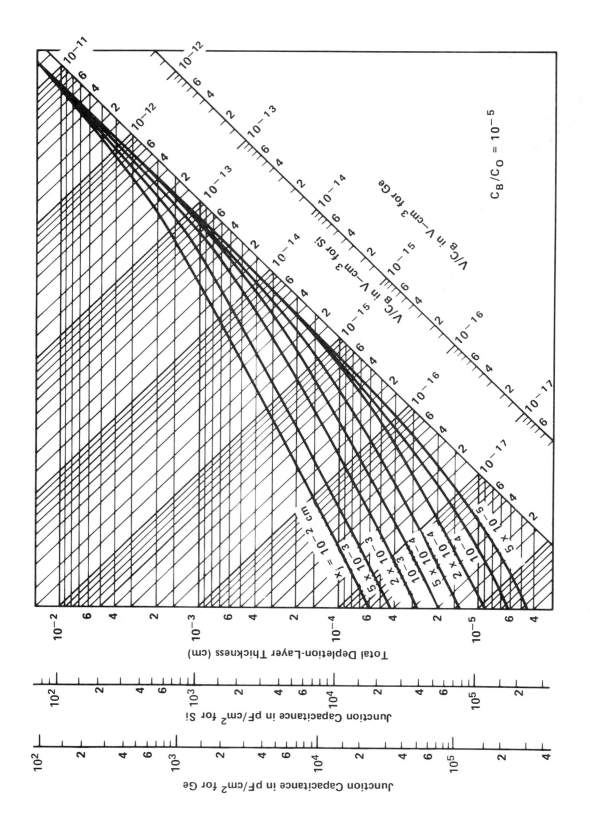

Figure 11-17 — Chart for Use in Range 3×10^{-6} to 3×10^{-5}, Erfc Distribution

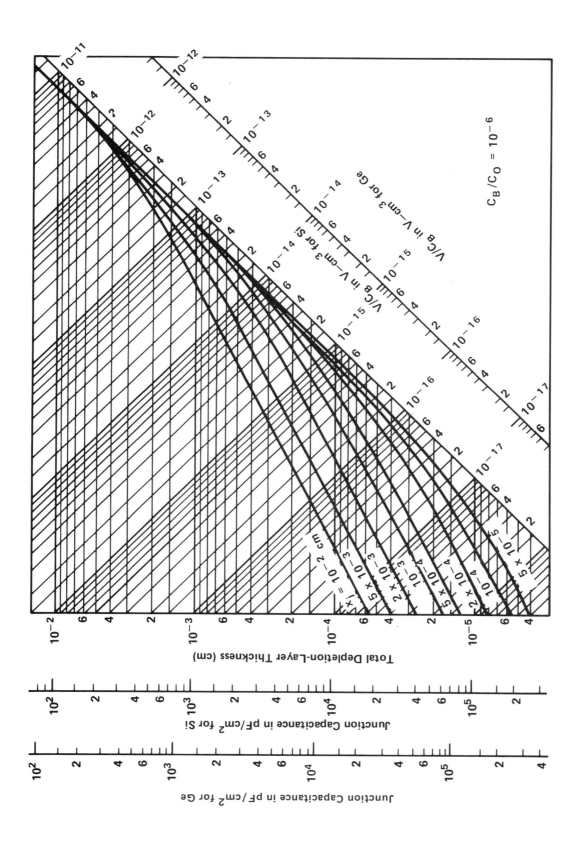

Figure 11-18 — Chart for Use in Range 3×10^{-7} to 3×10^{-6}, Erfc Distribution

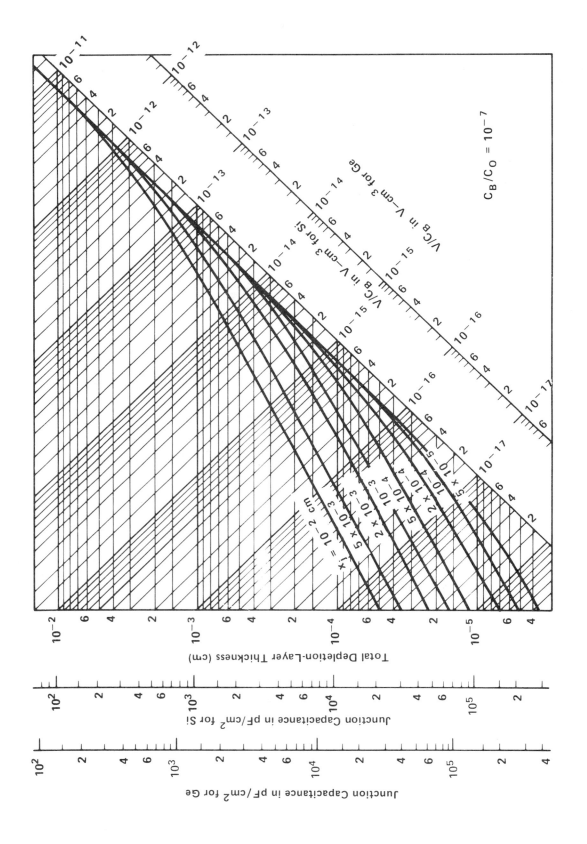

Figure 11-19 — Chart for Use in Range 3×10^{-8} to 3×10^{-7}, Erfc Distribution

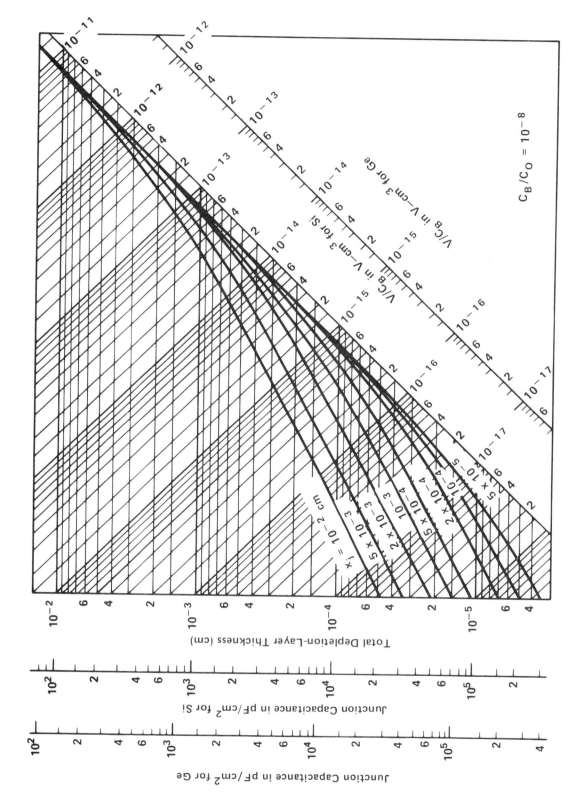

Figure 11-20 — Chart for Use in Range 3×10^{-9} to 3×10^{-8}, Erfc Distribution

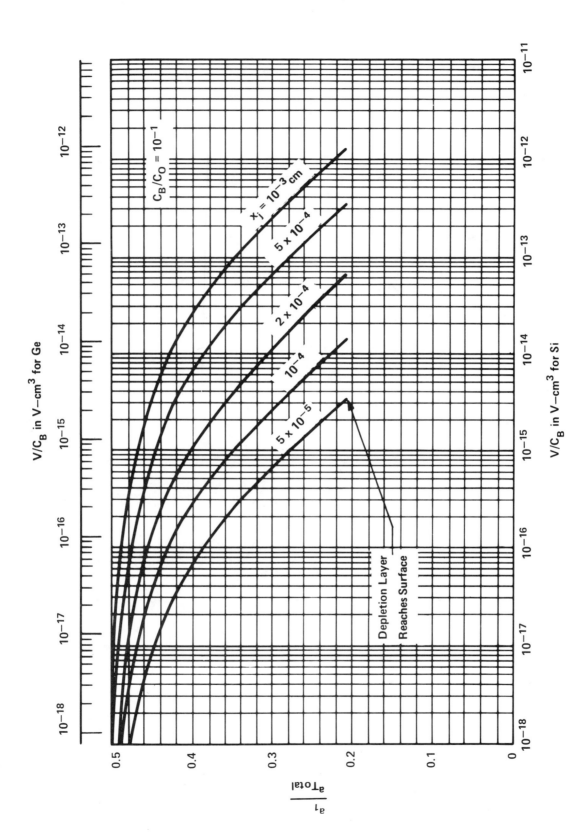

Figure 11-21 — Chart for Use in Range 3×10^{-2} to 3×10^{-1}, Erfc Distribution

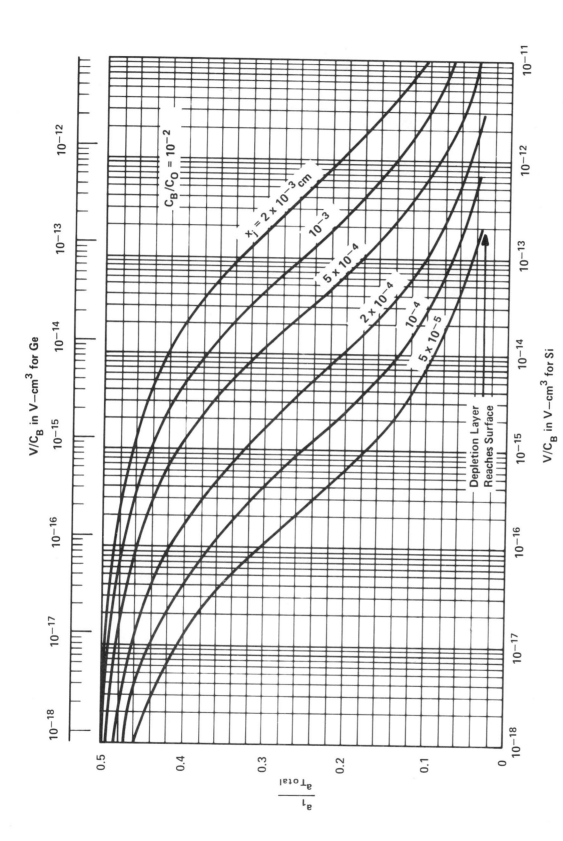

Figure 11-22 — Chart for Use in Range 3×10^{-3} to 3×10^{-2}, Erfc Distribution

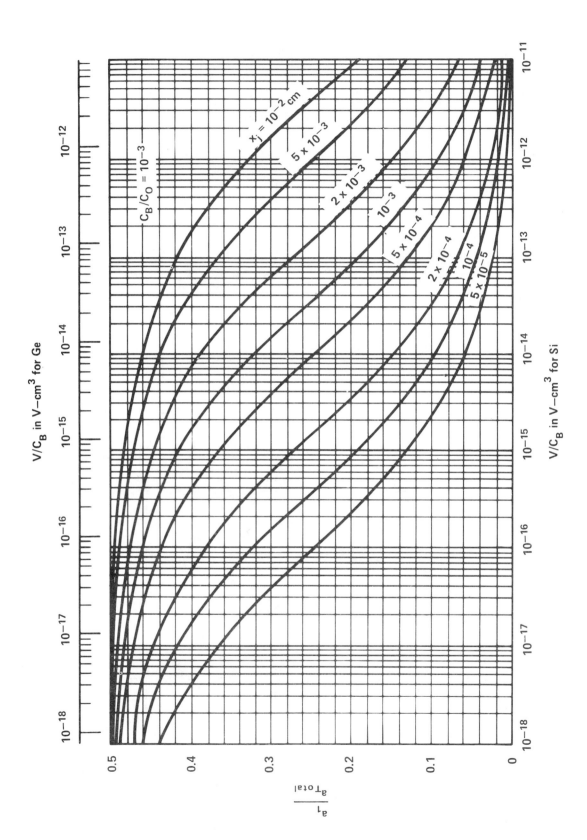

Figure 11-23 — Chart for Use in Range 3×10^{-4} to 3×10^{-3}, Erfc Distribution

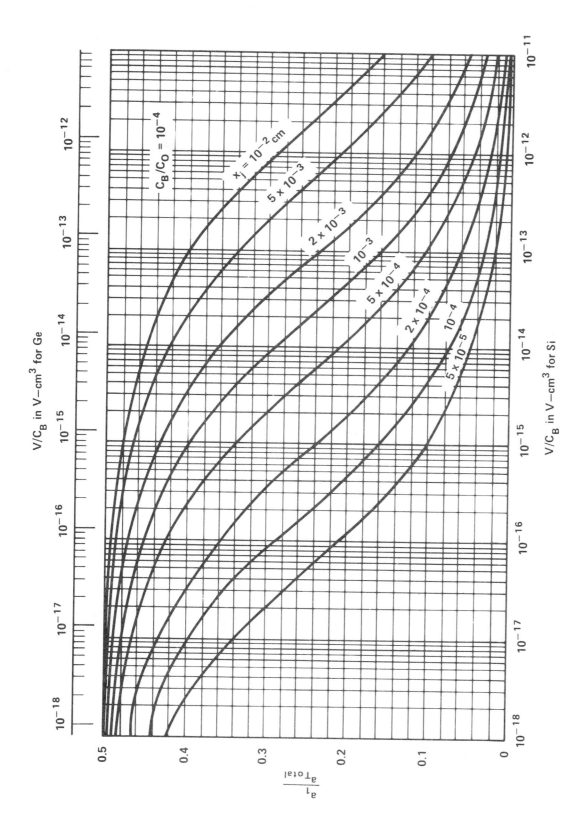

Figure 11-24 — Chart for Use in Range 3×10^{-5} to 3×10^{-4}, Erfc Distribution

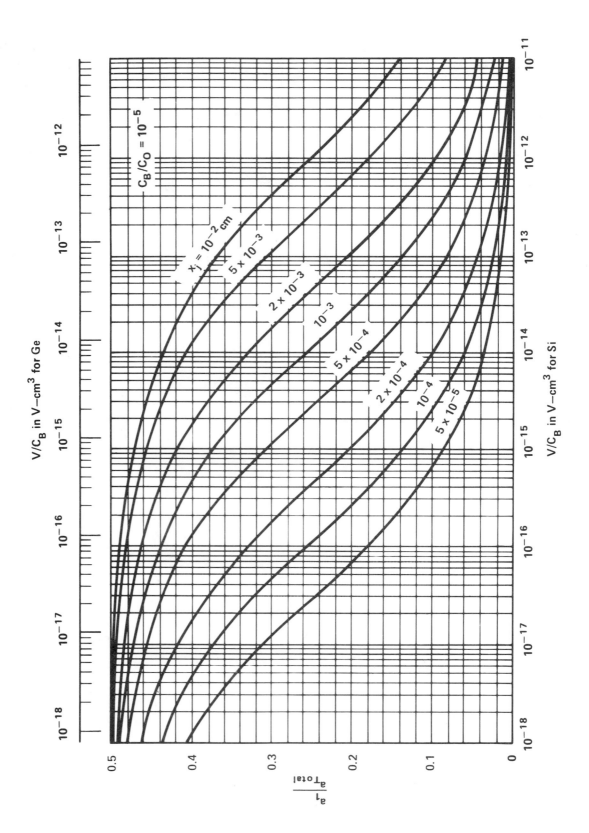

Figure 11-25 — Chart for Use in Range 3×10^{-6} to 3×10^{-5}, Erfc Distribution

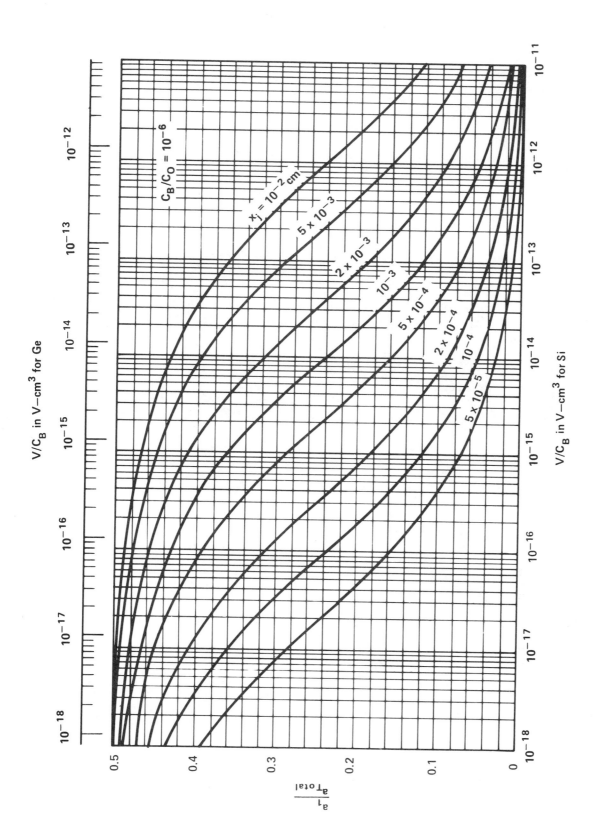

Figure 11-26 — Chart for Use in Range 3×10^{-7} to 3×10^{-6}, Erfc Distribution

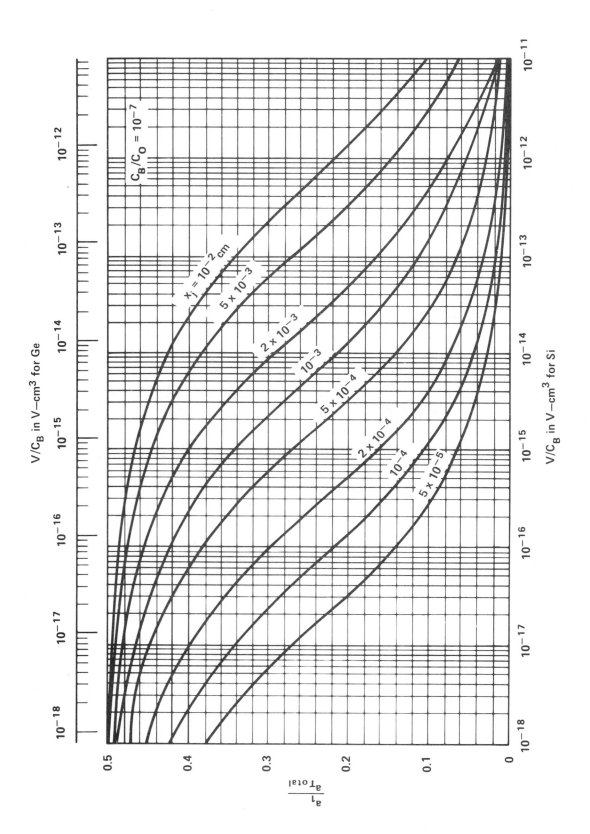

Figure 11-27 — Chart for Use in Range 3 X 10^{-8} to 3 X 10^{-7}, Erfc Distribution

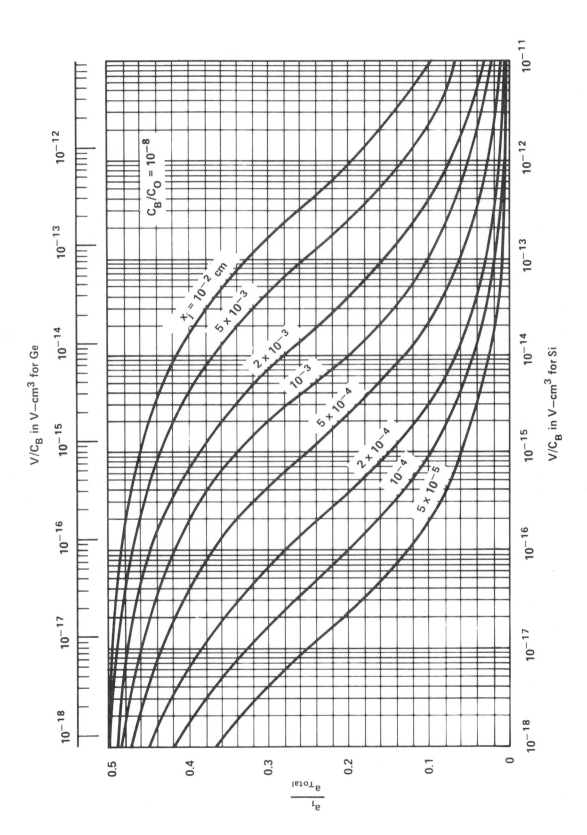

Figure 11-28 — Chart for Use in Range 3×10^{-9} to 3×10^{-8}, Erfc Distribution

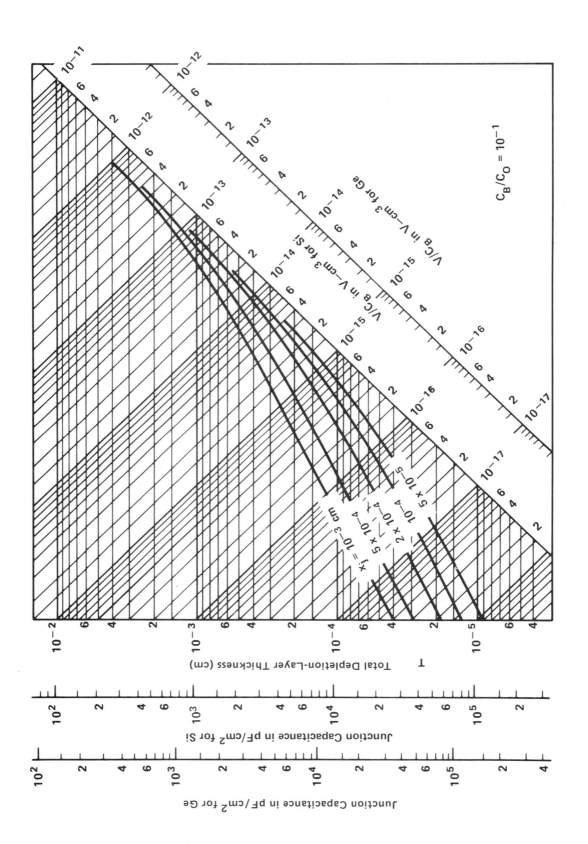

Figure 11-29 — Chart for Use in Range 3×10^{-2} to 3×10^{-1}, Gaussian Distribution

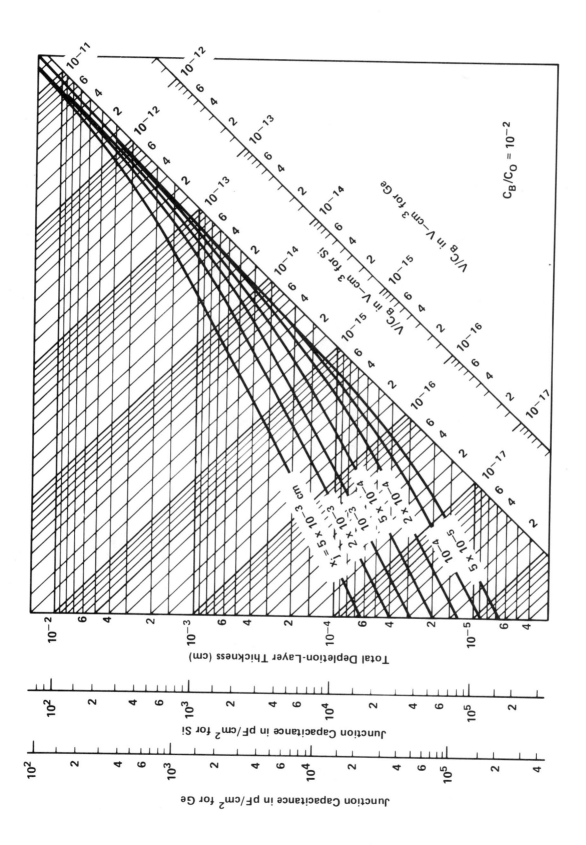

Figure 11-30 — Chart for Use in Range 3×10^{-3} to 3×10^{-2}, Gaussian Distribution

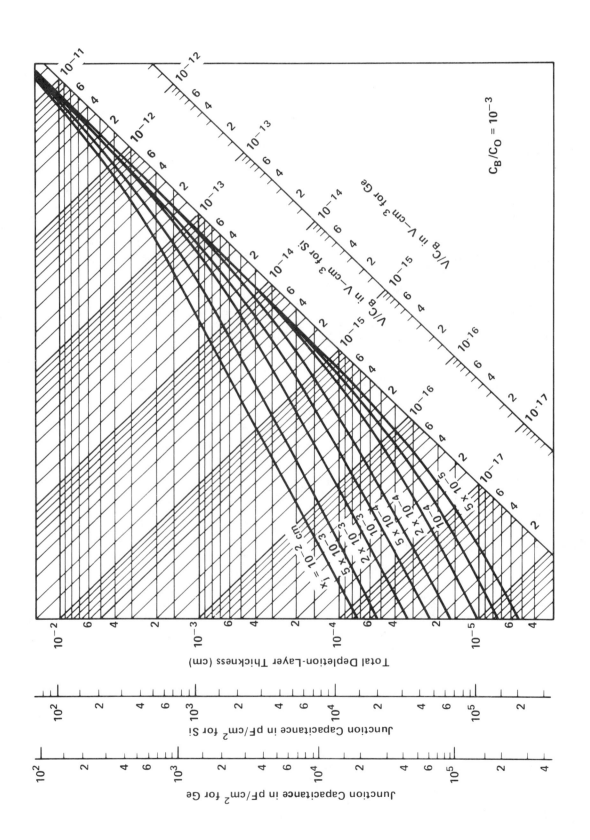

Figure 11-31 — Chart for Use in Range 3×10^{-4} to 3×10^{-3}, Gaussian Distribution

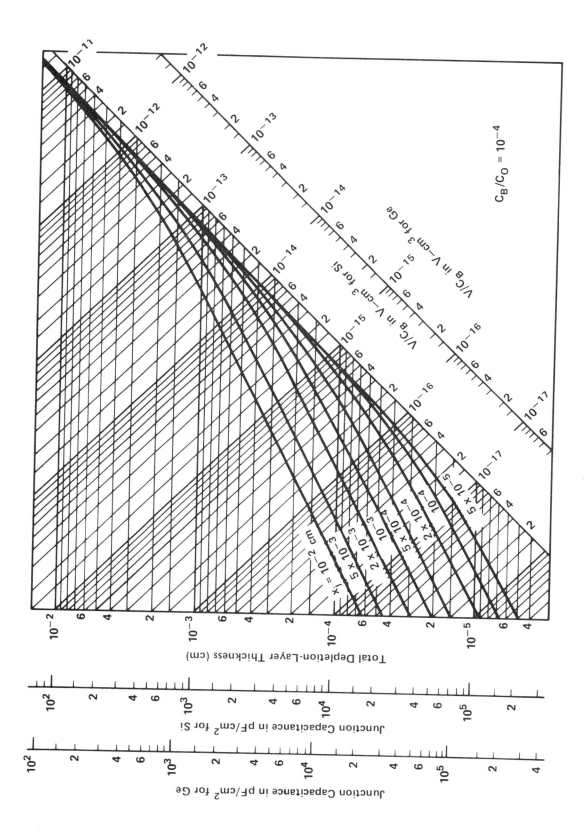

Figure 11-32 — Chart for Use in Range 3×10^{-5} to 3×10^{-4}, Gaussian Distribution

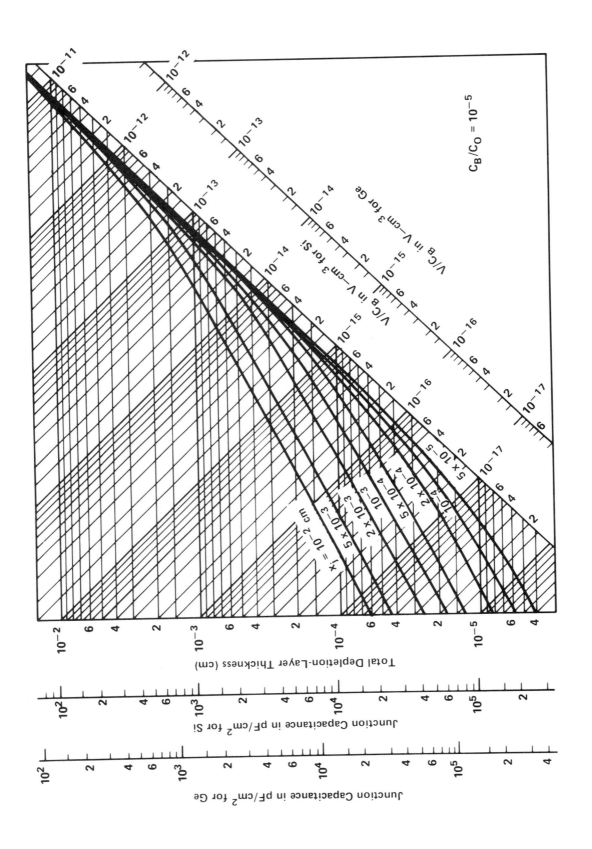

Figure 11-33 — Chart for Use in Range 3×10^{-6} to 3×10^{-5}, Gaussian Distribution

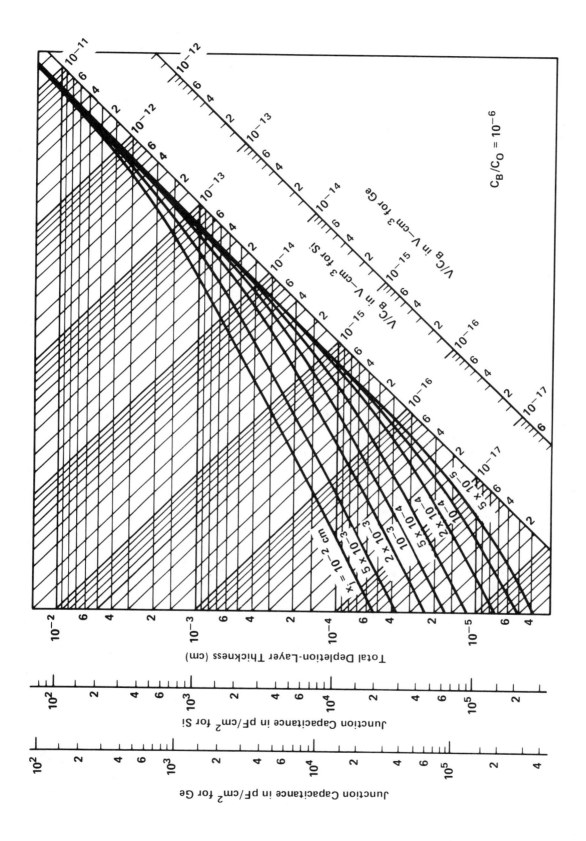

Figure 11-34 — Chart for Use in Range 3×10^{-7} to 3×10^{-6}, Gaussian Distribution

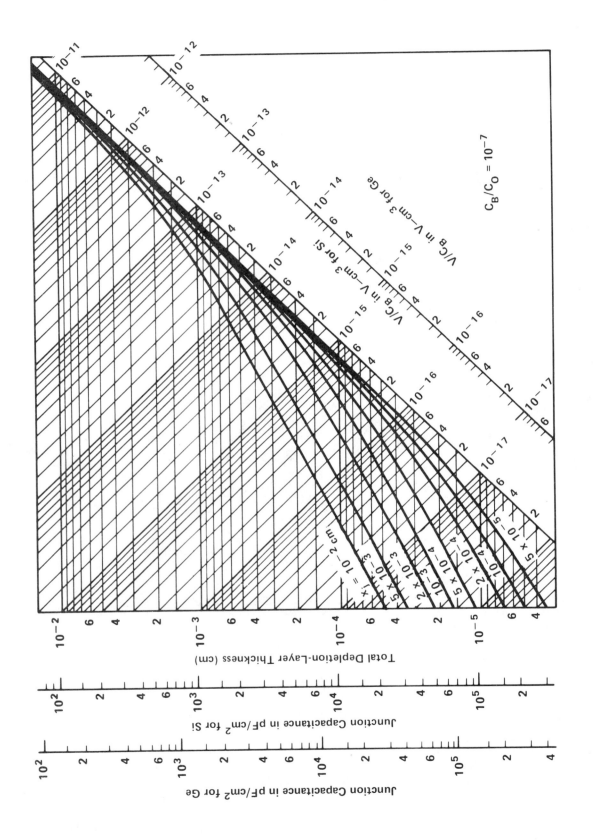

Figure 11-35 — Chart for Use in Range 3×10^{-8} to 3×10^{-7}, Gaussian Distribution

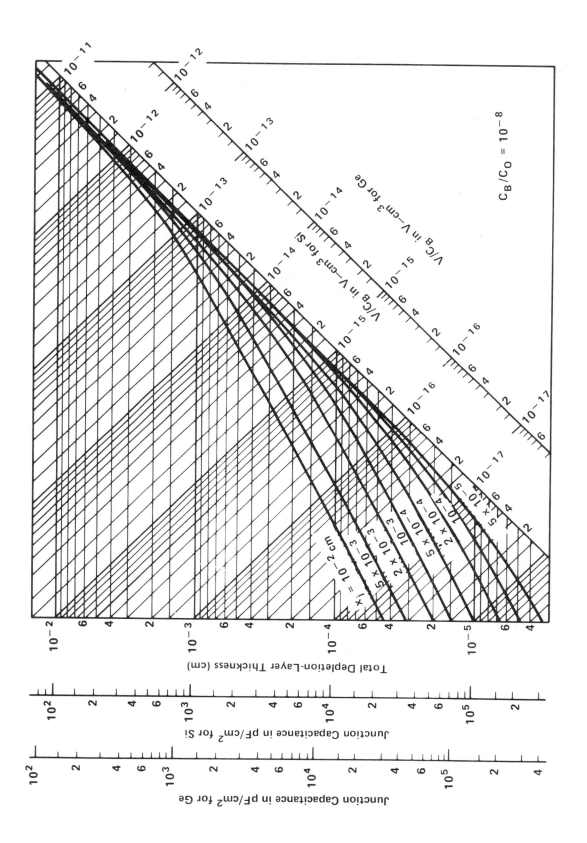

Figure 11-36 — Chart for Use in Range 3×10^{-9} to 3×10^{-8}, Gaussian Distribution

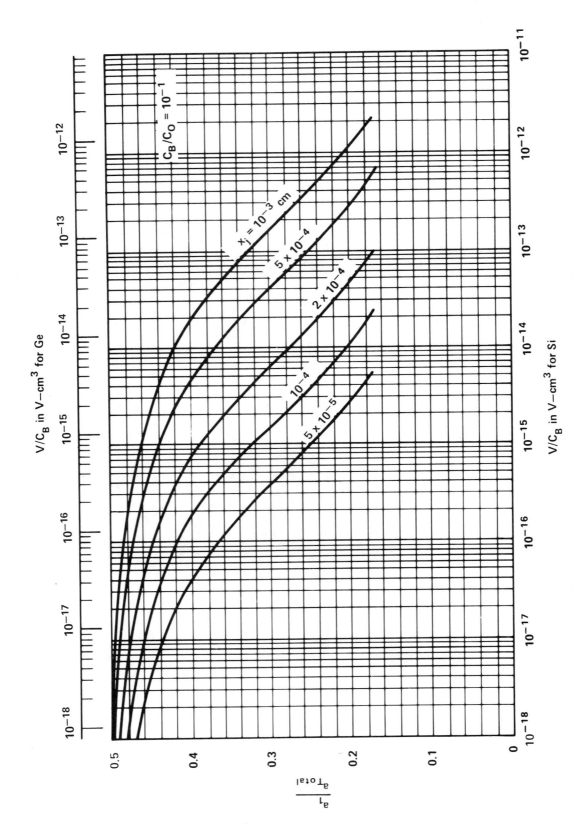

Figure 11-37 — Chart for Use in Range 3×10^{-2} to 3×10^{-1}, Gaussian Distribution

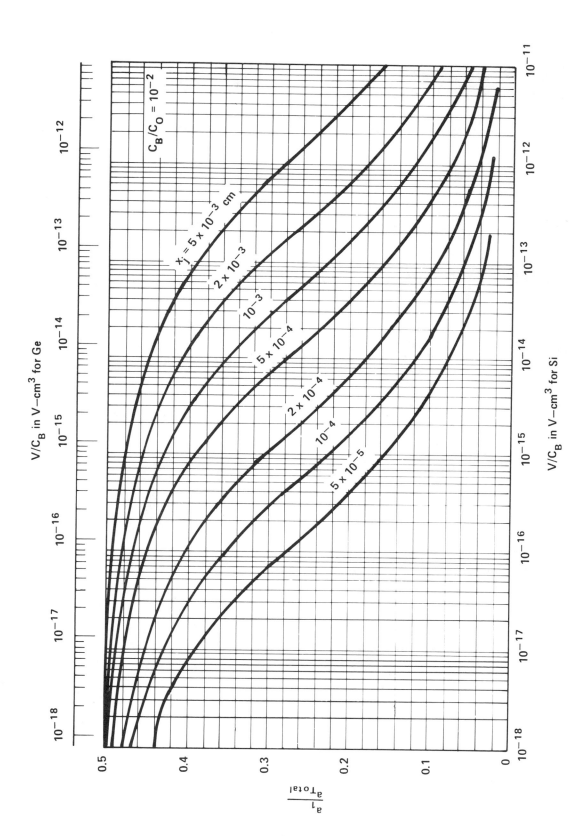

Figure 11-38 — Chart for Use in Range 3×10^{-3} to 3×10^{-2}, Gaussian Distribution

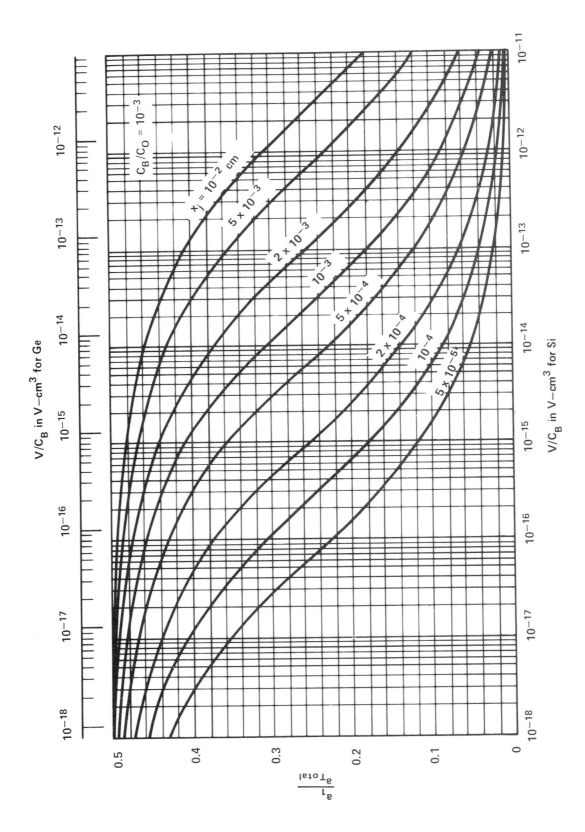

Figure 11-39 — Chart for Use in Range 3×10^{-4} to 3×10^{-3}, Gaussian Distribution

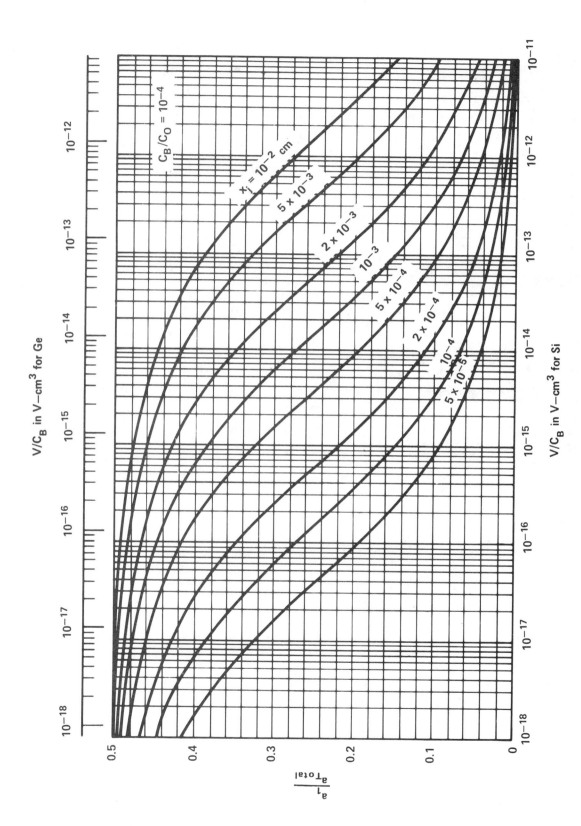

Figure 11-40 — Chart for Use in Range 3×10^{-5} to 3×10^{-4}, Gaussian Distribution

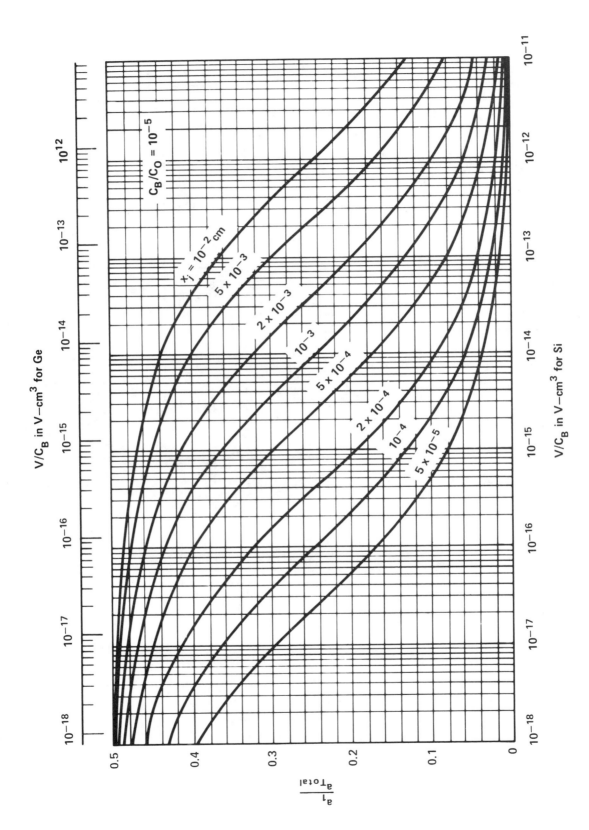

Figure 11-41 — Chart for Use in Range 3×10^{-6} to 3×10^{-5}, Gaussian Distribution

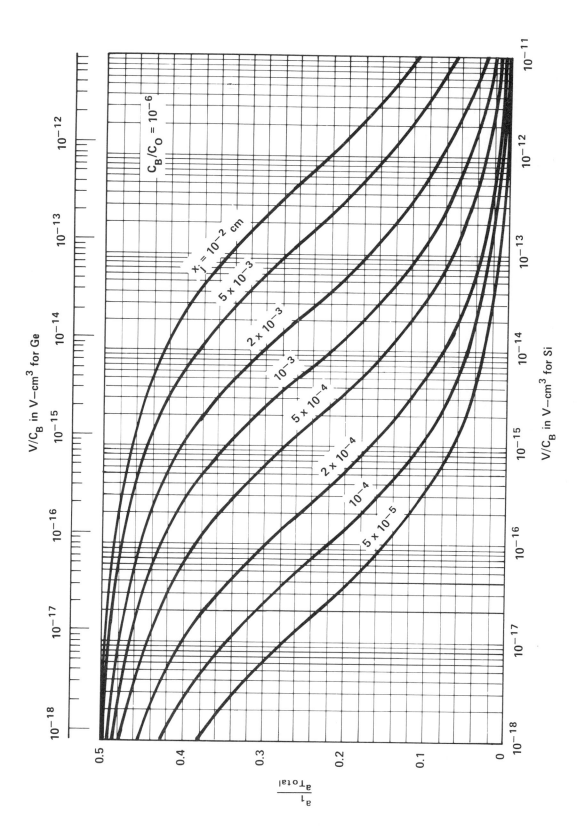

Figure 11-42 — Chart for Use in Range 3×10^{-7} to 3×10^{-6}, Gaussian Distribution

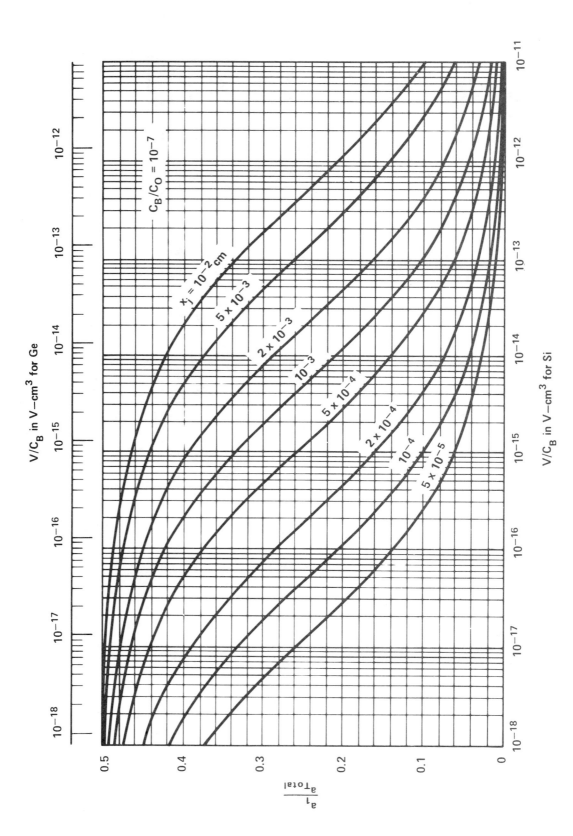

Figure 11-43 — Chart for Use in Range 3×10^{-8} to 3×10^{-7}, Gaussian Distribution

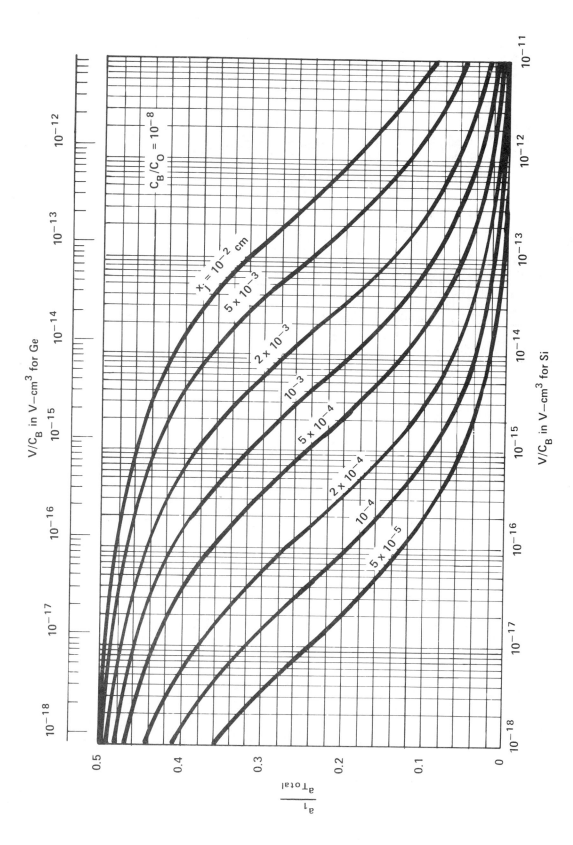

Figure 11-44 — Chart for Use in Range 3×10^{-9} to 3×10^{-8} , Gaussian Distribution

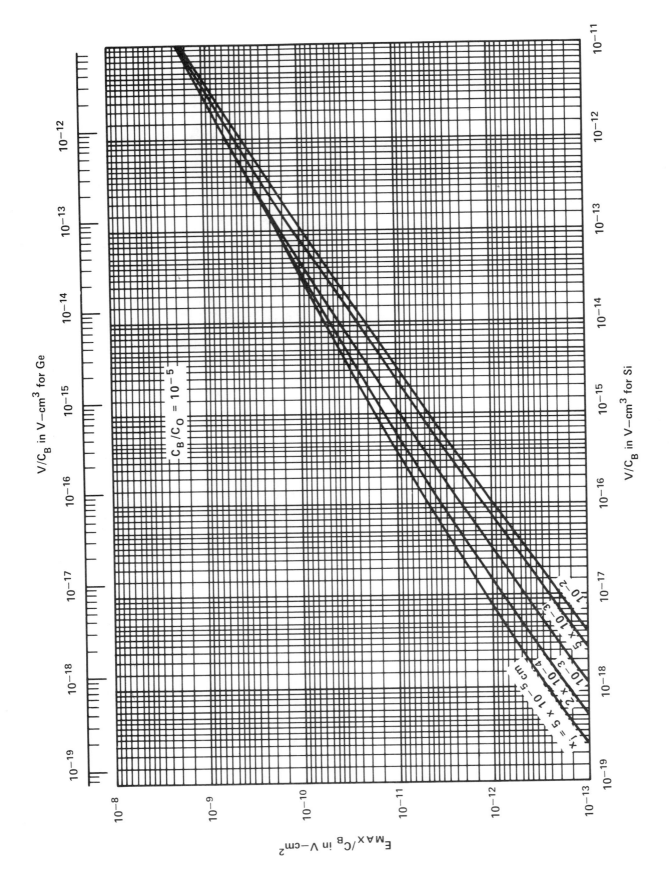

Figure 11-45 — Peak Electric Field Divided by C_B Versus V/C_B, for $C_B/C_O = 10^{-5}$. For Erfc Distribution, Use in Range 10^{-4} to 10^{-8}; for Gaussian Distribution, Use in Range 10^{-4} to 10^{-6}, $C_B = 10^{14}$

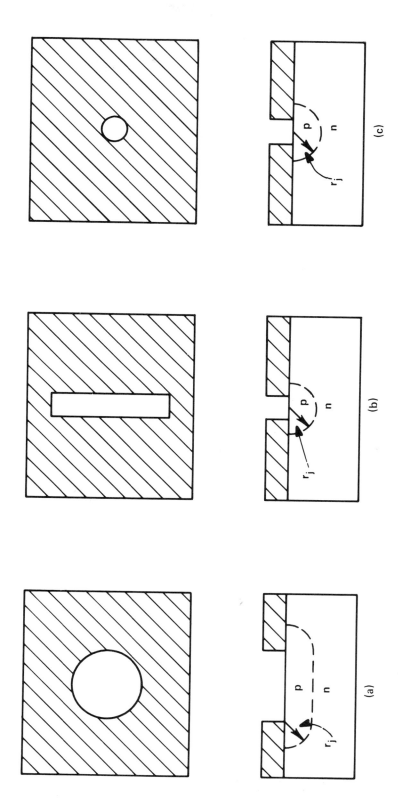

Figure 11-46— Definition of Planar (a), Semi-Cylindrical (b), and Hemispherical (c) Junctions— for Figures 11-47 through 11-50[3] (Reprinted with permission from Pergamon Press, Ltd.)

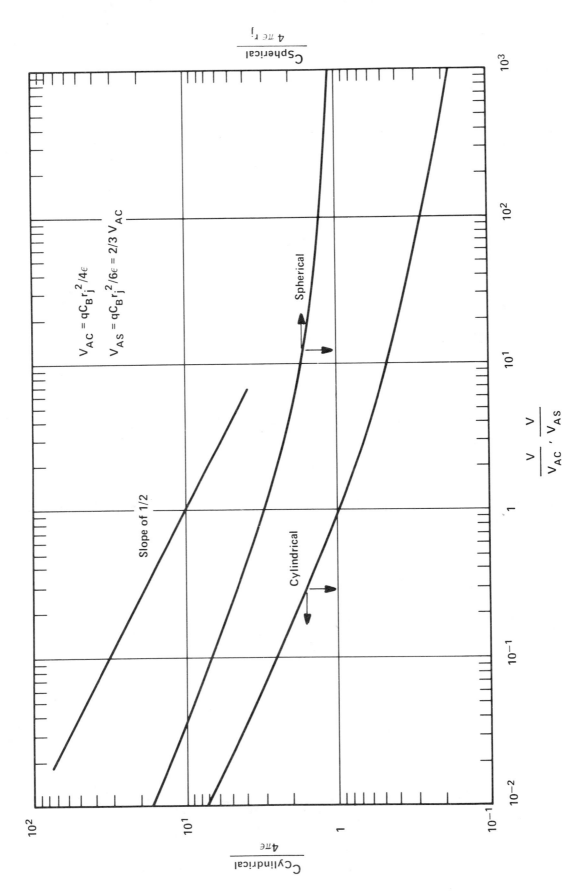

Figure 11-47—Normalized Capacitance Versus Normalized Voltage for Step-Junctions [3]
(Reprinted with permission from Pergamon Press, Ltd.)

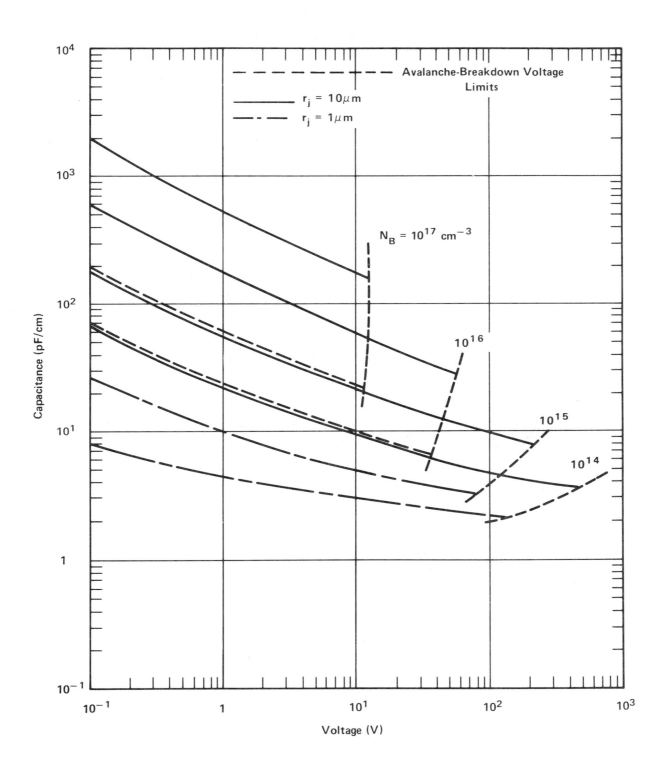

Figure 11-48—Capacitance of Cylindrical Step-Junction Versus Voltage for Si

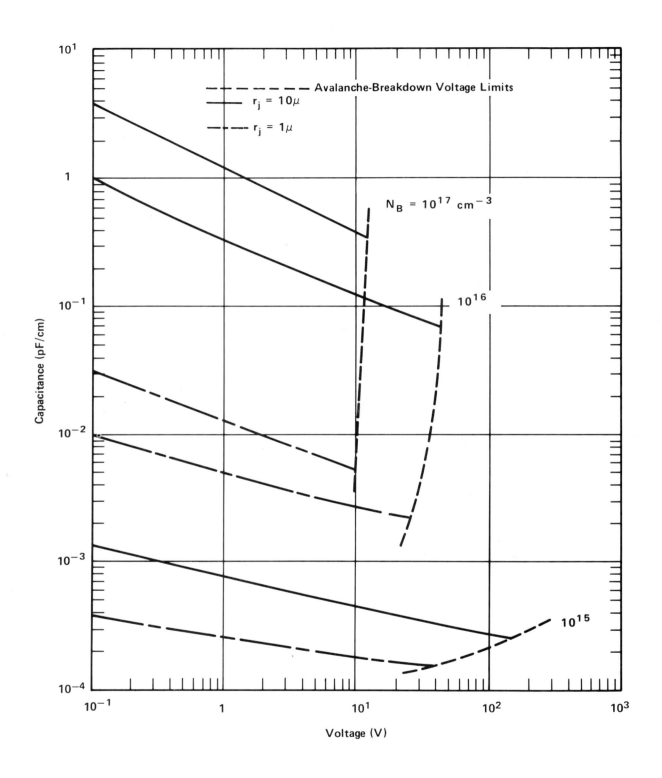

Figure 11-49 — Capacitance of Spherical Step-Junctions Versus Voltage for Si

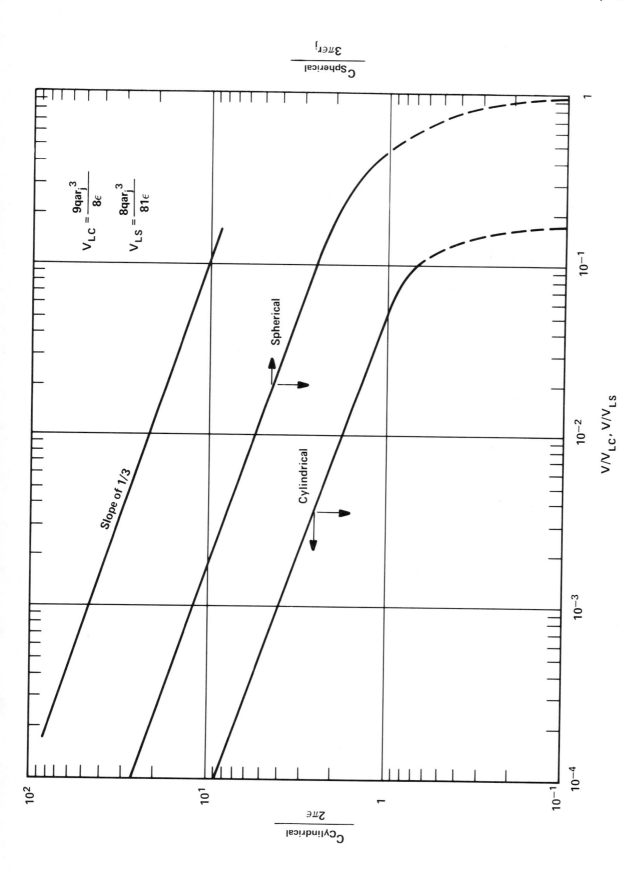

Figure 11-50 — Normalized Capacitance Versus Normalized Voltage for Linear-Graded Si Junction [3]
(Reprinted with permission from Pergamon Press, Ltd.)

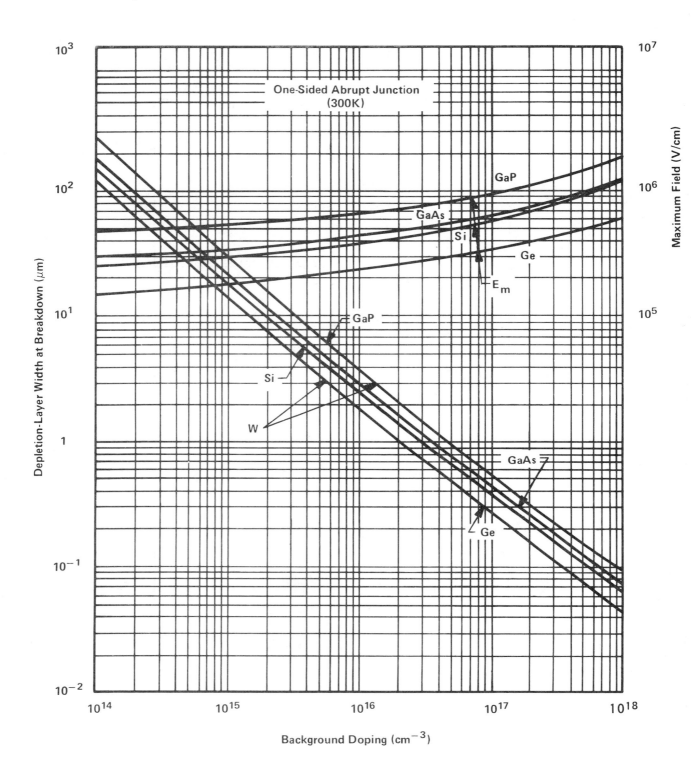

Figure 11-51 — Depletion-Layer Width and Maximum Field at Breakdown for
Step-Junctions in Ge, Si, GaAs, and GaP[7]

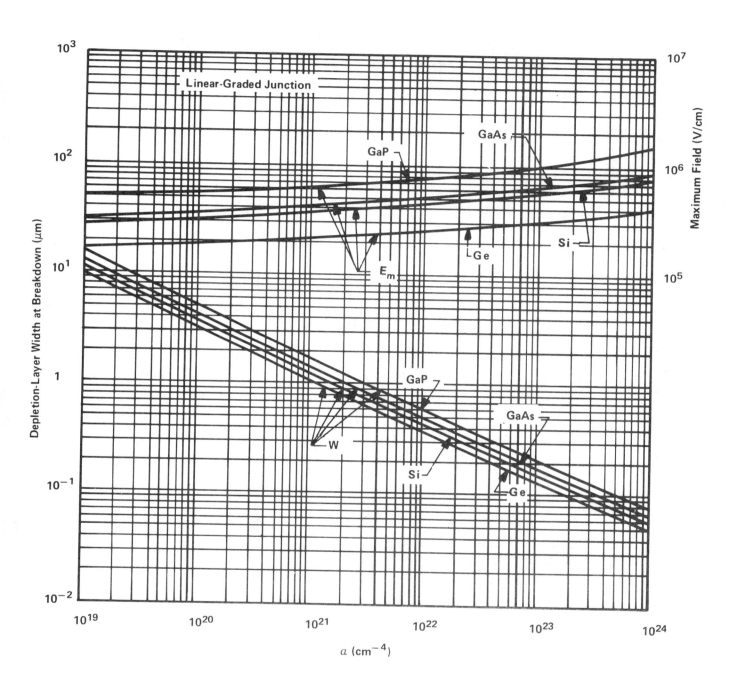

Figure 11-52—Depletion-Layer Width and Maximum Field at Breakdown for
Linear-Graded Junctions in Ge, Si, GaAs, and GaP[7]

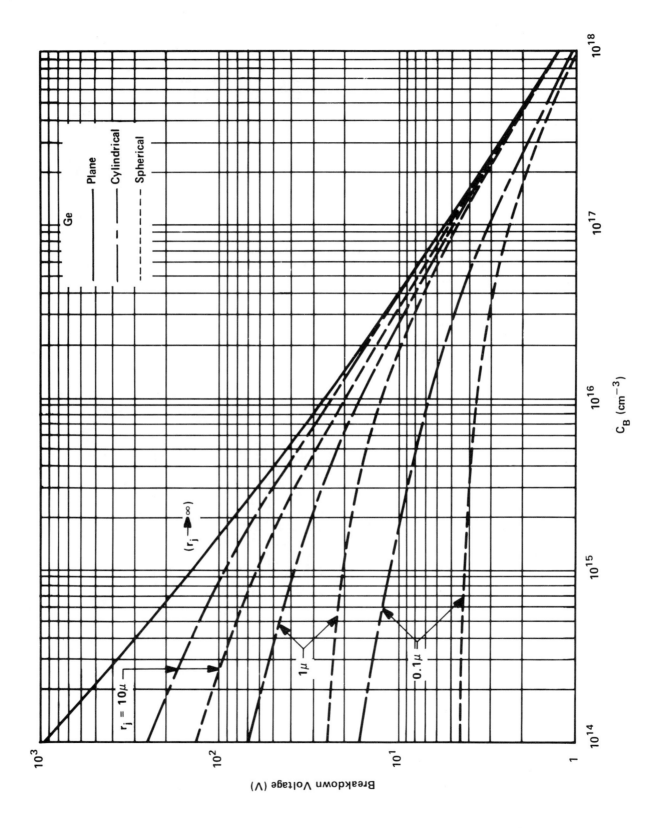

Figure 11-53—Ge Avalanche-Breakdown Voltage Versus Impurity Concentration for One-Sided Step-Junction, Where r_j Is Defined in Figure 11-46 [4] (Reprinted with permission from Pergamon Press, Ltd.)

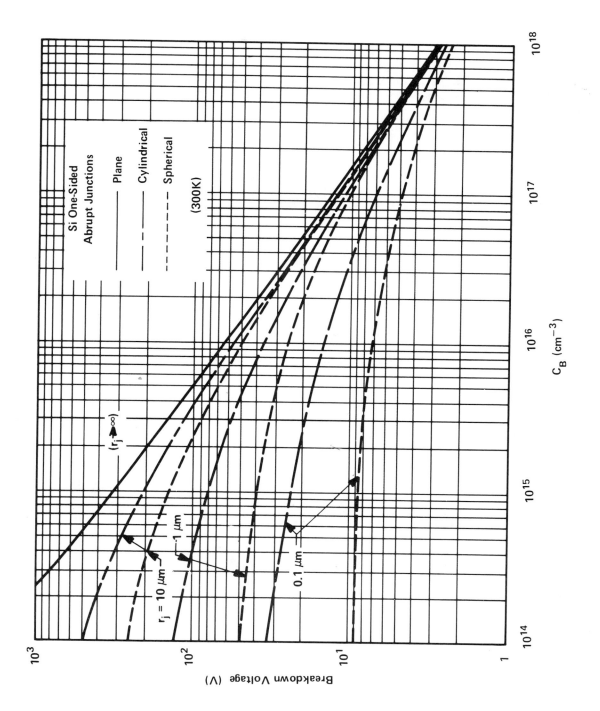

Figure 11-54 — Si Avalanche-Breakdown Voltage Versus Impurity Concentration for One-Sided Step-Junction, Where r_j Is Defined in Figure 11-46 [4] (Reprinted with permission from Pergamon Press, Ltd.)

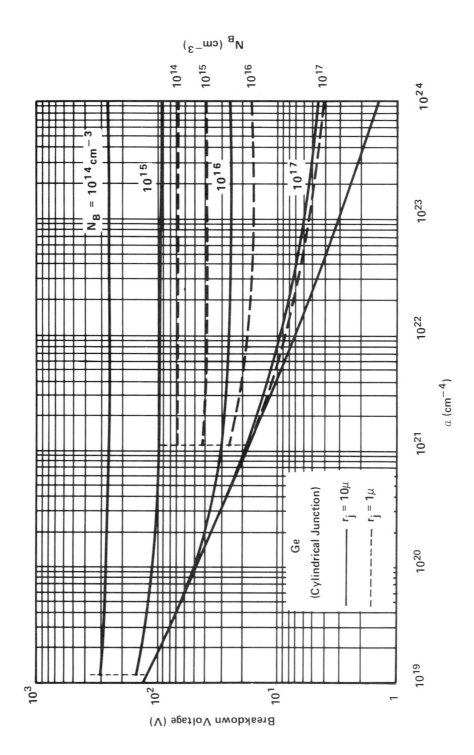

Figure 11-55 — Ge Avalanche-Breakdown Voltage Versus Impurity Gradient for Linear-Graded Junctions, Where r_j Is Defined in Figure 11-46[4] (Reprinted with permission from Pergamon Press, Ltd.)

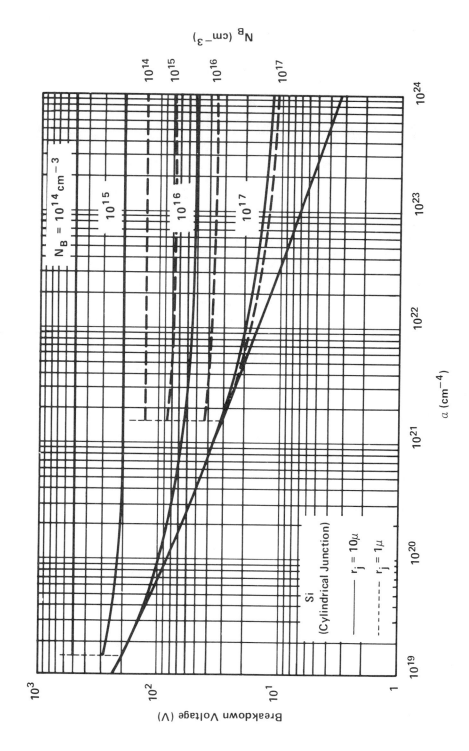

Figure 11-56 — Si Avalanche-Breakdown Voltage Versus Impurity Gradient for Linear-Graded Junctions, Where r_j Is Defined in Figure 11-46[4] (Reprinted with permission from Pergamon Press, Ltd.)

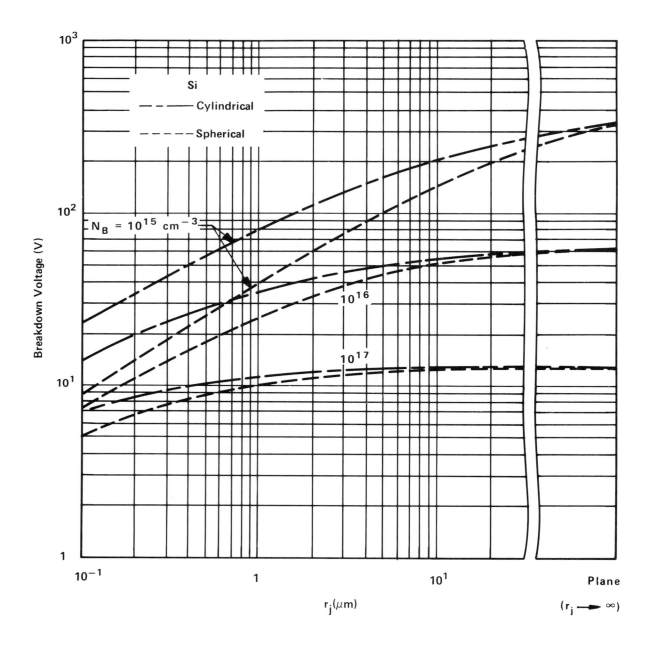

Figure 11-57—Si Avalanche-Breakdown Voltage Versus r_j for a Step-Junction, Where r_j Is Defined in Figure 11-46[4] (Reprinted with permission from Pergamon Press, Ltd.)

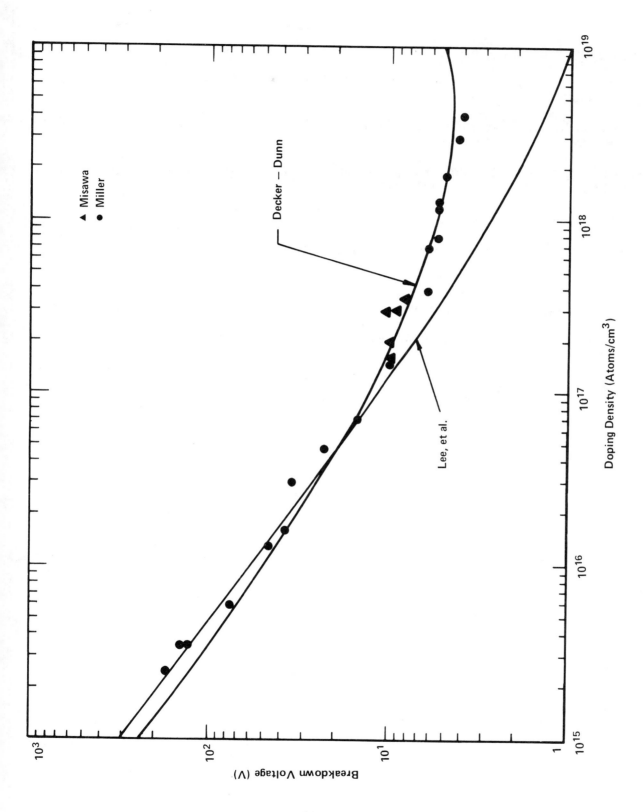

Figure 11-58—Si Breakdown Voltage Versus Substrate Doping for Step-Junctions [8]

REFERENCES

1. G. F. Foxhall, private communication.

2. H. Lawrence and R. M. Warner, Jr., "Diffused Junction Depletion Layer Calculations," B. S. T. J., *39* (March 1960), pp. 389-403.

3. T. P. Lee and S. M. Sze, "Depletion Layer Capacitance of Cylindrical and Spherical p-n Junctions," *Solid State Electron.,* Vol. 10 (1967), pp. 1105-1108.

4. S. M. Sze and G. Gibbons, "Effect of Junction Curvature on Breakdown Voltage in Semiconductors," *Solid State Electron.,* Vol. 9 (1966), pp. 831-845.

5. G. L. Miller and D. A. H. Robinson, private communication.

6. B. R. Chawla and H. K. Gummel, "Transition Region Capacitance of Diffused p-n Junctions," *IEEE Trans. on Electron. Dev.,* Vol. ED-18, No. 3 (March 1971), pp. 178-195.

7. S. M. Sze and G. Gibbons, "Avalanche Breakdown Voltages of Abrupt and Linearly Graded p-n Junctions in Ge, Si, GaAs, and GaP," *Appl. Phys. Letters,* Vol. 8, No. 5 (March 1, 1966), pp. 111-113.

8. D. R. Decker and C. N. Dunn, private communication.

12. METAL-SEMICONDUCTOR JUNCTIONS

METAL-SEMICONDUCTOR JUNCTIONS

LIST OF SYMBOLS

Symbol	Definition
E_C	Bottom of the conductive band
E_F	Fermi energy level
E_g	The semiconductor bandgap energy
E_V	Top of the valence band
k	Boltzman Constant
N	Impurity concentration
N_A	Acceptor impurity density
N_c	Effective density-of-states in the conduction band
N_D	Donor concentration in a semiconductor
N_v	Effective density-of-states in the valence band
q	Electron charge
Q_{sc}	Space charge in depletion layer
T	Absolute Temperature
V	Voltage
V_{bi}	Built-in potential
V_n	The Fermi potential relative to the conduction band edge
W	Depletion width
ε_s	Permittivity of the semiconductor
μ	Electron mobility
χ	Electron affinity measured from the bottom of the conduction band
Φ_m	Metal work function
Φ_{Bn}	Schottky barrier height on P-type semiconductor
Φ_{Bp}	Schottky barrier height on A-type semiconductor
ϕn, ϕp	Quasi Fermi level or imref for electrons and holes respectively

12.1 Introduction

Metal-semiconductor junctions are an integral part of every semiconductor device. The rectifying metal-semiconductor junction or Schottky barrier has numerous uses as a high-speed switching device, while the special case of the ohmic contact is used to interface semiconductor devices to the outside world. The earliest metal-semiconductor junction devices were probably point contact diodes first studied in the late 1800s. In 1938, Schottky[1] formulated the basic theoretical model for what has come to be called the Schottky barrier. The historical development of the metal-semiconductor contact has been dealt with in some detail by Henisch.[2]

12.2 The Schottky Barrier

Figure 12-1(a) shows the energy state for a metal n-type semiconductor system not yet in contact and not in equilibrium. If equilibrium is established by connecting the two materials, as by a wire contacting some other surface of the metal and semiconductor, then the two Fermi levels must be equal, as in Figure 12-1(b). Relative to the Fermi level in the metal, the Fermi level in the semiconductor has been lowered by an amount equal to the two work functions. This potential difference or contact potential is given by

$$q \phi_m - q (\chi + V_n).$$
12.2.1

As the metal is brought closer to the semiconductor, as in Figure 12-1(c), an increasing negative charge is built up at the metal surface, and an equal and opposite positive charge must exist in the semiconductor. When the gap between metal and semiconductor decreases to the order of interatomic distances as shown in Figure 12-1(d), we have formed the metal-semiconductor contact. The limiting barrier (neglecting image effects) is given by

$$q \phi_{Bn} = q (\phi_m - \chi).$$
12.2.2

Similarly, for a p-type semiconductor the barrier height is given by

$$q \phi_{Bp} = Eg - q (\phi_m - \chi).$$
12.2.3

The sum of the barrier heights on a given material is equal to the bandgap.

$$q (\phi_{Bn} + \phi_{Bp}) = E_g.$$
12.2.4

In the presence of surface states, the equilibrium relations at the contact are changed as indicated in Figures 12-1(e) through 12-1(h). In this case, a certain amount of the interface charge can be accommodated by ionizing the surface states. This will reduce the effect of the thermionic work functions on the barrier height and, in the extreme, the barrier can become almost independent of the work functions.

The effect of the image force on the barrier is to reduce the electron energy. This is usually called the Schottky effect. The resulting distribution of energy reaches a maximum at some distance X_m which is given by

$$X_m \cong \sqrt{\pi N_D W},$$
12.2.5

where W is the depletion width. The effective barrier lowering is generally small at low fields but can be significant at high fields.

12.3 Properties of the Schottky Barrier

The barrier height at the metal-semiconductor interface is determined by both the metal work function and the surface states. A detailed band diagram for the metal n-type semiconductor interface is shown in Figure 12-2. In general, the following properties of the junction will be of interest to the device designer.

METAL-SEMICONDUCTOR JUNCTIONS

Space Charge Q_{SC}

The space charge which forms in the depletion layer of the semiconductor at thermal equilibrium is given by

$$Q_{SC} = \sqrt{2q\,\epsilon_S\,N_D\,\left(V_{bi} - V - \frac{kT}{q}\right)}.$$

12.3.1

Depletion Width

The depletion width is identical to that for a one-sided, abrupt p-n junction and is given by

$$W = \sqrt{\frac{2\epsilon_S}{qN_D}\left(V_{bi} - V - \frac{kT}{q}\right)}.$$

12.3.2

Capacitance

The capacitance per unit area of the junction is given by

$$C = \sqrt{\frac{q\,\epsilon_S\,N_D}{2\left(V_{bi} - V - \frac{kT}{q}\right)}} = \frac{\epsilon_S}{W}.$$

12.3.3

Doping Level

The preceding equations can be differentiated and rewritten to give

$$N_D = \frac{2}{q\epsilon_S}\left[-\frac{dV}{d\,\frac{1}{C^2}}\right].$$

12.3.4

This is the basis for the use of the Schottky barrier in impurity profiling.

12.4 Current Transport in Schottky Barriers

Current transport in metal-semiconductor junctions is principally due to majority carriers. Carrier transport in the barrier is dominated by two mechanisms: thermionic emission and diffusion. A combined theory for carrier transport by these mechanisms has been developed by Crowell and Sze.[3] Consider the electron potential energy as is shown in Figure 12-3 for a metal-semiconductor contact. The origin of the barrier is due to a combination of surface states in the semiconductor and the metal work function as discussed in paragraph 12.2. For the case where the barrier is high enough so that the charge density between the metal surface and the edge of the depletion region at W is essentially that of the ionized donors, the current in the region between X_m and W is given by

$$J = -q\,\mu n\,\frac{d\phi_n}{d_x},$$

12.4.1

where

$n = N_c e^{\,q(\phi_n - \psi)/kT}$ is the electron density at X

12.4.2

μ = the electron mobility

and

N_c = the effective density of states in the conduction band

T = the electron temperature.

Between X_m and the metal surface, the potential changes rapidly with respect to the electron-mean free path; therefore, the current flow cannot be described by Equation 12.4.1. Rather, it is described in terms of the effective recombination velocity at the potential maximum V_R, as

$$J = q (n_m - n_o) V_R,$$

12.4.3

where n_m is the electron density at X_m and n_o is a quasi-equilibrium electron density at X_m (the density which would occur if it were possible to reach equilibrium without altering the magnitude or position of the potential energy maximum).

If we measure both ϕ and ψ with respect to the Fermi level in the metal, then

$$\psi_n(W) = -V,$$

12.4.4

$$n_o = N_c \exp [-q \phi_{Bn}/kT],$$

12.4.5

and

$$n_m = N_c \exp \left[\frac{-q \phi_n(X_m) - q \phi_{Bn}}{kT} \right],$$

12.4.6

where $q \phi_{Bn}$ is the barrier height and $q \phi_n(X_m)$ is the imref at X_m.

If we now combine Equations 12.4.1 and 12.4.2, integrate from X_m to W, and combine the result with Equations 12.4.3 and 12.4.6 it can be shown that

$$J = \frac{q N_c v_e}{1 + \dfrac{v_n}{v_D}} \exp \left[-\frac{q \phi_{Bn}}{kT} \right] \left[\exp \frac{-qV}{kT} - 1 \right],$$

12.4.7

where

$$v_D \equiv \left[\int_{X_m}^{w} \frac{q}{\mu kT} \exp \left[-\frac{q}{kT} (\phi_{Bn} + \psi) \right] dX \right]^{-1}$$

12.4.8

is an effective diffusion velocity associated with the transport of electrons from the edge of the depletion layer at W to the potential energy maximum.

If the electron distribution is Maxwellian in the depletion region, and if no electrons return from the metal other than those associated with recombination ($q n_o V_R$), then the semiconductor acts as a thermionic emitter and

$$V_R = \frac{A^* T^2}{qN_c},$$

12.4.9

where A^* is the effective Richardson constant. A^* is tabulated in Table 12-1.

In general, if $v_D \gg V_R$, the preexponential term is dominated by V_R and thermionic emission is the dominant transport process. In this case, Equation 12.4.7 reduces to approximately

$$J_n \cong \left\{ A^* T^2 \exp \left(-\frac{q\phi_{Bn}}{kT}\right) \left[\exp \frac{qV}{kT} - 1 \right] \right\}.$$

12.4.10

However, if $v_D \ll V_R$, the diffusion process is dominant and, neglecting image force effects and assuming that the electron mobility is independent of field,

$$J_n \cong q N_c \mu \epsilon \exp \left(-\frac{q\phi_{Bn}}{kT} \left[\exp \left(\frac{qV}{kT}\right) - 1 \right] \right).$$

12.4.11

For a more complete discussion of barrier height and current transport in Schottky barriers, including the effect of tunneling and minority carriers, see, for example, Reference 4.

Tables 12-2 through 12-5 contain useful data on metal-semiconductor contacts.

12.5 Ohmic Contacts

The flow of current in metal-semiconductor contacts is governed by Equation 12.4.10. (The I-V characteristic will be linear at voltages up to the order of 1 or 2 kT for current densities at the order of J_s.) If the barrier height is made sufficiently small by selection of the contact material, reasonably large current densities can flow in the contact in this range of voltages. Thus, metals with low barrier heights can be used to form ohmic contacts to semiconductors. However, if this was the only way to form ohmic contacts, it would severely limit the choice of contact materials for semiconductor devices. Fortunately, however, there is another approach to the formation of ohmic contacts.

If the carrier concentration in the semiconductor is increased, the width of the depletion region is decreased and, at very high carrier concentrations, the depletion layer becomes sufficiently narrow that quantum mechanical tunneling can take place. This is the more usual approach to ohmic contact formation on semiconductors. Typical ohmic contacts to silicon devices, for example, are formed on material which has a surface concentration $\geqslant 10^{19}/cm^3$. The band diagram for a typical ohmic contact to an n-type semiconductor and typical I-V characteristics are shown in Figure 12-4.

Typical ohmic contact systems for semiconductors make use of both the effect of low barrier height materials as well as tunneling wherever possible. Figures 12-5 through 12-8 show the properties of various contact systems on silicon.

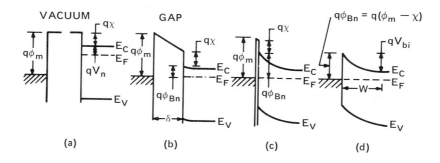

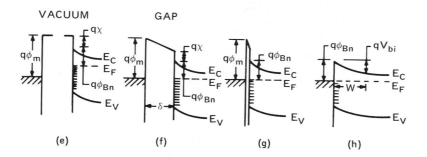

Figure 12-1 — Energy Band Diagrams of Metal-Semiconductor Contacts.[5]
(Reprinted with permission from Oxford University Press)

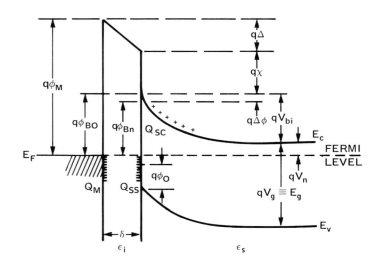

ϕ_m = WORK FUNCTION OF METAL
ϕ_{Bn} = BARRIER HEIGHT OF METAL-SEMICONDUCTOR BARRIER
ϕ_{BO} = ASYMPTOTIC VALUE OF ϕ_{Bn} AT ZERO ELECTRIC FIELD
ϕ_O = ENERGY LEVEL AT SURFACE
$\Delta\phi$ = IMAGE FORCE BARRIER LOWERING
Δ = POTENTIAL ACROSS INTERFACIAL LAYER
χ = ELECTRON AFFINITY OF SEMICONDUCTOR
V_{bi} = BUILT-IN POTENTIAL
ϵ_S = PERMITTIVITY OF SEMICONDUCTOR
ϵ_i = PERMITTIVITY OF INTERFACIAL LAYER
δ = THICKNESS OF INTERFACIAL LAYER
Q_{SC} = SPACE-CHARGE DENSITY IN SEMICONDUCTOR
Q_{SS} = SURFACE-STATE DENSITY OF SEMICONDUCTOR
Q_M = SURFACE-CHARGE DENSITY ON METAL

Figure 12-2 — Detailed Energy Band Diagram of a Metal n-Type Semiconductor Contact
With an Interfacial Layer of the Order of Atomic Distance.[6]

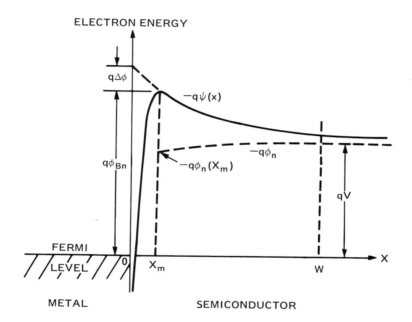

Figure 12-3 — Electron Potential Energy (qψ) Versus Distance for a
Metal-Semiconductor Barrier.[7]

BAND STRUCTURE

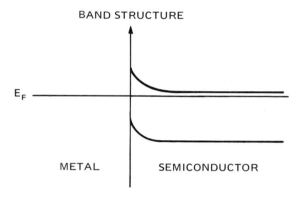

E_F

METAL   SEMICONDUCTOR

I-V CHARACTERISTICS

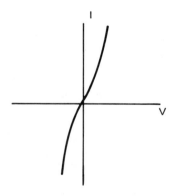

Figure 12-4 —Schematic Band Structure and I-V Characteristics for a Tunneling
Metal-Semiconductor Contact.

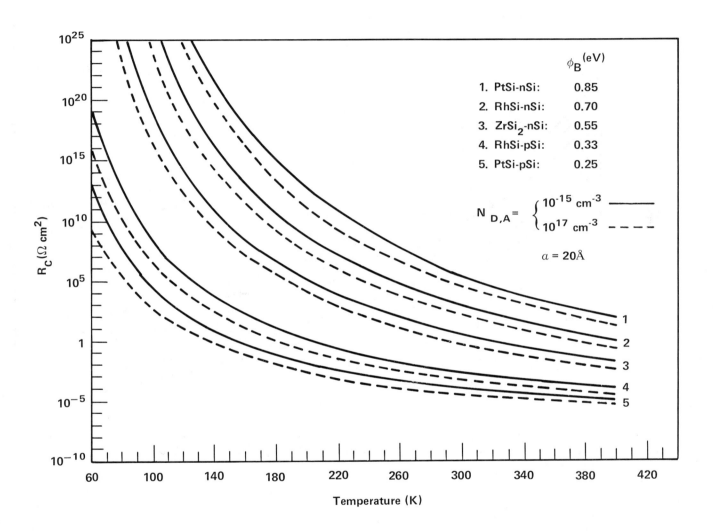

Figure 12-5 —Calculated Specific Contact Resistance Versus Temperature (10^{-10} to 10^{25}) [8]

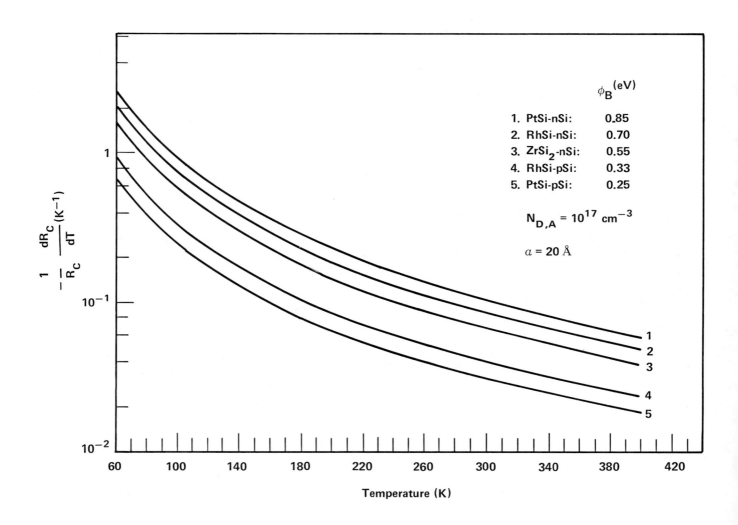

Figure 12-6 —Calculated Thermal Coefficient of Contact Resistance Versus Temperature (10^{-2} to 1)[8]

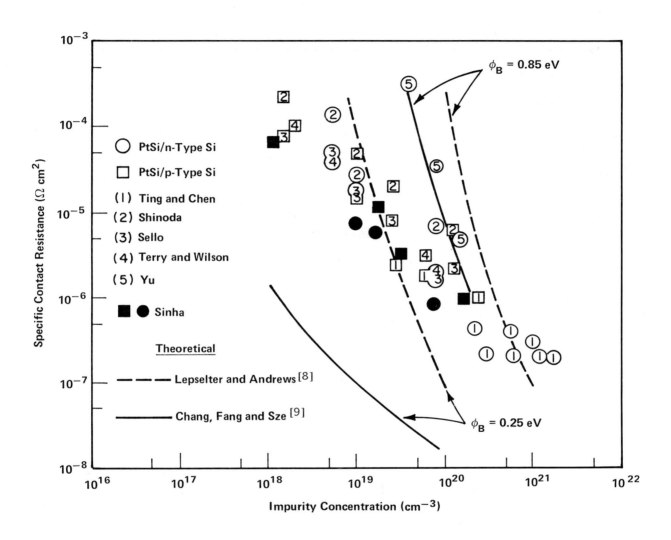

Figure 12-7 — PtSi Specific Contact Resistance Versus Doping Concentration [10]

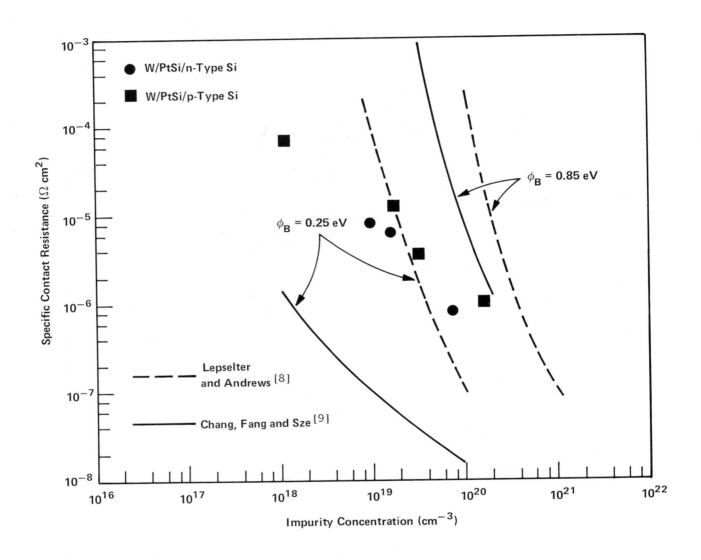

Figure 12-8 — W/PtSi Specific Contact Resistance Versus Doping Concentration

TABLE 12-1

VALUES OF A*/A[11]

SEMICONDUCTOR	Si	Ge	GaAs
p-type	0.66	0.34	0.62
n-type <111>	2.2	1.11	0.068 low field
n-type <100>	2.1	1.19	1.2 high field

$A \equiv 120$ amperes/cm^2/K^2

TABLE 12-2

BARRIER HEIGHTS ON n- AND p- TYPE SILICON*

METAL	ϕ_B (eV)		REF
	n-TYPE Si	p-TYPE Si	
Ag	0.78	0.54	12
Al	0.74	0.58	12
Au	0.80	0.32	12
Co	0.70		13
Cr	0.61	0.50	14
Cu	0.58	0.46	12
Hf	0.55	0.60	13
Mg	0.4		
Mo	0.68	0.42	12
Ni	0.61	0.51	12
Pb		0.55	12
Pd	0.77		13
Pt	0.90		
Ti	0.50	0.61	12
V	0.70		15
W	0.67		12
Zn	0.75	0.50	12

*E_g = 1.12 @ 300K (eV)

TABLE 12-3

METAL-SILICIDE SCHOTTKY BARRIERS

METAL	ϕ_B (n-TYPE Si) eV	ϕ_B (p-TYPE Si)	REF
CoSi	0.68		12
CoSi$_2$	0.64		12
CrSi$_4$	0.57		12
HfSi	0.4-0.5		12
IrSi	0.93		12
MnSi	0.76		12
Mn$_{11}$Si$_{19}$	0.72		12
MoSi$_2$	0.55		12
NiSi	0.70-0.75		12
Ni$_2$Si	0.70-0.75		12
NiSi$_2$	0.70		12
Pd$_2$Si	0.75	0.35	16
PtSi	0.85	0.25	8
RhSi	0.70	0.33	8
TaSi$_2$	0.59		12
TiSi$_2$	0.60		12
W Si$_2$	0.65		12
ZrSi$_2$	0.55	0.55	8

TABLE 12-4

CHARACTERISTICS OF METALS USED FOR Si METALLIZATION

METAL	LINEAR COEFFICIENT OF THERMAL EXPANSION $\frac{\nabla \iota}{\iota}$ ($^\circ C^{-1}$) 20°C	MELTING POINT OF METAL ($^\circ$C)	BULK RESISTIVITY (Ω — cm AT 20°C)	MELTING POINT OF Si EUTECTIC ($^\circ$C)
Si	4.15×10^{-6}	1430	—	—
Si	3.6×10^{-6}	—	—	—
Al	23.6×10^{-6}	660	2.65×10^{-6}	577.2
Au	14.2×10^{-6}	1063	2.35×10^{-6}	370
Mo	4.9	2610	5.2×10^{-6}	~ 1410
Ni	13.3×10^{-6}	1453	6.8×10^{-6}	964
Pt	8.9×10^{-6}	1769	10.6×10^{-6}	830
Ti	8.4×10^{-6}	1668	42×10^{-6}	1330
W	4.6×10^{-6}	3410	5.9×10^{-6}	1400

TABLE 12-5

METAL-SILICON BARRIER POTENTIALS AND METAL WORK FUNCTIONS FOR Ag, Aℓ, Au, Cu, Mg, Ni, Pd, AND Pt CONTACT ON ~0.5 Ω•cm n-TYPE SILICON

PARAMETER*	DEFINITION	Ag	Aℓ	Au	Cu	Mg	Ni	Pd	Pt
V_{int} (V) [17] †	Built-In Voltage	0.507	0.450	0.612	0.518	0.200	0.528	0.618	0.696
Φ_m (eV) [17] †	Metal Work Function	4.31	4.20	4.70	4.52	3.7	4.74	5.0	5.3
$\Phi_{b(I-V)}$ (eV) [18] †	Barrier from I-V Measurements	0.68	0.61		0.73				
Φ_{bc} (eV) [18] †	Barrier from Capacitance Measurements	0.79	0.70	0.82	0.75				
Φ_{bph} (eV) [18] †	Barrier from Photoemission Measurements	0.68	0.61	0.73	0.62				
Φ (eV) [17] ‡	Calculated Barrier Height	0.721	0.699	0.822	0.732	0.417	0.703	0.883	1.022

* See Figure 1-1 for illustration of terms.

† Experimental values from listed reference.

‡ Calculated values from listed reference.

REFERENCES

1. W. Schottky, *Naturwiss.,* Vol. 26 (December 30, 1938), p. 843.

2. H. K. Henisch, *Rectifying Semi-Conductor Contacts.* London: Oxford University Press, 1957.

3. C. R. Crowell and S. M. Sze, "Current Transport in Metal-Semiconductor Barriers," *Solid-State Electron.,* Vol. 9 (November/December 1966), pp. 1035-1048.

4. S. M. Sze, *Physics of Semiconductor Devices.* New York: John Wiley & Sons, 1969.

5. H. K. Henisch, op cit, p. 183.

6. A. M. Cowley and S. M. Sze, "Surface States and Barrier Height of Metal-Semiconductor Systems," *J. Appl. Phys.,* Vol. 36, No. 10 (October 1965), p. 3213.

7. C. R. Crowell and S. M. Sze, op cit, p. 1037.

8. M. P. Lepselter and J. M. Andrews, private communication.

9. C. Y. Chang, Y. K. Fang, and S. M. Sze, "Specific Contact Resistance of Metal-Semiconductor Barriers," *Solid-State Electron.,* Vol. 14, No. 7 (July 1971), pp. 541-550.

10. A. K. Sinha, private communication.

11. S. M. Sze, op cit, p. 380.

12. S. M. Sze, unpublished work.

13. W. D. Powell, private communication.

14. R. A. Zettler and A. M. Cowley, "p-n Junction-Schottky Barrier Hybrid Diode," *IEEE Trans. on Electron Devices,* Vol. ED-16 (January 1969), p. 60.

15. K. J. Miller, M. J. Grieco, and S. M. Sze, private communication.

16. C. J. Kircher, "Metallurgical Properties and Electrical Characteristics of Palladium Silicide-Silicon Contacts," *Solid State Electron.,* Vol. 14 (1971), pp. 507-513.

17. B. Pellegrini, "Properties of Silicon-Metal Contacts Versus Metal Work-Function, Silicon Impurity Concentration and Bias Voltage," *J. Phys. D: Appl. Phys.,* Vol. 9 (1976), pp. 55-68.

18. A. Thanailakis, "Contacts Between Simple Metals and Atomically Clean Silicon," *J. Phys. C: Solid State Phys.,* Vol. 8 (1975), p. 655.

13. SURFACE PROPERTIES

LIST OF SYMBOLS

Symbol	Definition
C_B	Concentration in the bulk or interior
C_{SC}	Surface concentration
E_c	(Energy of) the bottom edge of the conduction band
E_{cs}	The value of E_c at the surface
E_f	The Fermi level (Energy of)
E_{ib}	(Energy of) the Fermi level in the bulk of the intrinsic semiconductor
E_{is}	(Energy of) the Fermi level at the surface of the intrinsic semiconductor
E_v	(Energy of) the top edge of the valence band
E_{vs}	The value of E_v at the surface
E_{Si}	Electric field (in the silicon) at the silicon surface
F	(As in $F_{z(u_s, u_B)}$ number concentrations relating to surface charges
G	(As in $G_{(u_s, u_B)}$ Excess number concentrations relating to surface charges
G^+	Excess surface-carrier density, for Majority carriers in accumulation
G^-	Excess surface-carrier density, for Majority carriers in depletion or inversion
g^+	Excess surface-carrier density for minority carriers in depletion or inversion
g^-	Excess surface-carrier density for minority carriers in accumulation
G	A normalized form of G
k	Boltzmans constant
L	Extrinsic Debye length
L_C	Depth of charge centroid
L_D	Intrinsic Debye length
n	Electron density in number/cm3
n_B	Electron concentration n in the bulk
N_b	electron concentration n in the bulk
N	Carrier concentration in number/cm3
N_a	Acceptor concentration in number/cm3
N_d	Donor concentration in number/cm3
p	Hole density in number/cm3
p_b	Hole density p in the bulk
q	Electronic charge

Continued on next page

LIST OF SYMBOLS (Continued)

Symbol	Definition
Q	Charge in coulombs/unit area
Q_{SC}	Total surface charge in a surface layer/unit area
Q_{SC}/q	Number density of surface charge/unit area
T	Absolute temperature (Kelvin)
U	Normalized (dimensionless) parameter related to carrier concentration
U_b or u_B	Value of U in the bulk
u_s	Value of U at the surface
V	Variation of U from bulk value
V_s	Value of V at the surface
W	Width or thickness
W_a	Accumulation layer thickness
W_d	Depletion layer thickness
X_d	Depletion layer thickness
Δn	Excess electron concentration
Δp	Excess hole concentration
ϑ	Potential difference of a band structure feature (i.e., ψ_{ib}, ψ_c, etc.) from its value in the bulk
ϑ_{fb}	Potential difference between the Fermi potential and the intrinsic Fermi potential in the bulk
ϑ_s	The value of ϑ at the surface
ψ_c	Potential of the bottom edge of the conduction band in the bulk
ψ_f	Potential of the Fermi level
ψ_1, ψ_{ib}	Potential of the Fermi level of the INTRINSIC semiconductor
ψ_v	Potential of the top of the valence band
Ω	Symbol for ohms

The properties of semiconductor surfaces have a profound influence over the characteristics of semiconductor devices.[1] [2] [3] [4] Some of the most widely-used relationships among the variables that describe the semiconductor surface are shown graphically in this section.

The metal-insulator-semiconductor capacitor and field-effect transistor are of particular interest and have been given special treatment in Section 14, MOS.

13.1 Definitions of Semiconductor Surface Parameters

Figure 13-1 illustrates the interrelationships of some basic surface parameters. Note that the electron energy is considered a positive quantity and is plotted upward on the energy-band diagrams. To be consistent, positive potentials must be plotted downward because of the negative electron charge. The application of a positive potential to the surface results in the energy bands bending downward and a resulting accumulation of electrons at the surface as illustrated. It is sometimes convenient to use a normalized "potential," u or u_s, which is dimensionless and is measured in multiples of the thermal potential kT/q (volts); in this case, positive is scaled upwards.

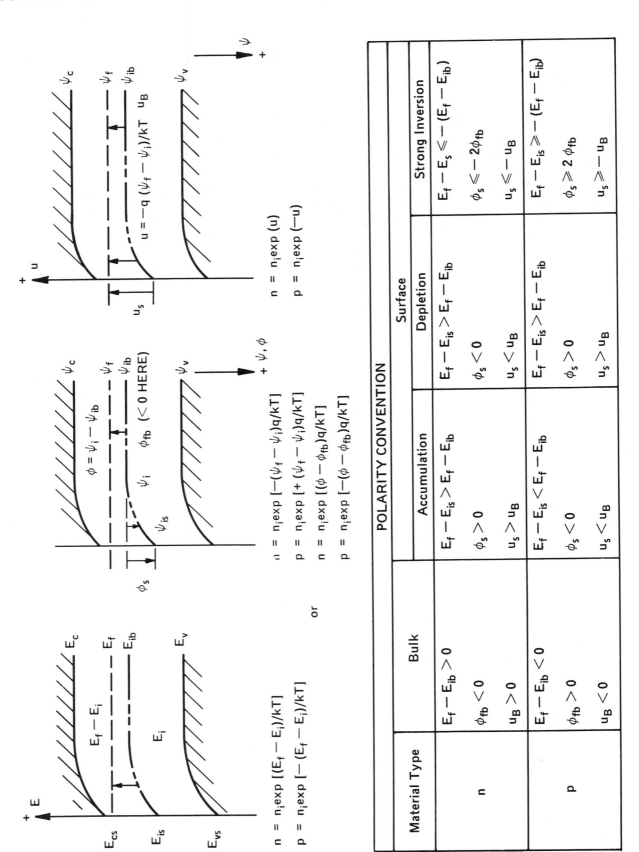

Figure 13-1 — Definition of Semiconductor Surface Parameters

13.2   Functions Related to Surface-Layer Thicknesses

Layer depths in semiconductors are frequently of interest. Because semiconductors have moderate carrier concentrations, accumulation or depletion-layer depths that commonly occur are much larger than inter-atomic spacing (in contrast to metals) and are comparable to dimensions used in device fabrication. Figure 13-2[5] shows the maximum accumulation width and depletion-layer widths and charge for a silicon-insulator surface at a given doping level N. A more detailed depletion-layer width for the doping range of greatest interest is given in Figure 13-3.[6]   The corresponding capacitance for the depletion layer at maximum width is given in Figures 13-4 through 13-6.

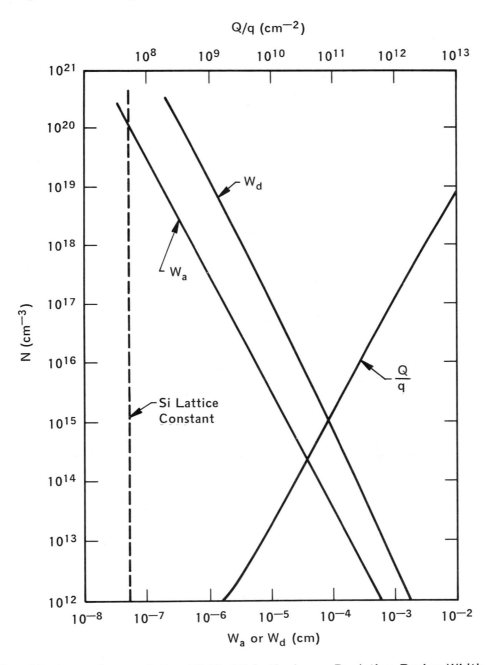

Figure 13-2 —   Maximum Accumulation Width (W_a), Maximum Depletion-Region Width (W_d), and Depletion-Region Charge (Q) Versus Impurity Concentration For Si[5]
(Reprinted by permission of Prentice-Hall, Inc., Englewood Cliffs, N.J.)

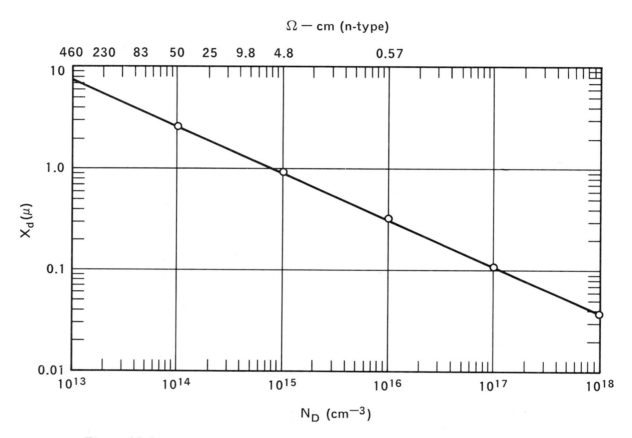

Figure 13-3 — Maximum Depletion-Layer Width for a Si-Insulator Structure[6]
(Reprinted by permission of John Wiley & Sons, Inc.)

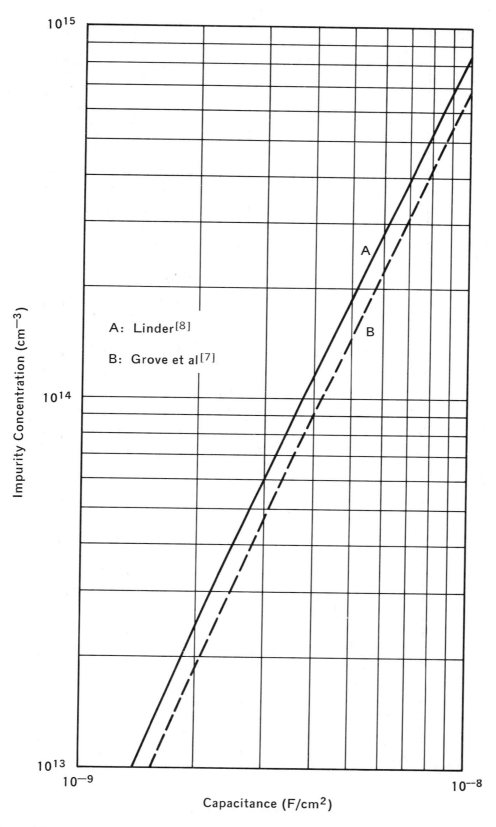

Figure 13-4 — Impurity Concentration Versus Minimum Si Capacitance in Inversion

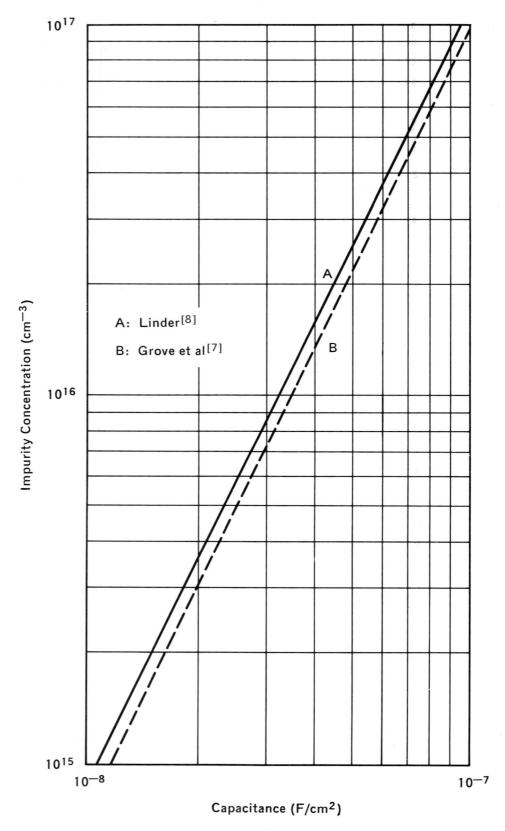

Figure 13-5 — Impurity Concentration Versus Minimum Si Capacitance in Inversion

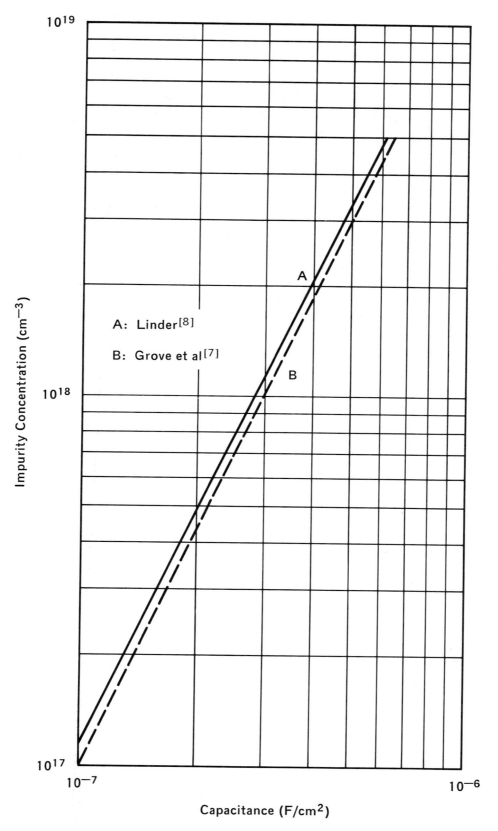

Figure 13-6 — Impurity Concentration Versus Minimum Si Capacitance in Inversion

13.3 Surface Charge

The surface charge and electric field at the silicon surface for two special cases of depletion are shown in Figures 13-7 and 13-8. Figure 13-9 shows the same situation for germanium. At the onset of inversion, the electron and hole densities are equal at the surface. The onset of strong inversion is a condition of considerable importance because this is the point where 1) the depletion-layer width approaches its maximum value, 2) the inversion layer starts its buildup, and 3) the lateral-channel conduction of the inversion layer starts to become significant and is the turn-on or threshold point for an enhancement-type field-effect device. The surface charges are the induced charges in the silicon for the condition given. For example, a positive oxide charge of $3 \times 10^{11}/cm^2$ (from whatever source—fixed charge, sodium ions, etc.) is almost sufficient to bring p-type silicon with $N=10^{16}/cm^3$ to the point of strong inversion. The depletion-charge density will be $-3 \times 10^{11}/cm^2$ (or -4.8×10^{-8} coul/cm^2) and the electric field in the silicon at the surface will be 4.6×10^4 volts/cm.

If the semiconductor in the above example was n-type with $N=10^{16}/cm^2$, the positive oxide charge would cause an accumulation layer to form. Figure 13-10 shows that the oxide charge of $3 \times 10^{11}/cm^2$ would cause the accumulation layer density to build up to a peak concentration (at the surface) of about $3 \times 10^{17}/cm^3$. Figure 13-11 shows the same situation for germanium. Remember that accumulation layers are very thin, and this concentration decays very rapidly to the bulk concentration.

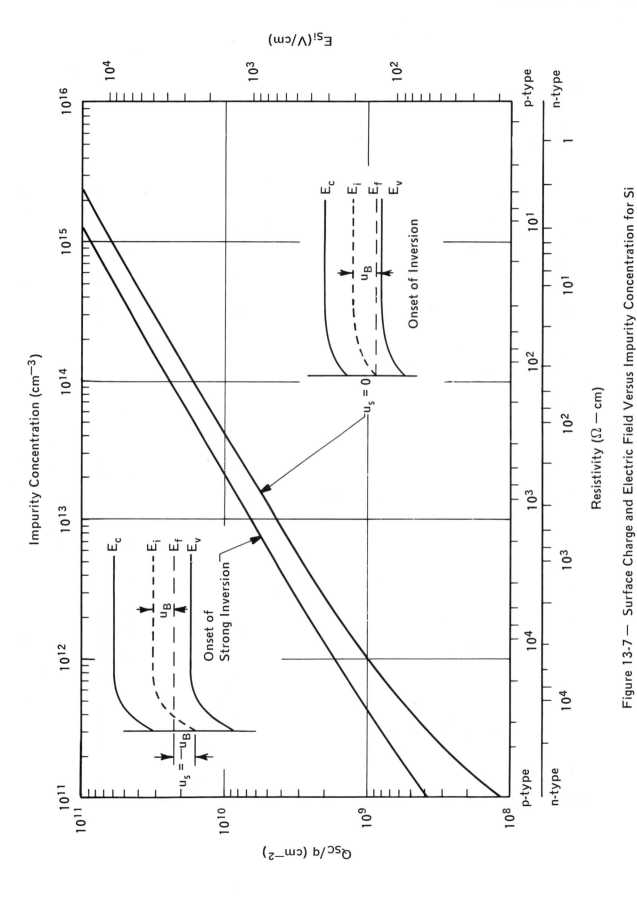

Figure 13-7 — Surface Charge and Electric Field Versus Impurity Concentration for Si

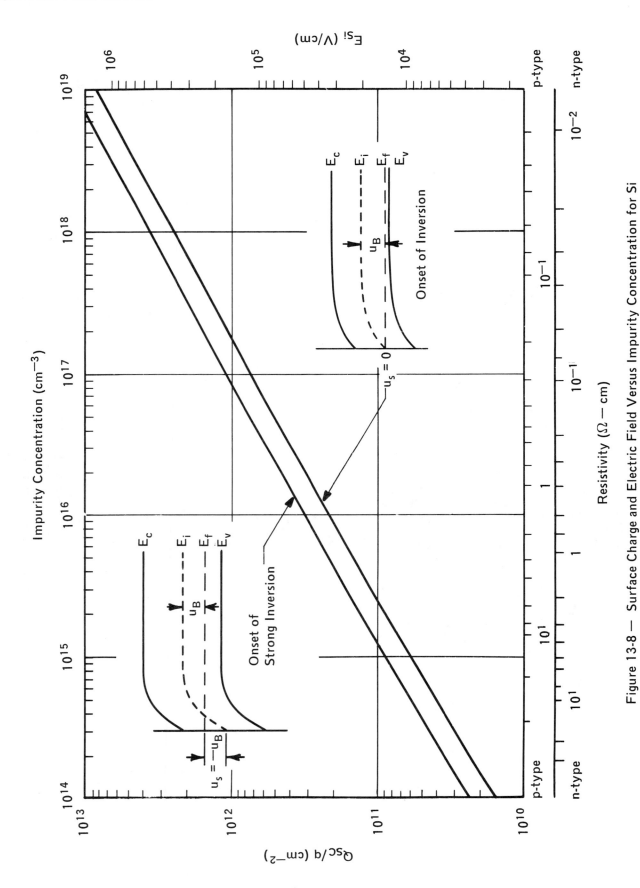

Figure 13-8 — Surface Charge and Electric Field Versus Impurity Concentration for Si

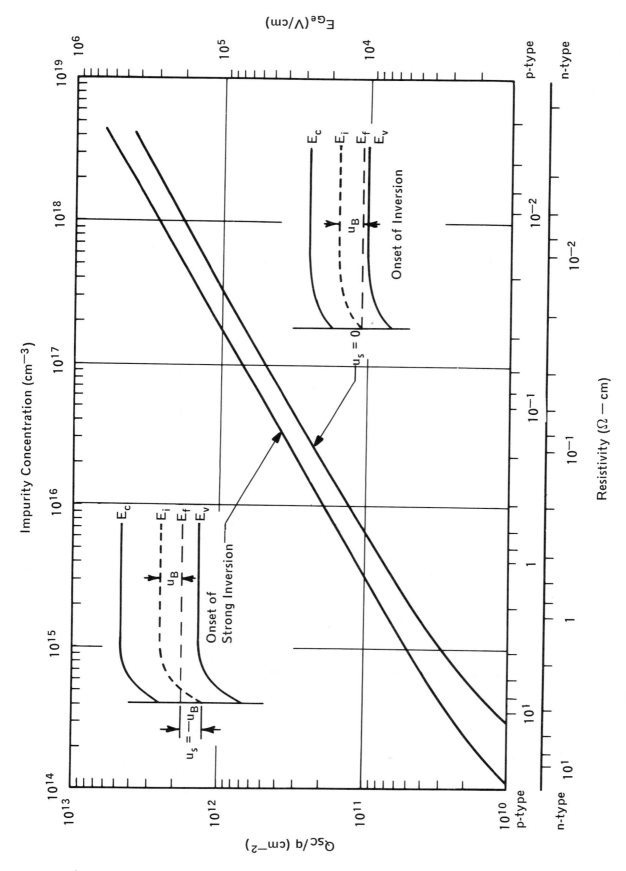

Figure 13-9 — Surface Charge and Electric Field Versus Impurity Concentration for Ge

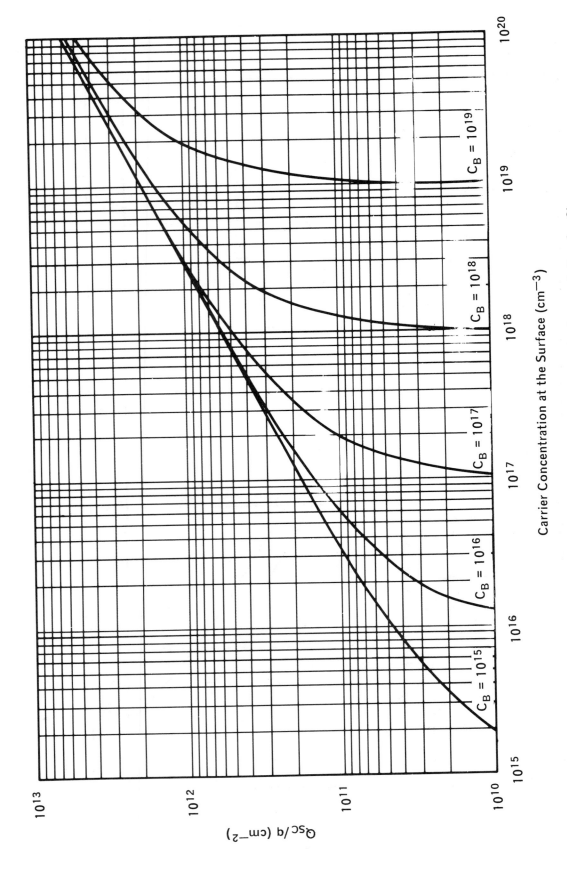

Figure 13-10 — Surface-Charge Density Variations in Accumulation for Si

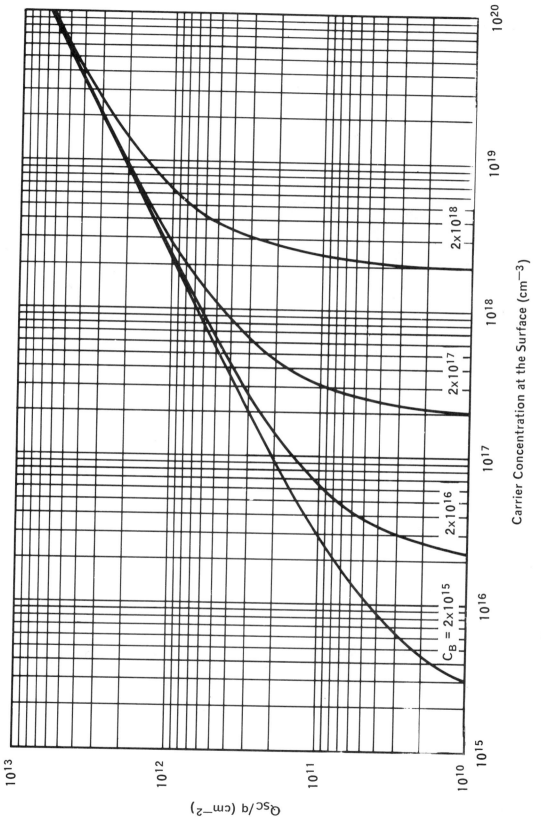

Figure 13-11 — Surface-Charge Density Variations in Accumulation for Ge

13.4 Extended Curves of Semiconductor Surface Parameters

This subsection graphically describes some of the functions used in calculations related to semiconductor surfaces. The curves taken from the work of Young[9] and of Many, Goldstein, and Grover[10] require further explanation, which follows.

Young's Curves

Young's curves are presented in Figures 13-12 through 13-21. They are plots of the functions $F(u_s, u_B)$ and $G(u_s, u_B)$. The expressions for the charge, electric field, and change in the electron and hole densities in terms of these functions are

$$Q_{SC} = K\epsilon_0 \, E_s = 2qn_i L_D \, F(u_s, u_B),$$

$$\Delta p = n_i L_D \, G(u_s, u_B),$$

$$\Delta n = n_i L_D \, G(-u_s, u_B),$$

where

Q_{SC} = space charge or surface charge

$K\epsilon_0$ = dielectric constant

Δp = change in hole density at the surface

Δn = change in electron density at the surface

q = electronic charge

n_i = intrinsic electron density

u_B = bulk potential $\dfrac{-q\phi_B}{kT}$

($u_B > 0$ for n-type, $n_B = n_i e^{u_B}$ and $p_B = n_i e^{-u_B}$)

u_s = surface potential $\dfrac{q(\phi_s - \phi_B)}{kT}$

($u_s > u_B$ implies $n > n_B$ and $p < p_B$ and conversely)

$L_D = \sqrt{\dfrac{K\epsilon_0 kT}{2q^2 n_i}}$ = intrinsic Debye length

$\quad = 2.4 \times 10^{-3}$ cm for Si

$\quad = 7.5 \times 10^{-5}$ cm for Ge.

Note: $F(u_s, u_B) \equiv -F(-u_s, -u_B)$.

Examples:

Determine the surface charge, Q_{SC}, of an n-type semiconductor (for instance $u_B = 10$) in inversion (for instance $u_s = -15$) as follows: from the curve in Figure 13-12 we see that $F(-15, 10) \cong +1950$. The positive sign applies for $u_s < u_B$ and the negative sign applies if $u_s > u_B$. $F(X, X) = 0$ at the cusp in the curves. This figure shows only positive values of u_B but the identity $F(u_s, u_B) \equiv -F(-u_s, -u_B)$ can be used to obtain the desired result for negative values of u_B. For instance, a p-type semiconductor ($u_B = -12$) in accumulation ($u_s = -15$) can be obtained by $F(-15, -12) = -F(15, 12) \cong -(-1618) = 1618$. Q_{SC} is then obtained by use of the appropriate values of K, n_i, and L_D for the particular semiconductor in question and the equations given above.

Figures 13-13 through 13-16 show the function $G(u_s, u_B)$ used in calculations of excess surface-carrier densities (number per unit surface area) of mobile electrons, Δp, in the space-charge layer with respect to their numbers at flat bands $V_s = u_s - u_B = 0$. If $\Delta n > 0$, then $\Delta p < 0$ and conversely. These quantities are related to the space-charge density $Q_{SC} = q(\Delta p - \Delta n)$. Using the values in the first example ($u_B = 10$, $u_s = -15$), we see from Figure 13-14 that $G(-15, 10) \cong 2480$ for calculating Δp and from Figure 13-13 $G(15, -10) \cong -1400$ for calculating Δn.

Finally, Figures 13-17 through 13-21 show curves of the normalized potential, u, as a function of depth (in units of intrinsic Debye length) into the semiconductor using five different scales to properly display all the curves. These curves use $u_s = -24$ at $x/L_D = 0$, but other values of $u_s > -24$ can be used by shifting the vertical axis to the right and assigning $x/L_D = 0$ at the desired value of u_s. For example, for $u_B = 16$ and $u_s = -5$ (referring to Figure 13-18), a shift of the vertical axis to $x/L_D = .001$ is required. At this point $u_s = -5$ for the $u_B = 16$ curve, and the remainder of the curve represents the variation of u into the semiconductor. These curves show u for n-type ($u_B > 0$) in inversion or depletion and p-type ($u_B < 0$) in accumulation. All signs of u can be changed for p-type curves in depletion or inversion and n-type in accumulation.

NOTE:

Jindal and Warner have published a series of papers containing normalized curves to determine the potential, electric field, charge, change in electron and hole densities, and capacitance at a semiconductor surface. These calculations cover a wider range than Young's curves and are also applicable to step junctions under equilibrium conditions. Refer to the original papers listed below for further details.

1. R. P. Jindal and R. M. Warner, Jr. "A General Solution for Step Junctions with Infinite Extrinsic End Regions at Equilibrium," *IEEE Trans. on Electron Dev.,* Vol. ED-28, No. 3 (March 1981), pp. 348-351.

2. R. P. Jindal and R. M. Warner, Jr. "An Extended and Unified Solution for the Surface Problem at Equilibrium," *J. Appl. Phys.,* Vol. 52, No. 2 (December 1981), pp. 7427-7432.

3. R. M. Warner, Jr. and R. P. Jindal, "Replacing the Depletion Approximation" *Solid-State Electron.,* Vol. 26, No. 4 (April 1983), pp. 335-342.

4. R. P. Jindal, "A Normalized Analytical Solution for the Capacitance Associated with Uniformly Doped Semiconductors at Equilibrium," *Solid-State Electron.,* Vol. 26, No. 10 (October 1983), pp. 1005-1008.

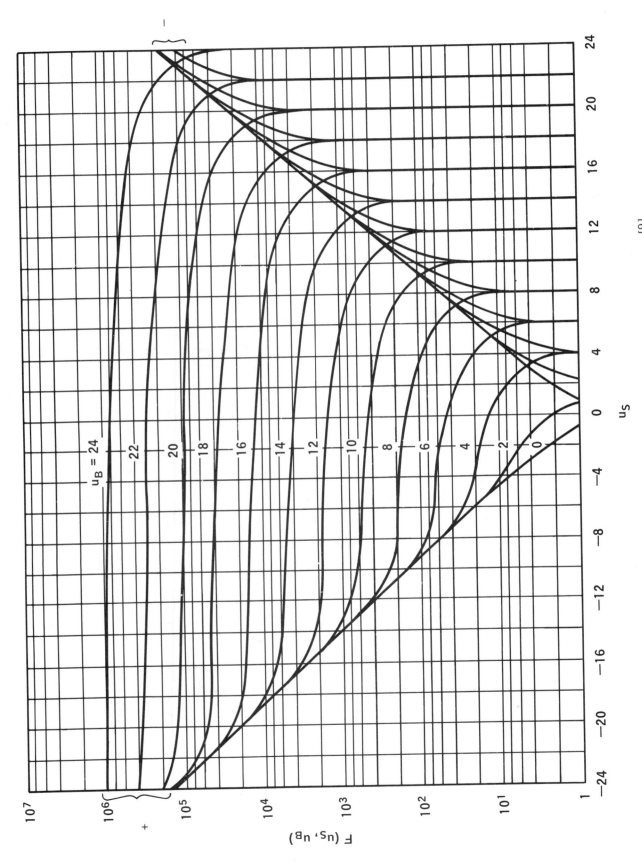

Figure 13-12 — $F(u_S, u_B)$ Versus u_S for Various Values of u_B [9]

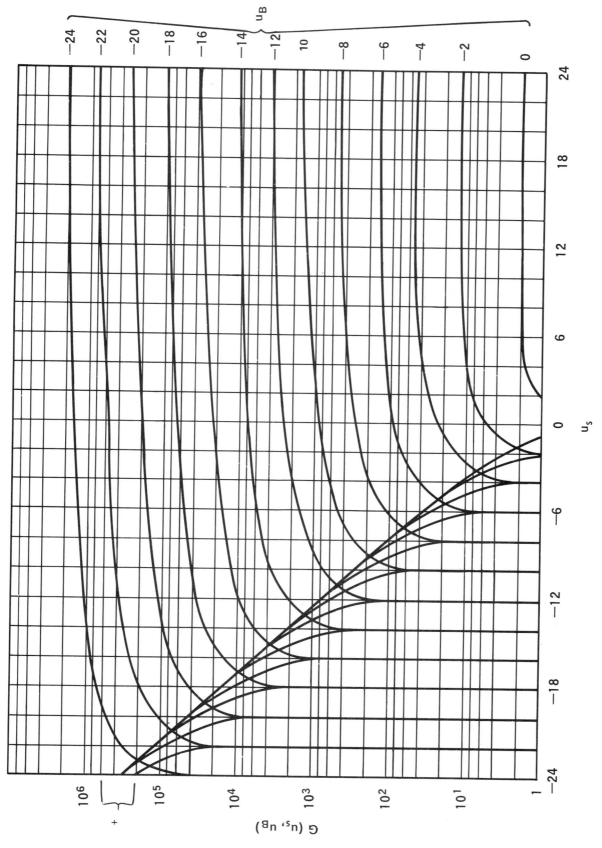

Figure 13-13 — $G(u_s, u_B)$ Versus u_s for Various Values of u_B [9]

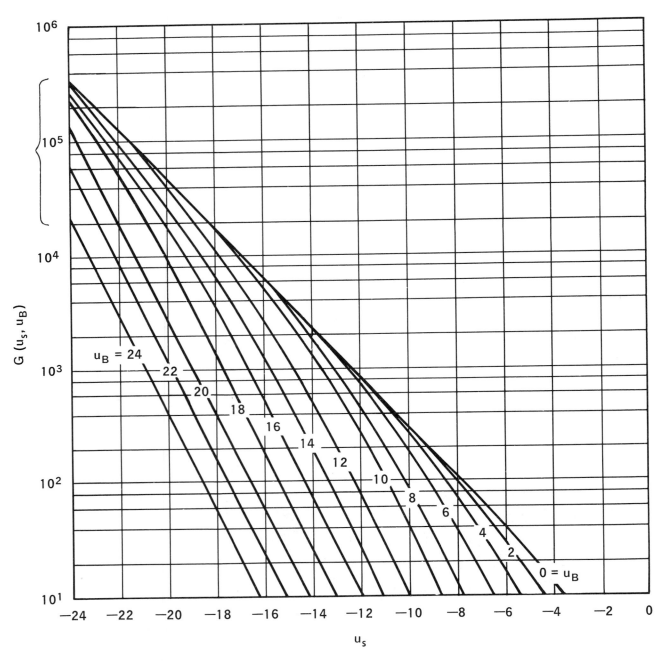

Figure 13-14 — $G(u_s, u_B)$ Versus u_s for Various Values of u_B [9]

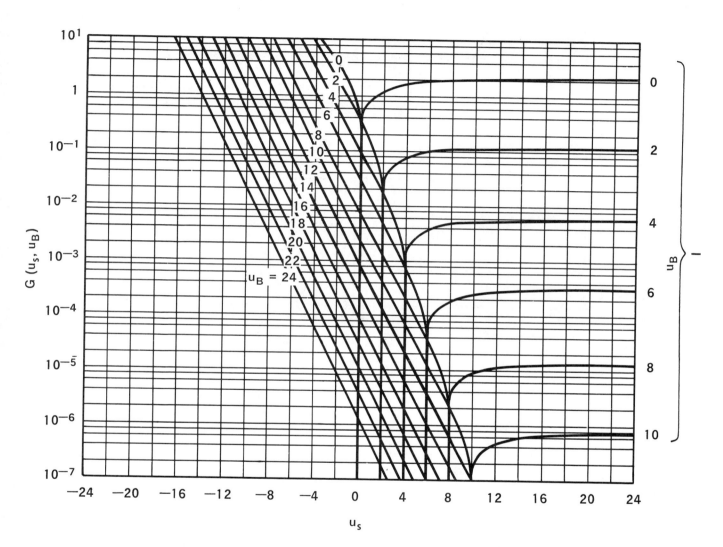

Figure 13-15 — G (u_S, u_B) Versus u_S for Various Values of u_B [9]

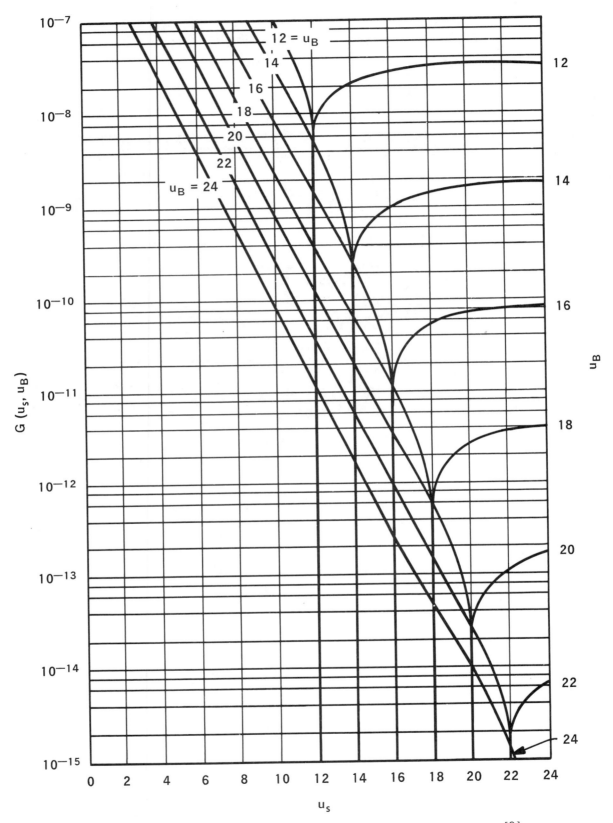

Figure 13-16 — G (u_S, u_B) Versus u_S for Various Values for u_B [9]

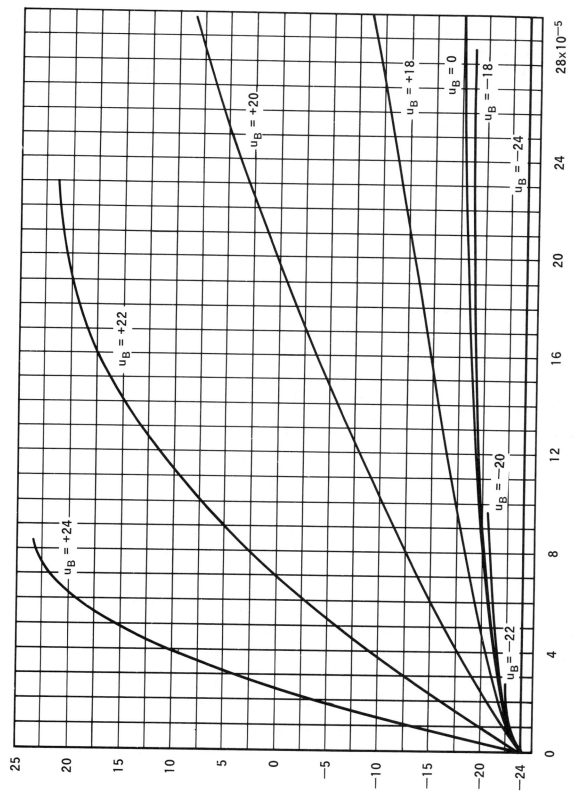

Figure 13-17 — u Versus x/L_D for $u_S = -24$ and Various Values of u_B [9]

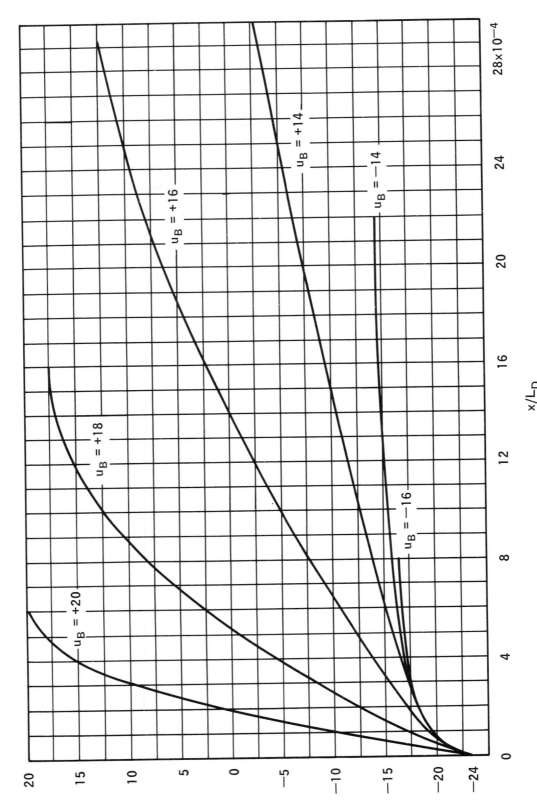

Figure 13-18 — u Versus x/L_D for $u_S = -24$ and Various Values of u_B [9]

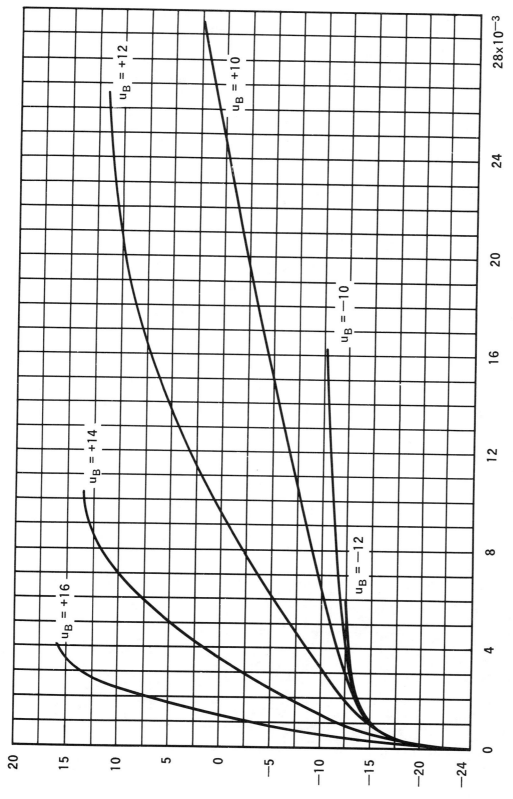

Figure 13-19 — u Versus x/L_D for $u_S = -24$ and Various Values of u_B [9]

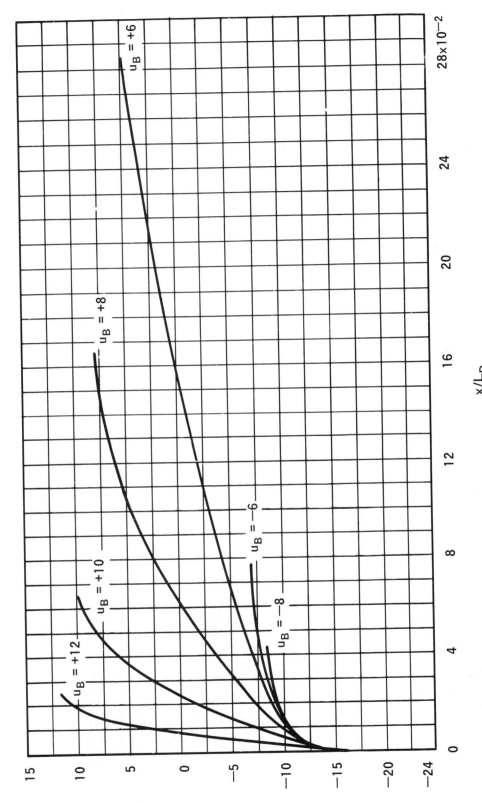

Figure 13-20 — u Versus x/L_D for $u_S = -24$ and Various Values of u_B [9]

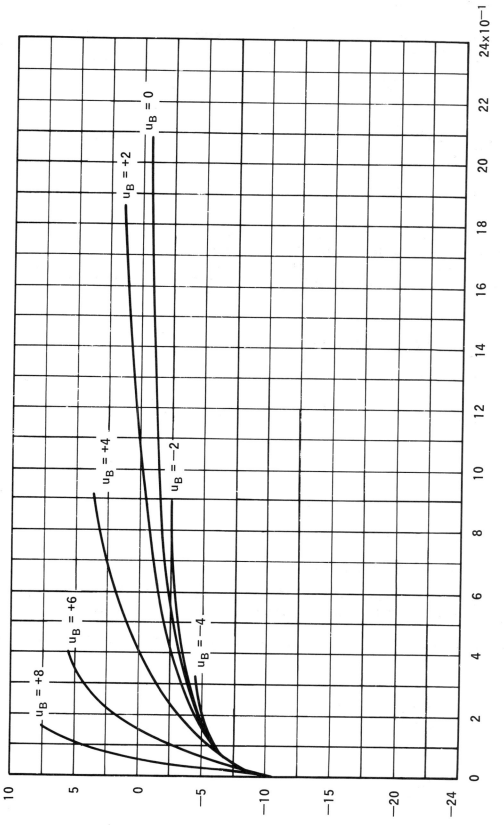

Figure 13-21 — u Versus x/L_D for $u_S = -24$ and Various Values of u_B [9]

Many, Goldstein, and Grover Curves

The curves of Many, Goldstein, and Grover (Figures 13-23 through 13-29) are presented next. Many, Gold-stein, and Grover use an effective Debye length defined as

$$L = \sqrt{\frac{K\epsilon_0 kT}{q^2(n_B + p_B)}} \quad ,$$

where n_B and p_B are the bulk electron and hole densities instead of the "intrinsic" Debye length used by Young. See Figure 13-22.

$$L_D = \sqrt{\frac{K\epsilon_0 kT}{2q^2 n_i}} \quad .$$

The effective Debye length reduces to L_D for intrinsic material, but the point here is that L, and not L_D, is best used in characterizing the width of space-charge regions in intrinsic material. This difference leads to different definitions of the functions F and G. The forms of these functions used by Many, Goldstein, and Grover will be denoted by italic letters F and G to differentiate them from the form used by Young.

The space-charge density is given by

$$Q_{SC} = K\epsilon_0 E_s = \pm q(n_B + p_B) L \, F_s(u_B, v_s),$$

where the sign is — or + as v_s is + or — respectively, and

$$F_s(u_B, v_s) = \sqrt{2} \left[\frac{\cosh(u_B + v_s)}{\cosh u_B} - V_s \tanh u_B - 1 \right]^{1/2} \quad ,$$

where

$$v = u - u_B \text{ and } v_s = u_s - u_B.$$

This differs from Young by a factor of $[\cosh u_B]^{1/2}$.

That is, $F(u_s, u_B) = [\cosh(u_B)]^{1/2} \, F_s(u_B, v_s)$.

The function F_s is plotted in Figure 13-23 for accumulation and in Figure 13-25 for inversion. These figures contain other useful information which is examined in the following paragraphs.

The effective-charge distance, L_C, is defined as the distance measured from the surface to the center of the space-charge.

$$L_C = \frac{\int_0^\infty \rho x dx}{\int_0^\infty \rho dx} = \frac{|v_s| L}{F_s} \quad [9] \quad .$$

It also happens that

$$C_{SC} \equiv \left| \frac{Q_{SC}}{V_s} \right| = \frac{K\epsilon_0}{L_c} \quad .$$

The parameter L_c/L is plotted in Figure 13-23 for accumulation and in Figure 13-24 for inversion layers. Note that in strong accumulation or inversion the charge center draws nearer the surface.

Degenerate surface conditions are also illustrated in these figures. The additional parameter, W_b, is the energy distance (in units of kT) between the Fermi level and the majority-carrier band edge in bulk. For inverted surfaces the important parameter is $(e_g - W_b)$, the energy distance (in units of kT) between the Fermi level and the minority-carrier band edge in the bulk. When the space-charge region becomes degenerate, the space charge increases more slowly than for the nondegenerate case, as v_s is changed.

Consider the same examples used in explaining the use of Young's curves; that is, determine the surface charge, Q_{SC}, of an n-type ($u_B = 10$) semiconductor in inversion ($u_s = -15$). In this case $v_s = u_s - u_B = -25$. So we need to determine $F(10, -25)$ when referring to Figure 13-25. Noting that $|v_s| - 2|u_B| = 5$, we find $F_s \cong 18.6$. This differs from F by a factor of $(\cosh 10)^{1/2}$; $F_s > 0$, since $u_s < 0$.

We must also note another important fact that cannot be obtained from Young's curves. If the semiconductor in the above example has a small bandgap, such as Ge (where $e_g \cong 26$), then $W_b \cong 3$ and $(e_g - W_b) -2|u_b| \cong 3$, and a degenerate surface condition will exist for this surface potential; $F_s \cong 16$ in this case.

In the example for a p-type semiconductor ($u_B = -12$) in accumulation ($u_s = -15$) we have $u_s = -3$ and $F_s \cong 5.7$. Again for Ge($e_g \cong 26$), degenerate surface conditions would have set in. There would also be bulk degeneracy in the latter example. These curves assume no bulk degeneracy, and, in addition, Young's curves assume no surface degeneracy. See R. Seiwartz and M. Green[11] for the case of the degenerate bulk semiconductor.

The excess surface-carrier densities are obtained using the functions $G^{\pm}$ for majority carriers and $g^{\pm}$ for minority carriers as follows:

Recall that

$$v_s = u_s - u_B .$$

In accumulation:

for n-type $v_s, \geqslant 0, u_B \geqslant 0,$

where

$$\Delta n = n_B L G^+(u_B, v_s) > 0$$

and

$$\Delta p = n_B L g^-(u_B, v_s) < 0 ;$$

for p-type, $v_s \leqslant 0, u_B \leqslant 0,$

where

$$\Delta p = p_B L G^+(u_B, v_s) > 0$$

and

$$\Delta n = p_B L g^-(u_B, v_s) < 0 .$$

In depletion or inversion:

$$\text{for n-type, } v_s \leqslant 0, u_B \geqslant 0,$$

where

$$\Delta n = n_B L G^-(u_B, v_s) < 0$$

and

$$\Delta p = n_B L g^+(u_B, v_s) > 0;$$

$$\text{for p-type, } v_s \geqslant 0, u_B \leqslant 0$$

where

$$\Delta p = p_B L G^-(u_B, v_s) < 0$$

and

$$\Delta n = p_B L g^+(u_B, v_s) > 0.$$

An overall plot of the four functions G^+, G^-, g^+, g^- is given in Figure 13-26. More detailed plots are given of G^+ and g^- in Figure 13-27 for accumulation layers and of g^+ and G^- in Figure 13-28 for inversion layers.

Finally, the shape of the potential barrier as a function of the normalized depth x/L is shown in Figure 13-29 for a surface potential of $|v_s| = 20$. The shape for values less than this can be obtained by translation along the x/L axis.

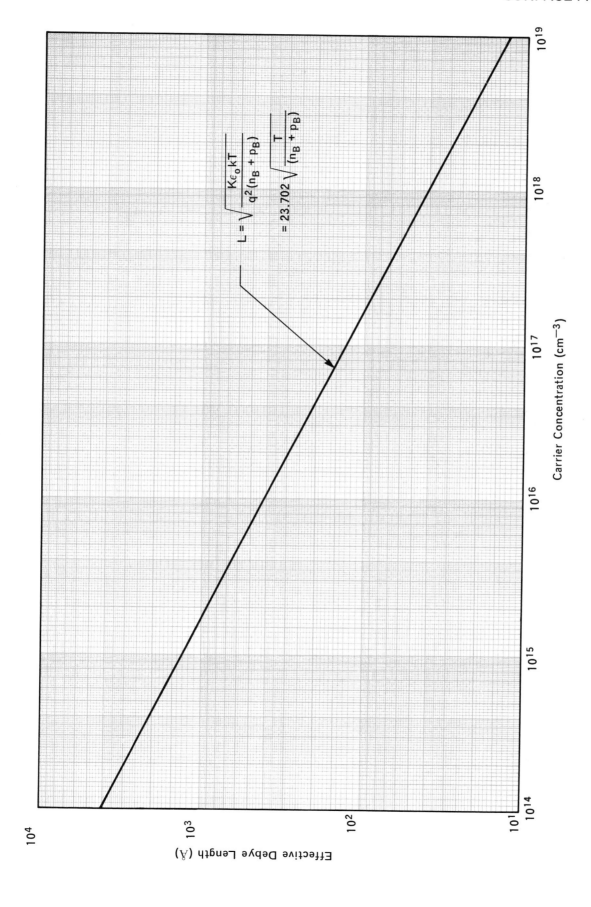

$$L = \sqrt{\frac{K\epsilon_0 kT}{q^2(n_B + p_B)}}$$

$$= 23.702 \sqrt{\frac{T}{(n_B + p_B)}}$$

Carrier Concentration (cm^{-3})

Effective Debye Length (Å)

Figure 13-22 — Debye Length in Si Versus Doping (T = 300 K)

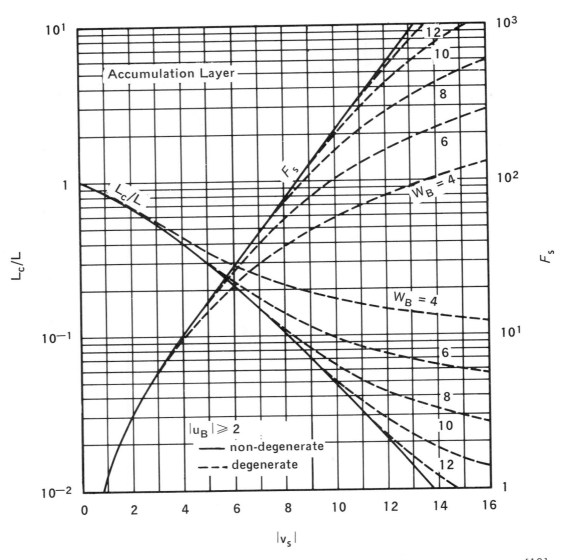

Figure 13-23 — F_s and L_c/L Versus Barrier Height $|v_s|$ in Accumulation Layers.[10]

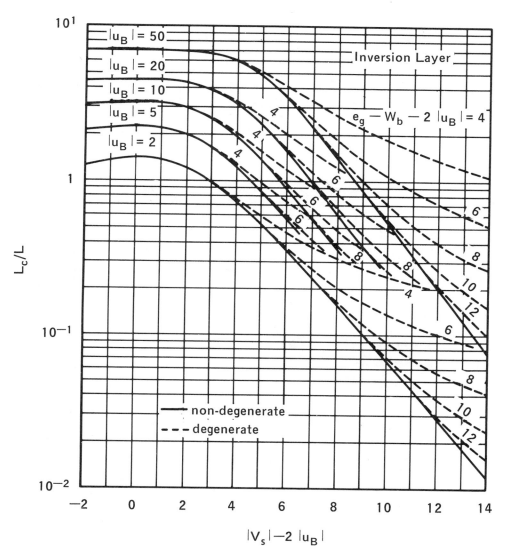

Figure 13-24 — L_c/L Versus $(V_s - 2 u_B)$ in Inversion Layers for Various Values of u_B [10]

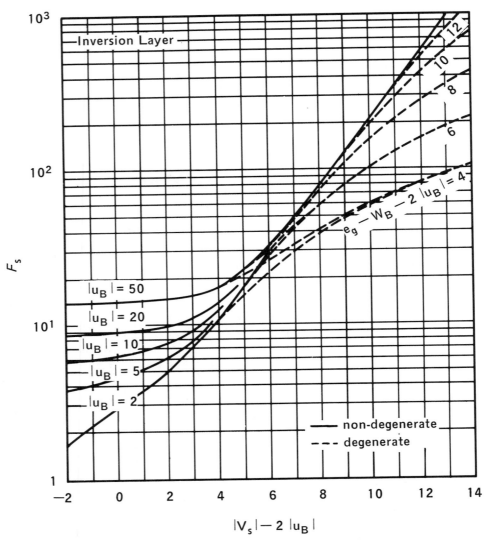

Figure 13-25 — F_s Versus $(V_s - 2 u_B)$ in Inversion for Various Values of u_B[10]

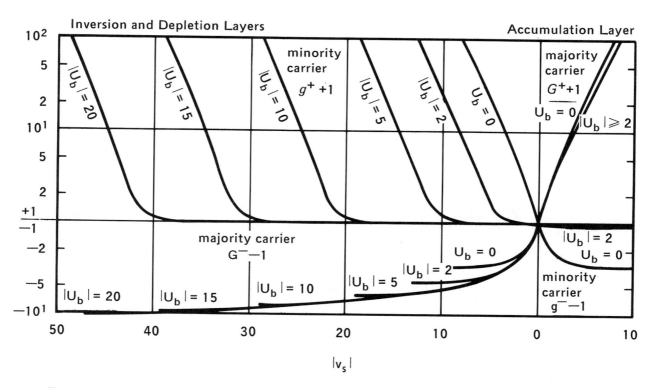

Figure 13-26 — Excess Surface-Carrier Densities Versus Barrier Height $|v_s|$ for Various Values of the Bulk Potential $|u_b|$[10]

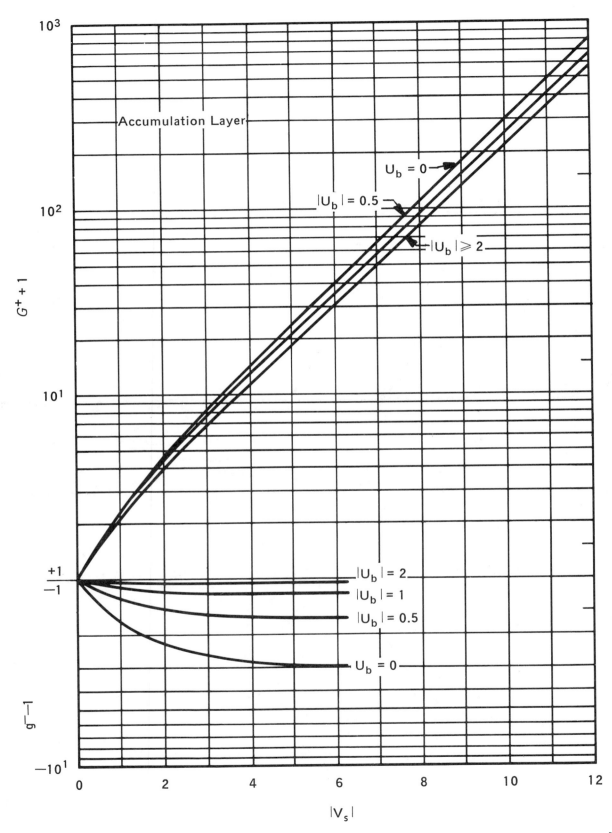

Figure 13-27 — An Expanded Plot of the Accumulation-Layer Region of Figure 13-26 [10]

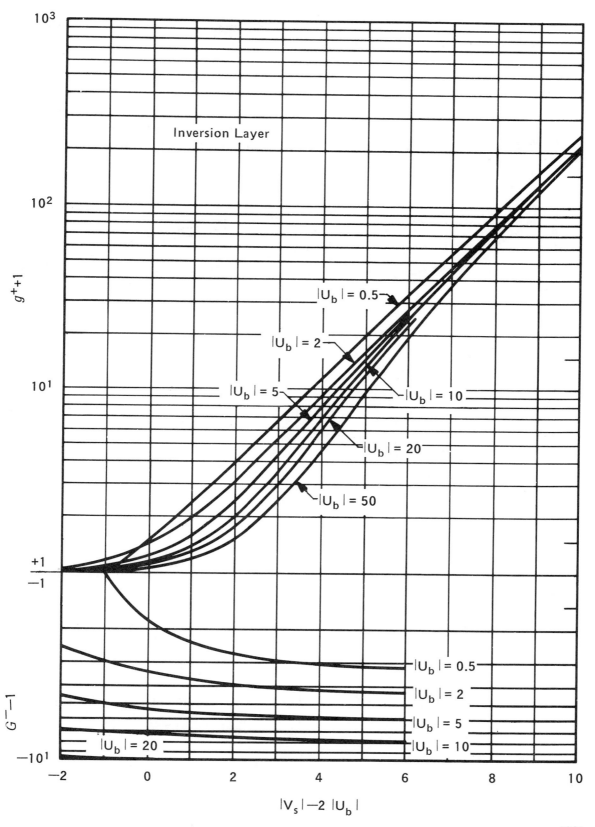

Figure 13-28 — An Expanded Plot of the Inversion-Layer Region of Figure 13-26[10]

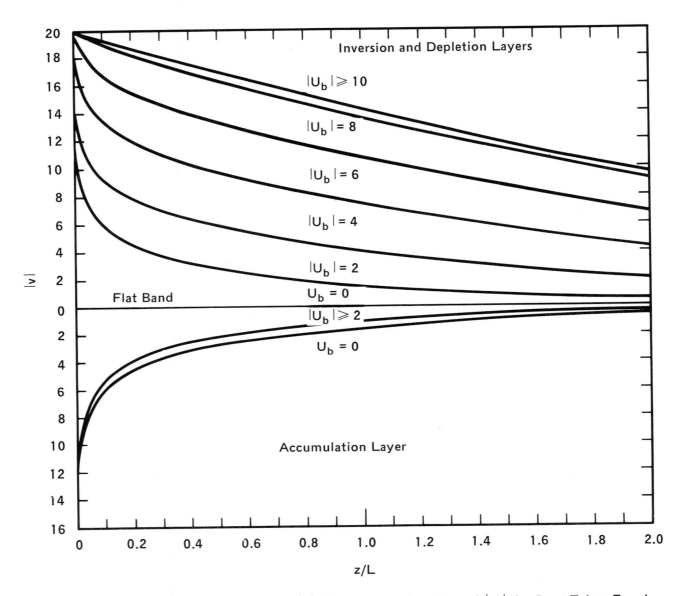

Figure 13-29 — Potential Barrier, $|v|$, Versus z/L (The Value of $|v_s|$ Has Been Taken Equal to 20)[10]

REFERENCES

1. A. Many, Y. Goldstein, and N. B. Grover, *Semiconductor Surfaces.* Holland: North-Holland Publishing, division of John Wiley & Sons, 1965.

2. D. R. Frankl, *Electrical Properties of Semiconductor Surfaces.* New York: Pergamon Press, 1967.

3. A. S. Grove, *Physics and Technology of Semiconductor Devices.* New York: John Wiley & Sons, 1967, pp. 263-355.

4. S. M. Sze, *Physics of Semiconductor Devices.* New York: John Wiley & Sons, 1969, pp. 425-566.

5. *Fundamentals of Silicon Integrated Device Technology,* ed. R. M. Burger and R. P. Donovan, Englewood Cliffs, New Jersey: Prentice-Hall, 1968, Vol. II, p. 357.

6. A. S. Grove, op cit, p. 270.

7. *Ibid.,* pp. 263-288.

8. R. Linder, "Semiconductor Surface Varactor," B.S.T.J., *41,* No. 3, (May 1962), pp. 803-831.

9. C. E. Young, "Extended Curves of the Space Charge, Electric Field, and Free Carrier Concentration at the Surface of the Electrostatic Potential Inside a Semiconductor," *J. Appl. Phys.* Vol. 32, No. 3 (March 1961), p. 330.

10. A. Many, Y. Goldstein, and N. B. Grover, op cit, pp. 140, 144-155.

11. R. Seiwatz and M. Green, "Space Charge Calculations for Semiconductors," *J. Appl. Phys.,* Vol. 29, No. 7, (July 1958), pp. 1034-1040.

14. MOS

14.1 Ideal MOS System

Energy Band and Charge Density Relationships

The energy bands and charge distribution in an ideal metal-oxide p-type semiconductor structure under various bias conditions are shown in Figure 14-1.[1] The conditions for an n-type semiconductor are similar except that bias polarities and carrier type are reversed for each surface condition. An ideal MOS structure is one without oxide charges (subsection 14.3), work function differences (subsection 14.2), or dc carrier transport through the insulator.[2]

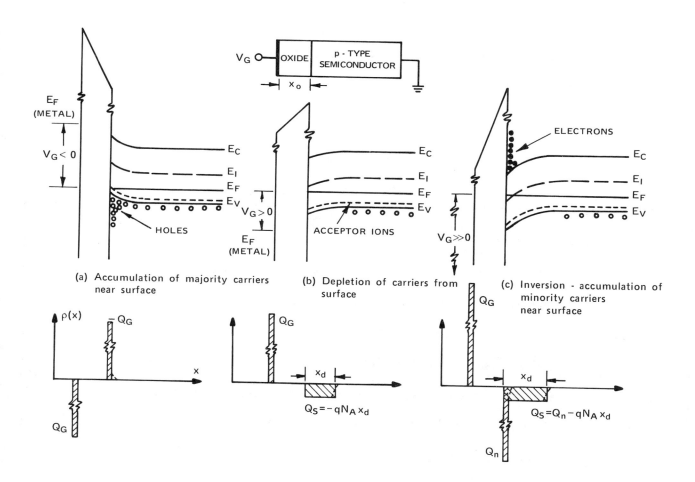

Figure 14-1 — Energy Bands and Charge Distribution in an MOS Structure Under Various Bias Conditions, in the Absence of Surface States and Work-Function Difference[1]
(Reprinted with permission from Metallurgical Transactions, a Publication of the Metallurgical Society of AIME)

The equilibrium space charge density in the semiconductor is

$$Q_{SC} = \epsilon_S E_S = \pm q (n_B + p_B) L F_S (u_B, v_S) \quad (C/cm^2),$$ 14.1

where

$$v_S = u_S - u_B,$$

$$L = \text{effective Debye length} = \left(\frac{\epsilon_S kT}{q^2 (n_B + p_B)} \right)^{1/2} \quad (cm),$$ 14.2

and

$$F_S(u_B, v_S) = \sqrt{2} \left[\frac{\cosh (u_B + v_S)}{\cosh u_B} - v_S \tanh u_B - 1 \right]^{1/2}.$$ 14.3

This treatment follows that shown in Section 13. All other terms are defined in Table 14-1 and Figure 14-2.

TABLE 14-1

DEFINITIONS OF TERMS

TERM	DEFINITION
A	Area of capacitor (cm^2)
A_p	Area of I-V peak corresponding to ionic motion (Eqtn 14.52)
a	Voltage ramp rate applied to capacitor (V/s)
β	Gain of a MOSFET (Eqtn 14.68)
C	Total device capacitance (F/cm^2)*
C_F	Final device capacitance after C-t transient (= C_{MIN}(HF)) (Fig 14-14)
C_{FB}	Total device capacitance at flatband (F/cm^2) (Eqtn 14.21)
C_{HF}	Total device capacitance at high frequency (F/cm^2)
C_I	Insulator capacitance (F/cm^2)
C_{LF}	Total device capacitance at low frequency (F/cm^2)
C_{MIN}(HF)	Minimum total device capacitance at high frequency (F/cm^2) (Eqtn 14.22)

* All quantities of charge and capacitance are used in terms of per unit area unless otherwise specifically indicated.

TABLE 14-1 (Contd)

DEFINITION OF TERMS

TERM	DEFINITION
$C_{MIN}(LF)$	Minimum total device capacitance at low frequency (F/cm^2) (Fig 14-12)
C_{OX}	Oxide capacitance (F/cm^2)
C_{SC}	Space-charge capacitance (F/cm^2)
$C_{SC}(FB)$	Space-charge capacitance at flatband (F/cm^2) (Eqtn 14.20)
$C_{SC}(MIN, HF)$	Minimum high-frequency space-charge capacitance (F/cm^2) (Eqtn 14.24)
$C_{SC}(MIN, LF)$	Minimum low-frequency space-charge capacitance (F/cm^2) (Eqtn 14.25)
C_i	Initial device capacitance after voltage step applied (Fig 14-14)
C_{it}	Capacitance associated with interface traps (F/cm^2) (Eqtn 14.58)
D_{it}	Interface trap density (number per cm^2 per eV)
Δ	Additive constant in calculation of ϕ_S from quasi-static C-V_G curve (V) (Eqtn 14.62)
E_A	Activation energy (eV) (Eqtn 14.60)
E_G	Semiconductor bandgap energy (eV)
E_I	Dielectric constant of insulator (F/cm)
E_S	Surface electric field (V/cm)
E_{eff}	Effective perpendicular electric field (V/cm)
ϵ_{OX}	Dielectric constant of SiO_2 = 3.45×10^{-13} F/cm
ϵ_S	Dielectric constant of semiconductor = 1.0443×10^{-12} F/cm for Si
F_S	Function used to calculate space-charge (Eqtn 14.3)
I_{MAX}	Maximum of $I(V_G)$ (Eqtn 14.53)
$I(V_G)$	Quasi-static current (A) (Eqtn 14.52)
k	Boltzmann constant = 1.38×10^{-23} joule/K = 8.6164×10^{-5} eV/K
L	Effective Debye length (cm) (Eqtn 14.2)
L'	Channel length of a MOSFET (cm)
λ_D	Extrinsic Debye length (cm) (Eqtn 14.4)
μ_n	Effective surface mobility ($cm^2/V{\bullet}s$)

TABLE 14-1 (Contd)

DEFINITION OF TERMS

TERM	DEFINITION
N	Number of charges per unit area (cm^{-2}) (See Sect 14.3, Nomenclature)
$\tilde{N}$	Effective doping concentration from ion-implantation (cm^{-3}) (Eqtn 14.73)
N_A	Acceptor concentration in the bulk (cm^{-3})
N_B	$\lvert N_D - N_A \rvert$
N_D	Donor concentration in the bulk (cm^{-3})
N_{EPI}	Doping concentration of an epitaxial layer (cm^{-3})
N_{SUB}	Doping concentration of a substrate (cm^{-3}) (Eqtn 14.38)
N_f	Effective number of fixed charges per unit area (cm^{-2})
N_m	Effective number of mobile charges per unit area (cm^{-2})
n_B	Bulk electron concentration $= n_i e^{u_B}$ (cm^{-3})
n_i	Intrinsic carrier concentration cm^{-3} (Fig 2-4)
N_{it}	Effective number of interface trapped charges per unit area (cm^{-2})
N_{ot}	Effective number of oxide trapped charges per unit area (cm^{-2})
n_S	Surface electron concentration $= n_i e^{u_S}$ (cm^{-3})
p_B	Bulk hole concentration $= n_i e^{-u_B}$ (cm^{-3})
p_S	Surface hole concentration $= n_i e^{-u_S}$ (cm^{-3})
Φ	Ion-implantation dose (cm^{-2})
Φ_M	Metal work function for metal on insulator (V)
Φ_{MS}	$\Phi_M - \Phi_S$ (V) (Eqtn 14.40)
Φ_S	Semiconductor work function for insulator on semiconductor (V)
ϕ_F	Bulk Fermi potential $= \pm \dfrac{kT}{q} \ln \dfrac{N_B}{n_i} = \phi_{fb}$ in Fig 14-2 (ϕ_F is positive for p-type semiconductor)
ϕ_{MI}	Metal-insulator barrier potential (V)
ϕ_S	Surface potential (band-bending) relative to the bulk (V)
ϕ_{SI}	Semiconductor-insulator barrier potential (V)

TABLE 14-1 (Contd)

DEFINITION OF TERMS

TERM	DEFINITION
Q	Net effective charge density at the Si-SiO$_2$ interface (C/cm^2) (See Sect 14.3, Nomenclature)
Q_B	Charge density in a depletion layer (C/cm^2)
Q_{FB}	Total effective oxide charge at the Si-SiO$_2$ interface when $\phi_S = 0$ (C/cm^2) (Eqtn 14.37)
Q_{SC}	Space charge density in the semiconductor (C/cm^2) (Eqtn 14.1)
Q_{eff}	Total effective oxide charge for $V_G = V_{th}$ (C/cm^2)
Q_f	Net effective fixed charge density (C/cm^2)
Q_{it}	Net effective interface trapped charge density (C/cm^2)
Q_m	Net effective mobile charge density (C/cm^2)
Q_n	Charge density in an inversion layer (C/cm^2)
Q_{ot}	Net effective oxide trapped charge density (C/cm^2)
q	Electronic charge = 1.602×10^{-19} C
R_{it}	Resistance associated with interface traps (ohm)
$\rho(X)$	Insulator space-charge distribution (C/cm) (Eqtn 14.37)
S	Surface generation velocity (cm/s) (Sect 14.1, Minority Carrier Generation Lifetime)
σ_n	Interface trap cross-section for electron capture (cm) (Eqtn 14.57)
T	Temperature (K)
t	Time (s)
t_F	Capacitor recovery time after a depleting voltage pulse (s) (Eqtn 14.36)
t_I	Insulator thickness (cm)
t_{OX}	SiO$_2$ thickness (cm)
τ_g	Generation lifetime of minority carriers (s) (Eqtn 14.35)
τ_M	Interface trap emission time constant (s) (Eqtn 14.57)
u_B	Bulk potential = $\dfrac{-q\phi_F}{kT}$ (See Fig 14-2)
u_S	Potential at the semiconductor surface (See Fig 14-2)
v_S	$u_S - u_B = q\phi_S/kT$

TABLE 14-1 (Contd)

DEFINITION OF TERMS

TERM	DEFINITION
$\overline{V}$	Mean electron thermal velocity (cm/s)
V_B	Negative voltage applied during bias-stress aging (V) (Eqtn 14.60)
V_{BB}	Voltage applied to substrate of a MOSFET
V_D	Voltage applied to drain of a MOSFET
V_{FB}	Flatband voltage (V) (Eqtn 14.37)
V_G	Voltage applied to metal gate of a capacitor or MOSFET (V)
V_S	Voltage applied to source of a MOSFET
V_{th}	Threshold or turn-on voltage of a capacitor or MOSFET (V) (Eqtn 14.26)
W	Width of a MOSFET (cm)
W_d	Depletion layer width in the semiconductor (cm)
$W_{d\ max}$	Maximum equilibrium depletion layer width (cm)
W_p	Width of $I(V_G)$ peak at $I = I_{MAX}/e$ (V) (Eqtn 14.53)
x	Distance (cm)
x_i	Effective depth of channel implant (cm)
χ_S	Electron affinity of a semiconductor (V)

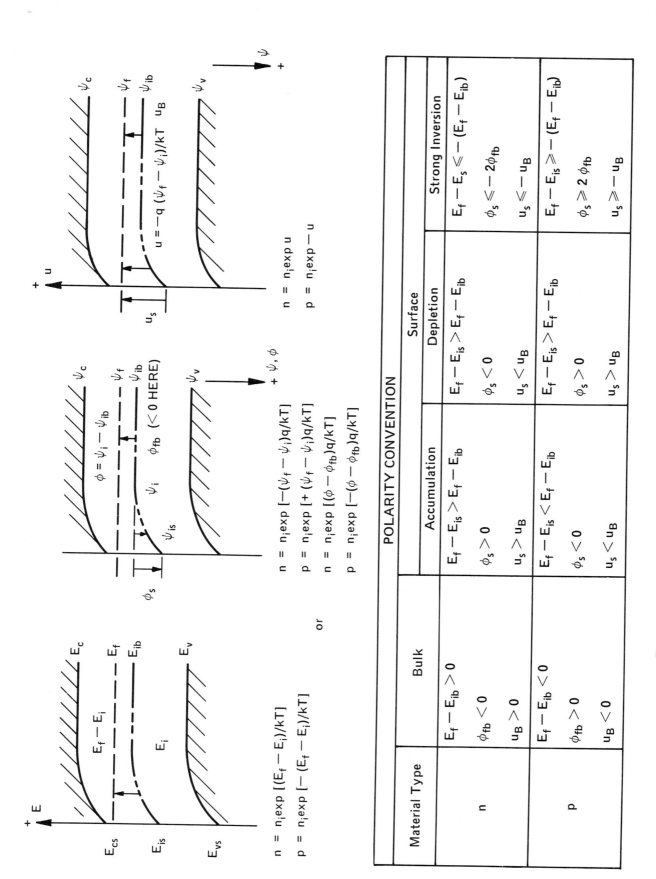

Figure 14-2 — Definition of Semiconductor Surface Parameters

The sign in Equation 14.1 is minus for positive v_S and vice versa. Note that an alternate expression for Debye length is the extrinsic value which, for holes, is

$$\lambda_D = \sqrt{2} L = \left(\frac{2\epsilon_S kT}{q^2 p_B}\right)^{1/2} \approx \frac{581.5}{\sqrt{p_B}} \text{ (cm) at 300 K.[2]} \qquad 14.4$$

A plot of Debye length versus free carrier concentration is given in Figure 2-11. In inversion,

$$Q_{SC} = Q_n + Q_B \quad (C/cm^2), \qquad 14.5$$

where

$$Q_n = \text{charge in the inversion layer,}$$

and

$$Q_B = \text{charge in the depletion layer.}$$

The relationship between Q_{SC} and Q_n for Si with varying bulk doping and surface potential is shown in Figure 14-3.[3] The effective width, W_d, of the depletion region induced in Si is depicted in Figure 14-4[3], where

$$W_d = \frac{(Q_{SC} - Q_n)}{q(N_D - N_A)} = \frac{(Q_{SC} - Q_n)}{qN_B} \quad (cm). \qquad 14.6$$

The surface potential, ϕ_S, is related to doping N_B and W_d by

$$|\phi_S| = \frac{qN_B W_d^2}{2 \epsilon_S} \quad (V). \qquad 14.7$$

When the surface becomes strongly inverted ($|\phi_S| \geq 2 \phi_{fb}$), a maximum depletion layer width is attained,[2] as

$$W_{d \; max} \approx \left[\frac{2\epsilon_S \phi_S(inv)}{qN_B}\right]^{1/2} = \left[\frac{4\epsilon_S kT \ln\left(\frac{N_B}{n_i}\right)}{q^2 N_B}\right]^{1/2} \quad (cm). \qquad 14.8$$

Figure 14-5[4] depicts $W_{d \; max}$ versus N_B for various semiconductors.

Capacitance Voltage Relationships

The capacitance per unit area associated with space charge Q_{SC} is

$$C_{SC} = -\frac{dQ_{SC}}{d\phi_S} = -\frac{dQ_{SC}}{dv_S} \quad (F/cm^2),$$

$$= \frac{\epsilon_S F (u_B, v_S)}{v_S L} \quad (F/cm^2). \qquad 14.9$$

For a uniformly doped semiconductor in accumulation or depletion[5]

$$C_{SC} = \frac{\epsilon_S}{\lambda_D} \frac{(1 - \exp v_S)}{(\exp v_S - v_S - 1)} \quad (F/cm^2). \qquad 14.10$$

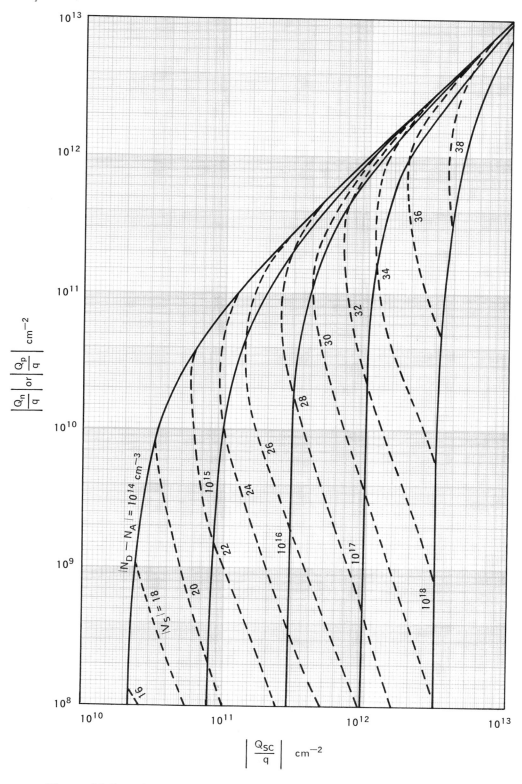

Figure 14-3 — Inversion Region Minority Carrier Charge Versus Total
Semiconductor Charge (300K)[3] (Reprinted with permission
from Pergamon Press, Ltd.)

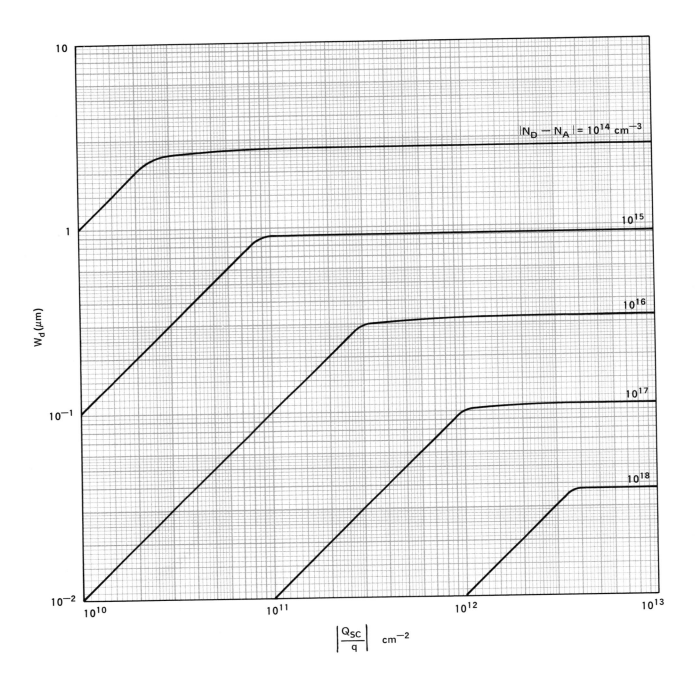

Figure 14-4 — The Width of the Depletion Region as a Function of the Total Charge Induced Within the Semiconductor (Silicon at 300K)[3] (Reprinted with permission from Pergamon Press, Ltd.)

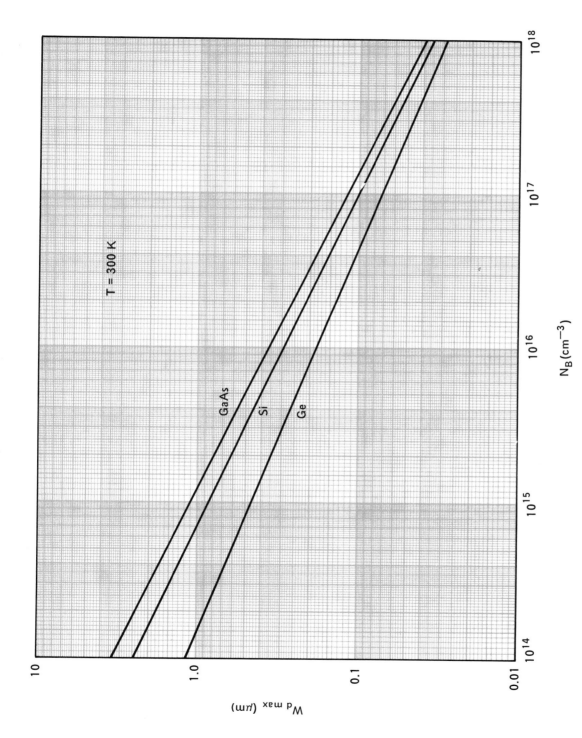

Figure 14-5 — Maximum Depletion Layer Width Versus Impurity Concentration
of the Semiconductor for Ge, Si, and GaAs Under Heavy Inversion Condition[4]
(Reprinted by permission of John Wiley & Sons, Inc.)

The total capacitance per unit area, C, of the ideal MOS capacitor is given by the series combination of C_{SC} with the insulator capacitance C_I, as

$$C = \frac{1}{\frac{1}{C_I} + \frac{1}{C_{SC}}} \quad (F/cm^2),$$
14.11

where

$$C_I = \epsilon_I / t_I \quad (F/cm^2),$$

and

$$\epsilon_I = \text{dielectric constant of the insulator}$$

$$t_I = \text{insulator thickness (cm)}.$$

The bias dependence of C is due to variation of the surface potential, ϕ_S, which affects C_{SC} via Equation 14.9 or 14.10.

When a dc bias, V_G, is applied to the metal in Figure 14-1 the following relationship holds:[2]

$$V_G = \phi_S - \frac{Q_{SC}}{C_I} \quad \text{(ideal MOS)}.$$
14.12

At flatband (for the ideal case), $V_G = \phi_S = \dfrac{Q_{SC}}{C_I} = 0.$
14.13

Modifications to Equation 14.12 to allow for oxide charges and work function differences are described in subsection 14.2. Note that in strong accumulation, $C_{SC} \to \infty$ and $C \approx C_I$.

Ideal Low-Frequency C-V_G Curves

At low measurement frequencies (~ 1 Hz), both majority and minority carriers can equilibrate with the signal. Thus, Equations 14.9, 14.11, and 14.12 are relevant in determining the C-V_G curve. An example of an ideal low-frequency C-V_G curve is shown in Figure 14-6[6], curve (a). In terms of charge, C_{SC} at low frequency[3] is

$$C_{SC} = \epsilon_S \frac{q(p_S - n_S + n_B - p_B)}{Q_{SC}} \quad (F/cm^2),$$
14.14

where p_S and n_S are surface carrier concentrations. (See Table 14-1).

Ideal High-Frequency C-V_G Curves (Slow DC Bias Ramp)

At high frequencies, minority carrier densities cannot change rapidly enough to follow the measurement signal. This is true for Si having a minority carrier lifetime $> 10^{-9}$ second and resistivity <1000 ohm-cm and ac signals having >10 mV magnitude and $\gtrsim 10$ kHz frequency.[5] In this case, changes in charge associated with minority carriers in the inversion layer do not contribute to the space-charge capacitance. Thus, [3]

$$C_{SC} = q \, (p_B - n_B) \, \frac{dW_d}{d\phi_S} \quad (F/cm^2).$$

14.15

For the idealized charge distribution of Figure 14-1,

$$C_{SC} \approx \frac{\epsilon_S}{W_d} \quad (F/cm^2),$$

14.16

where $W_d = W_{d \; max}$ (Equation 14.8) for biases corresponding to inversion. Equation 14.16 is not valid for accumulation. Curve (b) in Figure 14-6 shows the high-frequency, slow-ramp case.

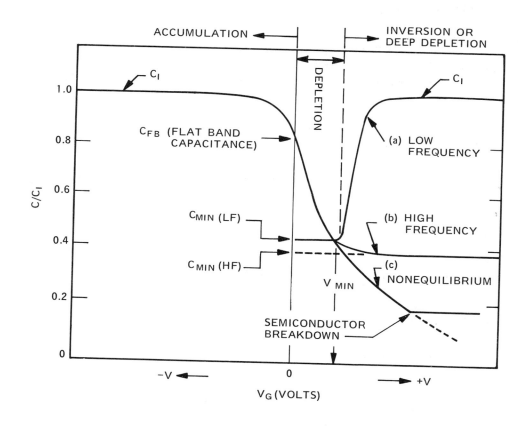

Figure 14-6 — MIS Capacitance-Voltage Curves. (a) Low Frequency, (b) High Frequency, (c) Nonequilibrium Case.[3] [6] (Reprinted by permission of John Wiley & Sons, Inc.)

Ideal Deep Depletion Curves (Fast Ramp and/or Low Temperatures)

The deep depletion C-V_G curve [Figure 14-6, curve (c)] is obtained when minority carriers cannot accumulate near the insulator-semiconductor interface, even for biases corresponding to inversion. This occurs for leaky insulators, for structures with p-n junctions under the metal plate, for rapid changes in dc bias from accumulation to inversion, or for low temperatures. The latter two situations may or may not result in deep depletion curves depending on the minority carrier lifetime in the semiconductor (i.e., long lifetime favors deep depletion). For deep depletion, $Q_n = 0$ and

$$Q_{SC} \approx q\,(n_B - p_B)\,W_d \quad (C/cm^2).[3]$$

14.17

The space-charge capacitance is still given by Equation 14.16 except W_d continues to increase beyond $W_{d\,max}$ for more depleting gate voltages. The total capacitance is now:[4]

$$\frac{C}{C_I} \approx \left[1 + \frac{2\,C_I^2\,V_G}{q(n_B - p_B)\epsilon_S}\right]^{1/2},$$

14.18

where C_I is given by Equation 14.11.

Calculations of Exact, Ideal C-V_G Curves

Iterative computer programs have been written to generate ideal curves for any metal-insulator-semiconductor system. Irvin[7] provided a FORTRAN program listing, and Goetzberger[8] provided C-V_G and $\phi_S - V_G$ curves for the Si-SiO$_2$ system. Some examples of Goetzberger's curves are shown in Figures 14-7 and 14-8. For n-type Si, reverse the sign of the voltage axis. For other insulators, use an effective thickness, as

$$t_{eff} = t_I \frac{\epsilon_{OX}}{\epsilon_I} \quad (cm),$$

14.19

where t_I is the true insulator thickness. This expression also holds for Figures 14-9 through 14-11.

Other Useful Expressions

Flatband Capacitance C_{FB}

For uncompensated semiconductors, the space-charge capacitance at flatband is

$$(C_{SC})_{FB} = \frac{\epsilon_S}{L} \quad (F/cm^2).$$

14.20

Thus, the total device capacitance becomes[2]

$$C_{FB} = \frac{\epsilon_I}{t_I + L\left(\frac{\epsilon_I}{\epsilon_S}\right)} = \frac{\epsilon_I}{t_I + \left(\frac{\epsilon_I}{\epsilon_S}\right)\left(\frac{kT\epsilon_S}{N_B q^2}\right)^{1/2}} \quad (F/cm^2),$$

14.21

where N_B = bulk doping concentration and L is given by Equation 14.2. Figure 14-9 shows the variation in C_{FB} versus t_I with bulk doping concentration as a parameter for the MOS system at 300K. Equation 14.21 and Figure 14-9 assume no contribution to the capacitance from interface trapped charge (See Section 14.3, Interface Trapped Charge). The flatband voltage (V_{FB}) is approximately offset by $\pm$ 1V from the point at which $C/C_I = 0.95$ (+1V for p-Si, $-$1V offset for n-Si).[12]

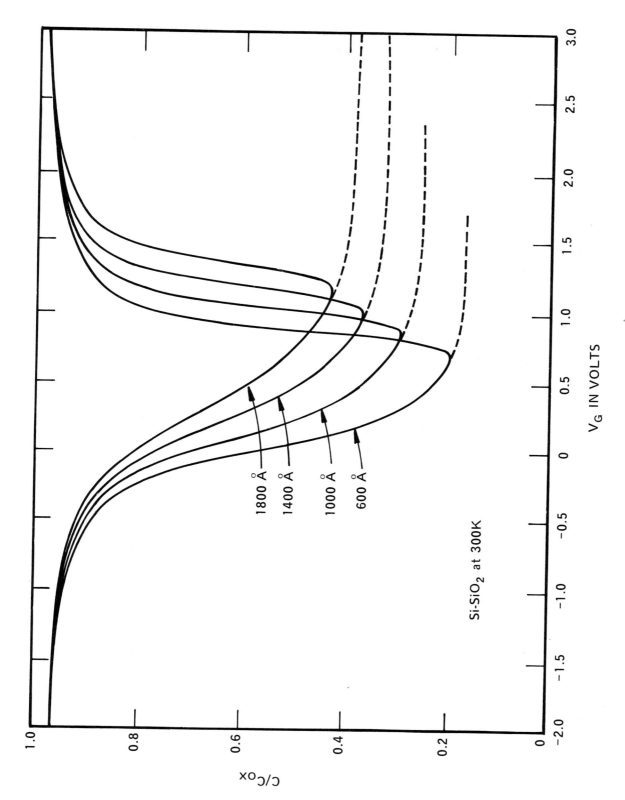

Figure 14.7 — MOS Capacity Versus Voltage. Oxide Thickness 600-1800Å. $(N_A = 1 \times 10^{15} \text{ cm}^{-3})$[8]

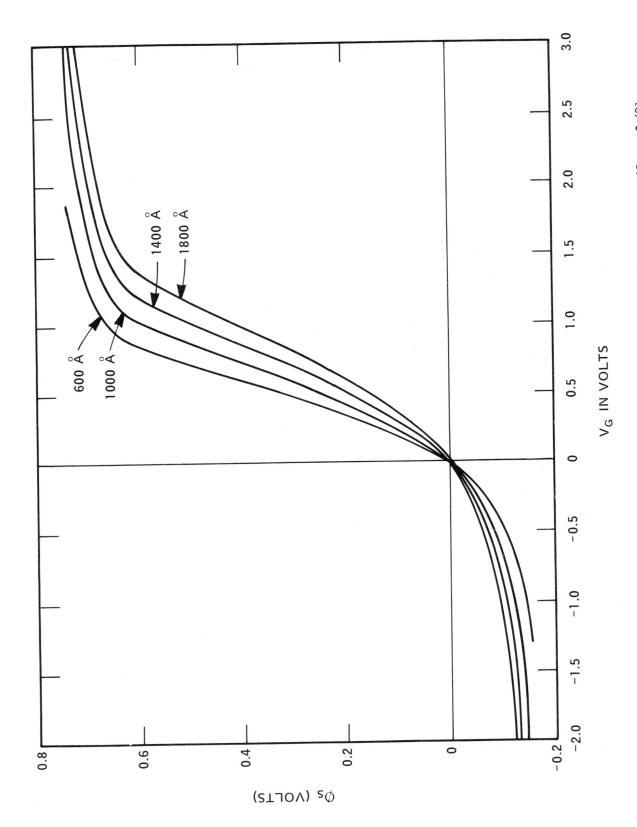

Figure 14-8 — Surface Potential Versus Voltage. Oxide Thickness 600-1800Å. ($N_A = 1 \times 10^{15}$ cm^{-3})[8]

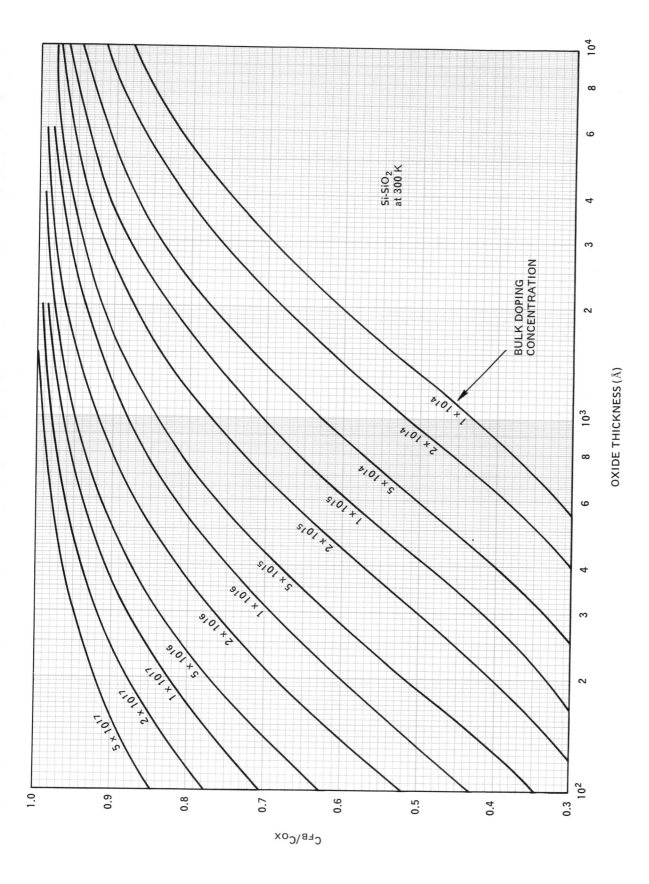

Figure 14-9 — Flatband Capacity Versus Oxide Thickness.[8]

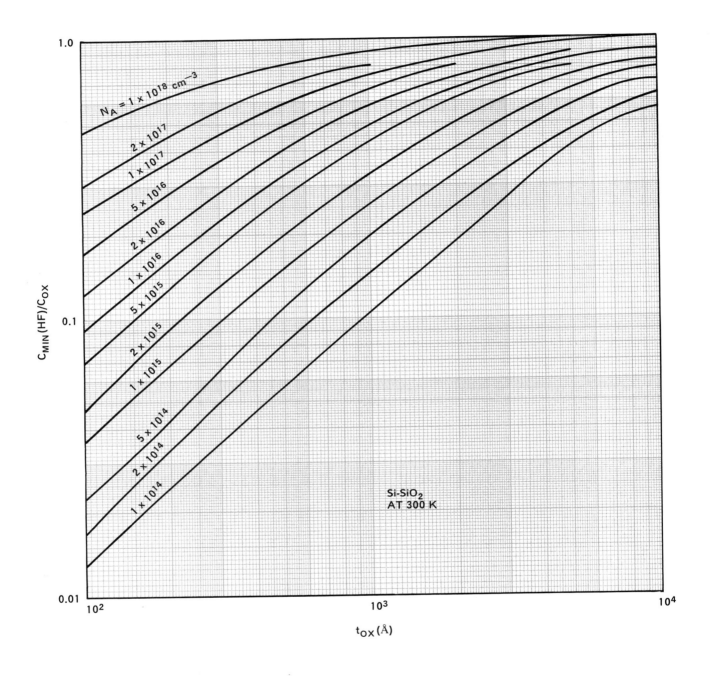

Figure 14-10 —Normalized Minimum Capacitance Versus Oxide Thickness for Ideal MIS Diodes Under High-Frequency Condition.[8] [9] (Reprinted by permission of John Wiley & Sons, Inc.)

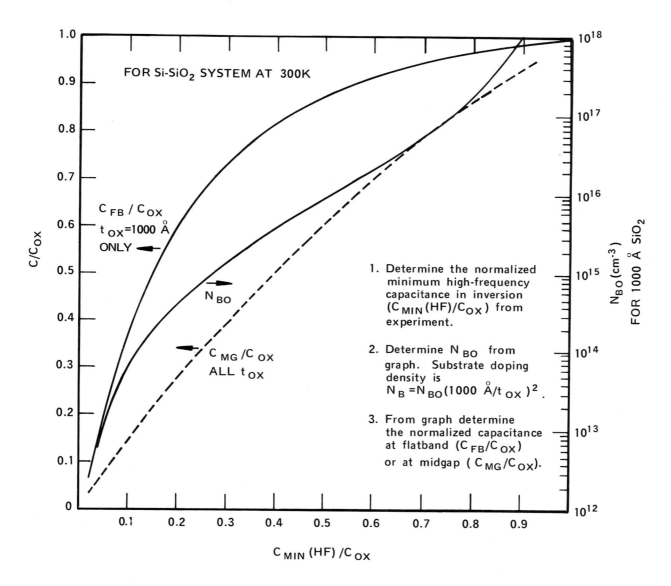

Figure 14-11 —Flatband and Midgap Capacitance and Substrate Doping From Experimental Curves[10] [11]

Minimum Capacitance, C_{MIN} (HF), C_{MIN} (LF)

The minimum capacitance for the high-frequency case is[2]

$$C_{MIN}(HF) = \frac{\epsilon_I}{t_I + \frac{\epsilon_I}{\epsilon_S} W_{d\ max}} \quad (F/cm^2) \qquad 14.22$$

or

$$C_{MIN}(HF)/C_I = \frac{C_{SC}(min, HF)}{C_I + C_{SC}(min, HF)}, \qquad 14.23$$

where

$$C_{SC}(min, HF) = \epsilon_S/W_{d\ max} \quad (F/cm^2). \qquad 14.24$$

Equation 14.23 is plotted in Figure 14-10 (Si-SiO$_2$ at 300K). In Figure 14-11, a plot of C_{FB}/C_I versus $C_{MIN}(HF)/C_I$ is provided.

For the low-frequency case,

$$C_{SC}(min, LF) \approx \frac{\epsilon_S \sqrt{2}}{5L} \quad (F/cm^2). \qquad 14.25$$

The total minimum low-frequency capacitance is plotted in Figure 14-12. Note that Equation 14.25 and Figure 14-12 apply only for oxides free of interface trapped charge.

Threshold or Turn-on Voltage, V_{th}

An approximate general expression for the applied voltage required for strong inversion is

$$V_{th} = V_{FB} + 2\phi_F - Q_B/C_I \quad (V), \qquad 14.26$$

where

$$Q_B \approx \pm (2\epsilon_S q N_B |2\phi_F|)^{1/2} \quad (C/cm^2) \qquad 14.27$$

V_{FB} = flatband voltage, i.e., voltage corresponding to

$$C = C_{FB} \text{ (Equation 14.21).}$$

In Equation 14.27, Q_B is positive for n-type Si and negative for p-type Si. As discussed in Section 14.2, subsections General Expression for Flatband Voltage and Threshold Voltage Considering Oxide Charge, and in Section 14.3, V_{FB} is usually nonzero for real MOS devices. Figure 14-13 gives curves of $|V_{th} - V_{FB}|$ versus N_B for Si-SiO$_2$ for various oxide thicknesses. (Use Equation 14.19 for other insulators). Equation 14.26 and Figure 14-13 are valid for uniform doping profiles only. See Section 14.5 for nonuniformly doped silicon.

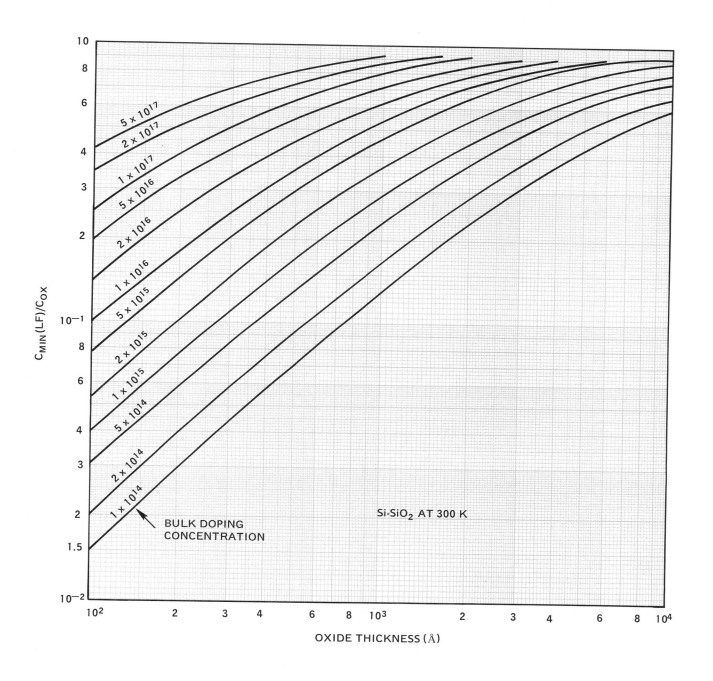

Figure 14-12 —Normalized Minimum Capacitance Versus Oxide Thickness With Silicon Doping
as the Parameter for Ideal MIS Diodes Under Low-Frequency Condition.[8] [13]
(Reprinted by permission of John Wiley & Sons, Inc.)

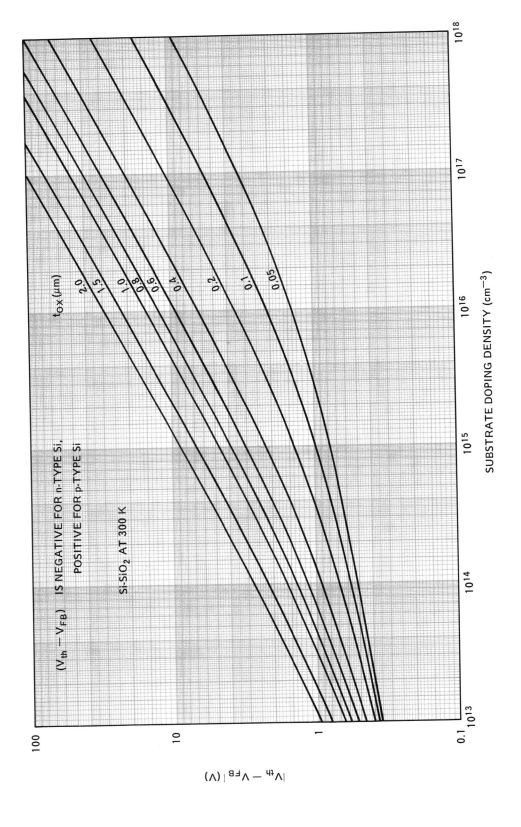

Figure 14-13 — Variation of Threshold Voltage for Si (300K) Versus Doping and Oxide Thickness

Doping Concentration from C-V_G Curves

Equations 14.22 and 14.23 will hold even when oxide fixed charge (See Section 14.3, Fixed Charge) is present, provided high enough voltages can be applied to measure C_I and C_{MIN}(HF) accurately. A useful algorithm for bulk doping concentration (Si at 299 K) is[14]

$$\log_{10} N_B = 30.38759 + 1.68278 \log_{10} C_{SC} \text{ (min, HF)}$$

$$-0.03177 \left[\log_{10} C_{SC} \text{(min, HF)} \right]^2, \tag{14.28}$$

where

$$C_{SC} \text{(min, HF)} = \frac{C_{MIN}(HF) C_I}{C_I - C_{MIN}(HF)} \quad (F/cm^2), \tag{14.29}$$

and

$$N_B \text{ is in units of } cm^{-3}.$$

Equation 14.28 assumes uniform doping. For nonuniform doping profiles, N_B is an average concentration within the depletion region of a width $W_{d \, max}$.

An alternative general expression for N_B is[10]

$$N_B = \frac{2\epsilon_I^2}{q\epsilon_S} \frac{\left| 2\phi_F + 3 \frac{kT}{q} \right|}{t_I^2 \left(\frac{C_I}{C_{MIN}(HF)} - 1 \right)^2} \quad (cm^{-3}). \tag{14.30}$$

For the Si-SiO$_2$ system at 300 K, this reduces to[10] [11]

$$N_B \approx 7.45 \times 10^4 \times \frac{\ln N_B - 24.93}{t^2_{OX} \left(\frac{C_{OX}}{C_{MIN}(HF)} - 1 \right)^2} \quad (cm^{-3}), \tag{14.31}$$

where N_B is solved for by iteration.

Doping Profiles from MOS C-V_G Curves

Assuming negligible influence from interface trapped charge (See Section 14.3, Interface Trapped Charge), high-frequency MOS C-V_G measurements can be used to determine doping profiles in the semiconductor. When the MOS capacitor is in depletion and deep depletion, the doping concentration, $N_B(W_d)$ is given by[15]

$$N_B(W_d) = \frac{2}{q\epsilon_S} \left[\frac{d(1/C)^2}{dV_G} \right]^{-1} \quad (cm^{-3}), \tag{14.32}$$

and

$$W_d = \frac{\epsilon_S}{C_{SC}} = \epsilon_S \left(\frac{1}{C} - \frac{1}{C_I} \right) \quad (cm). \tag{14.33}$$

These expressions are valid for W_d greater than 2 to 3 Debye lengths (λ_D = 58 nm for $10^{16} cm^{-3}$ doped Si at 300 K).[16] However, profiling can still be done to much smaller depletion depths compared to that for junction CV measurements.[17]

The maximum depletion depth (MOS C-V) is limited by inversion layer formation for slow voltage ramps. However, the capacitor can be driven into deep depletion by a voltage pulse, in which case the measurement becomes avalanche-breakdown limited.[15] Alternatively, the measurement depth can be extended using slow ramps if the sample is cooled to low temperature.[16] This also allows profiling closer to semiconductor surface since $\lambda_D \propto \sqrt{T}$. Both the pulsed and low-temperature ramp methods minimize inaccuracies from interface states.[17]

Brews[18] extended the usefulness of Equation 14.32 to situations where interface states are not negligible. Using both low- and high-frequency C-V_G measurements, then

$$N_B(W_d) = \frac{2}{q\epsilon_S} \left(\frac{1 - C_{LF}/C_{OX}}{1 - C_{HF}/C_{OX}}\right) \left[\frac{d\left(\frac{1}{C}\right)^2}{dV_G}\right]^{-1} \quad (cm^{-3}). \qquad 14.34$$

Ziegler et al[19] discussed a method which allows profiling right to the surface of the semiconductor.

Minority Carrier Generation Lifetime

When an MOS capacitor is pulsed from accumulation or inversion to deep depletion, a large depletion region is formed almost instantaneously.[20] Minority carriers generated in the depletion region are then swept toward the Si-SiO$_2$ interface where they accumulate in an inversion layer. Majority carriers flow to the edge of the depletion region reducing its width. The rate at which the depletion region relaxes depends on the bulk and surface generation rates. For bulk-limited cases (low interface trap density D_{it} is in the low $10^{10}\,cm^{-2}eV^{-1}$ range), the process can be described by a relaxation time Θ as

$$\Theta = \gamma\,\tau_g\,\frac{N_B}{n_i} \quad (s), \qquad 14.35$$

where

τ_g = generation lifetime of minority carriers   (s)

N_B = doping density (cm^{-3})

γ = 2 for midgap generation center energy levels.

The Zerbst analysis[21] consists of plotting $-d\left[\frac{C_{OX}}{C}\right]^2 \bigg/ dt$ versus $\frac{C_F}{C} - 1$, where C_F and C_{OX} are defined in Figure 14-14. The linear portion of such a plot has a slope A and intercept B such that:

$$\tau_g = \frac{2}{A}\frac{n_i}{N_B}\frac{C_{OX}}{C_F} \quad (s),$$

and

$$S = B\,t_{OX}\,\frac{N_B}{2n_i}\frac{\epsilon_{Si}}{\epsilon_{OX}} \quad (cm/s),$$

where S is the surface generation velocity for a depleted surface. The term S is proportional to interface-trap density. (See Section 14.3, Interface Trapped Charge.)

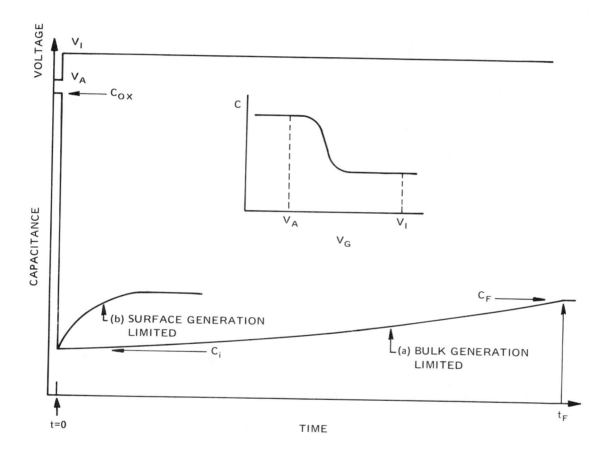

Figure 14-14 —Capacitance-Time Curves for MOS Capacitors Pulsed From Accumulation
to Deep Depletion

References [20] and [22] through [25] give refinements of the Zerbst technique. Reference [24] shows that care must be taken when surface-dominated C-t curves [curve (b) in Figure 14-14] are observed resulting in totally nonlinear Zerbst plots. Surface generation along the periphery of the capacitor must be considered for accurate measurements on long-lifetime material. Best results are obtained when pulsing from inversion to deep depletion.

If surface effects are negligible, an approximate expression for τ_g is[24]

$$\tau_g \approx m \; \frac{n_i}{8N_B} \; \frac{C_F}{C_{OX}} \; t_F \left(1 + \frac{C_i}{C_F}\right)^2 \quad \text{(s)}, \qquad\qquad 14.36$$

where t_F, C_i, and C_F are defined in Figure 14-14 and m = 1 from Reference [24] or m = 0.25 to 0.40 from Reference[26].

Pierret provided a technique for rapid C-t curve evaluation[22] and also discussed a linear sweep technique[23] which involves only the use of slow voltage ramps.

14.2 Practical MOS C-V_G Curves

General Expression for Flatband Voltage

In practice, oxide charges and work function differences contribute to the surface potential leading to nonzero flatband voltage. A common expression is[27]

$$V_{FB} = \Phi_{MS} - \frac{Q_{FB}}{C_I} - \frac{1}{C_I} \int_0^{t_I} \frac{x}{t_I} \rho(x) dx \quad \text{(V)}, \qquad\qquad 14.37$$

where

$\quad$ x = 0 and x = t_I at the metal insulator and semiconductor-insulator interfaces, respectively

$\quad \rho(x)$ = charge distribution (C/cm) in the insulator

$\quad Q_{FB}$ = total oxide charge (C/cm^2) located at the semiconductor insulator interface when ϕ_S = 0

$\quad \Phi_{MS}$ = metal-semiconductor work function difference $\quad$ (V)

$\quad C_I$ = insulator capacitance per unit area $\quad$ (F/cm^2).

The term Q_{FB} includes both fixed and interface trapped charge. (See Section 14.3.) Although commonly used, Equation 14.37 ignores possible effects of laterally nonuniform charge for which a flatband distribution function is appropriate.[28] Also, this expression holds only for gate voltages referenced to the back contact of a uniformly doped sample. For epitaxial structures having epitaxy and substrate of the same doping type but different concentrations (N_{EPI} and N_{SUB}), the measured V_{FB} referenced to the back contact is offset by the built-in potential of the epi-substrate interface.[29] Thus,

$$V_{FB} \text{ (EPI wafer)} = V_{FB} \text{ (uniform wafer)} \pm \frac{kT}{2q} \ln \frac{N_{SUB}}{N_{EPI}} \quad \text{(V)}, \qquad\qquad 14.38$$

where epitaxial doping is assumed equal to the doping of the corresponding uniformly doped wafer. The plus sign in Equation 14.38 is for p-type material (minus for n-type).

Threshold Voltage Considering Oxide Charge

Using Equation 14.37, the expression for threshold voltage (Equation 14.26) becomes

$$V_{th} = \Phi_{MS} - \frac{Q_{eff}}{C_I} + 2\phi_F - \frac{Q_B}{C_I} \quad \text{(V)},$$

14.39

where

Q_{eff} = total effective oxide charge per unit area for $V_G = V_{th}$.

Note that $Q_{eff} = Q_{FB} + \int_0^{t_I} \frac{x}{t_I} \rho(x)dx$ (C/cm^2) $\underline{only}$ if interface trapped charge is negligible, i.e., if total oxide charge does not change in magnitude with surface potential. (See Section 14.3, Interface Trapped Charge.)

Equation 14.39 is plotted in Figures 14-15 through 14-19 for various values of Q_{eff}, t_{OX}, and N_B. For other insulators, use Equation 14-19.

Work Function Differences

An energy band diagram for a metal-insulator p-type semiconductor is shown in Figure 14-20.[30] If oxide charge is neglected, a gate voltage $V_G = \Phi_{MS}$ must be applied to achieve flatband. The metal semiconductor work function difference is

$$\Phi_{MS} = \Phi_M - \Phi_S = \phi_{MI} - (\phi_{SI} - \frac{E_G}{2q} + \phi_F) \quad \text{(V)},$$

14.40

where all terms are defined in Figure 14-20. The commonly used value for ϕ_{SI} = 4.35eV for Si-SiO$_2$.[31] The metal-insulator barrier potential ϕ_{MI} has been determined for several metals on SiO$_2$ as listed in Table 14-2. For the Si-SiO$_2$ system at 300 K,

$$\Phi_{MS} = \phi_{MI} - 3.80 - \phi_F \quad \text{(V)}.$$

14.41

For p-type Si

$$\Phi_{MS} = \phi_{MI} - 3.19 - 0.0259 \ln N_A \quad \text{(V)},$$

14.42

and for n-type Si

$$\Phi_{MS} = \phi_{MI} - 4.41 + 0.0259 \ln N_D \quad \text{(V)}.$$

14.43

Equations 14.42 and 14.43 are plotted in Figure 14-21 for several MOS structures.

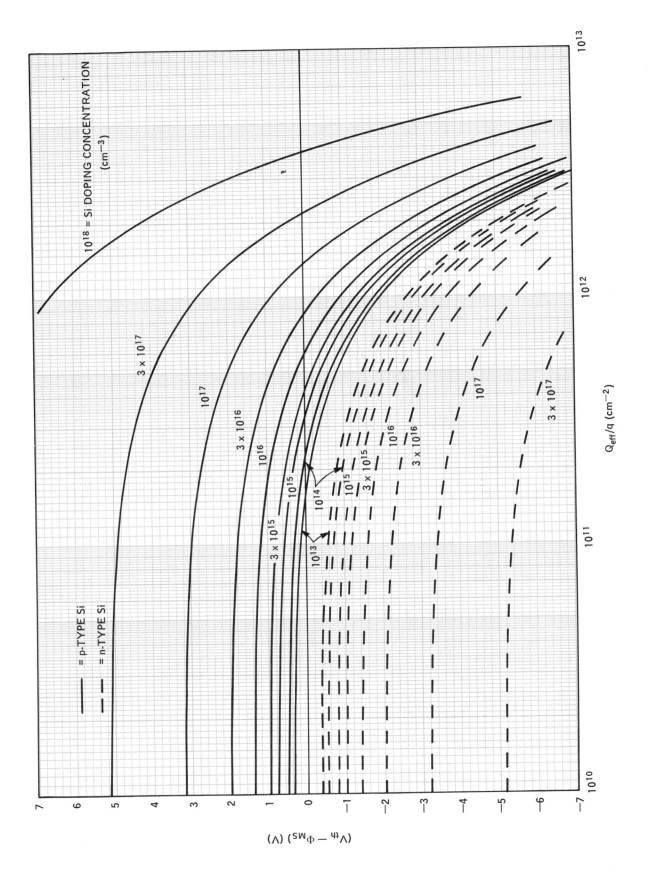

Figure 14-15 —Effect of Oxide Charge on Threshold Voltage for t_{OX} = 50 nm (0.05 μm)

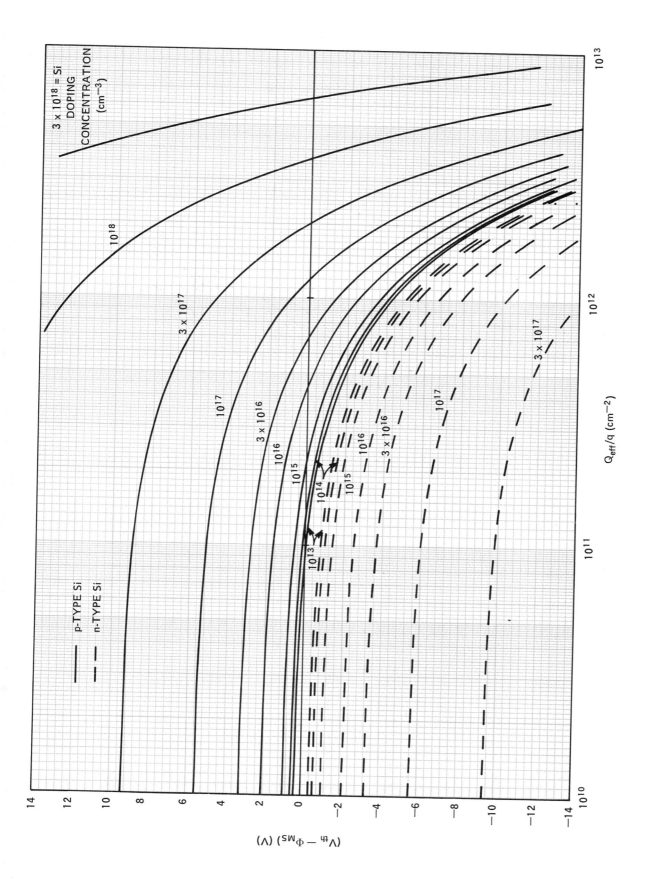

Figure 14-16 — Effect of Oxide Charge on Threshold Voltage for t_{OX} = 100 nm (0.1 μm)

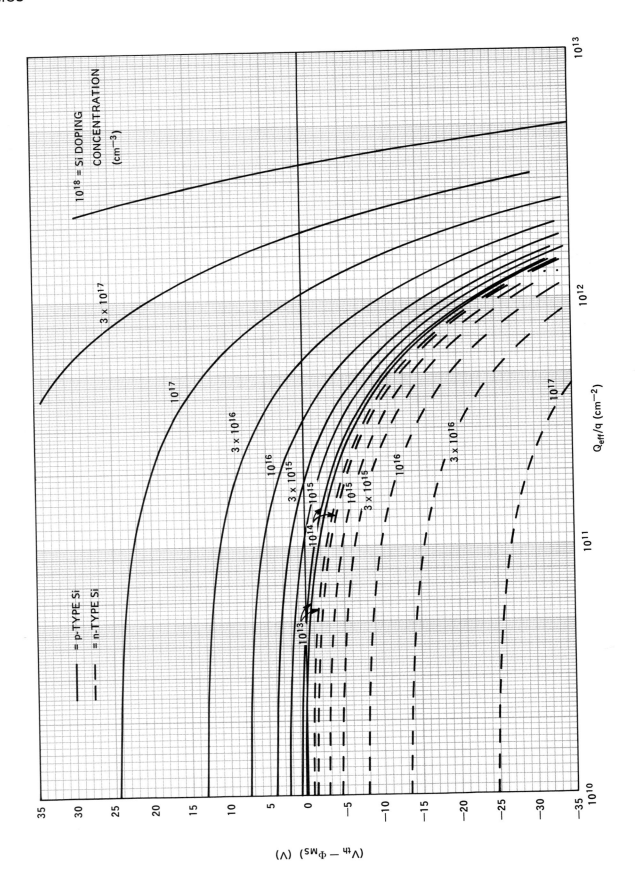

Figure 14-17 —Effect of Oxide Charge on Threshold Voltage for t_{OX} = 500 nm (0.5μm)

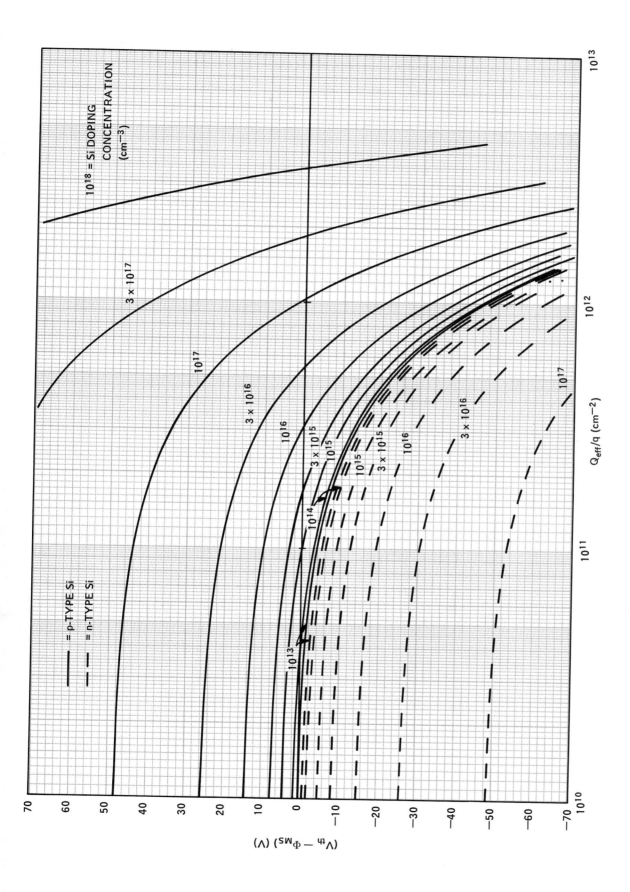

Figure 14-18 — Effect of Oxide Charge on Threshold Voltage for t_{OX} = 1.0 μm

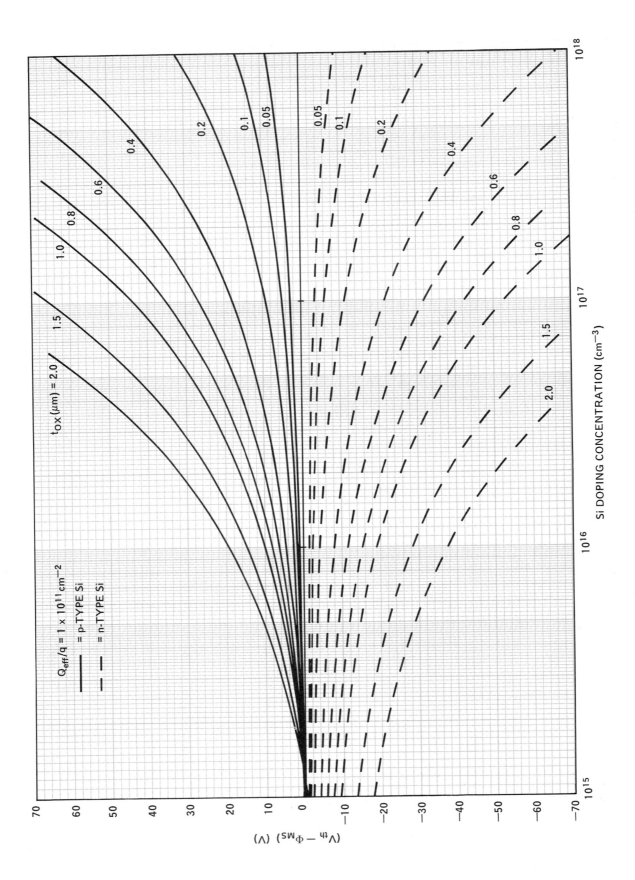

Figure 14-19 —Effect of Bulk Doping Density on Threshold Voltage for a Typical Oxide Charge Level

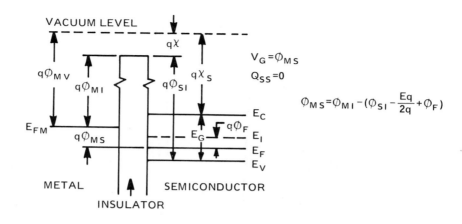

Figure 14-20 —Work Function Difference of MOS System; Energy Band Diagram in
Flatband Condition[30] (Reprinted with permission from Pergamon Press, Ltd.)

TABLE 14-2

METAL-SiO_2 BARRIER ENERGIES (eV)

METAL	BARRIER ENERGY (ϕ_{MI})*			
	C-V	REF	C-V[31]	PHOTO-RESPONSE[31]
Ag	4.07	32	4.2	4.15
Al	3.23	32	3.2**	3.2
Au	4.26	32	4.1	4.1
Cr	3.28	32		
Cu	3.97	32	3.8	3.8
Hg	4.3	34		
Hg:Na	2.1	34		
Mg	2.29	32	2.45	2.25
Ni		32	3.65	3.7
Sn	2.52	32		
WSi_2	3.9	33		

*   Assuming electron affinity of SiO_2 = 0.9V.

**   Value of ϕ_{MI} for aluminum (3.2V) is based on photoresponse measurements.

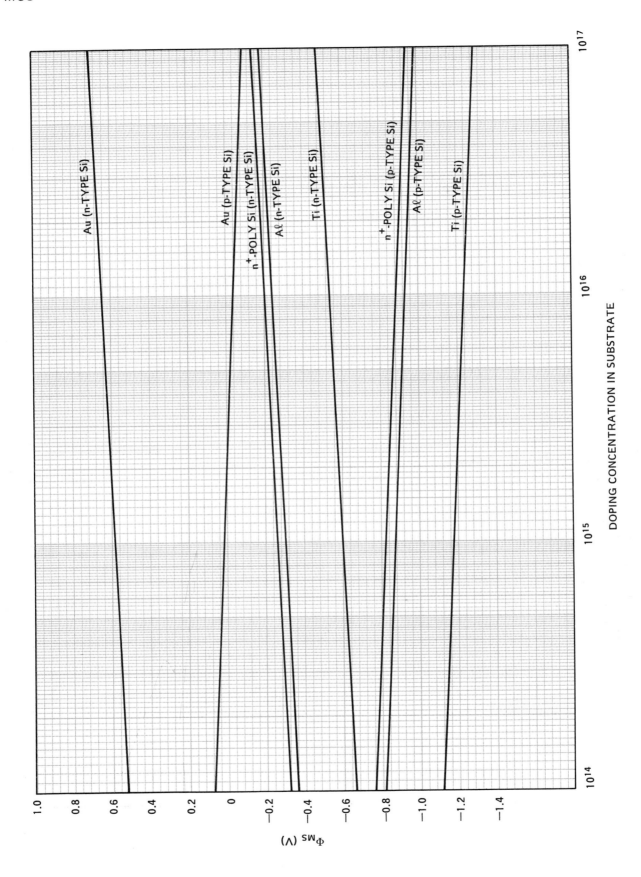

Figure 14-21 —Work Function Difference (Φ_{MS}) for Several Common MOS Structures (Si-SiO₂ at 300K)[31]

For poly-Si field plates, Φ_{MS} is dependent on the doping concentration in both the poly-Si and the substrate. Thus,

$$\Phi_M \text{ (poly)} = \phi_{SI} + \frac{E_G}{2q} + \phi_F \text{ (poly)} \quad \text{(V)},$$

14.44

and

$$\Phi_S = \phi_{SI} + \frac{E_G}{2q} + \phi_F \text{ (substrate)} \quad \text{(V)},$$

leading to

$$\Phi_{MS} = \phi_F \text{ (poly)} - \phi_F \text{ (substrate)}. \quad \text{(V)}.$$

14.45

For an n^+ (degenerate) doped poly-Si gate and p-substrate

$$\Phi_{MS} \approx 0.06 - 0.0259 \ln N_A \quad \text{(V)},$$

and for n-substrates,

$$\Phi_{MS} \approx -1.16 + 0.0259 \ln N_d \quad \text{(V)}.$$

14.46

Werner's values[30] for Φ_{MS} in Al-SiO$_2$-Si structures are $\sim 0.35V$ more positive than those of Deal et al[31] and Kar.[32] Haberle and Fröschle[35] found that, for the Al-sputtered SiO$_2$-Si system, ϕ_{SI} is a function of substrate orientation. Their data for <100> Si agrees with Werner's results while that for <111> Si agrees with Deal et al and Kar. However, Deal et al found no orientation dependence for thermal SiO$_2$. Kasprzak and Gaind[36] claimed that the electron affinity, χ_S, is sensitive to hydrogen concentration at the Si-SiO$_2$ interface. Thus, MOS structures containing wet O$_2$ or steam-grown oxides (like those of Kar) would exhibit $\sim 0.4V$ more negative Φ_{MS} than dry oxide structures.

14.3 Oxide Charges in the Si-SiO$_2$ System

Nomenclature

The four generally accepted types of charge in thermally oxidized Si are shown schematically in Figure 14-22. The standardized nomenclature described by Deal[37] is used here.

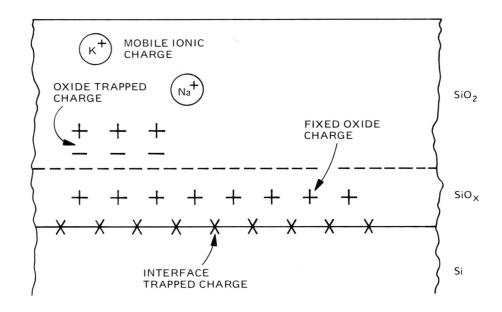

Figure 14-22 — Names and Locations of Charges in Thermally Oxidized Silicon[37]
(Reprinted by permission of the Electrochemical Society, Inc.)

The symbols for each type of charge may be either $Q_{\underline{\ \ }}$ (e.g., Q_m) or $N_{\underline{\ \ }}$ (e.g., N_m), where

Q = Net effective charge per unit area at the Si-SiO$_2$ interface

N = Net number of charges per unit area at the Si-SiO$_2$ interface*.

Note that $|Q/q| = N$ and that Q and N refer to the effective net charge at the Si-SiO$_2$ interface. Actual charge density in the oxide, but located at some distance from the Si, may be larger. The third term on the right in Equation 14.37 describes the effect of oxide space charge on V_{FB}.

The origins of most of the different types of charge are still in controversy. The following paragraphs will deal primarily with empirical observations accumulated to date. For recent detailed studies on the structure of SiO$_2$ and its interfaces, the following references are recommended:

The Physics of SiO$_2$ and its Interfaces. ed. S.T. Pantelides, New York: Pergamon Press, 1978.

The Physics of MOS Insulators, ed. G. Lucovsky, S.T. Pantelides, and F.L. Galeener, New York: Pergamon Press, 1980.

---

*Here, the symbol N is in units of cm^{-2} rather than units of cm^{-3} as is applicable to concentrations (e.g., N_B).

Mobile Ionic Charge: Q_m, N_m

Characteristics

Some of the known characteristics of mobile ionic charge are shown below.

(a) May be Na^+, K^+, Li^+, or H^+. Na^+ is most common.

(b) Mobility or drift rate decreases as ionic radius increases. (See Figures 14-23 and 14-24.)

(c) Drift rate is thermally activated; $E_{ACT} \approx 1.3$ to 1.4 eV for Na^+ and ~ 0.9 to 1.0 eV for Li^+.[40]

(d) Estimated time for drift saturation of Na^+ in SiO_2 is shown in Figure 14-25 (calculated from data of Snow et al[40]).

(e) Interface trapping (not mobility) of Na^+ ions is rate-controlling mechanism[41] (the reverse is true for K^+ ions[39]). Calculated transit times are much lower than total drift times. (See Figure 14-26.)

(f) Na^+ moves more readily (and at lower temperatures) from the Si-SiO_2 interface than a metal (Al or Au) interface.[42]

(g) Effect on Si surface potential (hence V_{FB}) depends on location of ions in oxide. (See Equation 14.37.)

(h) Only some fraction of total number of Na^+ ions is mobile.[38]

(i) Negative ions and heavy metals may contribute to Q_M even though they may be $\sim$ immobile below 500°C.[40]

(j) Drift rates for alkali ions are field-dependent.[40] (See Figure 14-27.)

When located near the Si-SiO_2 interface, N_m will cause parallel shifts of C-V_G curves along the voltage axis, as shown below for t_{OX} in μm.

$$\Delta V_{FB} = \frac{N_m \, t_{OX} (\mu m)}{2.15 \times 10^{10}} \quad (V).$$

14.47

Note that ΔV_{FB} from mobile charge will cause equal ΔV_{th} (Equation 14.26). Equation 14.47 is plotted in Figure 14-28.

Process Dependence

Device instabilities from mobile ions are minimized by avoiding contamination during processing. Sources of contamination include chemicals used for cleaning (e.g., hydrogen ions in acid residues), wafer handling equipment,[43] contaminated gloves,[43] furnace boats and tubes,[44] and metallization systems (e.g., Na from W filaments). Regarding metallization, electron gun evaporation has been shown to be much less prone to introducing sodium contamination on oxides than is filament evaporation. Chlorine-containing ambients are widely used to getter Na^+ from furnaces and oxides. (See Section 14.4.)

The use of phosphorus glass (PSG) in MOS devices to getter alkali metals is also widespread.[45] [46] [47] The same glass can be flowed at elevated temperature to achieve a smooth, contoured inter-level dielectric with good step coverage.[48] Figure 14-29 shows flatband voltage stability for phosphorus-doped glasses on 500Å thermal SiO_2. The rise in ΔV_{FB} for greater products of thickness x mol percentage of P_2O_5 arises from polarization effects in the PSG. The latter effect may not be a problem for MOS devices in which the gate metal is interposed between the PSG and the gate oxide. Seven more orders of magnitude of time are required to drift Na^+ ions through a PSG/SiO_2 composite layer (PSG layer consisting of 3.5 percent P_2O_5, 125Å thick; SiO_2 layer 2025Å thick) than through a 2150Å pure SiO_2 layer.[46]

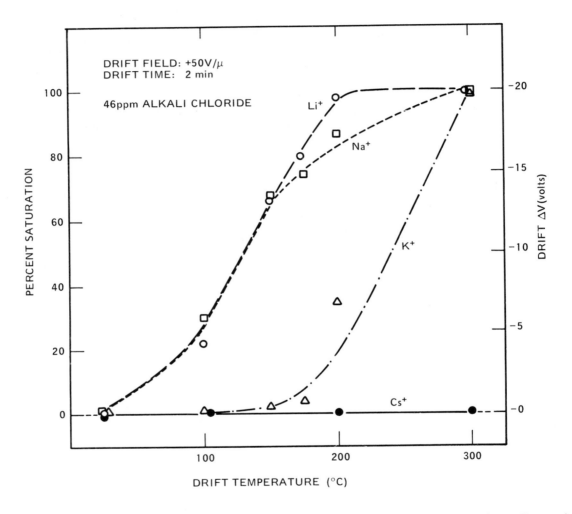

Figure 14-23 —Comparison of Drift Rates for Various Alkali Ions Through 0.2μ Thermal
Silicon Dioxide Over the Temperature Range 25° — 300°C [38] (Reprinted by
permission of the Electrochemical Society, Inc.)

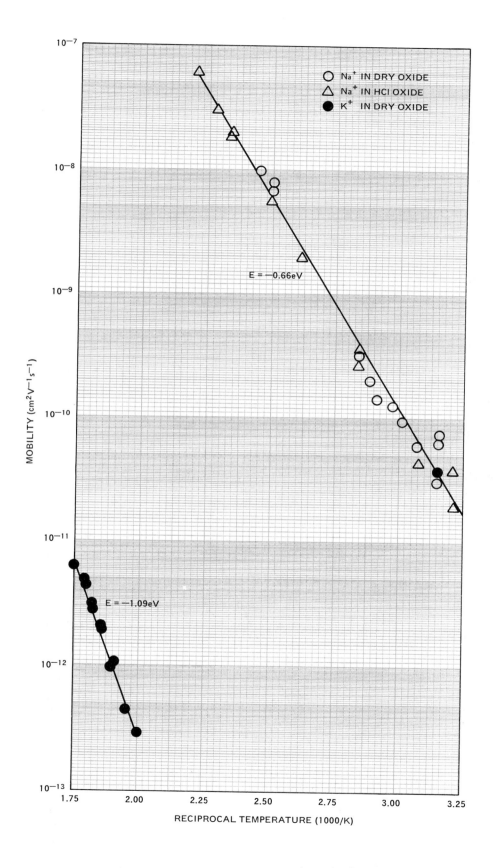

Figure 14-24 —Experimental Values of Na[+] and K[+] Ion Mobilities
Versus Reciprocal Temperature.[39]

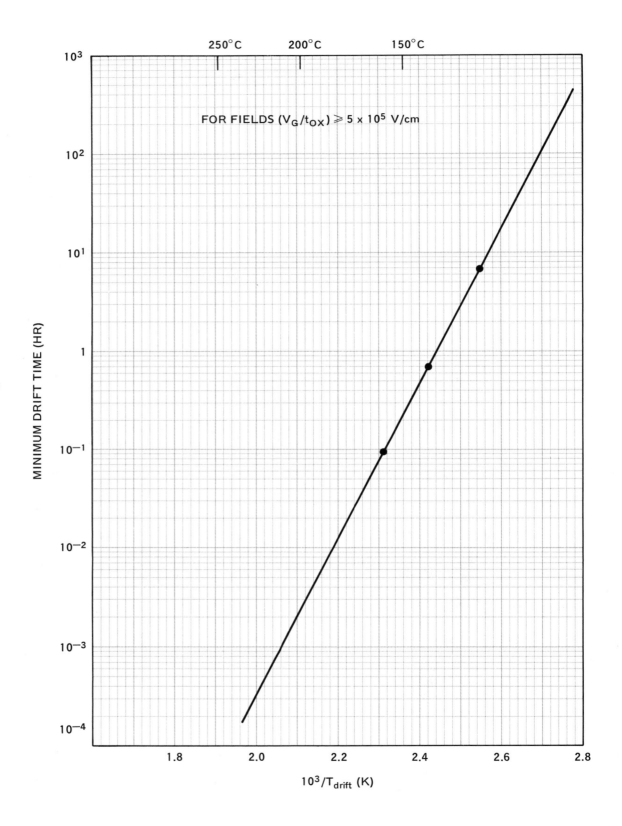

Figure 14-25 —Minimum Bias-Temperature Drift Times for Saturated
Drift of Sodium[40]

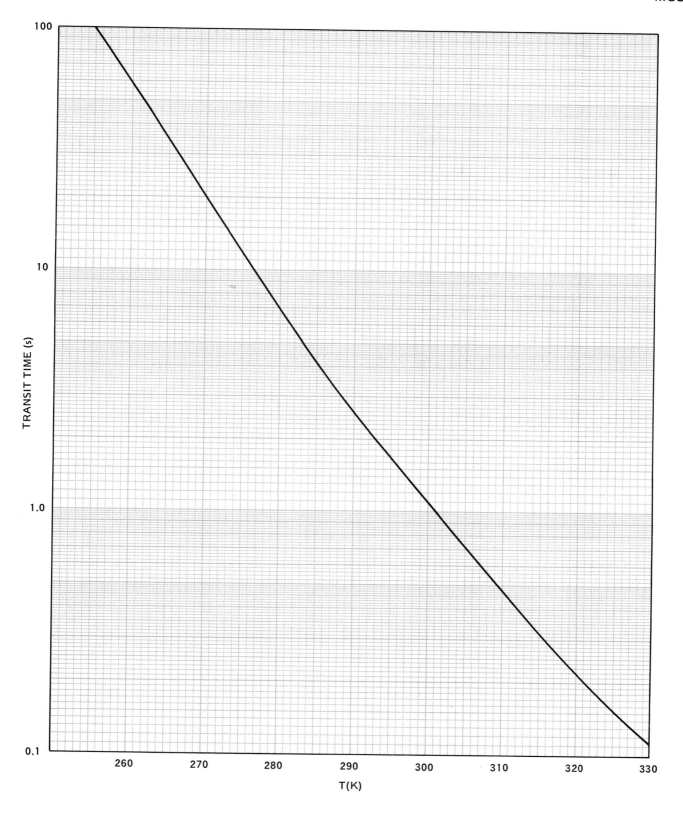

Figure 14-26 — Transit Time for Na^+ Ions Across $1000Å$ of SiO_2 Driven by
by a Field of 10^6 V cm^{-1}.[41]

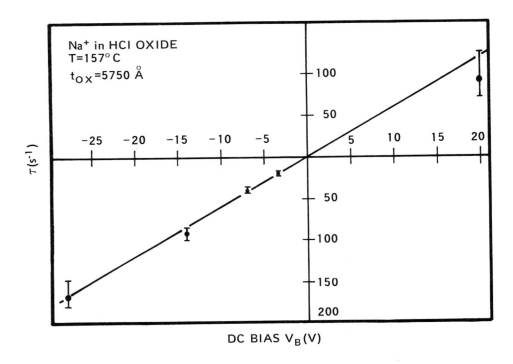

Figure 14-27 —Plot of Inverse Transit Times Versus Bias Voltage[39]

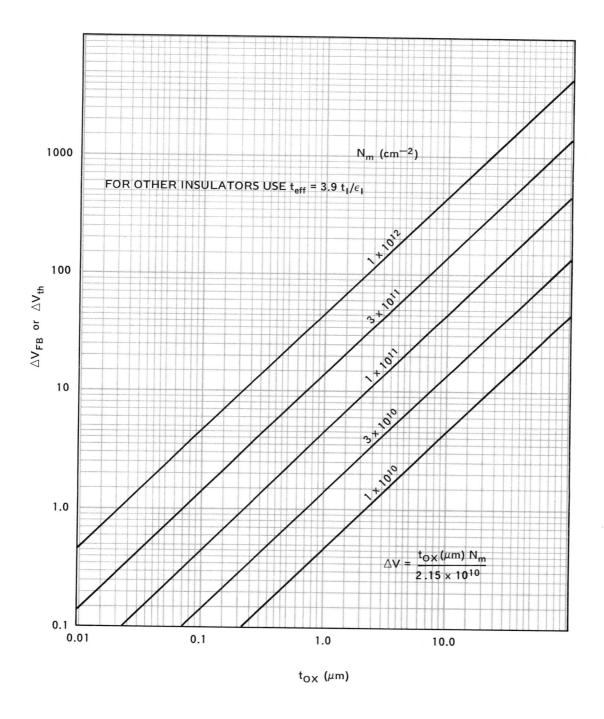

Figure 14-28 —Potential Maximum Flatband or Threshold Voltage Shift
due to Mobile Ion Contamination in Si-SiO$_2$ Structures

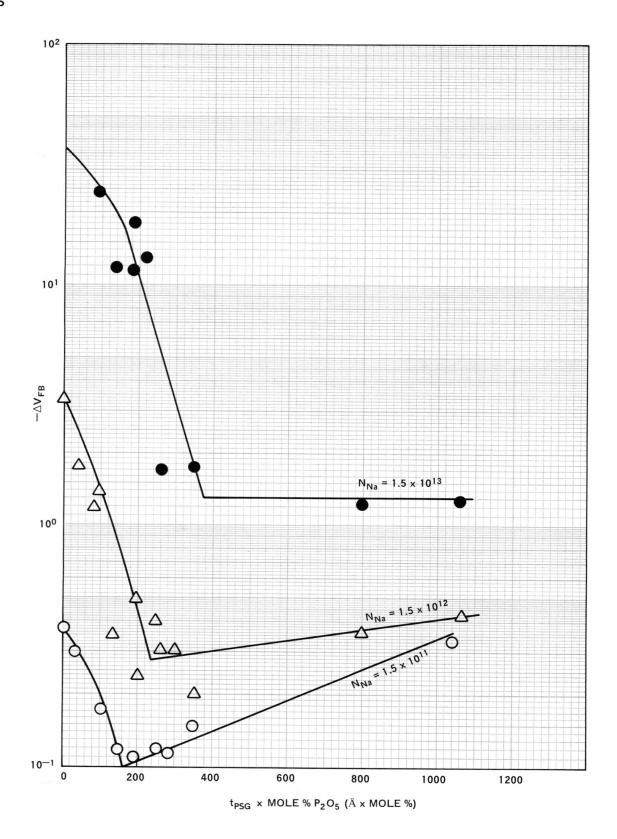

Figure 14-29 —Variation of Thermal Bias Flatband Voltage Shift With
Thickness x m/o P_2O_5. (Thickness Measured in A.)[47]
(Reprinted by permission of the Electrochemical Society, Inc.)

Measurement Techniques

A. Bias-Temperature Drift[40]

    (1) <u>Apparatus:</u> See Figure 14-30

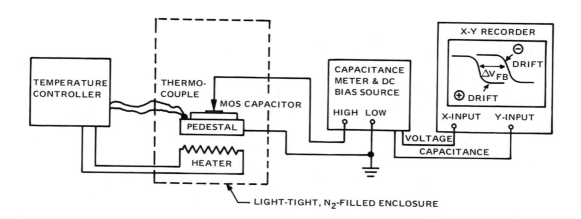

Figure 14-30 —C-V_G Apparatus for Bias-Temperature Drift

  (2) <u>Procedure</u>

    (a) Measure initial C-V_G curve of capacitor.

    (b) Apply $\geq + 1 \times 10^6$ V/cm field to metal for 10 to 15 minutes at 200°C (or equivalent time-temperature; see Figure 14-25).

    (c) Cool sample under bias.

    (d) Measure C-V_G curve to obtain $V_{FB}(+)$ and $Q_{FB}(+)$. (See Equation 14.37.)

    (e) Apply same field but negative polarity for same temperature and time as above.

    (f) Cool sample under bias.

    (g) Measure C-V_G curve $\rightarrow V_{FB}(-)$ and $Q_{FB}(-)$.

  (3) <u>Calculation:</u> In the general case, for any insulator

$$N_m = \frac{|Q_{FB}(+) - Q_{FB}(-)|}{q} \quad (cm^{-2}) \tag{14.48}$$

$$= |V_{FB}(-) - V_{FB}(+)| \times \frac{C_I\ (F)}{q \times area\ (cm^2)} \quad (cm^{-2}), \tag{14.49}$$

or, for SiO_2,

$$N_m = |V_{FB}(-) - V_{FB}(+)| \times \frac{2.15 \times 10^{10}}{t_{OX}\ (\mu m)} \quad (cm^{-2}). \tag{14.50}$$

Equation 14.50 is plotted in Figure 14-31.

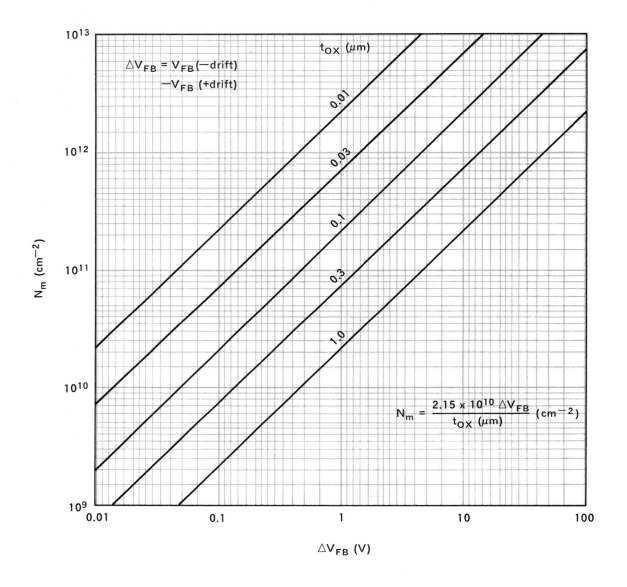

Figure 14-31 —Mobile Charge Versus Flatband Voltage Shift From BT Stress on
MOS Capacitors

(4) <u>Limitations</u>: Changes in fixed and interface trapped charge due to prolonged negative bias drift can distort the C-V_G curve.[49] (See Section 14.3, Fixed Charge and Interface Trapped Charge and Section 14.4, Oxide Charges.) If this occurs, only the difference between N_m and these changes in charge can be resolved (short drift times minimize this problem).

(5) <u>Resolution</u>: $N_m \gtrsim 10^{10}$ cm^{-2}

B. Quasi-Static (Ramp, TVS) Technique[49]

(1) <u>Apparatus</u>: See Figure 14-32.

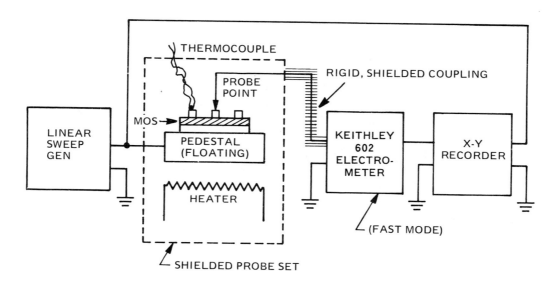

Figure 14-32 —Schematic Diagram for Mobile Ion Drift Test Set[49] (**Reprinted by permission of the Electrochemical Society, Inc.**)

(2) <u>Procedure</u>

(a) Heat sample to as high a temperature as possible [$\gtrsim 300°$C for Al field plates, see comments (a) and (b) below].

(b) Apply to the metal field plate a linear, noise-free ramp, dV/dt $\lesssim$ 100 mV/s from positive to negative voltage and then reverse direction, recording I-V_G curves.

(c) Record the zero baseline current with probe lifted.

(3) <u>Calculation</u>: Using the positive-to-negative bias curve,

$$N_m = A_p/(aqA) \quad (cm^{-2}),$$

14.51

where

A_p = area of peak in I-V_G curve due to ionic motion

$$A_p = \int_{-V_o}^{V_o} [I(V_G) - a\,C(V_G)]\,d\,V_G \quad \text{(volt-amperes)}$$

14.52

V_o = starting voltage   (volts)

$I(V_G)$ = absolute current measured   (amperes)

a = |ramp rate| (volts/second)

$C(V_G)$ = MOS capacitance   (F)

$aC(V_G)$ = electronic displacement current   (amperes)

A = capacitor area (cm^2).

Assuming a Gaussian approximation for the ionic component,

$$A_p \approx \frac{\sqrt{\pi}}{2} I_{MAX}\,W_p \quad \text{(volt-amperes)},$$

14.53

where

I_{MAX} = peak current due to ionic drift

W_p = full width of the peak (in volts) at $I_{MAX}/e = 0.37\,I_{MAX}$. (See Figure 14-33.)

The MOS component $[aC(V_G)]$ must be estimated for a baseline before I_{MAX} can be determined.

From Equations 14.51 through 14.53,

$$N_m \approx \frac{\sqrt{\pi}}{2}\,\frac{I_{MAX}\,W_p}{aqA} \quad (cm^{-2}),$$

14.54

or

$$N_m\,a\,A \approx \frac{\sqrt{\pi}}{2q}\,I_{MAX}\,W_p \quad \text{(volts/second)}.$$

14.55

Equation 14.55 is plotted in Figure 14-33, using W_p as a parameter.

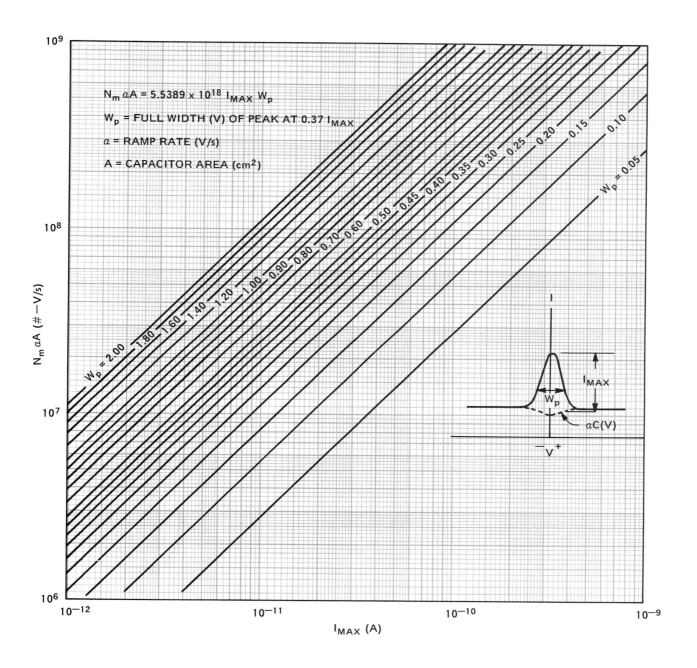

Figure 14-33 — Curves for Quasi-Static Method of Mobile Ion Measurement

(4) Limitations

(a) Low currents ($\sim 10^{-11}$ A) must be measured.

(b) MOS C-V_G curve must be known at high temperature before it can be subtracted from total $I(V_G)$ curve.

(5) Resolution

(a) Depends on background noise and shielding; typically $N_m \gtrsim 2 - 4 \times 10^9$ cm^{-2}.

(b) Larger capacitors yield higher current.

(6) Comments

(a) Calculated N_m versus increasing measurement temperature saturates at $\sim 225°$C for poly-Si field plates but does not saturate for Al field plates.[50]

(b) An additional ionic peak (Al field plate) appears for $T > 300°$C which may be due to detrapping.[41]

(c) Activation energies for A_p versus T^{-1} were 0.6 eV and 0.2 eV for 10^{11} and 10^{13} Na$^+$ ions/cm^2 at 300°C (Au on Cr field plates).[49]

C. Thermally Stimulated Ionic Current (TSIC) Method. See References[39], [40], and [42].

D. Automatic Registration of C-V_G Curve Shifts (ACVS). See Reference[51].

Fixed Oxide Charge: Q_f, N_f

Characteristics

Some characteristics of fixed oxide charge are shown below.

(a) Is positive.[38]

(b) Is located within 25Å from the Si-SiO$_2$ interface.[38]

(c) May arise from structural damage associated with oxidation, i.e., ionized silicon in the transition layer between Si and thermal SiO$_2$,[38], [52] or from various impurity atoms.[52]

(d) Does not communicate electrically with the Si, i.e., its magnitude is independent of surface potential.[38]

(e) Is immobile.[38]

(f) Magnitude depends on Si orientation:[53] Q_f (111) > (110) > (100).

(g) Magnitude depends on oxidation ambient and temperature (Figures 14-34 and 14-35).

(h) Is independent of doping type and concentration in the Si.[52] [53]

(i) Is independent of oxide thickness and oxidation time.[53]

(j) Is lowest after anneals for <1 hour in inert ambients, but increases for longer anneal times at 1100-1200°C.[38], [55] (Figure 14-36).

(k) Can be increased in magnitude ("slow trapping") by application of large negative fields at elevated temperatures (100-400°C) (Figures 14-37 and 14-38).[52] [53] See Section 14.4 for effect on Cl-containing oxides.

The introduction of Q_f into an otherwise charge-free system will cause a parallel shift of the C-V_G curve by an amount $-Q_f/C_I$. (See Equation 14.37.)

Process Dependence

The effects of several processing variables on Q_f are shown in Figures 14-34 through 14-36. Slowly cooled samples exhibit Q_f characteristic of temperatures lower than the oxidation temperature since Q_f equilibrates in minutes even down to $\sim 800°C$. The cooling ambient is similarly important.[54]

Measurement Techniques

Q_f is usually taken to equal Q_{FB} after all other types of charge have been minimized (i.e., anneal out interface traps and space charge, drift mobile ions to metal). Then,

$$Q_f = Q_{FB} = (\Phi_{MS} - V_{FB}) C_I \quad (C/cm^2),$$ 14.56

$$N_f = N_{FB} = (\Phi_{MS} - V_{FB}) C_I/q \quad (cm^{-2}).$$

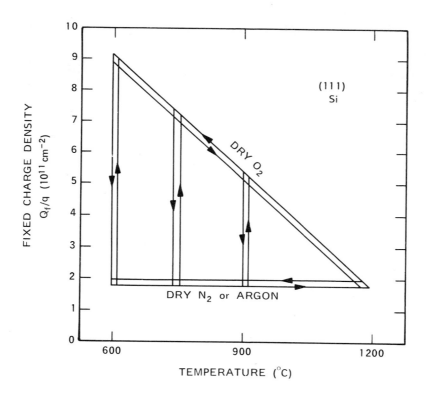

Figure 14-34 — Illustration of the Reversibility of Heat Treatment Effects on the Fixed Charge Density Q_f[53]

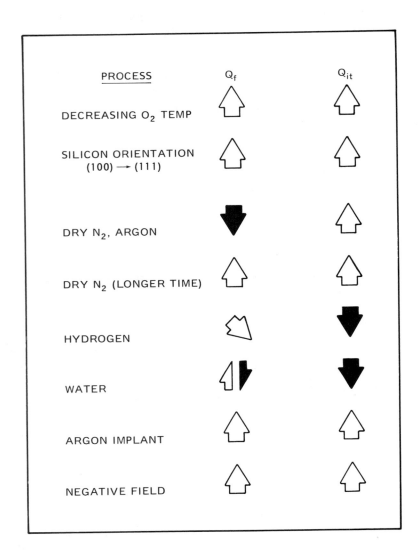

Figure 14-35 —Effect of Processing Variables on Densities of Oxide Fixed Charges (Q_f) and Oxidation-Induced Interface States (Q_{it}). Direction of Arrow Indicates Increase (Up) or Decrease (Down) in the Magnitude of the Charges. In the Case of Annealing in Hydrogen, Decrease in Q_f is for High Temperatures Only (>800°C).[54] (Reprinted by permission of the Electrochemical Society, Inc.)

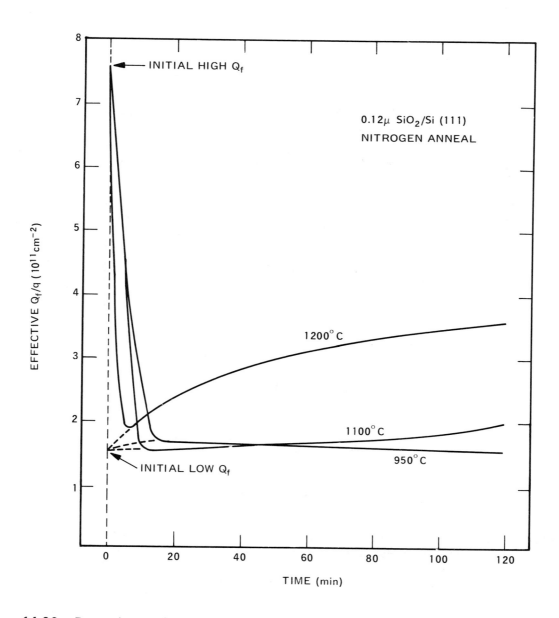

Figure 14-36 —Dependence of Effective Q_f/q on Anneal in Nitrogen for Times of 0-120 Minutes at Various Temperatures[38] (Reprinted by permission of the Electrochemical Society, Inc.)

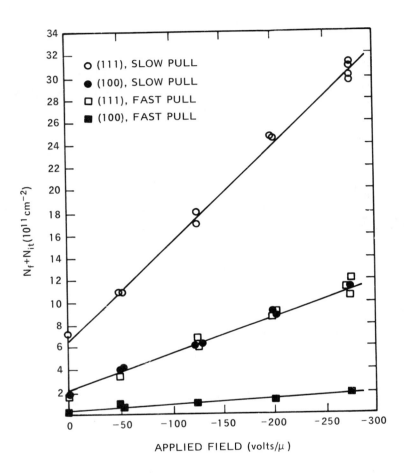

Figure 14-37 —Steady-State Values of $N_f + N_{it}$ as a Function of Applied Field for Four Different MOS Structures With Various Initial N_f Values [1200°C Dry O_2 Oxidation; $x_0 = 0.20\mu$; p-Type, $N_A = 1.4 \times 10^{16}$ cm^{-3}].[53] (Reprinted by permission of the Electrochemical Society, Inc.)

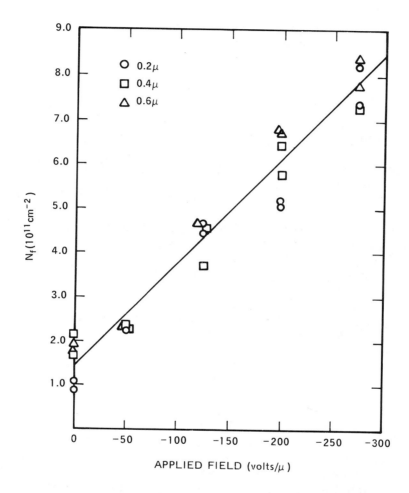

Figure 14-38 —Steady-State Values of N_f as a Function of Applied Field for MOS Structures With Various Oxide Thicknesses [$1200°C$ Dry O_2 Oxidation; (111) Orientation; n-Type, N_D = 1.4 x 10^{16} cm^{-3}] .[53] (Reprinted by permission of the Electrochemical Society, Inc.)

Interface Trapped Charge: Q_{it}, N_{it}

<u>Characteristics</u>

Some characteristics of interface trapped charge are shown below.

(a) May be either positive (donor-like) or negative (acceptor-like).

(b) Is associated with both discrete energy levels and with a continuum of energy levels in the band-gap of the Si[38] (Figure 14-39).

(c) May arise from

(1) Structural, oxidation-induced defects such as trivalent silicon centers.[56]

(2) Metal impurities (including Na).

(3) Other defects caused by radiation or similar bond-breaking processes.[38]

(d) Exhibits the same orientation dependence as fixed charge[52] [54] (Figure 14-40).

(e) Exhibits trapping time constants which vary exponentially with surface potential (Figure 14-41). These time constants determine the frequency response of Q_{it}.[52] [58]

(f) Can exhibit trap time-constant dispersion for laterally nonuniform distribution of charges (band-bending, varying laterally).[52] [58]

(g) Can be increased in magnitude by application of negative fields to metal at elevated temperatures[38] [53] [59] (Figures 14-42 and 14-43).

(h) Magnitude can be reduced significantly by annealing at 400° to 500°C in hydrogen-containing ambients[54] or in N_2 ambients if Al is used for the metal.[60]

(i) Density increases drastically with small amounts of hot electron trapping.[61]

This type of charge has been called surface states, fast states, and interface states. The introduction of Q_{it} into an otherwise oxide charge-free system will usually result in C-V_G curves which are distorted or smeared out along the voltage axis relative to the ideal case. This is due to the variation in the total number of empty or filled traps (or states) as (effectively) the Fermi level is swept across the bandgap by changing the surface potential (or applied voltage).

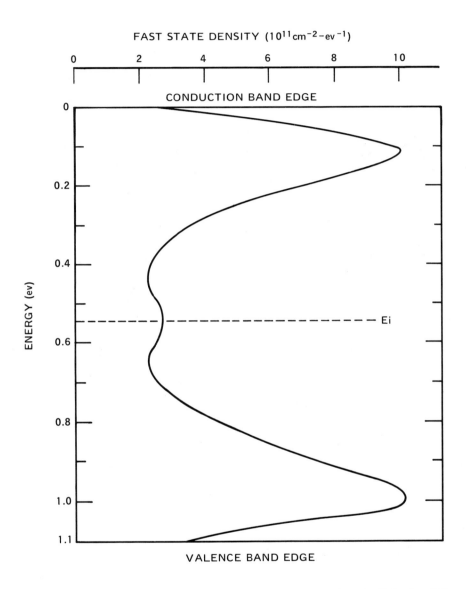

Figure 14-39 —Approximate Distribution of Fast States Associated With the Silicon-Silicon
Dioxide Interface Produced by Thermal Oxidation in Dry Oxygen[38]
(Reprinted by permission of the Electrochemical Society, Inc.)

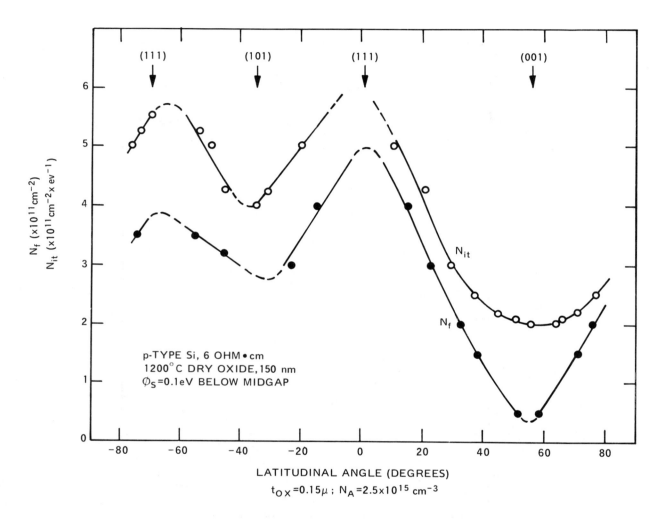

Figure 14-40 — Fixed Charge and Interface Trap Density as a Function of Silicon Surface Orientation[52] [57]

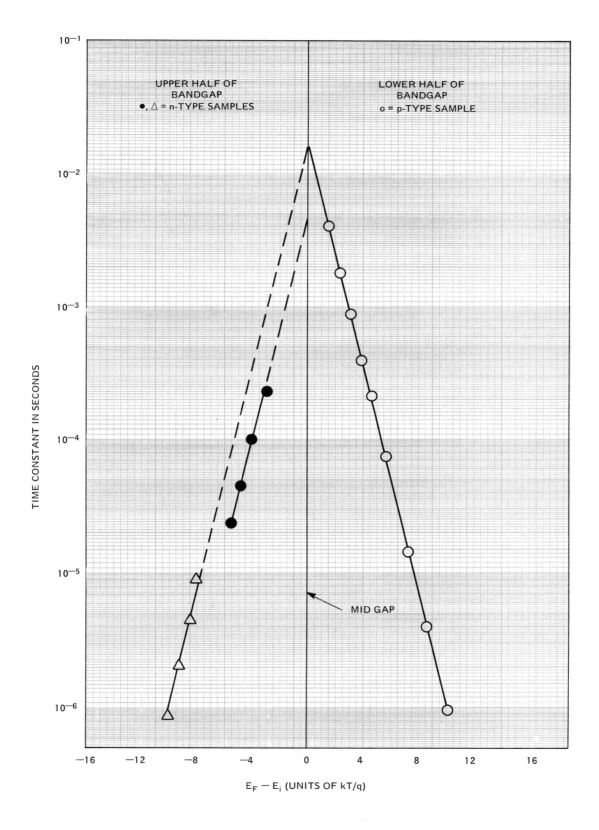

Figure 14-41 —Interface Trap Time Constants Versus Energy For Steam-Grown Oxide on <111> Si[52] [58]

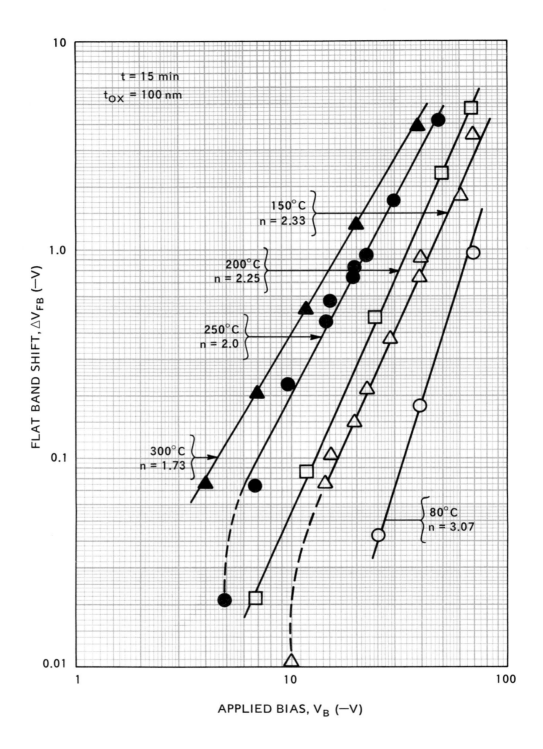

Figure 14-42 — Variation of Slow-Trapping Induced Flatband Shift With Applied Bias at Different Temperatures, and For Times of 15 Min.[59] (Reprinted by permission of the Electrochemical Society, Inc.)

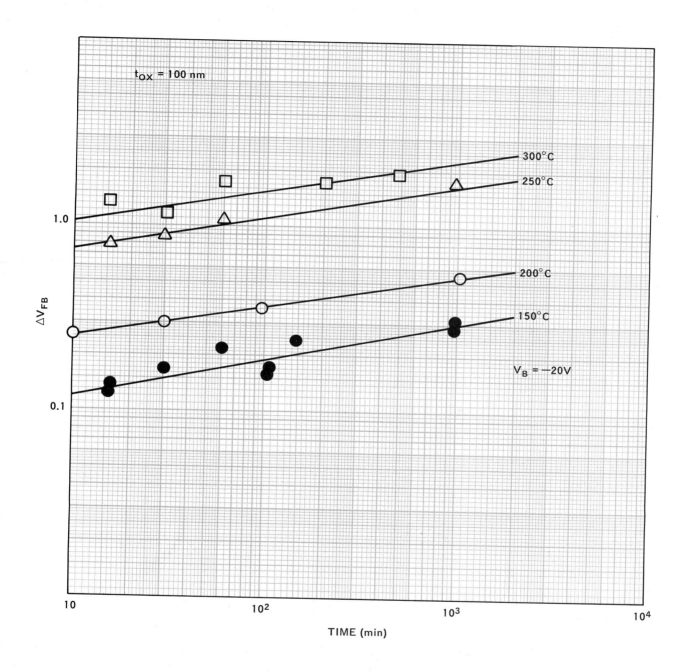

Figure 14-43 — Time Dependence of Flatband Shift due to Slow Trapping at Various Temperatures and for an Applied Bias of —20V.[59] (Reprinted by permission of the Electrochemical Society, Inc.)

Q_{it} will contribute to the total device capacitance if the measurement frequency is lower than the inverse of the trap time constants. The time constants vary with trap energy level in the Si bandgap (Figure 14-41). The time constant for emission of majority carriers from a uniform distribution of interface traps in the bandgap is given by[16] [58]

$$\tau_M = \frac{1}{\overline{V}\sigma_n n_i} \exp\left[\frac{q(\phi_B - \phi_S)}{kT}\right] \quad (s),$$ 14.57

where

ϕ_B = bulk potential (V)

ϕ_S = surface potential (V)

$\overline{V}$ = mean electron thermal velocity (cm/s)

σ_n = interface trap cross section for electron capture $\approx 6 \times 10^{-16}$ cm^{-2}.[58]

Flatband voltage shifts from Q_{it} will not necessarily be the same as threshold voltage shifts. Figure 14-44 shows the effect of hypothetical interface trap distributions on C-V_G curves allowing for both Q_{it} and Q_f. Distorted C-V_G curves may also be caused by lateral nonuniformities near the Si-SiO$_2$ interface.[52]

For a continuous and uniform interface trap distribution and neglecting time-constant dispersion, the capacitance associated with N_{it} is[16]

$$C_{it} = q\,\frac{N_{it}}{\omega\tau_M}\,\arctan\,(\omega\tau_M) \quad (F/cm^2),$$ 14.58

where

$\omega = 2\pi f$ = angular measurement frequency   (rad/s).

For total device capacitance, Equation 14.11 becomes:

$$C = \frac{1}{\dfrac{1}{C_I} + \dfrac{1}{(C_{SC} + C_{it})}} \quad (F/cm^2).$$ 14.59

Figure 14-45 shows the equivalent circuit for an MOS capacitor in depletion containing interface trapped charge.[58] Statistical fluctuations of surface potential cause time-constant ($R_{it}C_{it}$) dispersion represented by an infinite number of series RC networks connected in parallel. Traps are assumed to be in equilibrium with the measurement signal.

Interface trapped charge density is usually expressed in terms of number per unit area and energy in the bandgap, symbolized here by D_{it} (cm^{-2}eV^{-1}).[37] Note that N_{it} is the integral of D_{it} across the bandgap.

Cheng[63] presented a comprehensive survey of the characteristics and process dependence of interface trapped charge.

Process Dependence

The effect of some processing variables on D_{it} was shown qualitatively in Figure 14-35.[54] Typical distributions of D_{it} for different post-oxidation anneals are shown in Figure 14-46. The dependences of N_f and D_{it} on Si orientation, oxidation temperature, and ambient, cooling ambient, and hydrogen anneal are depicted in Figures 14-47 through 14-49.

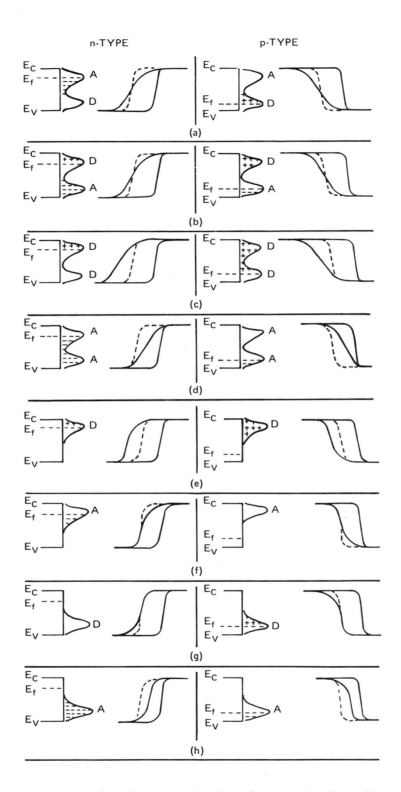

Figure 14-44 —Hypothetical Interface-State Distributions for p- and n-Type Silicon and Resulting High-Frequency C-V Curves of MOS Capacitors. In Each Case, the Right-Hand Continuous Curve is the Preirradiation Curve, the Parallel Shifted Dashed Curve is Postirradiation Considering Only Oxide Charge, and the Left-Hand Continuous Curve is Postirradiation With Oxide Charge and Interface States. States Labeled A are Acceptors and States Labeled D are Donors.[62]

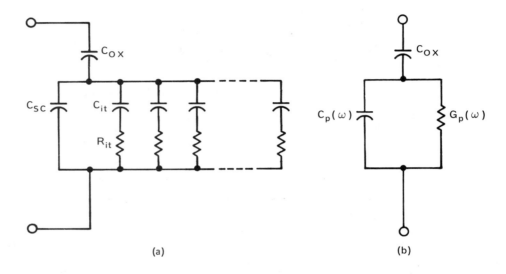

Figure 14-45 — (a) Equivalent Circuit for Depletion Region Showing Time Constant Dispersion Caused Primarily by Statistical Fluctuations of Surface Potential. (b) Simplified Version of (a). $C_p(\omega)$ is the Capacitance per Unit Area at a Given Bias and Frequency of the Distributed Network in Parallel With C_{SC}. $G_P(\omega)$ is the Equivalent Parallel Conductance per Unit Area.

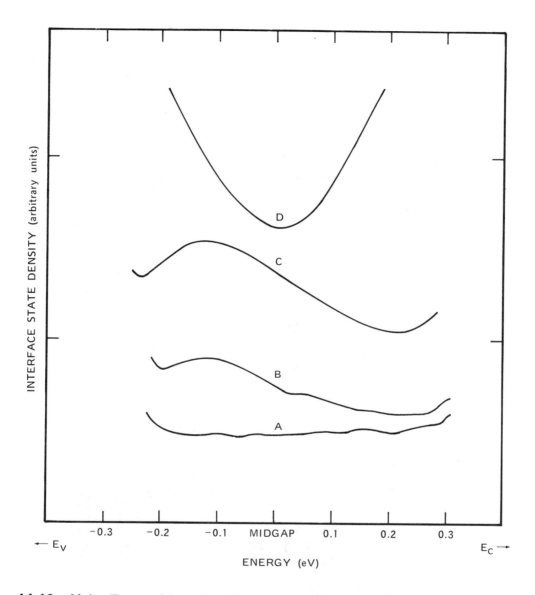

Figure 14-46 —Major Types of Interface State Density Distributions Obtained by Different Processing Methods for n-Type (100) and (111) Silicon Oxidized in Dry O_2, H_2O, and HCl. (A) Sample With Near Optimum H_2 Anneal, (B) Sample With Nonoptimum H_2 Anneal, (C) Unannealed Sample Pulled in Dry O_2, and (D) Unannealed Sample Pulled in N_2 [54] (Reprinted by permission of the Electrochemical Society, Inc.)

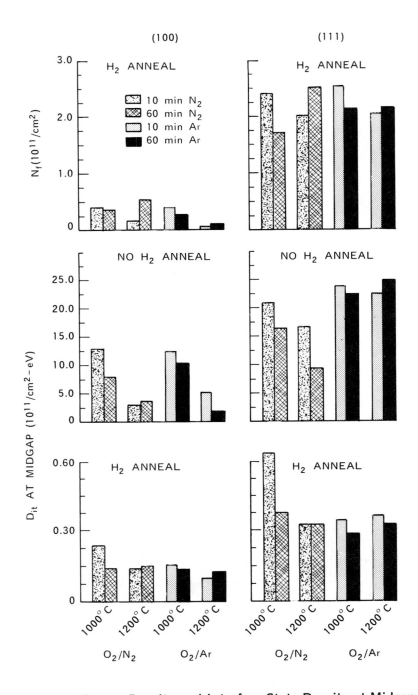

Figure 14-47 — Oxide Fixed Charge Density and Interface State Density at Midgap for n-Type and p-Type, (100) and (111), 4-9 $\Omega \bullet$CM Silicon Samples, Oxidized in Dry O_2 at 1000°C or 1200°C and Annealed at Temperature in Nitrogen or Argon. (N- and p-Type Data Have Been Averaged.) H_2 Anneal: 500°C, 10 Minutes, 10 Percent H_2 in N_2.[54] (Reprinted by permission of the Electrochemical Society, Inc.)

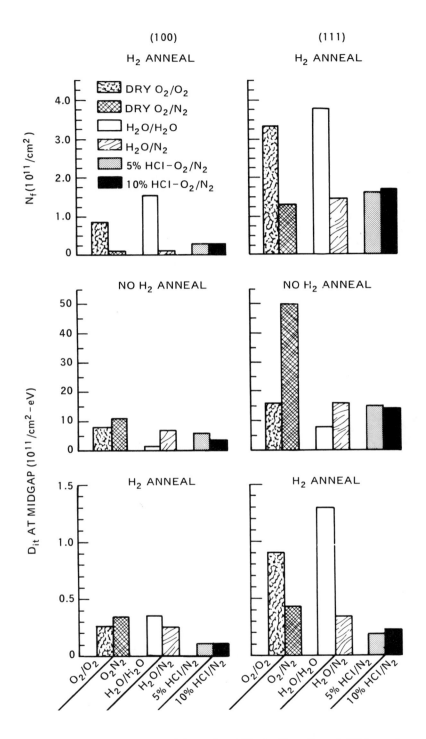

Figure 14-48 —Oxide Fixed Charge and Interface State Density at Midgap for n- and p-Type Silicon, (100) and (111), 4-9 $\Omega\bullet$CM, Oxidized at 1000°C in Various Ambients and Cooled Either in the Oxidizing Ambient or in N_2. (N- and p- Type Data Have Been Averaged.)[54] (Reprinted by permission of the Electrochemical Society, Inc.)

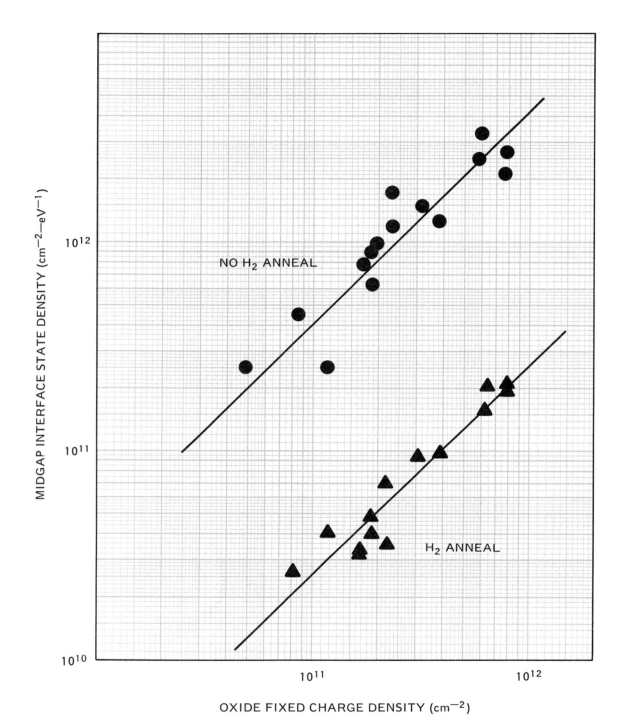

Figure 14-49 —Midgap Interface State Density Versus Oxide Fixed Charge
for Unannealed and H$_2$-Annealed n-Type and p-Type Silicon
Samples, (100) (111) Orientation, Oxidized in dry O$_2$ at
1000° and 1200°C. (Hydrogen Anneal: 2 Liters/Min, 10%
H$_2$ in N$_2$, 10 Min, 500°C)[54] (Reprinted by permission of the
Electrochemical Society, Inc.)

The "slow trapping" instability[38] [53] [59] can result in large increases in D_{it} and Q_f if negative fields are applied to MOS structures at elevated temperatures. Sinha and Smith[59] observed that for Al-gate devices on n-Si, $\Delta V_{th} > \Delta V_{FB}$, whereas for p-Si, $\Delta V_{th} < \Delta V_{FB}$. This behavior is due to generation of an interface trapped charge with a distribution peaking near midgap. A semiempirical expression for slow trapping flatband shift prior to saturation is[59]

$$\Delta V_{FB} \approx Y |V_B|^n t^{0.2} \exp\left(\frac{-E_A}{kT}\right) \quad (V), \tag{14.60}$$

where

$\qquad V_B$ = applied negative bias (10 to 40V for 1000Å dry oxide)

$\qquad t$ = time (10 to 1000 min)

$\qquad T$ = 373 to 573K

$\qquad n$ = 4.76 to 5.3 x 10^{-3} T (dimensionless for T in K)

$\qquad Y \sim 1.6$ x 10^3 (volts^{-n}min^{-2})

$\qquad E_A$ = thermal activation energy ~ 0.6 eV.

Measurement Techniques

A. Quasi-Static Method[64] [65] [66]

This technique measures the density of interface trapped charge that contributes to low-frequency (quasi-static) capacitance but not to high-frequency capacitance (~ 1 MHz). (See (3) below.)

    (1) Apparatus: The equipment shown in Figure 14-32 (less the hot chuck) is sufficient for the low-frequency measurements. A suitable high-frequency C-V set (Figure 14-30) is also required.

    (2) Procedure

        (a) Record the displacement current versus bias (low-frequency curve) at a low sweep rate, typically $\leqslant 100$ mV/s for both increasing and decreasing ramp voltages (current should be $< 10^{-8}$ A/cm^2).

        (b) Record the "stray capacitance" current in both ramp directions with the probe lifted slightly above the field plate.

        (c) Plot the high-frequency curve, superimposing the capacitance in strong accumulation (C_I) and the zero on the corresponding quasi-static plot.

        (d) The resulting curves should resemble those in Figure 14-50. For long lifetime material, the quasi-static curve capacitance may not recover to the C_I value in inversion.[67]

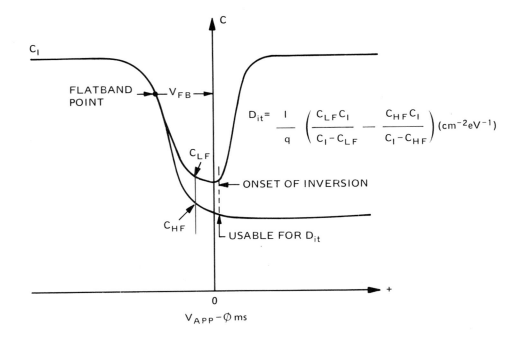

Figure 14-50 — Quasi-Static and High-Frequency C-V Curves Used to Determine Fast-State Density in the Depletion Region[65]

(3) <u>Calculation</u>: At any value of surface potential (ϕ_S), the interface trap density is given by[68]

$$D_{it}(\phi_S) = \frac{C_I}{q} \left(\frac{C_{LF}}{C_I - C_{LF}} - \frac{C_{HF}}{C_I - C_{HF}} \right) \quad (cm^{-2} \, eV^{-1}),$$

14.61

where all capacitances are measured in farads/cm^2. Equation 14.61 is valid only for ϕ_S from inversion threshold ($\phi_S = 2\phi_F$) to a value where the surface state time constant becomes equal to the ac signal period.[66] At 500 kHz, this potential is $\sim$250 mV from the majority carrier bandedge. At higher frequencies, the data are valid closer to the bandedge. (See Figure 14-41.)

The surface potential as a function of V_G is obtained by integrating the low frequency curve[68]

$$\phi_S(V_G) = \int_{V_{FB}}^{V_G} \left[1 - \left(\frac{C_{LF}}{C_I} \right) \right] dV_G + \Delta \quad (V).$$

14.62

The additive constant Δ can be obtained by comparing C_{LF} versus ϕ_S with an ideal C_{LF} versus ϕ_S curve. The voltage offset between the two curves is equal to Δ. The data should be considered unreliable if the integral in Equation 14.62, evaluated from strong accumulation to strong inversion, is larger than the bandgap of the semiconductor.[8] [65]

(4) <u>Resolution</u>: The resolution of $D_{it} \approx$ low 10^{10} cm^{-2} eV^{-1} range.

(5) <u>Quick Graphical Approach</u>: See Reference[69].

B. Conductance Method

The frequency-dependent conductance technique, described in detail by Nicollian and Goetzberger,[58] yields accurate information about D_{it} with improved resolution ($\sim 10^9$ cm^2 eV^{-1}) as compared to the quasi-static method. The measurements are rather tedious. Hill and Colemen[70] described a less elaborate approximation technique that combines g-V_G and C-V_G data at one frequency to calculate a minimum value of D_{it} between midgap and flatband.

C. Temperature Variable C-V_G Method

See Reference[71]

D. Transient Capacitance Spectroscopy

See Reference[72]. MOSFET structures have been used to measure D_{it} with high sensitivity throughout the bandgap. Capacitors can be used but with more limited range of ϕ_S.

Johnson et al[73] used this method to measure D_{it} and capture cross-sections ($\sim 10^{-18}$ cm^2) of MOS structures.

Oxide Trapped Charge: Q_{ot}, N_{ot}

Characteristics

Some of the characteristics of oxide trapped charge are shown below.

(a) Is positive (trapped hole) or negative (trapped electron) charge located in the oxide, often near the Si-SiO$_2$ interface.[38] Hole traps may be trivalent silicon donors or excess oxygen centers.[56]

(b) Can be caused by ionizing radiation such as implanted ions,[74] X-rays,[3] [75] electron beams,[56] neutrons, and gamma rays.[76]

(c) Magnitude is a function of radiation dose and energy and the field across oxide during irradiation.[38] Saturation of induced charge occurs with increasing dose.[77]

(d) Radiation probably induces electron-hole generation in the oxide and subsequent hole trapping at sites where Si-O bonds are broken.

(e) Trapping and detrapping may occur during C-V_G measurements leading to hysteresis at room temperature. The magnitude of the hysteresis (ΔV_{FB}) is proportional to the maximum fields applied.[78]

(f) Trapped positive charge can be nearly eliminated by low-temperature ($\geq 300°C$) anneals in inert ambients.[38]

(g) Annealing is hindered by dense dielectric layers such as Si$_3$N$_4$ over the SiO$_2$.[38]

(h) Low-temperature anneals ($< 600°C$) do not remove trap sites in oxides, but cause neutralization or compensation of trapped charge. These trap sites can be refilled with holes or electrons.[77] Higher temperatures are required to remove trap sites.

(i) Radiation can induce electron traps (capture cross-sections $\sim 10^{-15} cm^{-2}$)[75] that are neutral until filled by electrons injected into the oxide (hot electron trapping).

(j) For different conditions, RF plasmas can either generate or anneal out electron traps.[79]

(k) Capture cross-sections for X-ray induced electrons traps are 10^{-14} to 10^{-15} cm^2 at low fields (Figure 14-51).[75] [79] The low-field value for positively charged traps is larger, $\sim 3 \times 10^{-13}$ cm^2.[75]

Q_{ot} resembles fixed charge in that its magnitude is not a function of Si surface potential, and there is no capacitance associated with it. However, in this case, charges may exist throughout the oxide. During a C-V_G measurement, electrons or holes may be ejected from the Si to fill oxide traps after sufficient positive or negative bias is applied.[78] This will cause hysteresis in the C-V_G curve. The introduction of Q_{ot} into an oxide charge-free system will not distort the C-V_G curve but will cause shifts on the voltage axis (i.e., third term in Equation 14.37). However, phenomena which generate Q_{ot}, such as radiation, usually also cause increases in interface trapped charge and its associated C-V_G curve distortion.[56] Electron-gun induced soft X-rays are one of the more common sources of radiation-induced Q_{ot}.[75]

In previously unbiased devices, Q_{ot} is usually positive. After electron-hole pairs are generated by the ionizing radiation, the electrons can move rapidly to the metal-oxide or Si-SiO$_2$ interfaces; electron mobility in SiO$_2$ $\approx$ 30 cm^2/V•sec.[80] Hole mobility is lower by a factor of 10^{-5} due to the presence of shallow traps in the SiO$_2$ forbidden gap near the top of the SiO$_2$ valence band. Some of the positively charged holes remain bound at deep trap sites near the interface in the oxide, thereby shifting flatband voltage. The low hole mobilities are characteristic of small polarons.[81]

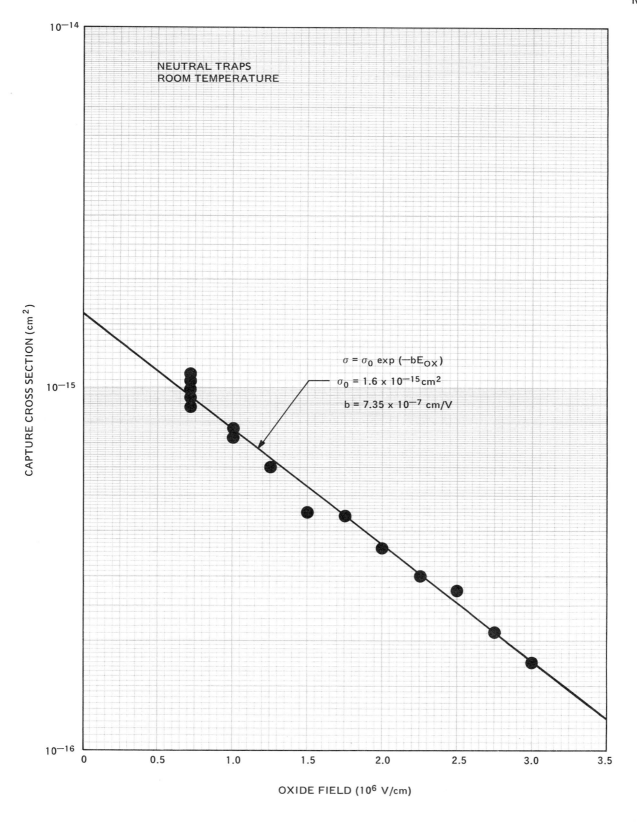

Figure 14-51 —Dependence of the Capture Cross Section of the Neutral Centers on the Average Oxide Field. The Solid Line Represents an Empirical Fit.[75]

Process Dependence

Q_{ot} is induced by exposure to beams of ions or ionizing radiation during and after device processing. The charge, but not necessarily the trap sites induced during processing, can usually be annealed out at $300°$ to $500°C$ in inert or reducing ambients.[38] [77]   The neutral trap centers remaining after annealing can become charged again (e.g., by hot-electron or hot-hole injection) during device operation, thereby causing threshold voltage shifts. Also, devices may become exposed to radiation during operation with the same effect on stability. So called "radiation-hard" oxides are less sensitive to these effects.

Radiation sensitivity is improved by

(a) Minimizing OH^- ion concentration, e.g., using dry oxides or following steam oxidation with high temperature anneals in neutral ambients.[56] [82] [83] [84]

(b) Using processes which also minimize fixed charge.[84] [85]   (See Section 14.3, Fixed Oxide Charge.)

(c) Using alkali-ion free process conditions, e.g., HCl-gettered furnace tubes.   (See Section 14.4).[83]

Incorporation of Cl in oxides seems to be detrimental to radiation hardness.[83]   (See Section 14.4).

Figures 14-52 and 14-53 show the dependence of radiation-induced threshold voltage shift on gate bias during irradiation for Al-gate p-channel MOSFETs. The total voltage shift is the sum of the shifts due to induced space charge (ΔV_{ot}) and interface traps (ΔV_{it}). Poly-Si gates inhibit the formation of interface traps but not oxide space charge.[84]

DiMaria's review[80] is recommended for detailed information on electron and hole trapping including observed trap cross sections, types of traps, and their origins.

Measurement Techniques

A. Capacitance-Voltage Method

This technique involves measurement of flatband voltage, as

$$\Delta N_{FB} = N_{ot} + (\Delta N_{it})_{FB} = \frac{C_I \Delta V_{FB}}{q},$$

14.63

where

ΔN_{FB} = total change in flatband charge (cm^{-2})

$(\Delta N_{it})_{FB}$ = change in interface trapped charge density at $\phi_S = 0$ (cm^{-2})

C_I = insulator capacitance per unit area.

The term ΔN_{FB} is the difference in flatband charge before and after irradiation or that between an irradiated device and a control. To obtain N_{ot}, the magnitude of $(\Delta N_{it})_{FB}$ must be determined by an independent measurement (See Section 14.3, Interface Trapped Charge.)   Resolution of N_{ot} is in the low $10^{10} cm^{-2}$ range. MOSFET threshold voltage shifts can also be used, substituting the subscripts "th" for "FB" in Equation 14.63.

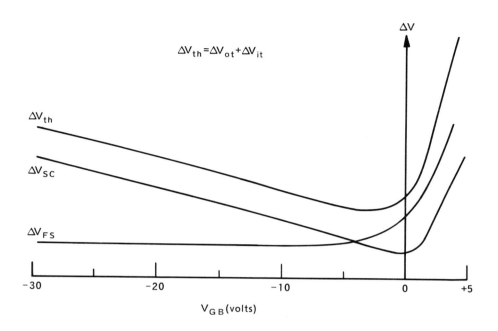

Figure 14-52 —Typical Dependence of Radiation-Induced Threshold Shift on Gate Bias During Irradiation, Showing the Individual Contributions From Space Charge and Fast Interface States[84] © 1971 IEEE.

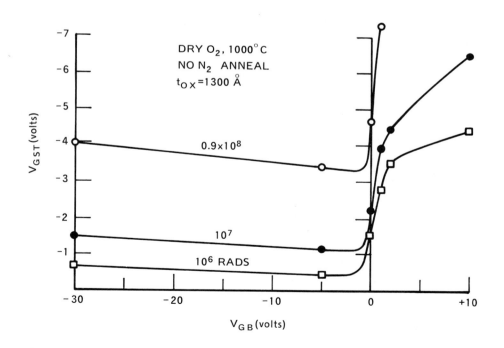

Figure 14-53 —Threshold Shifts Versus Gate Bias During Irradiation for Dry Oxide Transistors After Various Doses of Co^{60} Radiation[84] © 1971 IEEE.

B. Electron Avalanche Injection and Photo I-V Methods[74]

This approach yields information on neutral trap density. Electrons (or holes) are injected from p-type (or n-type) substrates to charge the traps. C-V_G curve shifts, as in Equation 14.63, can be used to calculate the density of the trapped charge. Also, photo-IV measurements[74] are often done at the same time in order to yield trap distribution centroids. Optically induced hot-electron injection in MOSFETs has also been employed.[86]

The change in flatband voltage resulting from a constant current density, J, passing through an oxide containing traps of cross section σ is[87]

$$\Delta V_{FB}(t) = q\, N_{eff}\, (t_{ox}/\epsilon_{ox})\, [1 - \exp(-\sigma J t/q)] \quad (V), \qquad 14.64$$

where

$$t_{ox} = \text{oxide thickness} \quad (cm)$$

$$t = \text{time} \quad (s)$$

$$N_{eff} = \text{effective trap density} \equiv (\overline{x}/t_{ox})\, N_T \quad (cm^{-2})$$

$$\overline{x} = \text{centroid of a trap distribution } N(x) \quad (cm)$$

$$N_T = \int_0^{t_{ox}} N(x)\, dx \quad (cm^{-2}). \qquad 14.65$$

The trap time constant can be defined as

$$\tau = q/(J\sigma) \quad (s). \qquad 14.66$$

Some measurements have shown that ΔV_{FB} from avalanche injection is due to changes in oxide charges other than interface trapped charge.[87] However, other work showed significant increases in N_{it} due to photoinjection of electrons[88] and avalanche injection.[61]

14.4 Oxides Grown in Cl-Containing Ambients

For recent comprehensive reviews, References [89] and [90] are recommended.

Oxidation Rates and Gas Thermodynamics

Figures 14-54 through 14-57 provide data on the increases in oxidation rate when Cl-containing gases are added to dry O_2. The accelerated growth rate may be due to:[94]

    (a) Enhanced diffusion of oxygen in the oxide.

    (b) Enhanced reaction rate at the interface due to catalytic nature of the Cl.

    (c) Contribution of H_2O formed as a byproduct in the gas.

The oxidation rate in steam or wet O_2 ambients is not enhanced by Cl additions.[95]

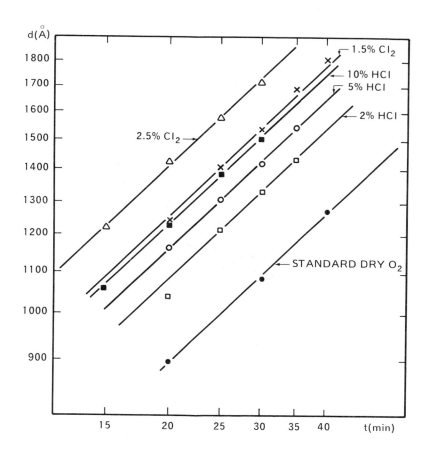

Figure 14-54 —Oxidation Rate of (100) Si in the Presence of HCl and Cl_2 [91]
(Reprinted by permission of the Electrochemical Society, Inc.)

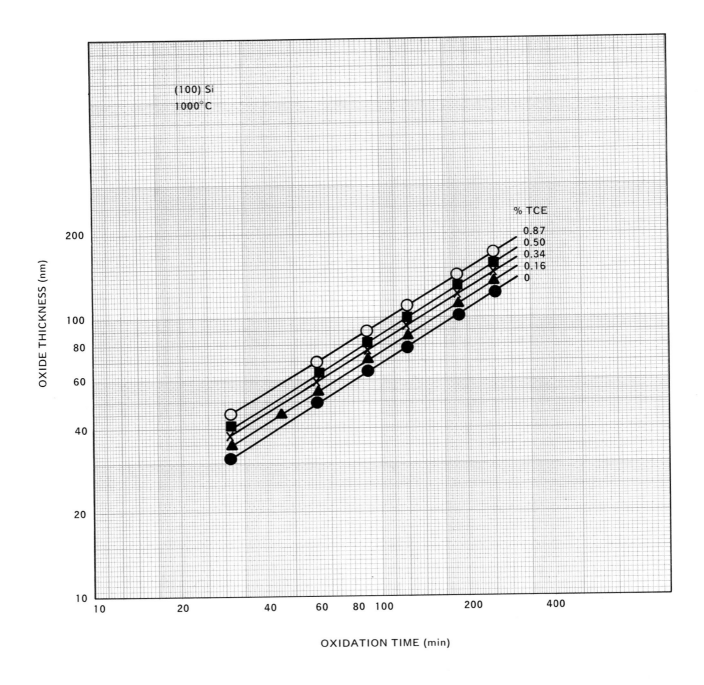

Figure 14-55 —Oxide Thickness Versus Oxidation Time for the Oxidation of Silicon in Dry Oxygen and in Various TCE/O$_2$ Mixtures [n-Type Si, (100), 7-13 Ω cm, T, 1000°C].[92] (Reprinted by permission of the Electrochemical Society, Inc.)

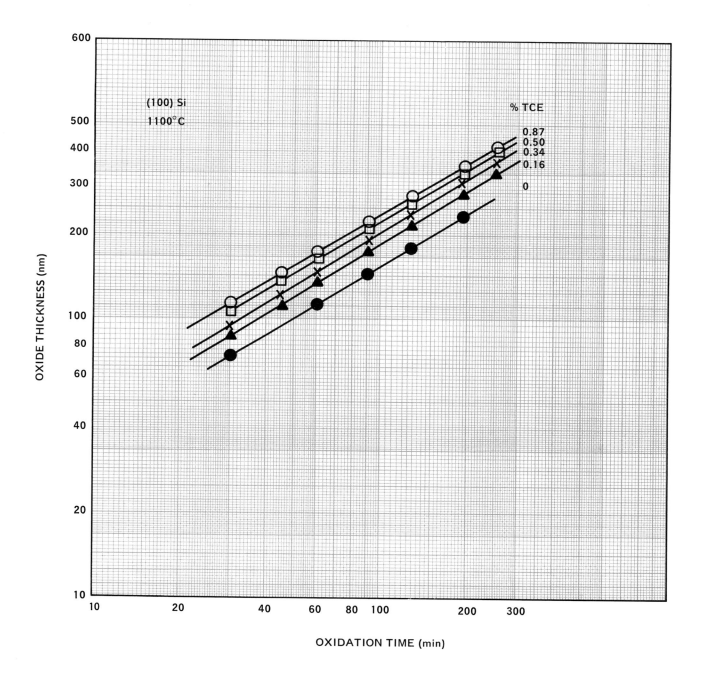

Figure 14-56 — Oxide Thickness Versus Oxidation Time for the Oxidation of Silicon in Dry
Oxygen and in Various TCE/O$_2$ Mixtures (n-Type Si, (100), 7-13 Ω cm,
T, 1100°C) [92] (Reprinted by permission of the Electrochemical Society, Inc.)

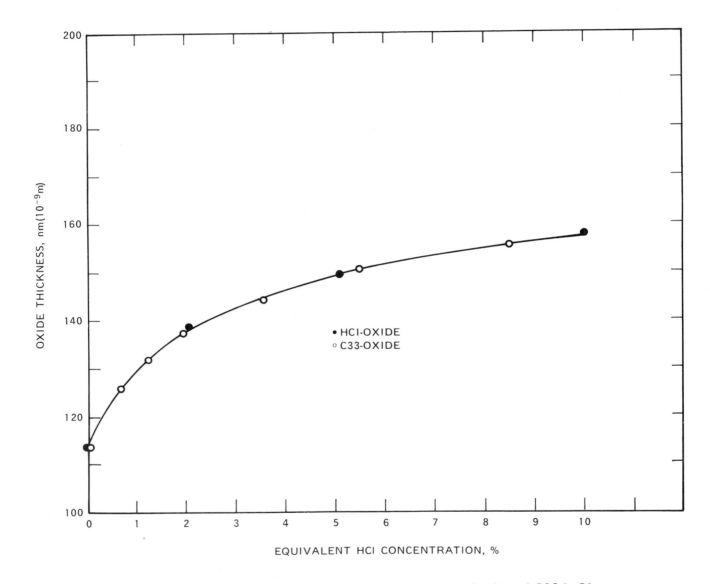

Figure 14-57 —The Increase in Oxide Thickness by the Use of HCl and C33 is Given as a
Function of Effective HCl Concentration (C33 Times 3), Oxides Grown
at 1150°C, 34 min on n-Type (100) Material[91] [93] (Reprinted by permission
of the Electrochemical Society, Inc.)

Figures 14-58 through 14-60 show the equilibrium concentrations of HCl, H_2O, and Cl_2 at high temperature relative to input additive concentration for O_2/TCE (trichloroethylene), O_2/C33 (trichloroethane), and O_2/HCl ambients. Carbon dioxide byproducts in O_2/C33 and O_2/TCE ambients do not seem to have any detrimental effect.* The oxide growth rate enhancement correlates well only with the equilibrium partial pressure of Cl_2 at the oxidation temperature.[90]

Sodium Passivation and Chlorine Content

Figures 14-61 and 14-62 depict the effect of O_2/HCl oxidations on mobile ionic sodium content in the oxide. The passivation effect is not observed for oxidation temperatures below 1050°C due to reduced levels of Cl incorporation. Trapping and neutralization occur only after the ions have migrated to the Si-SiO_2 interface.[95] The incorporated chlorine is located within 300Å of the interface.[95] [97] Passivation (fraction of mobile ions trapped) is directly proportional to the Cl content in the oxide for a given oxidation time and temperature and is independent of the total sodium level.[90] [96]

Figure 14-63 shows the chlorine content in oxides versus HCl partial pressure (P_{HCl}) for various temperatures (O_2/HCl oxidation).[98] To convert from P_{HCl} to volume fraction in the room temperature gas mixture in Figures 14-61 through 14-63, divide P_{HCl} by 0.678, 0.698, 0.700, 0.715, and 0.749 for 1100°, 1125°, 1150°, 1200°, and 1300°, respectively. Figure 14-64 shows passivation data for TCE and C33 oxidations. The rapidly increasing portions of the curves in Figures 14-61 through 14-64 correspond to the formation of a second phase (chlorosiloxane) which leads to a rough oxide surface. The same regions correspond to the Cl ambients required for complete passivation.[90] [99] At high temperatures and HCl flows (e.g., 1200°C, 10 vol % HCl, 6 hr), gas bubbles form due to reaction of the Cl with the Si and the oxide may lift from the surface.[100] High temperature (1150° to 1200°C) neutral ambient anneals after oxidation decrease the passivation effect.[97] [101] Passivation of previously grown oxides with subsequent HCl treatment occurs only when O_2 is also present.[90] Complete neutralization of sodium ions is possible only for concentrations up to $\sim 10^{12} cm^{-2}$.[89]

---

* Concentrations of TCE or C33 in O_2 should be kept at less than 10 vol % in order to avoid formation of the toxic gas phosgene ($COCl_2$).[93]

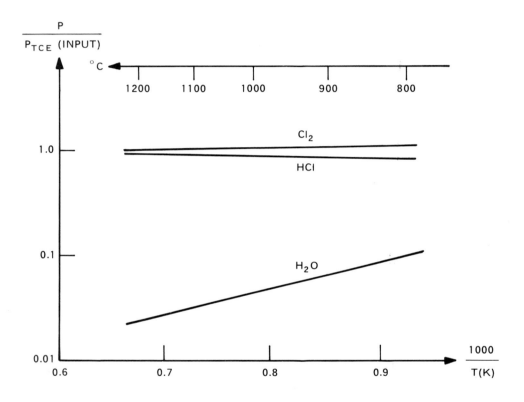

Figure 14-58 —Equilibrium Concentrations of HCl, H_2O, and Cl_2 Relative to Input
Additive Concentration at Various Temperatures for O_2/TCE[92]
(Reprinted by permission of the Electrochemical Society, Inc.)

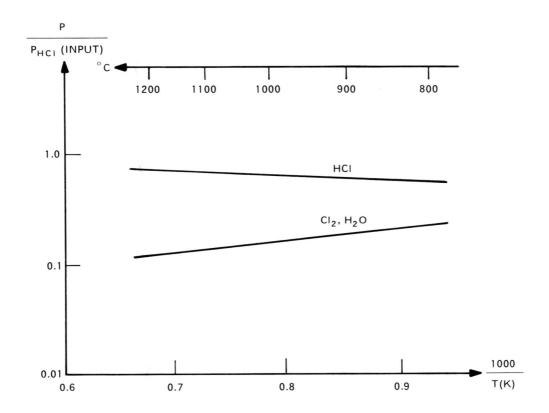

Figure 14-59 —Equilibrium Concentrations of HCl, H_2O, and Cl_2 Relative to Input
Additive Concentration at Various Temperatures for O_2/HCl[92]
(Reprinted by permission of the Electrochemical Society, Inc.)

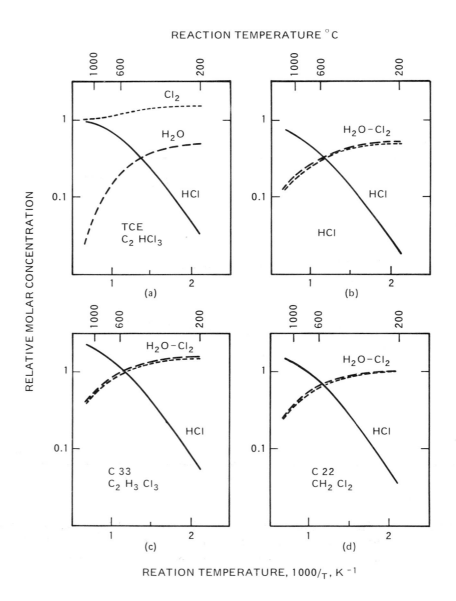

Figure 14-60 —The Concentrations of the Reaction Products HCl, H_2O, and Cl_2, Relative to the Additive Concentration Are Given as a Function of the Reaction Temperature: (a) the Pyrolysis of TCE (C_2HCl_3): (b) the Pyrolysis of HCl, (c) the Pyrolysis of C33 ($C_2H_3Cl_3$); (d) the Pyrolysis of C22 (CH_2Cl_2)[93] (Reprinted by permission of the Electrochemical Society, Inc.)

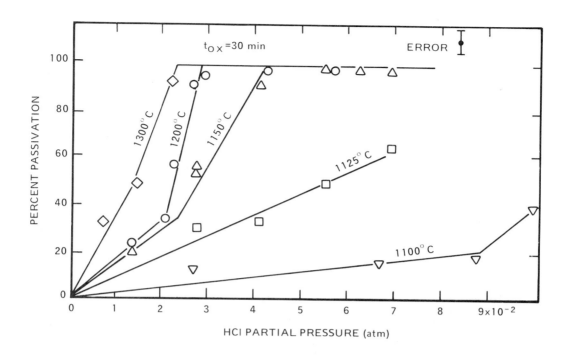

Figure 14-61 — Passivation P Versus Growth Ambient HCl Pressure P_{HCl}. All Oxides Were Grown for 30 Min. Different Growth Temperatures Are Indicated.[96]
(Reprinted by permission of the Electrochemical Society, Inc.)

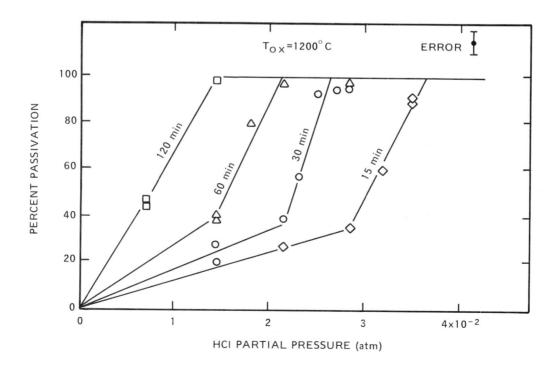

Figure 14-62 — Passivation P Versus Growth Ambient HCl Partial Pressure P_{HCl}. All Oxides Were Grown at 1200°C. Different Growth Times Are Indicated. To Convert P_{HCl} to Volume Fraction in the Room Temperature Gas Mixture, Divide by 0.721.[96]
(Reprinted by permission of the Electrochemical Society, Inc.)

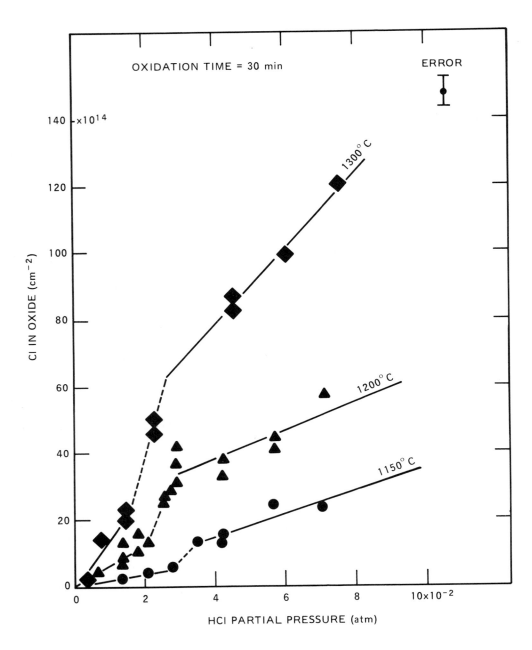

Figure 14-63 —Cl Content of SiO$_2$ Films as a Function of HCl Partial Pressure for 30 Min
Oxidation Time at 1150°, 1200°, and 1300°C. To Obtain the Volume
Fraction of HCl in O$_2$ (in the Original Gas Mixture), Divide the Value of
P$_{HCl}$ by: 0.700 for 1150°C, 0.715 for 1200°C, and 0.749 for 1300°C.[98]
(Reprinted by permission of the Electrochemical Society, Inc.)

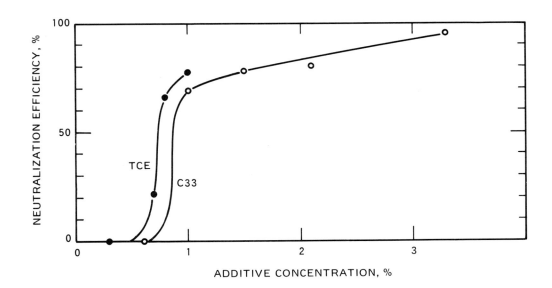

Figure 14-64 —The Neutralization Efficiency Calculated From Flatband Shift After BT Stress
Compared to the Flatband Shift With 0% Additive as a Function of Additive
Concentration. The Oxides Are Grown at 1150°C for 32 and 36 Min for TCE
and C33, Respectively. The TCE Oxide Contamination Level Is 2×10^{12} and
for the C33 Oxides Is 4.5×10^{11} Ion/cm^2.[93] (Reprinted by permission of the
Electrochemical Society, Inc.)

Gettering of Sodium

Ambients of HCl can remove Na from oxidation ambients and furnaces by causing the formation of
volatile chlorides.[44] [101] [102] Purging oxidation tubes for 20 hours at 1000°C in 6 percent HCl/N$_2$ is
required to deplete a typical quartz oxidation tube of sodium.[44] Frequent purges may be necessary if
wafers, boats, or furnace liners are contaminated. Additions of HCl to steam (900° to 1100°C) has re-
sulted in mobile ion densities of $<2 \times 10^{10}$cm^{-2}.[102]

Gettering of Heavy Metals: Lifetime Enhancement

Oxidation in Cl-containing ambients at >950°C can enhance minority carrier lifetimes in Si by 2 to 3
orders of magnitude.[93] [103] [107] Possible mechanisms include formation of volatile metal chlorides
allowing metal ion removal from both wafers and furnace tubes.[103] [105] Katz et al[105] showed that
thick (>20 nm) oxide formation on wafers should be avoided for maximum benefit. A typical 1000°C,
O$_2$/1 percent HCl oxidation does not getter wafers of heavy metals but does prevent further contamination
from the furnace.[102]

The lifetimes obtained increase with increasing temperature (850° to 1150°C) and additive concentration
(1 to 6 percent HCl) during oxidation.[106]

Stacking Faults

Additions of HCl,[108] TCE,[109] and C33[110] to dry O$_2$ tend to suppress formation of, or shrink, pre-
existing oxidation-induced stacking faults (OISF). Stacking fault nuclei can also be eliminated, resulting
in OISF-free wafers after subsequent oxidations. (See Figures 14-65 and 14-66) The effects on existing
OISF are also observed for wet O$_2$/TCE oxidation.[109]

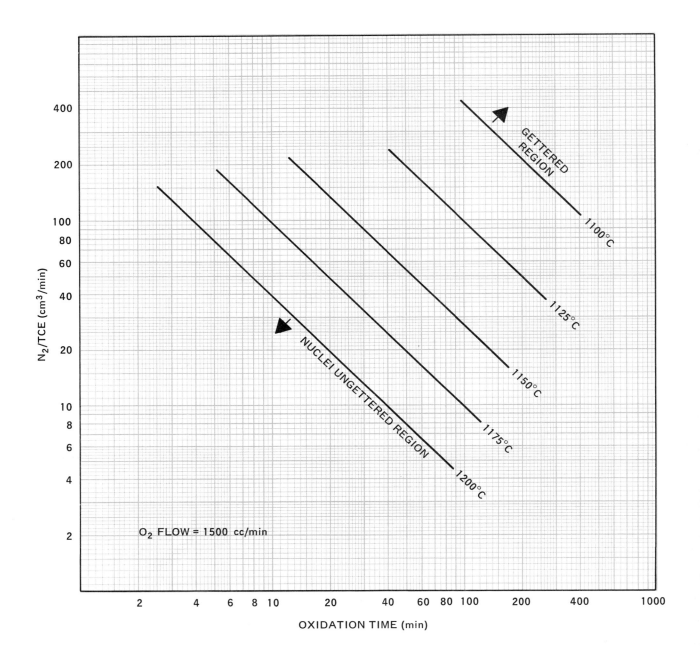

Figure 14-65 — Boundaries Between Two Regions Whereby Stacking-Fault Nuclei Are
Gettered and Ungettered for TCE Oxidation Temperatures Ranging
from 1100°C to 1200°C[109] (Reprinted with permission from Solid State
Technology, Technical Publishing, A Company of Dun & Bradstreet.)

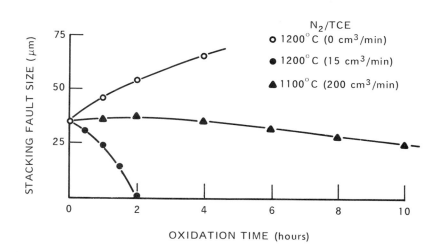

$O_2 = 1500$ cm^3/min

N$_2$/TCE
o 1200°C (0 cm^3/min)
● 1200°C (15 cm^3/min)
▲ 1100°C (200 cm^3/min)

Figure 14-66 — Shrinkage (or Expansion) of Already-Existing Stacking Faults During Additional Oxidation in the Presence (or Absence) of TCE[109]
(Reprinted with permission from Solid State Technology, Technical Publishing, A Company of Dun & Bradstreet.)

Oxide Charges

Fixed Charge

Oxidations O_2/HCl and O_2/Cl have little effect on fixed charge.[89] [91] But data for O_2/TCE at 1050°C suggests a factor-of-five reduction in Q_f.[111] The reduction in Q_f due to TCE oxidation occurs only if a high-temperature N_2 anneal is not done after oxidation.[93] If N_2 annealing is done, oxidations in O_2/TCE or C33 yield the same low Q_f as dry O_2 only (See Figure 14-67).

Interface Trapped Charge

D_{it} is typically reduced by a factor of 5 to 10 if chlorine-containing gases are added to dry O_2.[89] [93] [101] [107] $D_{it} \sim$ low 10^9 cm^{-2} eV^{-1} has been reported.[107] However, D_{it} increases for TCE or C33 additions greater than 0.6 and 1.0 vol %, respectively.[93]

Oxide Trapped Charge

Purging furnaces with HCl or oxidations in steam or dry O_2 that result in low Cl-content oxides seems to improve radiation hardness.[44] [83] [89] But oxides with high Cl concentrations show larger Q_{ot} compared to dry O_2 oxides, possibly due to chlorine ion migration.[105]

Negative Bias Instability

Cl-containing oxides are susceptible to this instability at high negative fields ($>5 \times 10^6$ V/cm) even at room temperature.[89] [97] [111] [112] [113] Negative ΔV_{FB} and C-V_G curve distortion increase with chlorine content. Thus, more stable oxides are grown at lower temperatures since less Cl is incorporated. O_2/Cl$_2$-grown oxides are less stable than either O_2/HCl[97] or O_2/TCE[113] oxides. At normal device operating fields ($\lesssim 2 \times 10^6$ V/cm), HCl and TCE oxides are as stable as dry O_2-only oxides. Negative bias instability is observed at lower fields as stressing temperature increases (Figure 14-68).[112]

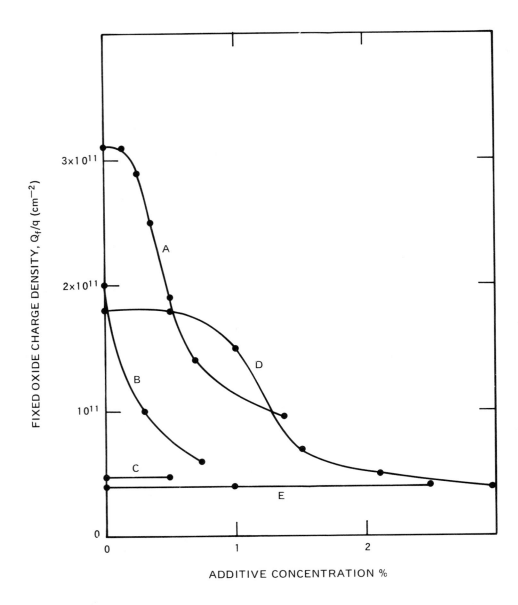

Figure 14-67 — The Reduction in Fixed Oxide Charge Density (N_{OX}) With Increasing
Chlorine Addition. A, (TCE) 1200°C, 14 Min Oxidation, no Anneal;
B, (TCE) 1200°C, 14 Min Oxidation, 30 Min 450°C Wet N_2 Anneal; C,
(TCE) 1150°C, 32 Min Oxidation, 20 Min N_2 in Situ, 450°C, 30 Min Wet
N_2 Anneal; D, C33, 1150°C, 36 Min Oxidation, 5 Min N_2 in Situ, 450°C,
30 Min Wet N_2 Anneal; E, Same as D but 20 Min N_2 in Situ Anneal. Sample
A Was Measured by Means of an Hg Probe; B, C, D, and E, After Metallization.[93]

(Reprinted by permission of the Electrochemical Society, Inc.)

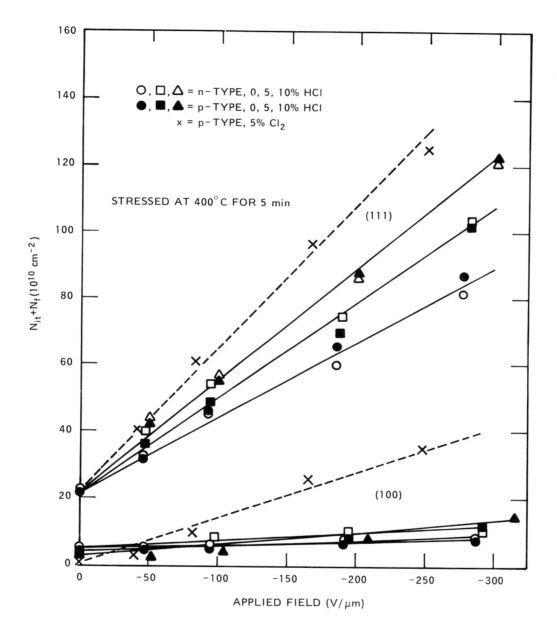

Figure 14-68 — Saturation Values of Effective Oxide Charge ($N_f + N_{it}$) as a Function of Applied Negative Fields for Oxides Thermally Grown on (100) and (111) Oriented Silicon at 1100°C in Various Chlorine-Containing Dry Oxygen Atmospheres. Solid Lines Are for Oxides Grown in HCl/O_2 Atmospheres, While the Dotted Lines Are for Oxides Grown in Cl_2/O_2 Atmospheres. The ($N_f + N_{it}$) Value at Zero Applied Field Is the Value Prior to Negative Bias Stressing.[112] (Reprinted by permission of the Electrochemical Society, Inc.)

Oxide Dielectric Breakdown

Oxides grown in HCl, Cl_2, or TCE ambients contain fewer of the types of defects that lead to premature breakdown, compared to dry O_2 oxides.[114] (See Figure 14-69). Oxide wearout times under accelerated stress conditions are also enhanced.

14.5 MOSFET

Structure and Nomenclature

The basic n-channel MOSFET (metal-oxide-semiconductor field effect transistor) structure is shown in Figure 14-70, with symbols and terms defined in subsection 14.6. Other commonly used MOSFET acronyms are defined in the glossary of Section 9. See page 9-25 for additional MOSFET cross sections.

In present day technology, the insulator is made of SiO_2 and the gate electrode is made of polycrystalline silicon, doped n-type. The conductance of a MOS transistor is modulated by applying a voltage (V_G) to the gate electrode. The presence of the gate voltage causes an inversion region (channel) to be formed under the insulator, thus allowing carriers to travel from the source to the drain.

First-Order Current-Voltage Relationships

A MOSFET has three basic modes of operation that depend upon bias conditions: subthreshold, linear, and saturation. The subthreshold region is characterized by a weakly inverted channel ($\phi_S < 2\phi_F$). In this region, the drain current (I_D) varies nearly exponentially with V_G. See References [115] and [116] for a discussion of subthreshold characteristics.

Figure 14-71 shows operation in the linear or triode region where the channel extends the entire distance between the source and drain diffusions. In this region, the current increases linearly with drain voltage.

Figure 14-72(a) shows operation in the saturation region. In saturation, the channel becomes "pinched-off" by the extension of the drain depletion region. The drain current is approximately constant in this region.

A first-order expression for the drain current of a MOSFET is given by[117]

$$I_D = \beta (V_G - V_{th} - \frac{V_D}{2})\ V_D.$$

14.67

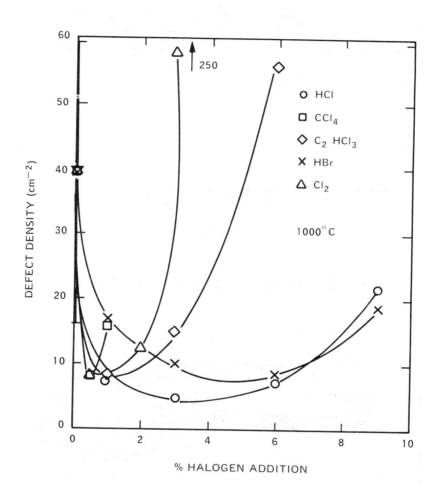

Figure 14-69 —Effect of Halogen Added During Oxidation on the Breakdown Defect Density[114]
(Reprinted by permission of the Electrochemical Society, Inc.)

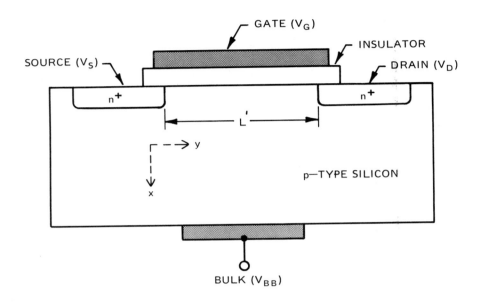

Figure 14-70 — Basic n-Channel MOSFET Structure

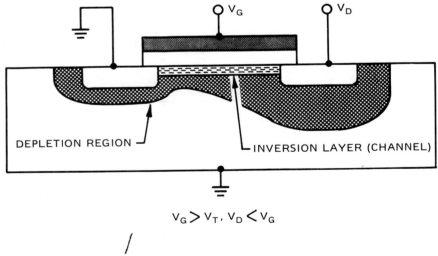

$V_G > V_T$, $V_D < V_G$

Figure 14-71 — Cross Section of MOSFET in Linear Region

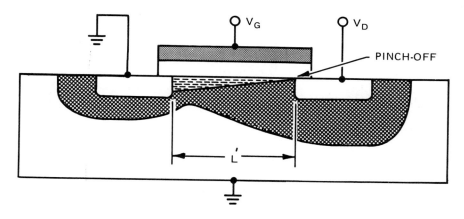

$$V_D = (V_G - V_T),\ V_G > V_T$$

(a) Cross Section of MOSFET at Onset of Saturation

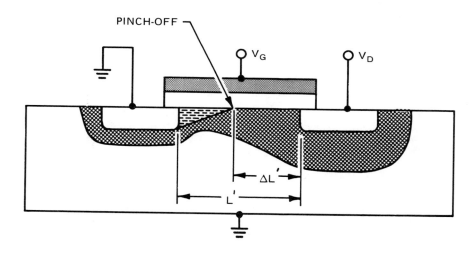

$$V_D > (V_G - V_T),\ V_G > V_T$$

(b) Cross Section of MOSFET in Strong Saturation

Figure 14-72 —Cross Sections of MOSFET in Saturation

The assumptions under which Equation 14.67 was derived are listed below.

(a) Depletion approximation: $W_{d\ max} = \left[\dfrac{2\epsilon_S (V_{BB} + 2\phi_F)}{qN_A}\right]^{1/2}$.

(b) One-dimensional geometry.

(c) Channel is strongly inverted.

(d) Total oxide charge is constant.

(e) Surface mobility, μ_n, is constant.

(f) I_D is drift current.

(g) Channel depletion region is rectangular in shape.

(h) Gradual channel approximation — the electric fields in the direction of current flow are much smaller than the fields in the direction perpendicular to the silicon surface.

In Equation 14.67, β is called the gain of the device and is given by

$$\beta = \mu_n \frac{W}{L'} C_{OX}, \qquad\qquad 14.68$$

and V_{th}, the threshold voltage, is given by

$$V_{th} = V_{FB} + |V_S| + 2\ |\phi_F| + \frac{[2\epsilon_S qN_A (2|\phi_F| + |V_S - V_{BB}|)]^{1/2}}{C_{OX}}. \qquad 14.69$$

For drain voltages, $V_D > V_G - V_{th}$, the MOSFET is in saturation and, for long channel devices, the drain current remains constant or saturates.

Because of this, it is well to define a saturation voltage as

$$V_{D,\ sat} = V_G - V_{th}. \qquad\qquad 14.70$$

By substituting Equation 14.70 in Equation 14.67, the following expression for saturation current is obtained:

$$I_{D,\ sat} = \frac{\beta}{2} (V_G - V_{th})^2. \qquad\qquad 14.71$$

Figure 14-73 shows a plot of I_D versus V_D for a long channel device, indicating $V_{D,\ sat}$ for each gate voltage.

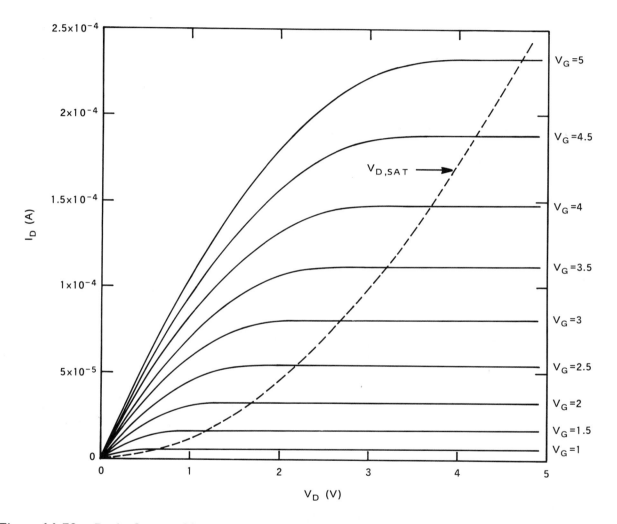

Figure 14-73 —Drain Current Versus Drain Voltage for MOSFET With Channel Length of 200 μm

Equations 14.70 and 14.71 are valid approximations for long channel devices only. As the drain voltage is increased above $V_{D, \, sat}$, the drain depletion region becomes larger and larger, the point of "pinch-off" moves farther from the drain [Figure 14-72(b)], and the effective channel length becomes shorter by an amount $\Delta L'$. An approximate expression for $\Delta L'$ is given by

$$\Delta L' = \left[\frac{2\epsilon_S (V_D - (V_G - V_{th}))}{qN_A} \right]^{1/2},$$

14.72

and the drain current of the device is increased by a factor of $\frac{L'}{L' - \Delta L'}$. Figure 14-74 shows a plot of I_D versus V_D for a short channel device, indicating $V_{D, \, sat}$ and the slight rise in I_D with V_D above $V_{D, \, sat}$.

Threshold Control

The threshold voltage is an important characteristic of a MOSFET since it determines at what gate voltage the device will begin to conduct current. Equation 14.69 shows which physical parameters effect the threshold voltage. The bulk or substrate bias, V_{BB}, is one parameter that can be modified in circuit application to dynamically change V_{th}. Thus, it is useful to have an expression for the change in V_{th} as a function of V_{BB}, as

$$\Delta V_{th} = V_{th} (V_{BB}) - V_{th} (V_{BB} = 0).$$

14.73

From Equation 14.69 one can derive the following:

$$\Delta V_{th} = \frac{(2\epsilon_S qN_A)^{1/2}}{C_{OX}} \left[(2|\phi_F| + |V_S - V_{BB}|)^{1/2} - (2|\phi_F|)^{1/2} \right].$$

14.74

Figure 14-75 shows a plot of ΔV_{th} versus V_{BB} for a given oxide thickness and several substrate doping concentrations. To calculate ΔV_{th} for a different oxide thickness, multiply the plotted value by the ratio t_{OX}/t_{OX} (plotted).

In circuit applications, it is sometimes useful to have MOSFETs with different values of V_{th}. To accomplish this, shallow channel implants are done to locally change the substrate doping. Table 14-3 indicates the direction V_{th} will move for n- and p-channel MOSFETs for both n- and p-type implants.

TABLE 14-3

V_{th} SHIFTS AS A FUNCTION OF CHANNEL IMPLANT

DEVICE TYPE	n-IMPLANT	p-IMPLANT
n-Channel	− (Depletion)	+ (Enhancement)
p-Channel	− (Enhancement)	+ (Depletion)

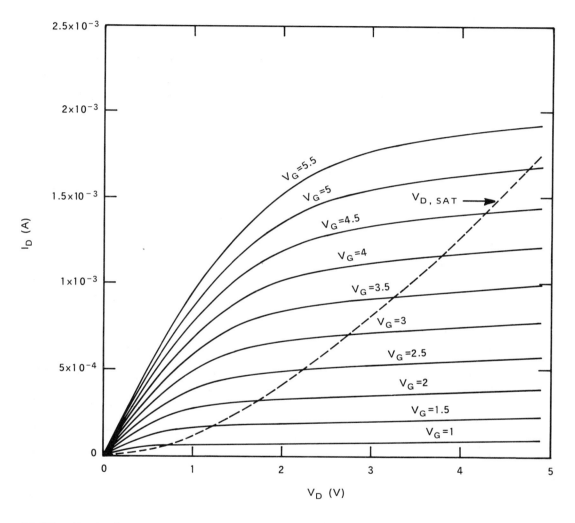

Figure 14-74 —Drain Current Versus Drain Voltage for MOSFET With Channel Length of 2.5 μm

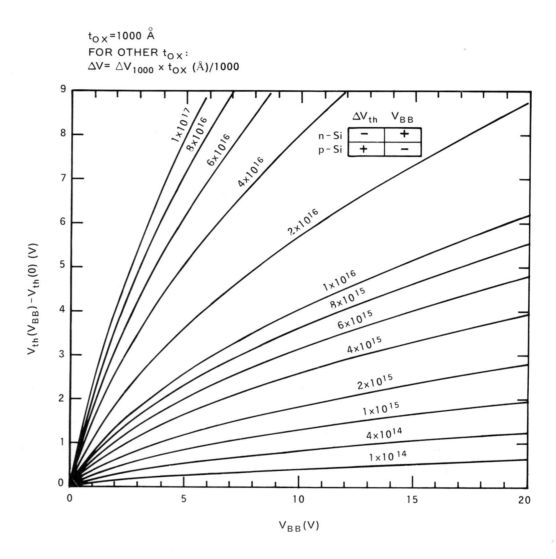

Figure 14-75 — Threshold Voltage Shift Due to Back-Gate Bias

To calculate the threshold voltage for an ion-implanted device, one must consider the final depth of the implant profile and the channel impurity concentration.

To simplify the calculations, the actual impurity distribution is approximated via a "box" distribution, and an effective uniform concentration[117], $\widetilde{N}$, is defined as

$$\widetilde{N} = \frac{\Phi}{x_i},$$
14.75

where

Φ = total implant dose per unit area,

x_i = junction depth of channel implant after drive-in.

To calculate V_{th} shifts as a function of implant parameters, two cases should be considered:[117]

(a) Effective depth of implant is less than the depletion region width at threshold.

(b) Effective depth of implant is greater than the depletion region width at threshold.

Using the depletion approximation, V_{th} for case (a) is given by[117]

$$V_{th} = V_{FB} + V_S + |\phi_F| + |\phi_{FS}| + \frac{q\Phi}{C_{OX}}$$

$$+ \frac{1}{C_{OX}} [2qN_A\epsilon_S (|\phi_{FS}| + |\phi_F| + V_S - V_{BB}) - q^2 x_i N_A \Phi]^{1/2},$$
14.76

where

$$\phi_{FS} = \left(\frac{kT}{q}\right) \ln [(\widetilde{N} + N_A)/n_i].$$

For case (b), V_{th} can be calculated by replacing N_A with $\widetilde{N} + N_A$ in Equation 14.69.

The difference in threshold with and without an ion implant, $\Delta V_{th} = V_{th}(\Phi) - V_{th}(\Phi = 0)$,

can be written for case (a) as

$$\Delta V_{th} = |\phi_{FS}| - |\phi_F| + \frac{q\Phi}{C_{OX}} + \frac{(2qN_A E_S)^{1/2}}{C_{OX}} \left\{ [|\phi_{FS}| + |\phi_F| + V_S - V_{BB}) - \frac{q}{2E_S} x_i \Phi]^{1/2} \right.$$

$$\left. - [2|\phi_F| + V_S - V_{BB}]^{1/2} \right\}.$$
14.77

A plot of Δ_{th} versus Φ/x_i for different substrate doping concentrations is given in Figure 14-76. For a more rigorous treatment of the channel implant effect on V_{th}, see Reference 118.

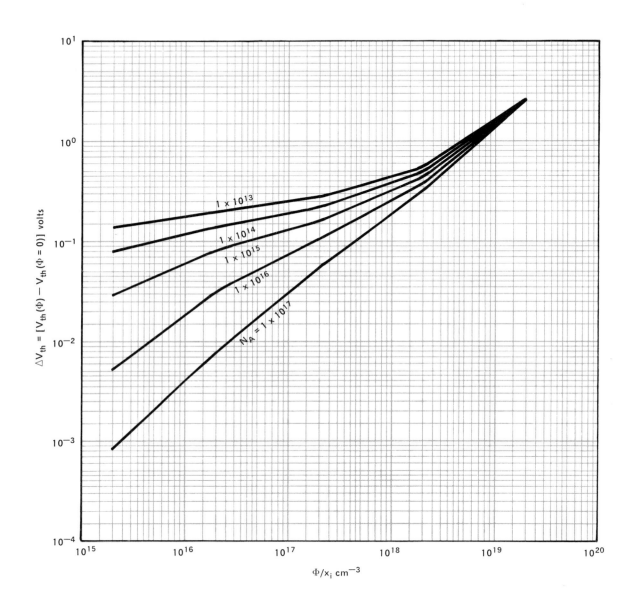

Figure 14-76 —Effect of Channel Implant on Threshold Voltage

Short Channel Length and Width Effects

In the analysis in the previous paragraphs of subsection 14.5, it was assumed that the channel length and width only affected the MOSFET drain current in a linear fashion as shown in Equation 14.67, and did not modify the V_{th} of the transistor.

However, it was found by Poon[119] and Yau[120] that V_{th} decreases as the channel length decreases. Likewise, Noble[121] showed that V_{th} increases as the channel width decreases. Both of these effects require at least a two-dimensional analysis to be treated rigorously. However, it is useful to obtain approximate one-dimensional solutions in order to have a feel for the magnitude of the effects.

Short Channel Effects

In the analysis mentioned above, it was assumed that the bulk charge that must be compensated for in order to form an inversion region under the gate extends from drain to source diffusions and is rectangular in shape. [See Figure 14-77(a).] Due to the source and drain depletion regions, the channel actually looks more like a trapezoid as shown in Figure 14-77(b) (assuming drain and source depletion regions are symmetrical). For long channel devices, the difference in area (bulk charge) between a rectangle and a trapezoid is small. However, for short channel devices, this difference becomes significant. For very short channel devices, the shape of the channel may even resemble a triangle.

In essence, the short channel effect can be thought of as an effective decrease in the bulk charge under the gate. Therefore, a smaller gate voltage is required to cause complete inversion. For a first-order approximation, we can use the geometry of the channel region to predict the V_{th} dependence on channel length.

The bulk charge term of Equation 14.69, Q_B, may be defined as

$$Q_B = \frac{[2\epsilon_S q N_A \, (2|\phi_F| + |V_S - V_{BB})]^{1/2}}{C_{OX}} \, .$$

14.78

This can be modified by a factor f[117], the ratio of the charge contained in the trapezoid to the charge contained in a rectangle. Using Figure 14-77(b) and some simple geometry

$$f = 1 - \frac{r_j}{L'} \, [(1 + \frac{2W_{d\,max}}{r_j})^{1/2} - 1].$$

14.79

Therefore, Equation 14.69 becomes

$$V_{th} = V_{FB} + V_S + 2 \, |\phi_F| + \frac{Q_B f}{C_{OX}} \, .$$

14.80

It can be seen from Equation 14.79 that the short channel effect can be minimized by decreasing the source and drain junction depths. It should be noted that Equation 14.80 is only a good approximation as the drain-to-source voltage goes to zero, since, for large drain-to-source voltages, the depletion regions at the source and drain are not equal.

Short Width Effects

Short width effects arise from two factors that act to increase the bulk charge in the channel. The first factor is related to fringing gate fields and thickening gate oxide ("bird's beak") at the channel edge. The fringing fields cause the depletion region to extend under the insulating field oxide that surrounds the MOSFET. (See Figure 14-78.) The added depletion region increases the bulk charge and, thus, V_{th} increases.

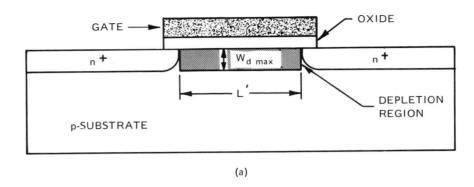

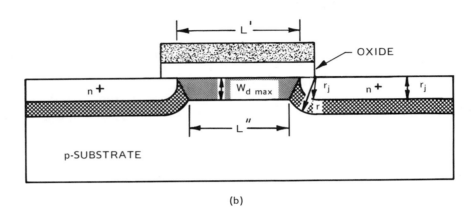

(a)

(b)

Figure 14-77 —Cross-Sectional View of MOSFET Showing (a) Channel and (b) Source and Drain
Depletion Regions

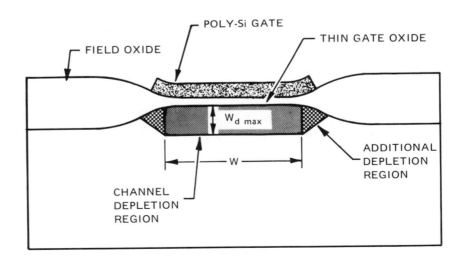

Figure 14-78 —Cross-Sectional View of MOSFET Showing Additional Depletion
Regions Caused by the Bird's Beak

The second factor is the channel stop implant. This implant is done under the field oxide to increase the substrate doping in this region so that the region cannot be easily inverted. After implant and subsequent heat treatments, the regions near the edge of the gate electrode become more highly doped due to the lateral diffusion of the channel-stop impurity. The increased doping gives rise to a larger bulk charge in these regions and, thus, V_{th} increases.

The short width effect becomes noticeable at widths <30 μm.[122] The decrease in V_{th} is parabolic as a function of width. The actual parameters of the parabola are very much dependent on the specific MOS process.

Surface Mobility

The channel conductance of a MOSFET depends upon the surface mobility of the carriers in the inversion layer. As the mobility increases, the gain and the drain current of a MOSFET also increase. It has been found empirically, that the surface mobility is limited to about one-half the value of the bulk mobility.[27] Furthermore, the surface mobility decreases as V_G is increased.

Even though an exact theory for surface mobility does not exist, it is generally accepted that the reduced mobility is due to loss mechanisms associated with collisions at the Si-SiO$_2$ interface.[117] The reduction in mobility as a function of increasing V_G is attributed to the stronger attraction of carriers to the surface at high fields.

Sabnis and Clemens[123] developed a model that shows that the surface mobility is not a function of doping density in the range ($N_A < 1.0 \times 10^{17}$ cm^{-3}). In their model, they define an effective electric field as

$$E_{eff} = \frac{C_{OX}}{\epsilon_S} \; [1/2 \, (V_G + V_T) - V_{FB} - 2\phi_F].$$
(14.81)

When channel mobility is plotted against E_{eff}, all data for $N_A < 1.0 \times 10^{17}$ cm^{-3} falls on one universal curve, as shown in Figure 14-79.

Cooper and Nelson[124] have measured the high-field drift velocity of electrons in silicon inversion layers. The technique used involved the measuring of the time-of-flight of a packet of carriers (created by a pulsed laser) at the semiconductor surface. The results of their measurements are shown in Figure 14-80, where the electric field normal to the surface is given by

$$E_N = -[qN(x)/2 + Q_B(x)]/\epsilon_S$$
(14.82)

where $Q_B(x)$ is the depletion layer charge at point x along the surface and $N(x)$ is the electron density at point x, which is controlled by the laser intensity. At low tangential fields, the electron velocity approaches a linear function of field and yields a mobility of 840 cm^2/V•s at a normal field of 30 kV/cm, which is in agreement with the work of Sabnis and Clemens.[123]

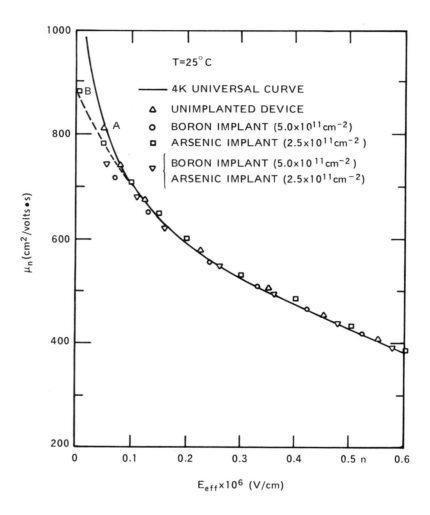

Figure 14-79 —Effective Channel Mobility as a Function of Effective Electric Field

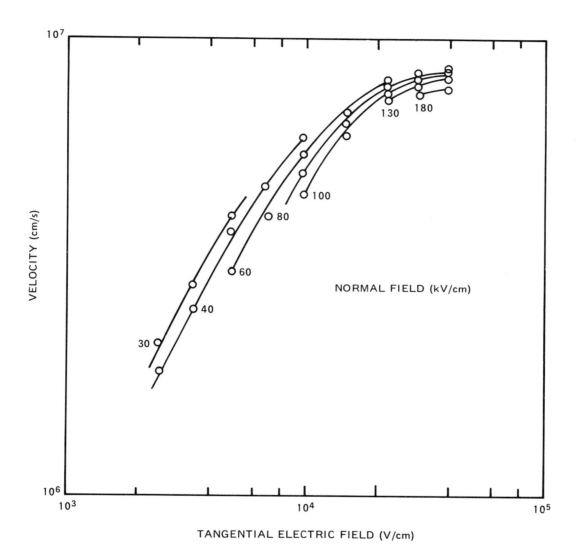

Figure 14-80 — A Plot of Velocity Versus Tangential Electric Field[124]
(© 1981 IEEE)

Summary of Useful Equations

$$V_{th} = V_{FB} + |V_S| + 2|\phi_F| + \frac{[2\epsilon_S q N_A (2|\phi_F| + |V_S - V_{BB}|)]^{1/2}}{C_{OX}} .$$

$$I_D = \mu_n \frac{W}{L'} C_{OX} (V_G - V_{th} - \frac{V_D}{2}) V_D .$$

$$V_{D, sat} = V_G - V_{th} .$$

$$I_{D, sat} = \mu_n \frac{W}{2L'} C_{OX} (V_G - V_{th})^2 .$$

$$\Delta V_{th} = V_{th}(V_{BB}) - V_{th}(V_{BB} = 0) = \frac{[2\epsilon_S q N_A]^{1/2}}{C_{OX}} \left\{ [2|\phi_F| + |V_S - V_{BB}|]^{1/2} - [2|\phi_F|]^{1/2} \right\}$$

$$Q_B = \frac{[2\epsilon_S q N_A (2|\phi_F| + |V_S - V_{BB}|)]^{1/2}}{C_{OX}}$$

MOSFET Parameter Extraction

Threshold Voltage, V_T

The threshold voltage, V_T, is defined as the gate voltage, V_G, at which the channel becomes strongly inverted, i.e., $\phi_S = 2\phi_F$.

V_T can be determined from Equation 14.67, which predicts a linear relationship between I_D and V_G (for a constant $V_D \leqslant 0.100V$) with an intercept at $V_G = V_0$ where

$$V_0 = V_T + 1/2 \, V_D . \tag{14.83}$$

In practice, due to the change of surface mobility with V_G, no linear behavior is found.[116] Due to this, V_0 is determined by first finding the maximum slope $M_i = \left(\frac{I_D}{V_G}\right)_{V_D}$, fitting a straight line through this point, and extrapolating to zero current, as

$$V_0 = V_{G_i} - \frac{I_{D_i}}{M_i} , \tag{14.84}$$

where I_{D_i} and V_{G_i} are the current and gate voltage at the point of maximum slope, $M_i = \left(\frac{\partial I_D}{\partial V_G}\right)_{V_D}$.

The above procedure is best applied to devices with interface state densities of $10^{10} cm^{-2}$ or less.[116]

Transconductance

The transconductance g_m is defined as the slope of the I_D versus V_G characteristic at constant drain voltage,

$$g_m = \left(\frac{\partial I_D}{\partial V_G}\right)_{V_D} . \tag{14.85}$$

Drain Conductance

The drain conductance, g_d, is defined as the slope of the I_D versus V_D characteristic at constant gate voltage,

$$g_d = \left(\frac{\partial I_D}{\partial V_D}\right) \tag{14.86}$$

14.6 Glossary of MOSFET Symbols and Terms

Channel — The surface region on an MOS device between source and drain that is inverted (converted to a conductivity type opposite that of substrate) and provides the current carrying path between source and drain.

Depletion Mode MOSFET — $V_{th} < 0$ for n-channel. ($V_{th} > 0$ for p-channel). MOSFET is on at $V_G = 0V$.

Drain — The electrode in an MOS device from which carriers exit to the external circuit.

Enhancement Mode MOSFET — $V_{th} > 0$ for n-channel ($V_{th} < 0$ for p-channel). MOSFET is off at $V_G = 0V$.

Gate — The metal- (or other conductor) to-dielectric contact that controls the flow of charge between source and drain by controlling the characteristics of the channel.

MISFET — A metal-insulator semiconductor field-effect transistor.

Source — The source of carriers that enter the channel and travel from source to drain.

Substrate (Bulk) — The semiconductor material in and on which the MOS device is fabricated.

C_{OX} — Oxide capacitance per unit area (F/cm^2).

L — The channel length—length of the physical path traveled by carriers from source to drain. This is shorter than the length of the gate electrode due to lateral diffusion of source and drain.

N_A — Substrate impurity concentration.

μ_n — Surface mobility, typically ~ 800 $cm^2/V \bullet s$ for n-channel MOSFETs at low fields.

V_{BB} — Voltage applied to substrate electrode.

V_D — Voltage applied to drain electrode.

V_G — Voltage applied to gate electrode.

V_S — Voltage applied to source electrode.

V_{th} — Threshold voltage.

W — Channel width—channel dimension perpendicular to L'.

REFERENCES

1. B. E. Deal et al, "Recent Advances in the Understanding of the Metal-Oxide-Silicon System," *Trans. AIME,* Vol. 233 (March 1965), pp. 524-529.

2. S. M. Sze, *Physics of Semiconductor Devices,* New York: John Wiley & Sons, 1969.

3. A. S. Grove et al, "Investigation of Thermally Oxidized Silicon Surfaces Using Metal-Oxide Semiconductor Structures," *Solid-State Electron.,* Vol. 8, (1965), pp. 145-163.

4. S. M. Sze, op cit, p. 437.

5. R. Lindner, "Semiconductor Surface Varactor," B.S.T.J., *41,* No. 3, (May 1962), pp. 803-831.

6. S. M. Sze, op cit, p. 436.

7. J. C. Irvin, private communication.

8. A. Goetzberger, "Ideal MOS Curves for Silicon," B.S.T.J., *45,* No. 9, (September 1966), pp. 1097-1122.

9. S. M. Sze, op cit, p. 442.

10. S. Wagner and C. N. Berglund, private communication.

11. S. M. Spitzer, private communication.

12. K. H. Zaininger and F. P. Heiman, "The C-V Technique as an Analytical Tool," *Solid State Technology,* (June 1970), pp. 46-55.

13. S. M. Sze, op cit, p. 441.

14. L. P. Adda, private communication.

15. W. van Gelder and E. H. Nicollian, "Silicon Impurity Distribution as Revealed by Pulsed MOS C-V Measurements," *J. Electrochem. Soc.,* Vol. 118, No. 1 (January 1971), pp. 138-141.

16. L. P. Adda, private communication.

17. E. H. Nicollian, M. H. Hanes, and J. R. Brews, "Using the MIS Capacitor for Doping Profile Measurements with Minimal Interface State Error," *IEEE Trans. Electron Dev.,* Vol. ED-20, No. 4 (1973), pp. 380-389.

18. J. R. Brews, "Correcting Interface-State Errors in MOS Doping Profile Determinations," *J. Appl. Phys.,* Vol. 44, No. 7 (July 1973), pp. 3228-3231.

19. K. Ziegler, E. Klausmann, and S. Kar, "Determination of the Semiconductor Doping Profile Right Up to its Surface Using the MIS Capacitor," *Solid-State Electron.,* Vol. 18 (1975), pp. 189-198.

20. F. P. Heiman, "On the Determination of Minority Carrier Lifetime from the Transient Response of an MOS Capacitor," *IEEE Trans. Electron Dev.,* Vol. ED-14, No. 11 (1967), pp. 781-784.

21. M. Zerbst, "Relaxation Effects at Semiconductor-Insulator Interfaces," *Z. angew. Phys.,* Vol. 22 (May 1966), pp. 30-36.

22. R. F. Pierret, "Rapid Interpretation of the MOS-C C-t Transient," *IEEE Trans. Electron Dev.,* Vol. ED-25, No. 9 (September 1978), pp. 1157-1159.

23. R. F. Pierret, "A Linear Sweep MOS-C Technique for Determining Minority Carrier Lifetimes," *IEEE Trans. Electron Dev.,* Vol. ED-19, No. 7 (July 1972), pp. 869-873.

24. D. K. Schroder and J. Guldberg, "Interpretation of Surface and Bulk Effects Using the Pulsed MIS Capacitor," *Solid-State Electron.,* Vol. 14 (1971), pp. 1285-1297.

25. D. W. Small and R. F. Pierret, "Separation of Surface and Bulk Components in MOS-C Generation Rate Measurements," *Solid-State Electron.,* Vol. 19 (1976), pp. 505-511.

26. L. P. Adda, private communication.

27. A. S. Grove, *Physics and Technology of Semiconductor Devices,* New York: John Wiley & Sons, 1967.

28. C. C. Chang and W. C. Johnson, "Distribution of Flat-Band Voltages in Laterally Nonuniform MIS Capacitors and Application to a Test for Nonuniformities," *IEEE Trans. Electron Dev.,* Vol. ED-25, No. 12 (December 1978), pp. 1368-1373.

29. J. T. Clemens, personal communication to C. A. Goodwin (December 1979).

30. W. M. Werner, "The Work Function Difference of the MOS-System with Aluminum Field Plates and Polycrystalline Silicon Field Plates," *Solid-State Electron.,* Vol. 17 (1974), pp. 769-775.

31. B. E. Deal, E. H. Snow, and C. A. Mead, "Barrier Energies in Metal-Silicon Dioxide-Silicon Structures," *J. Phys. Chem. Solids,* Vol. 27 (1966), pp. 1873-1879.

32. S. Kar, "Determination of Si-Metal Work Function Differences by MOS Capacitance Technique," *Solid-State Electron.,* Vol. 18 (1975), pp. 169-181.

33. F. Mohammadi and K. C. Saraswat, "Properties of Sputtered Tungsten Silicide for MOS Integrated Circuit Applications," *J. Electrochem. Soc.: Solid-State Science and Technology,* Vol. 127, No. 2 (February 1980), pp. 450-454.

34. G. A. Corker and C. M. Svensson, "Sodium-Induced Work Function Shift of Mercury as a Metal-Oxide-Semiconductor Electrode," *J. Electrochem. Soc.: Solid-State Science and Technology,* Vol. 125, No. 11 (November 1978), pp. 1881-1882.

35. K. Haberle and E. Fröschle, "On the Work Function Difference in the Al-SiO$_2$-Si System with Reactively Sputtered SiO$_2$," *J. Electrohcem. Soc.: Solid-State Science and Technology,* Vol. 126, No. 5 (May 1978), pp. 878-880.

36. L. A. Kasprzak and A. K. Gaind, "The Si-SiO$_2$ Interface: Oxide Charge, Electron Affinity and Fast Surface States," *The Physics of SiO$_2$ and Its Interfaces,* ed. T. Pantelides, New York: Pergamon Press, 1978, pp. 438-442.

37. B. E. Deal, "Standardized Terminology for Oxide Charges Associated with Thermally Oxidized Silicon," *J. Electrochem. Soc.: Solid-State Science and Technology,* Vol. 127, No. 4 (April 1980), pp. 979-981.

38. B. E. Deal, "The Current Understanding of Charges in the Thermally Oxidized Silicon Structure," *J. Electrochem. Soc.: Solid-State Science and Technology,* Vol. 121, No. 6 (June 1974), pp. 198C-205C.

39. J. P. Stagg, "Drift Mobilities of Na^+ and K^+ Ions in SiO_2 Films," *Appl. Phys. Lett.,* Vol. 31, No. 8 (October 1977), pp. 532-533.

40. E. H. Snow et al, "Ion Transport Phenomena in Insulating Films," *J. Appl. Phys.,* Vol. 36, No. 5 (May 1965), pp. 1664-1673.

41. M. R. Boudry and J. P. Stagg, "The Kinetic Behavior of Mobile Ions in the Al-SiO_2-Si System, *J. Appl. Phys.,* Vol. 50, No. 2, (February 1979), pp. 942-950.

42. T. W. Hickmott, "Thermally Stimulated Ionic Conductivity of Sodium in Thermal SiO_2," *J. Appl. Phys.,* Vol. 46, No. 6 (June 1975), pp. 2583-2598.

43. W. R. Knolle and H. D. Seidel, private communication.

44. S. Mayo and W. H. Evans, "Development of Sodium Contamination in Semiconductor Oxidation Atmospheres at $1000°C$," *J. Electrochem. Soc.,* Vol. 124, No. 5 (May 1977), pp. 780-785.

45. P. Balk and J. M. Eldridge, "Phosphosilicate Glass Stabilization of FET Devices," *Proceedings IEEE,* Vol. 57, No. 9 (September 1969), pp. 1558-1563.

46. J. M. Eldridge and D. R. Kerr, "Sodium Ion Drift through Phosphosilicate Glass-SiO_2 Films," *J. Electrochem. Soc.: Solid-State Science and Technology,* Vol. 118, No. 6 (June 1971), pp. 986-991.

47. L. H. Kaplan and M. E. Lowe, "Phosphosilicate Glass Stabilization of Mos Structures," *J. Electrochem. Soc.: Solid-State Science and Technology,* Vol. 118, No. 10 (October 1971), pp. 1649-1653.

48. J. Drobek and L. P. Adda, private communication.

49. M. Kuhn and D. J. Silversmith, "Ionic Contamination and Transport of Mobile Ions in MOS Structures," *J. Electrochem. Soc.: Solid-State Science and Technology,* Vol. 118, No. 6 (June 1971), pp. 966-970.

50. L. P. Adda, private communication.

51. W. Marciniak and H. M. Przewlocki, "On the Behavior of Mobile Ions in Dielectric Layers of MOS Structures," *J. Electrochem. Soc.: Solid-State Science and Technology,* Vol. 123, No. 8 (August 1976), pp. 1207-1212.

52. E. H. Nicollian, private communication.

53. B. E. Deal et al, "Characteristics of the Surface-State Charge (Q_{SS}) of Thermally Oxidized Silicon," *J. Electrochem. Soc.: Solid-State Science and Technology,* Vol. 114, No. 3 (March 1967), pp. 266-274.

54. R. R. Razouk and B. E. Deal, "Dependence of Interface State Density on Silicon Thermal Oxidation Process Variables," *J. Electrochem. Soc.: Solid-State Science and Technology,* Vol. 126, No. 9 (September 1979), pp. 1573-1581.

55. D. W. Hess and B. E. Deal, "Effect of Nitrogen and Oxygen/Nitrogen Mixtures on Oxide Charges in MOS Structures," *J. Electrochem. Soc.: Solid-State Science and Technology,* Vol. 122, No. 8 (August 1975), pp. 1123-1127.

56. C. T. Sah and L. C. Sah, "Origins of Interface States and Oxide Charges Generated by Ionizing Radiation in Metal-Oxide-Silicon Structures," U.S. Defense Nuclear Agency, *Report No. HDL-CR-76-164-1,* under Contract DAAG39-75-C-0164, January 1976.

57. E. Arnold, J. Ladell, and G. Abowitz, "Crystallographic Symmetry of Surface State Density in Thermally Oxidized Silicon," *Appl. Phys. Lett.,* Vol. 13, No. 12 (December 1968), pp. 413-416.

58. E. H. Nicollian and A. Goetzberger, "The Si-SiO$_2$ Interface-Electrical Properties as Determined by the Metal-Insulator-Silicon Conductance Technique," B.S.T.J., *46,* No. 6 (July-August 1967), pp. 1055-1133.

59. A. K. Sinha and T. E. Smith, "Kinetics of the Slow-Trapping Instability at the Si/SiO$_2$ Interface," *J. Electrochem. Soc.: Solid-State Science and Technology,* Vol. 125, No. 5 (May 1978), pp. 743-746.

60. P. L. Castro and B. E. Deal, "Low-Temperature Reduction of Fast Surface States Associated with Thermally Oxidized Silicon," *J. Electrochem. Soc.: Solid-State Science and Technology,* Vol. 118, No. 2 (February 1971), pp. 280-286.

61. A. K. Sinha et al, "Avalanche-Induced Hot Electron Injection and Trapping in Gate Oxides on p-Si," *J. Electrochem. Soc.: Solid-State Science and Technology,* Vol. 127, No. 9 (September 1980), pp. 2046-2049.

62. G. W. Hughes, "Interface-State Effects in Irradiated MOS Structures," *J. Appl. Phys.,* Vol. 48, No. 12 (December 1977), pp. 5357-5359.

63. Y. C. Cheng, "Electronic States at the Silicon-Silicon Dioxide Interface," *Progress in Surface Science,* Vol. 8, No. 5 (1977), pp. 181-218.

64. M. Kuhn, "A Fast, Quasi-Static Technique for Thermal Equilibrium MOS C-V and Surface State Measurements," *Solid State Electron.,* Vol. 13, (1970), pp. 873-885.

65. R. D. Plummer, private communication.

66. R. Castagne and A. Vapaille, "Description of the SiO$_2$-Si Interface Properties by Means of Very Low Frequency MOS Capacitance Measurements," *Surface Science,* Vol. 28 (1971), pp. 157-193.

67. M. Kuhn and E. H. Nicollian, "Nonequilibrium Effects in Quasi-Static MOS Measurements," *J. Electrochem. Soc.: Solid-State Science and Technology,* Vol. 118, No. 2 (February 1971), pp. 370-373.

68. C. Berglund, "Surface States at Steam-Grown Silicon-Silicon Dioxide Interfaces," *IEEE Trans. Electron Dev.,* Vol. ED-13, No. 13 (October 1966), pp. 701-705.

69. R. Van Overstraeten, G. Declerck, and G. Broux, "Graphical Technique to Determine the Density of Surface States at the Si-SiO$_2$ Interface Using the Quasi-Static Method," *J. Electrochem. Soc.,* Vol. 120, No. 12 (December 1973), pp. 1785-1787.

70. W. A. Hill and C. C. Coleman, "A Single-Frequency Approximation for Interface-State Density Determination," *Solid-State Electron.,* Vol. 23 (1980), pp. 987-993.

71. P. V. Gray and D. M. Brown, "Density of SiO$_2$-Si Interface States," *Appl. Phys. Lett.,* Vol. 8, No. 2 (January 1966), pp. 31-33.

72. K. L. Wang, "MOS Interface-State Density Measurements Using Transient Capacitance Spectroscopy," *IEEE Trans. Electron Dev.,* Vol. ED-27, No. 12 (December 1980), pp. 2231, 2239.

73. N. M. Johnson, D. J. Bartelink, and M. Schulz, "Transient Capacitance Measurements of Electronic States at the SiO$_2$-Si Interface," *The Physics of SiO$_2$ and Its Interfaces,* ed. S. T. Pantelides, New York: Pergamon Press, 1978, pp. 421-427.

74. D. J. DiMaria et al, "Centroid Location of Implanted Ions in the SiO$_2$ Layer of MOS Structures Using the Photo I-V Technique," *J. Appl. Phys.,* Vol. 49, No. 11 (November 1978), pp. 5441-5444.

75. T. H. Ning, "Electron Trapping in SiO$_2$ Due to Electron-Beam Deposition of Aluminum," *J. Appl. Phys.,* Vol. 49, No. 7 (July 1978), pp. 4077-4082.

76. A. Kamgar and A. K. Sinha, private communication.

77. R. A. Gdula, "The Effects of Processing on Radiation Damage in SiO$_2$," *Digest of International Electron Devices Meeting* (December 1977), pp. 148-150.

78. F. P. Heiman and G. Warfield, "The Effects of Oxide Traps on the MOS Capacitance," *IEEE Trans. Electron Dev.,* Vol. ED-12 (April 1965), pp. 167-178.

79. T-P Ma and W. H-L. Ma, "Low Pressure RF Annealing: A New Technique to Remove Charge Centers in MIS Dielectrics," *Appl. Phys. Lett.,* Vol. 32, No. 7 (April 1978), pp. 441-444.

80. D. J. DiMaria, "The Properties of Electron and Hole Traps in Thermal Silicon Dioxide Layers Grown on Silicon," *The Physics of SiO$_2$ and Its Interfaces,* ed. S. T. Pantelides, New York: Pergamon Press, 1978, pp. 160-178.

81. R. C. Hughes and D. Emin, "Small Polaron Formation and Motion of Holes in a SiO$_2$," *ibid,* pp. 14-18.

82. K. Saminadayar and J. C. Pfister, "Evolution of Surface-States Density of Si/Wet Thermal SiO$_2$ Interface During Bias-Temperature Treatment," *Solid-State Electron.,* Vol. 20 (1977), pp. 891-896.

83. B. L. Gregory, "Process Controls for Radiation-Hardened Aluminum Gate Bulk Silicon CMOS," *IEEE Trans. Nuc. Science,* Vol. NS-22, No. 6 (December 1975), pp. 2295-2302.

84. K. G. Aubuchon, "Radiation Hardening of P-MOS Devices by Optimization of the Thermal SiO$_2$ Gate Insulator," *IEEE Trans. Nuc. Science,* Vol. NS-18, No. 6 (1971), pp. 117-125.

85. T. H. Ning, C. M. Osburn, and H. N. Yu, "Effect of Electron Trapping on IGFET Characteristics," *J. Electronic Mater.,* Vol. 6, No. 2 (1977), pp. 65-76.

86. T. H. Ning and H. N. Yu, "Optically Induced Injection of Hot Electrons into SiO$_2$," *J. Appl. Phys.,* Vol. 45, No. 12 (December 1974), pp. 5373-5378.

87. J. M. Aitken and D. R. Young, "Electron Trapping by Radiation-Induced Charge in MOS Devices," *J. Appl. Phys.,* Vol. 47, No. 3 (March 1976), pp. 1196-1198.

88. S. Pang, S. A. Lyon and W. C. Johnson, "Generation of Interface States in the Si-SiO$_2$ System by Photoinjection of Electrons," *The Physics of MOS Insulators,* ed. G. Lucovsky, S. P. Pantelides, and F. L. Galeener, New York: Pergamon Press, 1980, pp. 285-289.

89. B. R. Singh and P. Balk, "Oxidation of Silicon in the Presence of Chlorine and Chlorine Compounds," *J. Electrochem. Soc.: Solid-State Science and Technology,* Vol. 125, No. 3 (March 1978), pp. 453-461.

90. J. Monkowski, "Role of Chlorine in Silicon Oxidation, Part I," *Solid State Technology,* Vol. 122, No. 7 (July 1979), pp. 58-61; Part II, *ibid,* Vol. 22, No. 8 (August 1979), pp. 113-119.

91. R. J. Kriegler, Y. C. Cheng, and D. R. Colton, "The Effect of HCl and Cl$_2$ on the Thermal Oxidation of Silicon," *J. Electrochem. Soc.: Solid-State Science and Technology,* Vol. 119, No. 3 (March 1972), pp. 388-392.

92. B. R. Singh and P. Balk, "Thermal Oxidation of Silicon in O$_2$-Trichloroethylene," *J. Electrochem. Soc.: Solid-State Science and Technology,* Vol. 126, No. 7 (July 1979), pp. 1288-1294.

93. E. J. Janssens and G. J. Declerck, "The Use of 1,1,1-Trichloroethane as an Optimized Additive to Improve the Silicon Thermal Oxidation Technology," *J. Electrochem. Soc.: Solid-State Science and Technology,* Vol. 125, No. 10 (October 1978), pp. 1696-1703.

94. K. Hirabayashi and J. Iwamura, "Kinetics of Thermal Growth of HCl-O$_2$ Oxides on Silicon," *J. Electrochem. Soc.: Solid-State Science and Technology,* Vol. 120, No. 11 (November 1973), pp. 1595-1601.

95. B. E. Deal, A. Hurrle, and M. J. Schulz, "Chlorine Concentration Profiles in O$_2$/HCl and H$_2$O/HCl Thermal Silicon Oxides Using SIMS Measurements," *J. Electrochem. Soc.: Solid-State Science and Technology,* Vol. 125, No. 12 (December 1978), pp. 2024-2027.

96. A. Rohatgi, S. R. Butler, and F. J. Feigl, "Mobile Sodium Ion Passivation in HCl Oxides," *J. Electrochem. Soc.: Solid-State Science and Technology,* Vol. 126, No. 1 (January 1979), pp. 149-154.

97. Y. J. van der Meulen, C. M. Osburn, and J. F. Ziegler, "Properties of SiO$_2$ Grown in the Presence of HCl and Cl$_2$," *J. Electrochem. Soc.: Solid-State Science and Technology,* Vol. 122, No. 2 (February 1975), pp. 284-290.

98. A. Rohatgi et al, "Chlorine Incorporation in HCl Oxides," *J. Electrochem. Soc.: Solid-State Science and Technology,* Vol. 126, No. 1 (January 1979), pp. 143-149.

99. J. Monkowski, J. Stach, and R. E. Tressler, "Phase Separation and Sodium Passivation in Silicon Oxides Grown in HCl/O$_2$ Ambients," *J. Electrochem. Soc.: Solid-State Science and Technology,* Vol. 126, No. 7 (July 1979), pp. 1129-1134.

100. J. Monkowski, R. E. Tressler, and J. Stach, "The Structure and Composition of Silicon Oxides Grown in HCl/O$_2$ Ambients," *J. Electrochem. Soc.: Solid-State Science and Technology,* Vol. 125, No. 11 (November 1978), pp. 1867-1873.

101. R. J. Kriegler, "The Uses of HCl and Cl$_2$ for Preparation of Electrically Stable SiO$_2$," *Semiconductor Silicon 1973,* ed. H. R. Huff and R. R. Burgess, Princeton: The Electrochemical Society, Inc., 1973, pp. 363-375.

102. B. E. Deal, "Thermal Oxidation Kinetics of Silicon in Pyrogenic H$_2$O and 5% HCl/H$_2$O Mixtures," *J. Electrochem. Soc.: Solid-State Science and Technology,* Vol. 125, No. 4 (April 1978), pp. 576-579.

103. R. S. Ronen and P. H. Robinson, "Hydrogen Chloride and Chlorine Gettering: An Effective Technique for Improving Performance of Silicon Diodes," *J. Electrochem. Soc.: Solid-State Science and Technology,* Vol. 119, No. 6 (June 1972), pp. 747-752.

104. P. H. Robinson and F. P. Heiman, "Use of HCl Gettering in Silicon Device Processing," *J. Electrochem. Soc.: Solid-State Science and Technology,* Vol. 118, No. 1 (January 1971), pp. 141-143.

105. L. E. Katz, P. F. Schmidt, and C. W. Pearce, "Neutron Activation Study of a Gettering Treatment for Czochralski Silicon Substrates," *J. Electrochem. Soc.: Solid-State Science and Technology,* Vol. 128, No. 3 (March 1981), pp. 620-624.

106. D. R. Young and C. M. Osburn, "Minority Carrier Generation Studies in MOS Capacitors on N-Type Silicon," *J. Electrochem. Soc.: Solid-State Science and Technology,* Vol. 120, No. 11 (November 1973), pp. 1578-1581.

107. G. J. Declerck et al, "Some Effects of Trichloroethylene Oxidation on the Characteristics of MOS Devices," *J. Electrochem. Soc.: Solid-State Science and Technology,* Vol. 122, No. 3 (March 1975), pp. 436-439.

108. H. Shiraki, "Suppression of Stacking Fault Generation in Silicon Wafer by HCl Added to Dry O_2 Oxidation," *Jap. J. Appl. Phys.,* Vol. 15, No. 1 (January 1976), pp. 83-86.

109. T. Hattori, "TCE Oxidation for the Elimination of Oxidation-Induced Stacking Faults in Silicon," *Solid State Tech.,* Vol. 22, No. 11 (November 1979), pp. 85-89.

110. C. L. Claeys et al, "Elimination of Stacking Faults for Charge-Coupled Device Processing," *Semiconductor Silicon 1977,* ed. H. R. Huff and E. Sirtl, Princeton: The Electrochemical Society, Inc., 1977, pp. 773-784.

111. M. C. Chen and J. W. Hile, "Oxide Charge Reduction by Chemical Gettering with Trichloroethylene During Thermal Oxidation of Silicon," *J. Electrochem. Soc.: Solid-State Science and Technology,* Vol. 119, No. 2 (February 1972), pp. 223-225.

112. D. W. Hess, "Effect of Chlorine on the Negative Bias Instability in MOS Structures," *J. Electrochem. Soc.: Solid-State Science and Technology,* Vol. 124, No. 5 (May 1977), pp. 740-743.

113. D. L. Heald, R. M. Das, and R. P. Khosla, "Influence of Trichloroethylene on Room Temperature Flatband Voltages of MOS Capacitors," *J. Electrochem. Soc.: Solid-State Science and Technology,* Vol. 123, No. 2 (February 1976), pp. 302-303.

114. C. M. Osburn, "Dielectric Breakdown Properties of SiO_2 Films Grown in Halogen and Halogen-Containing Environments," *J. Electrochem. Soc.: Solid-State Science and Technology,* Vol. 121, No. 6 (June 1974), pp. 809-815.

115. W. Fichtner and H. W. Pötzl, "MOS Modelling by Analytical Approximations. I. Subthreshold Current and Threshold Voltage," *Int. J. Electronics,* Vol. 46, No. 1 (1979), pp. 33-55.

116. J. R. Brews, "Physics of the MOS Transistor," *Applied Solid State Science, Supplement 2, Silicon Integrated Circuits, Part A,* ed. D. Kahng (1981), pp. 1-120.

117. R. S. Muller, T. I. Kamins, *Device Electronics for Integrated Circuits,* New York: John Wiley & Sons, 1977.

118. J. R. Brews, "Threshold Shifts Due to Nonuniform Doping Profiles in Surface Channel MOSFETs," *IEEE Trans Electron Dev.,* Vol. ED-26, No. 11 (November 1979), pp. 1696-1710.

119. H. C. Poon, L. D. Yau, and R. L. Johnston, "DC Model for Short-Channel IGFETs," *Digest of International Electron Devices Meeting,* (December 1973), pp. 156-159.

120. L. D. Yau, "A Simple Theory to Predict the Threshold Voltage of Short-Channel IGFETs," *Solid-State Electron.,* Vol. 17 (1974), pp. 1059-1063.

121. W. P. Noble and D. E. Cottrell, "Narrow Channel Effects in Insulated Gate Field Effect Transistors," *Digest of International Electron Devices Meeting,* (December 1976), pp. 582-586.

122. D. Beecham, private communication to W. G. Meyer (1981).

123. A. G. Sabnis and J. T. Clemens, "Characterization of the Electron Mobility in the Inverted <100> Si Surface," *1979 International Electron Devices Meeting,* Technical Digest, (New York: IEEE, 1979), p. 180.

124. J. A. Cooper, Jr. and D. F. Nelson, "Measurement of the High-Field Drift Velocity of Electrons in Inversion Layers on Silicon," *IEEE Electron Devices Letters,* Vol. EDL-2, No. 7 (July 1981), p. 172.

15. RELIABILITY

LIST OF SYMBOLS

Symbol	Definition
A	Acceleration factor
A_0	Reference condition acceleration factor
A_{AC}	Air conditioned ambient acceleration factor
A_{BR}	Baton Rouge, La. ambient acceleration factor
A_1	Acceleration factor — bare triple track on Al_2O_3
A_2	Acceleration factor — RTV encapsulated triple track on Al_2O_3
A_3	Acceleration factor — bare triple track on Si_3N_4
A_4	Acceleration factor — RTV encapsulated triple track on Si_3N_4
a	Duane shape parameter
E_a	Activation energy
F	Cumulative failure distribution
f	Failure probability distribution
k	Boltzmans Constant
ML	Median life
MTTF	Mean time to failure
N	Number of devices, components, etc.
R	Reliability
T	Temperature
t	Time
α	"Original Weibull" parameter
β	Weibull distribution shape parameter
λ	Hazard rate or instantaneous failure rate
λ_C	Cumulative (average) failure rate
λ_D	Duane intercept
λ_0	Failure rate at $t = to$
λ_1	Failure rate at $t = 1$ hour
θ	Weibull distribution scale parameter
μ	Log-normal distribution scale parameter
σ	Log-normal distribution shape parameter
τ	Time in hours

15.1 Definitions

Reliability

Reliability, R(t), is defined as the <u>probability</u> that a device will perform its required function under <u>stated operating conditions</u> for a stated period of <u>time.</u> The essential points in this definition are:

(a) <u>Probability</u>: Reliability is inherently a statistical concept. All measures of reliability are statements about the average behavior of groups of devices.

(b) <u>Stated operating conditions:</u> Reliability is a strong function of operating conditions. These must always be stated when reliability is specified.

(c) <u>Time:</u> Reliability is inherently a function of time. In addition to the monotonic decrease in the reliability of a given component with time, there is also a variation in the probability of failure for equal time intervals located at different times.

Probability

The commonly used description of reliability is, as noted above, R(t) = reliability = the probability that a device will not fail between times 0 and t. It is more common to discuss failures and use measures of failure rather than measures of the absence of failure. In this case it is convenient to use

$$F(t) = 1 - R(t) = \text{cumulative failure distribution function}$$

$$= \text{the probability that a device will fail between 0 and t.}$$

This is related to the failure probability density function by

$$F(t) = \int_0^t f(t) \, dt,$$

where

$$f(t) = \text{failure probability density function (time}^{-1}).$$

The most commonly used expression of failure behavior is

$$\lambda(t) = \frac{f(t)}{1 - F(t)} = \text{hazard rate or instantaneous failure rate (time}^{-1}).$$

These functions are plotted for some distributions of interest in Figures 15-1 (exponential, Weibull, log-normal).

Operating Conditions

The primary environmental factors which affect reliability are:

(a) Temperature.

(b) Humidity.

(c) Temperature cycling.

(d) Voltage

(e) Current.

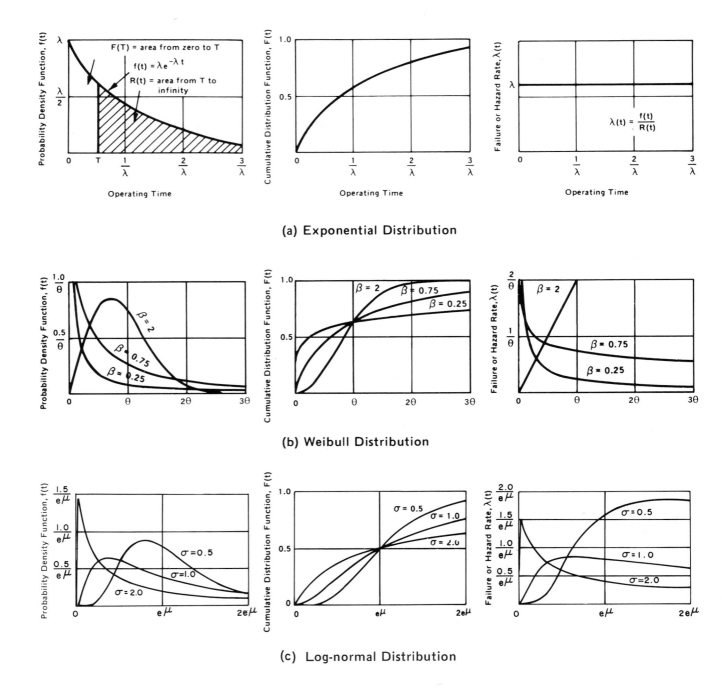

(a) Exponential Distribution

(b) Weibull Distribution

(c) Log-normal Distribution

Figure 15-1 — Typical Failure Distribution

Each of these factors is a stress that, if applied in sufficient magnitude, will cause rapid wearout failure of semiconductor devices. It is the goal of semiconductor device engineers to design devices that will be unaffected by these factors for any realistic period of use. This is done by studying the failure behavior as a function of stress at stress levels high enough to cause rapid failure. These accelerated stress test data are then used to estimate behavior under use conditions.

Time

The most common way of displaying the reliability of semiconductor devices is with a plot of failure rate, $\lambda(t)$, as a function of time. Such plots are shown in Figure 15-1 for some distributions of interest. However, no single distribution describes the behavior of typical semiconductor device populations. Appropriate modeling of these populations is discussed below.

15.2 Behavior of Device Populations

A single semiconductor device, like any other physical system, has a failure time that is a function of applied stress and time. A population of nominally identical devices will, however, exhibit a range of failure times due to normal device-to-device variation. It is precisely this variation which gives the failure distributions discussed above.

In addition to this variation, populations of devices often contain "freak" or "sport" devices that have substantially different reliability than the main device population. The overall behavior of such mixed populations will not be well-represented by a single failure distribution.

The result of this mixture of devices is the classic "bathtub curve" of failure rate as a function of time. (See Figure 15-2.) This curve is divided into three regions: an early failure or "infant mortality" region, a long-term or steady-state failure rate region, and a wearout region. The behavior in each of these regions has a different physical cause and, thus, is treated analytically in a different way.

Early Failures (Infant Mortality)

The high initial failure rate (infant mortality) is shown in Figure 15-2. This is caused by the presence of devices ("freaks" or "sports") with lives that are significantly shorter (under stress levels of normal use) than the main population. The overall population failure rate decreases rapidly in this region because there is only a small fraction of these devices in the population and they are being eliminated as they fail. The most common analytic description of this region is the Weibull model[1] with the β parameter of less than 1. Actual data are often analyzed by using a Duane Plot.[2] (See subsection 15.3, Weibull Distribution.) Sometimes it happens that a single failure mechanism is dominant in this region because all of the defects in the population are of the same type. In such a case, it is more appropriate to model the behavior as wearout of a subpopulation and use a log-normal distribution. (See subsection 15.3, Log-normal Distribution.)

Steady State Failures

When all the weak devices have failed, the failure rate falls to a very low level. Because there are very few failures even in large populations, it is difficult to measure any definite variation with time. The most common procedure is to fit the data to a constant failure rate model (exponential distribution) as described in subsection 15.3. This is the model used for virtually all system failure rate calculations.

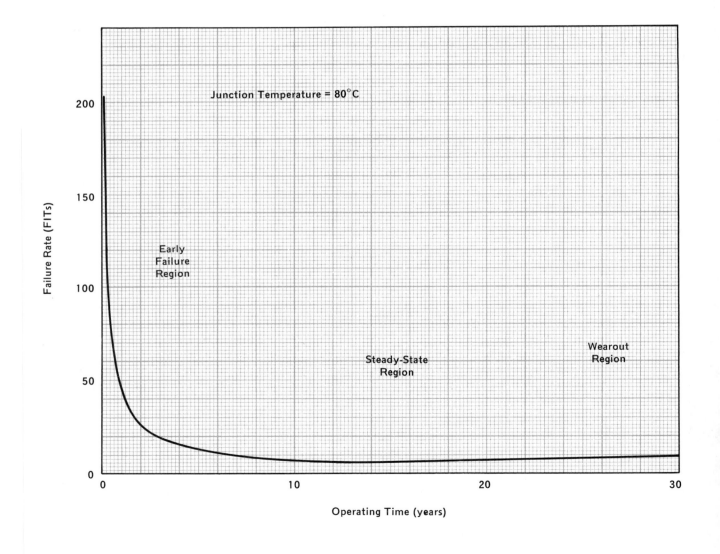

Figure 15-2 — Typical Failure Rate for Semiconductor Device

Wearout

For most semiconductor devices, the wearout region of the failure rate curve will occur at times that are much longer than any practical system life. Thus, this region can only be investigated by means of accelerated testing. In particular, the slight upturn in the curve of Figure 15-2 is not use condition data but is extrapolated from $200°C$ life test data. Proper analysis of accelerated life test data and its application to normal use conditions requires:

(a) A knowledge of the physical mechanism involved.

(b) Investigation of several stress levels to permit extrapolation to normal use stresses.

The most commonly used model for wearout is the log-normal distribution. (See subsection 15.3)

15.3 Failure Distributions and Units

The functional forms of the three commonly used distributions are presented in Figure 15-1. The corresponding analytic expressions are given in Table 15-1, and their parameters and some useful facts are summarized in Tables 15-2 and 15-3.

The failure or hazard rate is expressed as a proportion of the number of devices that fail per unit of time, or

$$\frac{\text{(number failed)}}{\text{(total number) (time)}} .$$

The expression employs units of time^{-1}. The expression is commonly stated as failures per device hour. In most cases this is a very small number, and it is convenient to define a unit called the FIT:

$$\text{FIT} = \text{one failure per } 10^9 \text{ device-hours}$$

This is the unit that is used throughout this section.

Exponential Distribution and Units

This constant failure rate distribution is perhaps the simplest model for failure. There is only one parameter, the scale parameter:

$$\lambda = \text{failure rate (time}^{-1}).$$

An equivalent parameter is the mean-time-to-failure (MTTF):

$$\text{MTTF} = 1/\lambda \text{ (time)}.$$

It is easily shown that a collection of devices, each with a different but constant failure rate, has an exponential distribution with:

$$\lambda \text{ (collection)} = \sum_{i=1}^{N} \lambda_i,$$

where

$$\lambda_i = \text{failure rate of } i^{th} \text{ individual component}$$

$$N = \text{number of components in collection}.$$

This ability to combine distributions by summing failure rates is the reason that the exponential model is so popular for system failure rate calculations.

Plots of this distribution are best done on ordinary linear graph paper.

TABLE 15-1

EXPRESSIONS FOR COMMON FAILURE DISTRIBUTIONS

DISTRIBUTION TYPE	FAILURE RATE $\lambda(t)$	CUMULATIVE DISTRIBUTION FUNCTION $F(t)$	PROPORTION FAILED FROM t_1 TO t_2 $\int_{t_1}^{t_2} f(t)dt$	PROBABILITY DENSITY FUNCTION $f(t)$
Exponential	λ	$1 - e^{-\lambda t}$	$e^{-\lambda t_1} - e^{-\lambda t_2}$	$\lambda e^{-\lambda t}$
Weibull	$\dfrac{\beta}{\theta}\left(\dfrac{t}{\theta}\right)^{\beta-1}$	$1 - e^{-\left(\frac{t}{\theta}\right)^{\beta}}$	$e^{-\left(\frac{t_1}{\theta}\right)^{\beta}} - e^{-\left(\frac{t_2}{\theta}\right)^{\beta}}$	$\dfrac{\beta}{\theta}\left(\dfrac{t}{\theta}\right)^{\beta-1} e^{-\left(\frac{t}{\theta}\right)^{\beta}}$
Log-normal	See Fig. 15-3	See Fig. 15-4	See Fig. 15-4	$\dfrac{1}{\sigma t \sqrt{2\pi}} e^{-\frac{1}{2}\left[\frac{\ln t - \mu}{\sigma}\right]^2}$

TABLE 15-2

COMMON FAILURE DISTRIBUTIONS

DISTRIBUTION	PARAMETERS*		FAILURE RATE PLOTS	USES
	SCALE (TIME)	SHAPE (DIMENSIONLESS)		
Exponential	$1/\lambda$	—	Constant with time	Steady state region
Weibull	θ	β	Straight line on log-log paper	Early failure region Life test data
Log-normal	e^μ	σ	Use Fig. 15-3	Wearout Life test data

*All distributions also have a location parameter that corresponds to the starting time of operation. This has been set to zero, as the absolute starting time (date, hour of day) of most experiments is immaterial. Of course, account must be taken of any previous operation of the devices.

TABLE 15-3

CENTRAL TENDENCIES OF COMMON FAILURE DISTRIBUTIONS

DISTRIBUTION TYPE	MEDIAN (TIME TO 50% FAILURE)	MEAN (OF TIMES TO FAILURE)	MODE [MAXIMUM OF f(t)]
Exponential	$\dfrac{\ln 2}{\lambda}$	$\dfrac{1}{\lambda}$	0
Weibull	$\theta [\ln 2]^{\frac{1}{\beta}}$	$\theta\,\Gamma\left(1+\dfrac{1}{\beta}\right)$	$\theta\left[\dfrac{\beta-1}{\beta}\right]^{\frac{1}{\beta}}$ If $\beta>1$ 0 If $\beta<1$
Log-normal	e^μ	$e^{\mu+\frac{\sigma^2}{2}}$	$e^{\mu-\sigma^2}$

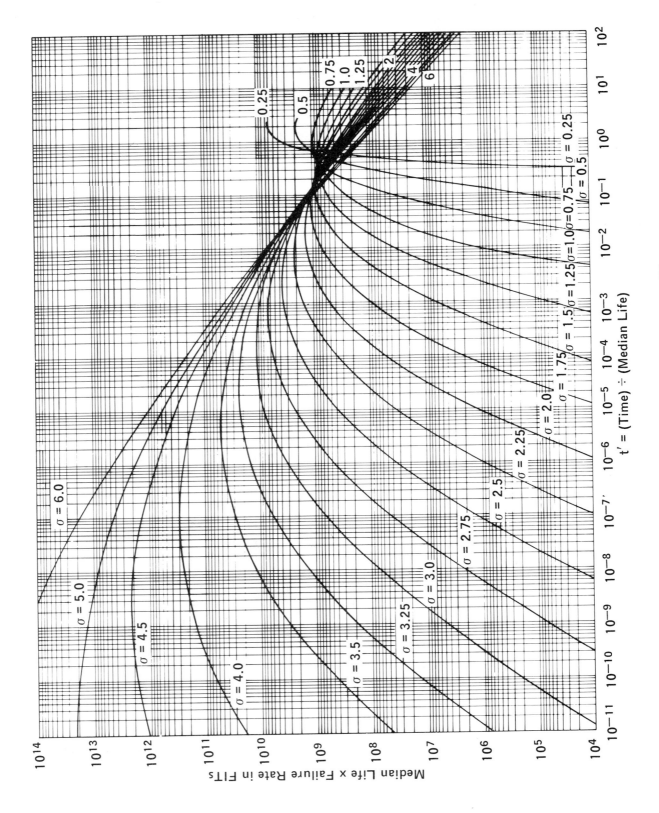

Figure 15-3 — Log-normal Failure Rates[3]   (© 1961 IEEE)

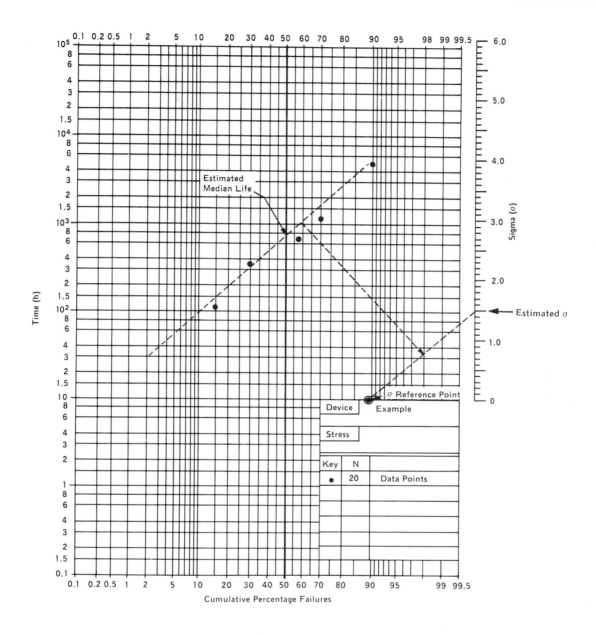

SAMPLE DATA, N = 20 DEVICES

TIME (HOURS)	110	330	700	1200	5000
NUMBER DEVICES FAILED	3	3	6	3	4
CUMULATIVE NUMBER DEVICES FAILED	3	6	12	15	19
CUMULATIVE PERCENT FAILED (% BASED ON N+1 = 21)	14	29	57	71	90

Figure 15-4 -- Log-normal Failure Distribution[4]

Weibull Distribution

This distribution is characterized by two parameters:

β = shape parameter (dimensionless)

θ = scale parameter (time).

The effect of these parameters is best seen from the cumulative distribution function

$$F(t) = 1 - e^{-(\frac{t}{\theta})^{\beta}}$$

The value of θ directly sets the time scale factor. At $t = \theta$, 63 percent of the devices have failed. The value of β determines whether the failure rate increases or decreases with time, as

$\beta < 1$ λ decreases with time,

$\beta = 1$ λ = constant (reduces to exponential distribution),

$\beta > 1$ λ increases with time.

A. Alternative Weibull Parameters

Although the above parameters are recommended, there are two other common ways to parameterize the Weibull distribution; the Duane Model and the "original" Weibull.

Duane Model

In this parameterization, the Weibull shape and scale parameters are replaced with new parameters as follows:

$$a = 1 - \beta = \text{Duane shape parameter (dimensionless)}$$

$$\lambda_D = \frac{\beta}{\theta^{\beta}} = \text{Duane intercept (time}^{-\beta}\text{).}$$

These parameters give a convenient form for the failure rate, as

$$\lambda(t) = \lambda_0 t^{-a} \ (\text{time}^{-1}).$$

This form displays the inherent logarithmic behavior of the Weibull distribution failure rate. The units of λ_D are awkward, but can be made convenient by introducing a reference time, t_0, and rewriting the failure rate as

$$\lambda_D t_0^{-a} \frac{t^{-a}}{t_0^{-a}} = [\lambda_0] [\frac{t}{t_0}]^{-a},$$

where

$$\lambda_D = \lambda(t_0) = \text{failure rate at } t = t_0.$$

Now if t_0 is chosen as 1 hour, it need not be written explicitly, and we have

$$\lambda(t) = \lambda_1 \tau^{-a},$$

where

$$\lambda_1 = \text{failure rate at 1 hour (time}^{-1})$$

$$\tau = \text{time in hours (dimensionless)}.$$

This has the same functional form and also has convenient units. It is sometimes used for modeling the infant mortality region of device behavior. It is interesting to note that this parameterization of the Weibull model was originally proposed to model an entirely different process: the learning curve behavior in aircraft production.[2]

This form is particularly convenient for graphical analysis using plots of the cumulative failure rate. This is discussed in Section C below and an example is shown in Figure 15-5.

"Original" Weibull

In this parameterization, the shape parameter is the same as β above, but the scale parameter is replaced; thus

$$\beta = \text{shape parameter (dimensionless)}$$

$$a = \theta^\beta = \text{"original Weibull" parameter (time}^\beta).$$

The cumulative distribution function then becomes

$$F(t) = 1 - e^{-\frac{t^\beta}{a}}.$$

This parameterization simplifies some functional forms and was used by Weibull in his original paper.[1] It is not recommended, however, because not only does the scale parameter, a, have awkward units, but its effect in scaling time depends on β. Table 15-4 gives a convenient method of converting between these parameters.

TABLE 15-4

WEIBULL PARAMETER RELATIONSHIPS

RECOMMENDED WEIBULL	DUANE	ORIGINAL WEIBULL	CONVERSION
θ β	$\left[\dfrac{1-a}{\lambda}\right]^{\frac{1}{1-a}}$ $1-a$	$a^{\frac{1}{\beta}}$ β	To find recommended Weibull from others
$\dfrac{\beta}{\theta^{\beta}}$ $1-\beta$	λ a	$\dfrac{\beta}{a}$ $1-\beta$	To find Duane from others
θ^{β} β	$\dfrac{1-a}{\lambda}$ $1-a$	a β	To find original Weibull from others

Note: If the Duane parameter, λ_1, is known rather than λ, these forms still hold, as long as λ_1 is the failure rate at time $t_1 = 1$ unit. For example:

$$\theta = \left[\frac{1-a}{\lambda}\right]^{\frac{1}{1-a}} = \left[\frac{1-a}{\lambda_1 t_1^{-a}}\right]^{\frac{1}{1-a}} = \left[\frac{1-a}{\lambda_1}\right]^{\frac{1}{1-a}}$$

B. Convenient Properties of Weibull Distribution

As can be seen from the Duane parameterization of the Weibull distribution, log-log plots of the Weibull failure rates against time are straight lines. This makes curve fitting and extrapolation straightforward. It is also simple to find the two parameter values, a and λ_1, from such a plot. An example is shown in Figure 15-5 as the lower, dashed line.

C. Cumulative Failure Rates

Cumulative failure rates are defined by

$$\lambda_c = \frac{N_c}{t\ N}$$

where

λ_c = cumulative failure rate (time^{-1}),

N_c = cumulative number failed from time 0 to t

N = total number of devices in the population

t = time.

The cumulative failure rate is easy to calculate and plot. It is particularly useful for the Weibull distribution because it is related to the instantaneous failure rate by:

$$\lambda_c = \frac{\lambda}{1-a},$$

(i.e., the cumulative failure rate is $\frac{1}{1-a}$ times larger than the instantaneous failure rate). Thus, it is easy to estimate λ by plotting λ_c and fitting a straight line. The instantaneous rate, λ, is then just a parallel straight line offset by $\frac{1}{1-a}$. An example of this type of analysis is shown in Figure 15-5.

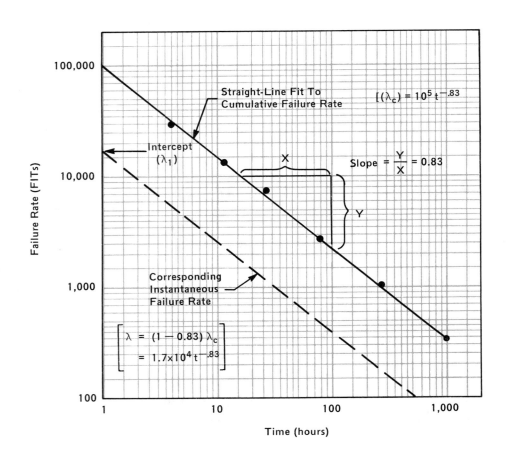

EXPERIMENTAL DATA, N = 25,000 DEVICES TOTAL

TIME (hours)	0	0.7	4	12	27	80	271	982
NUMBER DEVICES FAILED	0	2	1	1	1	1	1	1
CUMULATIVE DEVICE FAILURES	0	2	3	4	5	6	7	8
CUMULATIVE FAILURE RATE (FITs)	0	114,000	13,300	10,700	7,400	3,000	1,050	330

Figure 15-5 — Weibull Failure Rate Example

Caution should be exercised with this technique, however, as there is a very strong built-in smoothing effect. In fact, the cumulative failure rate $\lambda_c(t)$ is really an <u>average failure rate</u> over the interval 0 to t. This means that there is a strong correlation between points, and almost any data will appear to fit this model reasonably well. Whenever possible, a more detailed analysis should be done using a statistically rigorous technique. (See subsection 15.5.)

Log-normal Distribution

This distribution is characterized by a failure rate that increases with time to some maximum value and then decreases. It is useful for the analysis of life testing data where most devices survive the applied stress for some period of time and then fail over a relatively short time interval. The parameters that characterize this distribution are:

$$e^\mu = \text{median life} = \text{time for 50 percent of the devices to fail}$$

$$\sigma = \text{spread of failures around the median life.}$$

This distribution is useful primarily when the data can be fitted with a small σ (less than 4) that does not vary with the applied stress. In this case, only the median life is a function of stress. Such behavior can be supported theoretically for some types of stress effects. [5]

The most convenient way to analyze life test data and determine the e^μ and σ parameters is to plot the data on special log-normal paper. [4] An example is shown in Figure 15-4. A convenient blank form is included in subsection 15.5. If a straight line results, e^μ and σ may be read off as shown. These parameters can then be used with Figure 15-3 or the figures in reference 3, to determine the failure rate at any given time. It is important to note, however, that if the plot of Figure 15-4 shows a definite curvature, extreme care should be taken in deriving parameter values. Such a curvature often is caused by the presence of two distinct populations of devices in the experiment. Analysis of these cases is very complex as there are five parameters (e^μ and σ of each population plus the relative proportion). This requires a large amount of data and considerable knowledge of the failure modes and stress effects if it is to be more than a curve fitting exercise.

Another case that should be treated very carefully is a plot with a large σ (greater than 4). This type of data indicates that there is a very large device-to-device variation in life. This can be seen in Figure 15-3 where the extremely wide distribution for large σ values is apparent. Any conclusions about the behavior of such populations should be drawn with extreme care. For example, this type of data will fit both Weibull and log-normal models equally well for short-time failure rates, but will give significantly different long term failure estimates. A rigorous analysis using the statistical techniques of subsection 15.5 should always be made in such cases.

One cause of an apparent high σ on such plots is the presence of a subpopulation of "weak" devices which fail rapidly under the applied stress while the remaining devices are unaffected. The median life and σ of this subpopulation can be estimated by introducing the size of the subpopulation as a third parameter. Appropriate statistical techniques for this type of analysis are now available. [6]

15.4 Life Testing and Acceleration

In the following discussion, it is assumed that a log-normal failure distribution applies and the effects of stress can be represented by an acceleration factor A, such that

$$[ML]_0 = A [ML],$$

where

$$[ML] = \text{median life at accelerated condition}$$

$$[ML]_0 = \text{median life at use condition}$$

$$A = \text{acceleration factor.}$$

Temperature

Virtually all temperature effects are modeled using the Arrhenius Equation

$$[ML]_0 = \left[e^{\frac{Ea}{k}(\frac{1}{T} - \frac{1}{T_0})} \right] [ML],$$

where

ML = median life at temperature T

$[ML]_0$ = median life at use temperature T_0

E_a = activation energy (eV)

k = Boltzmann's Constant (eV/K)

T = temperature (K) (°C + 273)

This results from the assumption that the median life is inversely proportional to the chemical reaction rate that causes failure.[5]   The value of E_a, the activation energy, is characteristic of the mechanism that is causing the device to fail. An example of some typical values is given in Table 15-5. In general, the value will not be known, and data must be taken at several temperatures and then analyzed to determine if the activation energy model is appropriate. A convenient graph paper for doing this analysis is included in sub-section 15.5. An example of how this is used is given in Figure 15-6.[4]   Note that if two or more mechanisms are causing failures, they should be separated before the analysis of Figure 15-6 is carried out, or an average E_a will result. This can cause erroneous conclusions if this average is used to make estimates over a wide temperature range.

Humidity

Acceleration factors for humidity have been developed from surface conductance data for several metal-insulator systems.[9] [18] There is no simple analytic approach to estimating these accelerations, particularly when the effect of temperature is also important. Data analysis is carried out by using the appropriate acceleration factors from Figure 15-7 and by assuming that

    (a) Median life varies directly with surface conductance.

    (b) Median life varies inversely with voltage. [9] [11]

In practice, the device often has some power dissipation itself or is warmed by nearby equipment. This lowers the relative humidity (RH) at the chip surface and this must be considered in the calculation of acceleration factors. A plot for doing this for 85% RH, 85°C accelerated tests and for two use conditions is given in Figure 15-8. For example, consider a device whose chip runs at 5°C above ambient under life test conditions and 15°C above ambient under use conditions. If the life test gives a median life of 5000 hr, then referring to Figure 15-8 and using the Sbar curves we can calculate the values in Table 15-6.

Caution should be exercised for cases with

    (a) Very low operating voltage (<1.5 volts); corrosion may not occur at all.[13]

    (b) Severe contamination; electrochemistry is complex and surface conductance curves may greatly underestimate corrosion.

TABLE 15-5

ACTIVATION ENERGY FOR FAILURE MECHANISMS

VALUE (eV)	MECHANISM	REF.
0.4	Average of early failures due to mixture of defects	7
0.5 — 1.0	Surface charge motion on SiO_2 or Si_3N_4 due to applied electric field	8
1.0	Motion of Na in SiO_2 under applied electric field	10 (See also Section 6, page 35, this manual)
1.5 — 3.5	Diffusion of most metallic impurities in Si	See Section 6, Pages 30 & 32, this manual

TABLE 15-6

EXAMPLE OF USE OF FIGURE 15-8

CONDITION	CHIP TEMPERATURE RISE (°C)	ACCELERATION FACTOR (SYMBOL)	MEDIAN LIFE (HR)
85°C 85% RH Life Test	5	3.3 (A_O)	5000 (Observed)
25°C 50% RH Use	15	25,000 (A_{AC})	4×10^7
Baton Rouge Use	15	15,000 (A_{BR})	2×10^7

Since the Baton Rouge, La. use condition is a worst case outdoor humidity condition and an air conditioned ambient is a best case indoor humidity condition, acceleration factors for other use conditions will generally fall between these and can be estimated by interpolation.

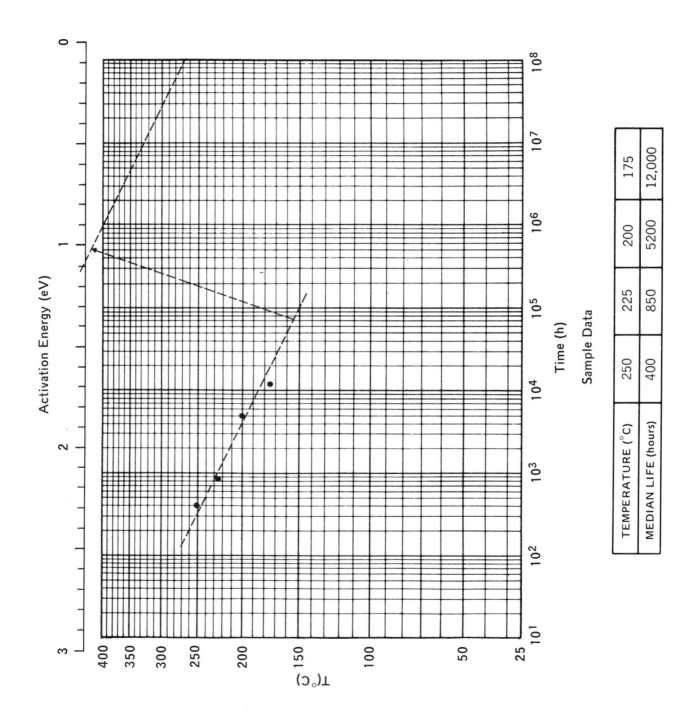

Sample Data

TEMPERATURE (°C)	250	225	200	175
MEDIAN LIFE (hours)	400	850	5200	12,000

Figure 15-6 — Effect of Temperature on Failure Time[4]

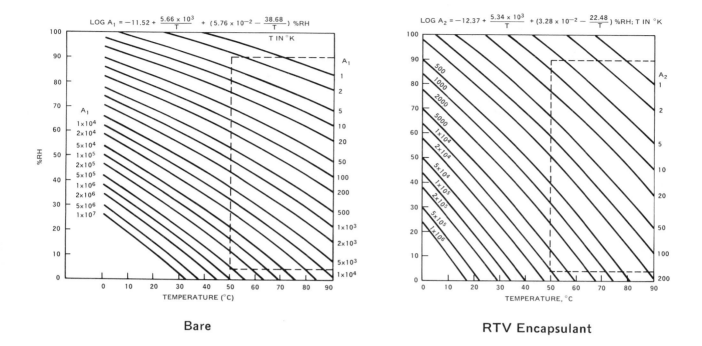

(a) Triple-Track Conductors On Al_2O_3 Ceramic

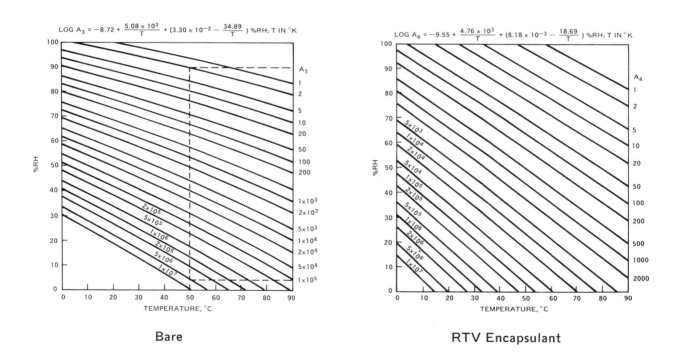

(b) Triple-Track Conductors On Si_3N_4-Covered Si

Figure 15-7 — Temperature, Humidity, and Bias Acceleration Factors Ti-Pd-Au Metallization[11]

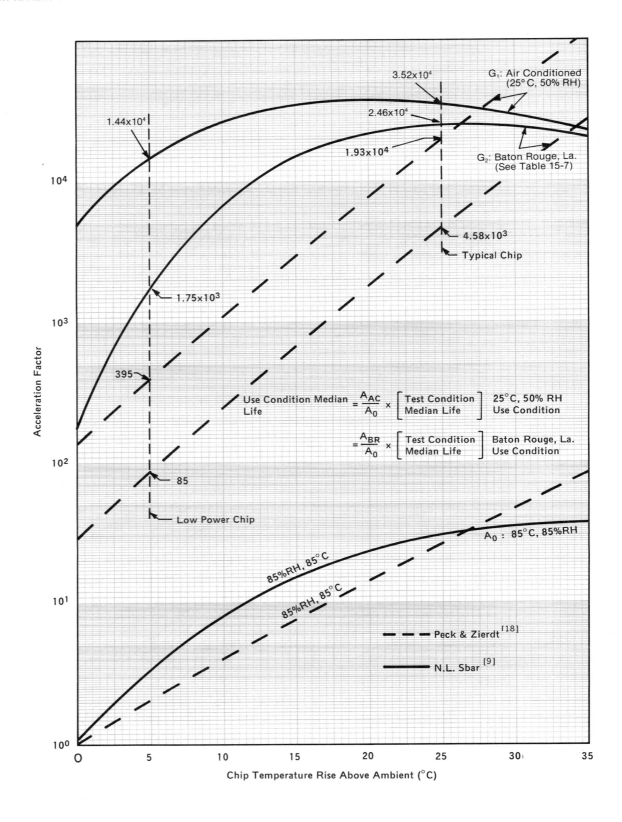

Figure 15-8 — Effect Of Chip Temperature Rise On Temperature, Humidity, and Bias Acceleration Factor [12]

TABLE 15-7

AVERAGED DAY AND NIGHT AMBIENT BY MONTH
BATON ROUGE, LA.[12]

	DAY		NIGHT	
	TEMPERATURE ($^\circ$C)	RELATIVE HUMIDITY (%)	TEMPERATURE ($^\circ$C)	RELATIVE HUMIDITY (%)
January	16	62	4	80
February	15	53	4	76
March	22	56	10	79
April	27	64	17	81
May	30	57	19	87
June	33	59	21	86
July	32	72	23	91
August	32	69	23	91
September	30	32	20	88
October	25	59	11	86
November	23	65	11	86
December	18	68	6	85

$$A_{BR} = \frac{1}{\sum_{i=1}^{24} \frac{1/24}{A(T_i, RH_i)}} = \text{Yearly Average Acceleration Factor}$$

Design of Life Tests

Techniques are now available for the optimum choice of stress levels and numbers of devices in life test experiments. References 14 and 15 provide a good introduction to this subject.

15.5 Data Analysis Aids

Computer Programs

In addition to the graphical data analysis techniques described previously in this section, a number of computer programs are available for analysis of life data. These programs have two major advantages over graphical techniques:

- Curve fitting is generally done with a maximum likelihood technique. This is statistically correct and eliminates the possible errors that may occur when relying on intuitive visual estimates.

- Confidence intervals are calculated for the model parameters. This gives the user a measure of the uncertainties in the model fitting.

Because of these advantages, one of these programs should always be used to supplement the standard graphical analysis.

Two programs which are available for this type of analysis are CENSOR and SURVREG. These programs share the following capabilities:

(a) They are available on many computer systems.

(b) The following parametric models are fitted using a maximum likelihood technique:

(1) Weibull   (4) logistic

(2) log-normal   (5) log logistic

(3) exponential (6) normal.

(c) Censored data is correctly handled (data where some items have been aged longer than others).

(d) They provide printouts of probability plots and survival plots of both input data and fitted curves.

(e) They calculate confidence interval based on assumed asymptotic normality of the maximum likelihood estimates.

(f) They can do comparative plots of populations with several failure modes (i.e., multiple probability plots).

(g) They provide correct handling of data where the failure time is not known exactly but is only known to lie between two measurement times.

A. CENSOR

This program is documented in Reference 16. In addition to the above features it also has:

   (a) Nelson's estimate of the cumulative hazard rate.

   (b) Random number generation which can be used for Monte Carlo simulation.

   (c) Plots of failure rate.

It is somewhat easier to learn than SURVREG, described below, but is presently limited to batch operation.

B. SURVREG

This program is documented in Reference 17. In addition to the general features, it also has:

   (a) Analysis of data and models with co-variates.

   (b) Plots of non-parametric analysis (Kaplan-Meier estimates and proportional hazard estimates).

A full interactive version is available, as well as batch mode job submission.

Worksheets

The following two worksheets for log-normal failure distribution curves and activation energy determination curves, representative of Figures 15-4 and 15-6, respectively, are included here for your own use.

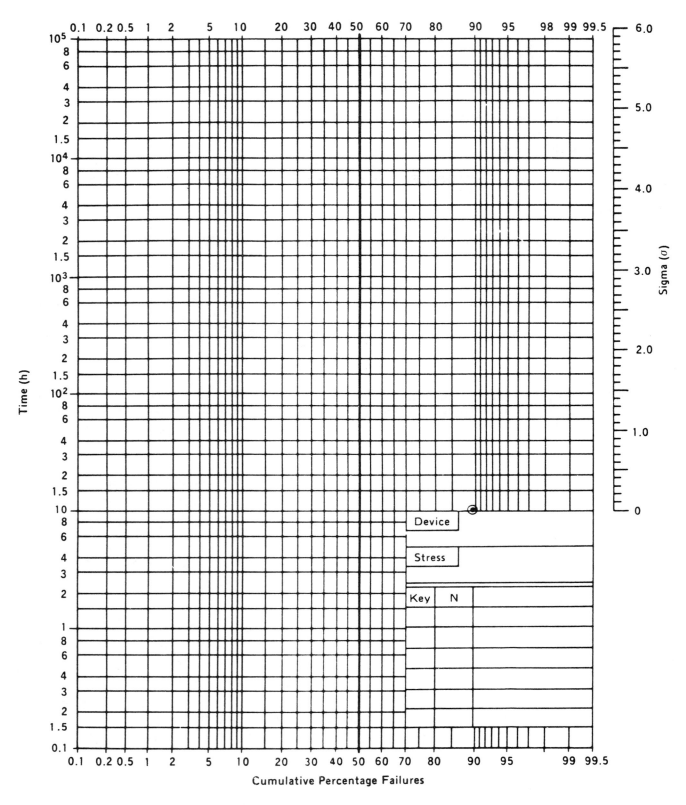

Log-Normal Failure Distribution Worksheet

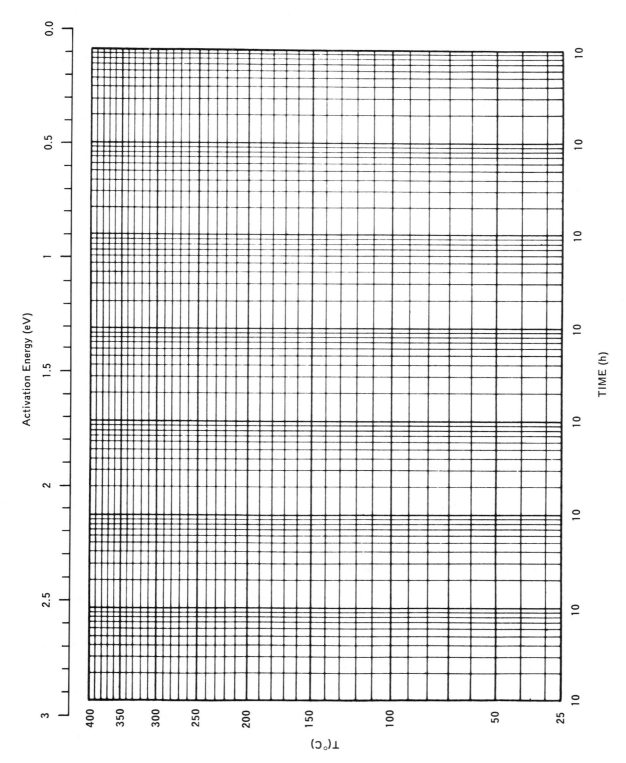

Activation Energy Distribution Worksheet

REFERENCES

1. W. Weibull, "A Statistical Distribution Function of Wide Applicability," *J. Appl. Mech.,* Vol. 18, (September 1951), pp. 293-297.

2. J. T. Duane, "Learning Curve Approach to Reliability Monitoring," *IEEE Trans. on Aerospace,* Vol. 2, No. 2 (April 1964), pp. 563-566.

3. L. R. Goldthwaite, "Failure Rate Study for the Log-normal Lifetime Model," *Proceedings of Seventh National Symposium on Reliability and Quality Control in Electronics (1961),* pp. 208-213.

4. G. A. Dodson, private communication.

5. G. A. Dodson and B. T. Howard, "High Stress Aging to Failure of Semiconductor Devices," *Proceedings of Seventh National Symposium on Reliability and Quality Control in Electronics (1961),* pp. 262-272.

6. W. Q. Meeker, Jr., "Limited Failure Population Life Tests: Application to Integrated Circuit Reliability," Statistics Department, Iowa State University, Ames, Iowa.

7. D. S. Peck, "New Concerns About Integrated Circuit Reliability," *16th Annual Proceedings Reliability Physics, 1978.* (New York: IEEE, 1978), pp. 1-6.

8. W. H. Schroen, "Process Testing for Reliability Control," Ibid., pp. 81-87.

9. N. S. Sbar and R. P. Kozakiewicz, "New Acceleration Factors for Temperature, Humidity Bias Testing," ibid., pp. 161-178.

10. C. Y. Bartholomew, private communication.

11. L. R. Hartman, private communication.

12. W. E. Beadle and A. A. Yiannoulos, private communication.

13. F. W. Hewlett and R. A. Pedersen, "The Reliability of Integrated Injection Logic Circuits for the Bell System," *14th Annual Proceedings Reliability Physics,* (New York: IEEE, 1976), pp. 5-10.

14. T. J. Kielpinski and W. Nelson, "Optimum Censored Accelerated Life Tests for Normal and Log-normal Life Distributions," IEEE Trans. on Reliability, Vol. R-24, (December 1975), pp. 310-320.

15. W. Q. Meeker, Jr., and W. Nelson, "Optimum Accelerated Life Tests for the Weibull and Extreme Value Distributions," ibid., pp. 321-332.

16. W. Q. Meeker, Jr. and S. D. Duke, "CENSOR — A User-Oriented Program for Life Data Analysis," American Statistical Association, 1980 Proceedings of the Statistical Computing Section, pp. 298-301.

17. D. L. Preston and D. B. Clarkson, "A User's Guide to SURVREG: Survival Analysis with Regression," ibid., pp. 195-197.

18. D. S. Peck and C. H. Zierdt, Jr., "The Reliability of Semiconductor Devices in the Bell System," Proc. IEEE, Vol. 62, No. 2 (February 1974), pp. 185-211.

Z